PROGRESS IN FUNCTIONAL ANALYSIS

NORTH-HOLLAND MATHEMATICS STUDIES 170

(Continuation of the Notas de Matemática)

Editor: Leopoldo NACHBIN

Centro Brasileiro de Pesquisas Físicas
Rio de Janeiro, Brazil
and
University of Rochester
New York, U.S.A.

NORTH-HOLLAND – AMSTERDAM • LONDON • NEW YORK • TOKYO

PROGRESS IN FUNCTIONAL ANALYSIS

Proceedings of the International Functional Analysis Meeting on the Occasion of the 60th Birthday of Professor M. Valdivia
Peñíscola, Spain, 22-27 October, 1990

Edited by

Klaus D. BIERSTEDT
University of Paderborn
Paderborn, Germany

José BONET
Technical University of Valencia
Valencia, Spain

John HORVÁTH
University of Maryland
College Park, MA, USA

Manuel MAESTRE
University of Valencia
Valencia, Spain

N·H

1992

NORTH-HOLLAND – AMSTERDAM • LONDON • NEW YORK • TOKYO

ELSEVIER SCIENCE PUBLISHERS B.V.
Sara Burgerhartstraat 25
P.O. Box 211, 1000 AE Amsterdam, The Netherlands

Distributors for the U.S.A. and Canada:

ELSEVIER SCIENCE PUBLISHING COMPANY, INC.
655 Avenue of the Americas
New York, N.Y. 10010, U.S.A.

Library of Congress Cataloging-in-Publication Data

International Functional Analysis Meeting (1990 : Peñíscola, Spain)
Progress in functional analysis : proceedings of the International Functional Analysis Meeting on the occasion of the 60th birthday of professor M. Valdivia, Peñíscola, Spain, 22-27 October 1990 / edited by Klaus D. Bierstedt ... [et al.].
p. cm. -- (North-Holland mathematics studies ; 170)
ISBN 0-444-89378-4
1. Functional analysis--Congresses. I. Valdivia, Manuel, 1928- . II. Bierstedt, Klaus-Dieter. III. Title. IV. Series.
QA319.I58 1990
515'.7--dc20 91-42139
CIP

ISBN: 0 444 89378 4

Printed in The Netherlands

PREFACE

When the 60th birthday of Professor Manuel Valdivia was approaching, some of his former students from the two universities of Valencia decided to hold an international functional analysis meeting in his honour. The Organizing Committee (consisting of K.D. Bierstedt, J. Bonet, J. Horváth and M. Maestre) was formed during the summer of 1989, more than one year ahead of the scheduled date of the congress.

Professor Valdivia is an internationally recognized functional analyst and one of the best known Spanish mathematicians. He has published more than 125 articles, most of them in renowned international journals (in English), and a monograph on locally convex spaces in the North-Holland series Mathematics Studies, he has been invited to many universities and has given invited talks at many meetings all over the world. Valdivia used to lecture abroad by writing in English on the blackboard, but talking in Spanish. In 1988, to the surprise of the audience, he finally started speaking English in his lectures at international conferences.

Professor Valdivia made important contributions to several areas of functional analysis. Among the topics which he became famous for, we only mention the closed graph and open mapping theorems, barrelled, bornological and ultrabornological spaces, Fréchet and (DF)-spaces, sequence space representations of spaces of functions and distributions, and Köthe sequence spaces. In particular, Valdivia proved that the spaces $\mathcal{D}$ and $\mathcal{D}'$ of Laurent Schwartz are not B_r-complete and that ultrabornological spaces (not having the finest locally convex topology) can always be represented as injective inductive limits of Banach spaces which are copies of an arbitrary (fixed) separable infinite dimensional Banach space. He gave the first examples of barrelled (DF)-spaces which are not bornological and of B_r-complete spaces which are not B-complete. He solved some problems of Grothendieck, in particular one on products of totally reflexive spaces, by giving a beautiful characterization of the totally reflexive Fréchet spaces and their duals. In recent years, he turned his interest to Banach space theory and, more specifically, to weak compactness and resolutions of the identity. The work of Valdivia has had a profound impact on functional analysis worldwide. It was used and mentioned in many articles and was referred to e.g. in the books of Köthe, Jarchow and Pérez Carreras/Bonet.

Valdivia came to Valencia as a Full Professor in the Faculty of Science of the University of Valencia in 1965. In the next year, he was one of the founders of the Section of Mathematics (later on, Faculty of Mathematics). The analysis group of Valencia originated with him. However, his influence in functional analysis in Spain went far beyond Valencia, and one may say that nowadays functional analysis is one of the most prominent topics in mathematics in Spain, due to Manuel Valdivia. In turn, the work of Spanish mathematicians in functional analysis is quite prominent internationally.

A thorough appreciation of Valdivia's mathematical work and precise references to his main theorems can be found in J. Horváth's article at the beginning of these Proceedings (which corresponds to his lecture during the opening ceremony of the Peñíscola meeting).

Recognizing Valdivia's importance for mathematics both in and out of Valencia, the two universities of Valencia were quite happy to sponsor the international functional analysis meeting on the occasion of his 60th birthday, and other sponsors joined them. We would like to acknowledge their support and, in particular, we would also like to thank the members of the Honour Committee in this respect. The "Centro de Estudios de la Función Pública" at Peñíscola, Castellón, about 130 km from Valencia was chosen as the site of the congress. This was a perfect choice in several respects, and we are very grateful to the people in charge of the Centro de Estudios for creating a good (working) atmosphere for the meeting.

The topics of the congress were centered around the (many) interests of Professor Valdivia in functional analysis. In September 1989, the organizing committee contacted 30 specialists (from several countries) and invited them to attend the Peñíscola meeting and to present the main talks. Practically everybody replied immediately and agreed to come. When the announcement of the meeting appeared (e.g. in the European Mathematical Newsletter and in the Notices of the American Mathematical Society), together with a list of the main speakers, the organizing committee received many letters with requests for participation in the congress and many abstracts for consideration in the section of short communications. A careful selection was made, and we thank several colleagues who helped us in the evaluation.

The last weeks before the meeting were the time for many preparations, and when the two foreign members of the organizing committee came to Valencia shortly before the congress, they realized how much work had already been done by their Spanish colleagues. During the congress, the amount of work connected with the organization was sometimes overwhelming. The organizing committee gratefully acknowledges that much of all this was done by the members of the Supporting Committee, and we thank them for the invaluable help.

During the week at Peñíscola (October 22–27, 1990), 28 invited lectures of 50 minutes, 7 invited lectures of 25 minutes and 38 short communications of 15 minutes were presented. (The Schedule of Lectures can be found in the editorial part of this book.) Probably none of the 96 participants will ever forget the spectacular beginning of the meeting, with a severe thunderstorm during the first lecture; we all admired Richard Aron for going on with some pertinent remarks when the world became dark and the lights went on and off several times. Fortunately enough, the weather improved, and the meeting turned out to be the success we had hoped and worked for. From the list of speakers and lectures and from the abstracts it had already been obvious that the mathematics would be interesting and very good. It remains to thank the speakers for their excellent, inspiring and well prepared talks, the chairpersons and all the participants in the meeting for their help, their interest and for many stimulating discussions. But also the social part of the congress turned out fine: The old city of Peñíscola is quite picturesque, and the visit to its castle provided us with beautiful sights and nice photos. The excursion on Wednesday afternoon showed another interesting place, Morella, not too far from Peñíscola.

All speakers of invited talks at the meeting had been invited to submit an article to the Proceedings volume. The papers were not to exceed 20 pages, unless authorized by the organizing committee, which also functioned as editors of the Proceedings. The original deadline was March 1, 1991. Indeed, most of the speakers contributed, but the last papers only arrived in May. The articles were refereed, and we thank the referees very much. Some of them took their job quite seriously, and many helpful comments and criticisms were made which definitely helped to improve several articles, mainly in the exposition and in the layout, or just by removing misprints. In some cases, however, new versions were requested, and with a small number of papers, final corrections had to be made at Paderborn (where the originals were sent), due to lack of time. (We would like to thank Mrs. Duddeck-Buijs for her competence.) At this point, we also thank the editor of "Notas de Matemática", Professor Leopoldo Nachbin, the publisher, Drs. Sevenster, and the Elsevier/North-Holland company for including this book in their series "Mathematics Studies".

As a glance at the table of contents shows, the present Proceedings volume contains 27 articles on various interesting areas of present-day functional analysis: Banach spaces and their geometry, weak compactness; operator ideals, tensor products and tensor norms; Fréchet, (DF)- and (LF)-spaces, barrelled spaces; sequence, function and distribution spaces, as well as infinite dimensional holomorphy (together with a "touch" of automatic continuity and approximation theory). Some of the reports we received from referees confirm our impression that the authors have taken much care with their articles and that many of the papers present important results and methods in active fields of research. We hope, and are quite confident, that this collection of papers, dedicated to Professor Manuel Valdivia on the occasion of his 60th birthday, with best wishes for the future, will prove helpful, valuable and inspiring for anybody interested in functional analysis and related fields.

College Park/Paderborn/Valencia, July 1991

K.D. Bierstedt, J. Bonet, J. Horváth, M. Maestre

INTERNATIONAL FUNCTIONAL ANALYSIS MEETING on the Occasion of the 60th Birthday of Professor M. Valdivia

Peñíscola, Castellón, 22 – 27 October 1990

SPONSORS

Presidencia de la Generalitat
Ministerio de Educación y Ciencia
Conselleria de Cultura Educació i Ciencia de la Generalitat Valenciana
Universidad Politécnica de Valencia
Universitat de Valencia
Facultad de Matemáticas, Universitat de Valencia
E.T.S.I. Arquitectura, Universidad Politécnica de Valencia
E.T.S.I. Telecomunicaciones, Universidad Politécnica de Valencia
Departamento de Análisis Matemático, Universitat de Valencia
Departamento de Matemática Aplicada, Universidad Politécnica de Valencia
Ayuntamiento de Peñíscola

HONOUR COMMITTEE

Molt Honorable President de la Generalitat D. Joan Lerma
Ilustrísimo Señor Conseller de Educació y Cultura D. Antonio Escarré
Ilustrísimo Señor Director General de Universidades D. Antonio Clemente
Magnífico Señor Rector de la Universitat de Valencia D. Ramón Lapiedra
Magnífico Señor Rector de la Universidad Politécnica de Valencia D. Justo Nieto

ORGANIZING COMMITTEE

Prof. Klaus D. Bierstedt, Univ. Paderborn, Federal Republic of Germany
Prof. José Bonet, Univ. Politécnica de Valencia, Spain
Prof. John Horváth, Univ. Maryland, College Park, Maryland, USA
Prof. Manuel Maestre, Univ. Valencia, Spain

SUPPORTING COMMITTEE

Prof. Pablo Galindo, Univ. Valencia
Prof. Domingo García, Univ. Valencia
Prof. Manuel López, Univ. Politécnica de Valencia
Prof. Vicente Montesinos, Univ. Politécnica de Valencia
Prof. Jose Luis Santos, Univ. Politécnica de Valencia

INTERNATIONAL FUNCTIONAL ANALYSIS MEETING on the Occasion of the 60th Birthday of Professor M. Valdivia

Peñíscola, Castellón, 22 – 27 October 1990

SCHEDULE OF LECTURES

Monday, October 22

Morning Session
Chair: K.D. Bierstedt

9.30 R.M. Aron, Weak-star continuous analytic functions

10.30 J. Schmets, On the extent of the (non-) quasianalytic classes

12.00 **Opening Ceremony**, including:
J. Horváth, The mathematical works of Manuel Valdivia
Reply of Professor M. Valdivia

Afternoon Session
Chair: J. Mujica

16.00 K. Floret, Continuity of tensor product operators between spaces $L_p(E)$ of Bochner integrable functions

17.00 A. Defant, Summing norms of finite rank operators

Chair: R. Aron

18.30 R. Meise, On the range of the Borel map for classes of nonquasianalytic functions

19.30 J.G. Llavona, Weak topologies on bounded sets of a Banach space and associated function spaces

20.30 L. Weis, On the inversion of the vector valued Laplace transform in $L_p(X)$-spaces

Tuesday, October 23

Morning Session
Chair: R. Meise

9.00 A. Pietsch, Eigenvalues of nuclear operators and geometry of Banach spaces

10.00 H. Jarchow, Some remarks on composition operators on classical Hardy spaces

11.15 S. Dineen, Complex geodesics on convex domains in Banach spaces

12.30 **Visit to the Castle of Peñíscola**

Afternoon Session
Chair: J. Bonet

17.00 P.J. Paul, Barrelled function spaces

Chairs: S. Dierolf (S = Salón de Actos), H. Jarchow (B = Aula B)

18.30 S.A. Saxon, The Pérez Carreras/Bonet problems (S)

19.05 E. Behrends, On points of symmetry of compact convex sets in the two-dimensional complex space (Yost problem) (S)

G. Metafune, Quojections and finite dimensional decompositions (B)

Short communications
Chairs: S. Dierolf (S), H. Jarchow (B), F. Cobos (A = Aula A)

19.40 J.A. López Molina, DF and quasibarrelled spaces of (q', p')-dominated operators (S)

A. Tonge, Another multilinear version of Grothendieck's Theorem (B)

S. Romanelli, Universal Korovkin closures related to positive multipliers on commutative Banach algebras (A)

20.00 J.C. Díaz, The problem of topologies of Grothendieck for quojections of Moscatelli type (S)

F.L. Hernández, Disjointly strictly-singular operators in Banach lattices (B)

V. Echandia, Some results on interpolation between dyadic Hardy spaces and dyadic Orlicz-Lorentz spaces (A)

20.20 C. Fernández, On distinguished Fréchet spaces of Moscatelli type (S)

M. González, Dual factorization of continuous operators (B)

E. Briem, Operating functions for subspaces of $C_0(X)$ (A)

20.40 A. Galbis, Tensor products of Fréchet or (DF)-spaces with a Banach space (S)

O. Blasco, Carleson measures and vector-valued Lipschitz functions (B)

J. Cerdá, Some remarks on quasi-Banach valued analytic functions (A)

Wednesday, October 24

Morning Session
Chair: K. Floret

9.00 D. Vogt, On inductive limits of Fréchet spaces

10.00 T. Terzioğlu, Decomposition method and imbeddings of smooth sequence spaces

Chair: G. Pisier

11.30 J. Diestel, Ülger's characterization of the relatively weakly compact subsets of the space of Bochner integrable functions

12.30 A. Pełzyński, Translation invariant projections in spaces of differentiable functions and Paley's effect

15.00 **Visit to Morella**

Thursday, October 25

Morning Session
Chair: A. Pełczyński

9.00 G. Pisier, Complete boundedness for Banach spaces

10.00 G. Godefroy, Smooth bumps and their applications

Chairs: A. Pietsch (S), M. Maestre (B)

11.30 W. Schachermayer, Slicings, selections and applications (S)

12.30 J. Orihuela, Compactness in function spaces (S)

13.30 W.R. Derrick, Interior properties and fixed points of certain discontinuous operators (S)

D.N. Zarnadze, James sup-theorem for Fréchet spaces and generalization of Grothendieck's homomorphism theorem (B)

Afternoon Session
Chair: J. Diestel

16.00 F. Bombal, Distinguished subsets in vector sequence spaces

17.00 K.B. Laursen, Automatic continuity of intertwiners in topological vector spaces

18.30 C. Bessaga, Functional analytic aspects of the theory of metric spaces. Linear extending of metrics and related problems

Short communications
Chairs: W. Schachermayer (S), M. Maestre (B), J.M. Ansemil (A)

19.45 R. Becker, Integral representation in locally convex spaces (S)

J. Prada, Operators commuting with differentiation (B)

L.D. Nel, Infinite dimensional calculus allowing nonconvex domains with empty interior (A)

20.05 W. Hensgen, Some properties of the vector-valued Banach ideal space $E(X)$ derived from those of E and X (S)

N. Robertson, The metrisability of precompact sets (B)

L.D. García-Tuñón, The Nachbin topology $\tau_{\omega,\vartheta}$ for a particular ϑ-holomorphy type (A)

20.25 A.J. Pallarés, Selection theorems in dual Banach spaces with the Radon Nikodym property (S)

R. Beattie, Multiplication and convolution of rapidly decreasing functions (B)

Y. Meléndez, Inductive limits of Moscatelli type (A)

20.45 P. Aiena, On Fredholm theory for multipliers on commutative Banach algebras (S)

J. Zafarani, Angelicity of spaces of vector-valued measures (B)

Friday, October 26

Morning Session
Chair: J. Horváth

9.00 F. Cobos, A class of approximation spaces

10.00 F. Altomare, Lototsky-Schnabl operators and degenerate diffusion equations

Chair: D. Vogt

11.30 S. Dierolf, Remarks on distinguished Fréchet spaces

12.30 J. Taskinen, Factorization of multilinear operators

13.30 L.A. de Moraes, On F-quotients and holomorphy

Afternoon Session
Chair: S. Dineen

16.00 J.M. Ansemil, Embeddings of spaces of holomorphic germs

17.00 J. Mujica, Linearization of holomorphic mappings of bounded type

Short communications
Chairs: G. Godefroy (S), J. Schmets (B), J. Orihuela (A)

18.15 P. Harmand, M-ideals and spaces of compact operators (S)

M. López Pellicer, On Valdivia's barrelledness property in $l_0^\infty(X, \mathcal{A})$ (B)

Fan Ming, Some problems on the joint spectrum and the joint numerical range (A)

18.35 R. Payá, Banach spaces with strongly subdifferentiable norm (S)

F. Balibrea, Permutation metric and weak topology (B)

J. Motos, A note on the vector-valued Fourier transform (A)

18.55 W. Werner, The asymptotic behaviour of BAP on subspaces of c_0 (S)

E. Martín-Peinador, On Pontryagin duality (B)

J.H. Fourie, Wedge spaces of operators on Hilbert spaces (A)

19.15 D. Yost, Convex sets and M-ideals in Banach spaces (S)

T. Gilsdorf, Regular inductive limits of (K)-spaces (B)

L.M. Sánchez, On the open mapping theorem without convexity conditions (A)

19.35 T. Kühn, Optimal asymptotic eigenvalue behaviour of weakly singular integral operators (S)

M.C. de las Obras, Relations between convergence in $G(H)$ (B)

J.F. Castillo, Sums and products of Banach spaces (A)

21.00 **Special dinner**

Saturday, October 27

9.30 **Departure to Valencia**

12.00 **Closing Ceremony** (Paraninfo de la Universidad)

The participants of the meeting during the visit to the castle of Peñíscola

INTERNATIONAL FUNCTIONAL ANALYSIS MEETING on the Occasion of the 60th Birthday of Professor M. Valdivia

Peñíscola, Castellón, 22 – 27 October 1990

LIST OF PARTICIPANTS

P. Aiena (Palermo, Italy)
F. Altomare (Potenza, Italy)
J.M. Ansemil (Universidad Complutense de Madrid, Spain)
J. Arias de Reyna (Sevilla, Spain)
R.M. Aron (Kent State University, USA)
J.L. Arregui (Zaragoza, Spain)
F. Balibrea (Murcia, Spain)
B.R. Beattie (Mount Allison University, Sackville, Canada)
R. Becker (Paris VI, France)
E. Behrends (Freie Universität Berlin, Germany)
C. Bessaga (University of Warszawa, Poland)
K.D. Bierstedt (Paderborn, Germany)
O. Blasco (Zaragoza, Spain)
F. Bombal (Universidad Complutense de Madrid, Spain)
J. Bonet (Universidad Politécnica de Valencia, Spain)
E. Briem (Reykjavik, Iceland)
B. Cascales (Murcia, Spain)
J.M.F. Castillo (Badajoz, Spain)
J. Cerdá (Universidad de Barcelona, Spain)
F. Cobos (Universidad Autónoma de Madrid, Spain)
A. Defant (Oldenburg, Germany)
W.R. Derrick (University of Montana, Missoula, USA)
J.C. Díaz (Córdoba, Spain)
S. Dierolf (Trier, Germany)
J. Diestel (Kent State University, USA)
S. Dineen (University College Dublin, Ireland)
V. Echandia (Budapest, Hungary)
C. Fernández (Universidad de Valencia, Spain)
L.M. Fernández Cabrera (Universidad Complutense de Madrid, Spain)
M. Florencio (Sevilla, Spain)
K. Floret (Campinas, Brazil/Oldenburg, Germany)
J.H. Fourie (Potchefstroom, South Africa)
F.J. Freniche (Sevilla, Spain)
A. Galbis (Universidad de Valencia, Spain)
P. Galindo (Universidad de Valencia, Spain)
D. García (Universidad de Valencia, Spain)

L.D. García Tuñón (Badajoz, Spain)
T. Gilsdorf (University of North Dakota, Grand Forks, USA)
G. Godefroy (Paris VI, France)
M. González (Santander, Spain)
P. Harmand (Oldenburg, Germany)
W. Hensgen (Regensburg, Germany)
F.L. Hernández (Universidad Complutense de Madrid, Spain)
F.J. Hernández García (Universidad Autónoma de Madrid, Spain)
J. Horváth (University of Maryland, College Park, USA)
H. Jarchow (Universität Zürich, Switzerland)
T. Kühn (Universidad Autónoma de Madrid, Spain/Leipzig, Germany)
K.B. Laursen (Copenhagen, Denmark)
J.G. Llavona (Universidad Complutense de Madrid, Spain)
J.A. López Molina (Universidad Politécnica de Valencia, Spain)
M. López Pellicer (Universidad Politécnica de Valencia, Spain)
S. Macario (Universidad de Valencia, Spain)
M. Maestre (Universidad de Valencia, Spain)
E. Martín-Peinador (Universidad Complutense de Madrid, Spain)
Y. Meléndez (Badajoz, Spain)
R. Meise (Düsseldorf, Germany)
G. Metafune (Lecce/Rome, Italy)
Fan Ming (Uppsala, Sweden)
A. Moltó (Universidad de Valencia, Spain)
V. Montesinos (Universidad Politécnica de Valencia, Spain)
L.A. de Moraes (Universidade Federal do Rio de Janeiro, Brazil)
J. Motos (Universidad Politécnica de Valencia, Spain)
J. Mujica (Campinas, Brazil)
L. Narici (St. John's University, New York, USA)
L.D. Nel (Ottawa, Canada)
M.C. de las Obras (Universidad de Valencia, Spain)
J. Orihuela (Murcia, Spain)
A.J. Pallarés (Murcia, Spain)
P.J. Paúl (Sevilla, Spain)
R. Payá (Granada, Spain)
A. Pelczyński (Polish Academy of Sciences)
A. Peris (Universidad Politécnica de Valencia, Spain)
A. Pietsch (Jena, Germany)
G. Pisier (Paris VI, France)
J. Prada (Salamanca, Spain)
M.J. Rivera (Universidad Politécnica de Valencia, Spain)
N. Robertson (University of Cape Town, South Africa)
L. Rodriguez Piazza (Sevilla, Spain)
S. Romanelli (Bari, Italy)
L.M. Sánchez (Universidad Politécnica de Valencia, Spain)
S.A. Saxon (University of Florida, Gainesville, USA)

W. Schachermayer (Wien, Austria)
J. Schmets (Liège, Belgium)
M. Simoes (Napoli, Italy)
J. Taskinen (Helsinki, Finland)
T. Terzioğlu (Middle East Technical University, Ankara, Turkey)
A. Tonge (Universidad Complutense de Madrid, Spain/Kent State University, USA)
Professor M. Valdivia (Universidad de Valencia, Spain)
G. Vera (Murcia, Spain)
A. Vignoli (Roma, Italy)
D. Vogt (Wuppertal, Germany)
L. Weis (Louisiana State University, Baton Rouge, USA)
W. Werner (Paderborn, Germany)
D. Yost (Granada, Spain/Freie Universität Berlin, Germany)
J. Zafarani (Isfahan, Iran)
D.N. Zarnadze (Georgian Academy of Science, Tblisi, USSR)

PROTOCOLO DEL ACTO DE APERTURA

Peñíscola, Centro de Estudios de la Función Pública, Avd. del Puerto. Lunes 22 de Octubre de 1990, 12 horas.

1. Apertura de la sesión por el Presidente.

2. El Presidente da la palabra al Profesor J. Bonet, de la Universidad Politécnica de Valencia, quien, en nombre del Comité Organizador, da la bienvenida a los participantes.

3. El Presidente da la palabra al Profesor J. Horváth de la Universidad de Maryland (EEUU), el cual dictará la Lección Inaugural Homenaje al Profesor M. Valdivia. (En este momento la mesa presidencial pasará a ocupar los asientos asignados en la primera fila, donde permanecerán hasta el término de la exposición del Profesor J. Horváth. Finalizada esta volverán a ocupar la mesa presidencial.)

4. El Presidente da la palabra al Profesor M. Valdivia.

5. Apertura del Congreso por el Presidente.

For the lecture of J. Horváth see his article "On the mathematical works of Manuel Valdivia" in these Proceedings. The English translation of the reply of Professor M. Valdivia follows on the next pages.

REPLY OF PROFESSOR M. VALDIVIA

The Honourable Minister of Culture, Education, Science of the Government of Valencia,
Your Excellency, the General Director of Universities in the Government of Valencia,
Your Excellency, Rector of the University of Valencia,
Your Excellency, Rector of the Technological University of Valencia,
Excellent Sirs, Honourable Ladies and Gentlemen, Dear Colleagues and Friends:

First of all, I want to thank the institutions who, through their official and financial support, have made possible this homage for which I am heartily indebted: the Government of the State of Valencia, the Ministry of Education and Science, the Valencian Ministry of Culture, Education and Science, the University of Valencia, the Technological University of Valencia, the Faculty of Mathematics, the School of Architecture, the School of Telecommunications, the Department of Applied Mathematics of the Technological University of Valencia and the Department of Mathematical Analysis of the University of Valencia.

I want to thank the official institutions and, more specifically, their appointed representatives at this event, not only for their collaboration in this celebration, but also for the enormous support they are giving to education and research for the benefit of University, Science and Culture.

For several years now the conditions under which research is being done have improved quite obviously and they continue to improve. Thanks to this new situation, mathematicians can establish contact and can cooperate with scientists in the whole world travelling here and there and doing joint work. In the State of Valencia we have now reached a mathematical level which is very respectable and widely respected. But we must recognize that this is not only due to the effort of the mathematicians, but also due to the strong support of the institutions. It is not by chance that some of the authorities who honour us today by their presence here are also professors and researchers. They have known very well how important research and education are for society and have acted accordingly, helping us in our tasks.

Twenty-five years ago I arrived in this hospitable land of Valencia and was received cordially and warmly by my colleagues in the former Faculty of Science. I came with the mission to explain mathematics in the Section of Physics and later on I was in charge of Mathematical Analysis in the Faculty. Those were very difficult times professionally. We did not have mathematical journals. The chair to which I was appointed could not afford to buy books. I went to Madrid to see the General Director of our Ministry to ask him for help, and he offered me 100.000 pesetas for books. For me, this was the actual beginning of mathematics in the Faculty.

During those twenty-five years, all my teaching and research was carried out in the University of Valencia as well as in the Technological University of Valencia which I joined at the same moment when I came here, first as an Associate Professor, later on as Full Professor for Mathematics 1 in the School of Agricultural Engineering. It was an honour for me to collaborate in the formation of the Technological University of Valencia, where my work was carried out in the beginning by giving lectures in the Schools of Architecture and Agriculture and then by directing the Department of Mathematics of that University.

I must say here that I am very proud of my original career as an agricultural engineer; as such, I remember with great pleasure the time when I worked in Madrid, giving classes of genetics, physics and chemistry, then in rational mechanics and mathematics. I used to say that I was a mathematician of late vocation: I was 31 years old when I enrolled in the Faculty of Mathematics in Madrid.

During my first years at the University of Valencia, an important event occurred, the foundation of the Faculty of Mathematics, in which I participated and contributed from that time on by explaining several topics with the aim of consolidating the Faculty. Some of you, who have worked together with me, studied in Valencia at that time and hence remember how many difficulties we used to have.

Concerning my research, the first years were a time of isolation and solitude. I had finished my Doctoral Thesis in Madrid, under the direction of D. Ricardo San Juan, on the classical theory of integration, but D. Ricardo insisted that I should change topics. Once I was in Valencia, I thought it would be interesting to do investigations in Functional Analysis. I got a book on locally convex spaces, the one of Professor Köthe, a person well-known to many of you, for whom I always had great respect and deep affection. I take this opportunity to say that I will always keep his memory in highest regard. Köthe's monograph, written in German, was difficult for me to read, and it took me a long time to study it with the help of a dictionary. Later on, there appeared, like a miracle, two books in English which were my last resort: the books of Professors Schaefer and Horváth, respectively. When I studied them, I started publishing some articles in the Revista de la Academia de Ciencias, in Madrid, yet without being able to check with anybody if they contained interesting results. But it was Professor Horváth who saved me from this situation. Before I worked with San Juan, Professor Horváth had visited Madrid, invited by D. Ricardo, and given a course on quasianalytic classes, and from that point on they both had kept in contact and exchanged letters. D. Ricardo wrote him about me, and the answer did not take long: I received from Liège, sent by De Wilde, his famous doctoral thesis. Later on, Professor Garnir, whose memory has always been in my highest regard, sent me an invitation to attend a congress which was organized at Liège. I went there and I met Professor Köthe and other mathematicians in the field. This was, in my opinion, the beginning of mathematics at an international level in our universities at Valencia. Consequently, we owe many thanks to Professor Horváth. We have just heard his lecture on my work. I cannot find the proper words to express how grateful I feel. I can only offer him my friendship, which he already has, even deeper and closer, if possible.

I feel the same way about the Professors and Friends José Bonet, Klaus Bierstedt and Manuel Maestre. I do not know how to address them at this occasion, but they should know that my thanks are sincere and cordial.

At the same time I should express my special gratitude for the members of the supporting committee: Professors Pablo Galindo, Domingo García, Manuel López Pellicer, Vicente Montesinos and José Luis Santos.

Many thanks to all those who are in the audience now: friends, companions, colleagues, prestigious mathematicians who honour me by their presence.

Before I finish I have to say that Functional Analysis in the Community of Valencia goes further and further thanks to those young mathematicians who work in our Universities: In spite of their youth, some of them are well-known internationally. In order to reach this situation, many hours of work, pondering and effort were necessary. Just as Laurent Schwartz said in the preface to one of his books: "There is no mathematics without tears."

I would also like to mention María Teresa, my wife, who has always been my helper, supporting me at each moment in my work and dedicating herself, together with me, to the service of the University of Valencia during these twenty-five years.

And that's all, I only want to express my emotion and gratitude with a very simple sentence: Thanks to all of you, my heart is with you.

Peñíscola, October 22, 1990

Professor M. Valdivia during the visit to the castle of Peñíscola

LIST OF CONTRIBUTORS

Franceso Altomare, Dipartimento di Matematica, Università di Bari
Traversa 200 Via Re David 4, I-70125 Bari, Italy

José M. Ansemil, Departamento de Análisis Matemático, Facultad de Matemáticas
Universidad Complutense, E-28040 Madrid, Spain

Richard M. Aron, Department of Mathematical Sciences, Kent State University
Kent, Ohio–44242, USA

Aydın Aytuna, Department of Mathematics, Middle East Technical University
06531 Ankara, Turkey

Ehrhard Behrends, Freie Universität Berlin, Institut für Mathematik I
Arnimallee 2 – 6, D-1000 Berlin 33, F.R. Germany

Cesław Bessaga, Institute of Mathematics, Warsaw University
ul. Banacha 2, 00-913 Warsaw, Poland

Klaus D. Bierstedt, FB 17 Mathematik, Universität-GH-Paderborn
Postfach 1621, D-4790 Paderborn, F.R. Germany

Fernando Bombal, Departamento de Análisis Matemático, Facultad de Matemáticas
Universidad Complutense, E-28040 Madrid, Spain

José Bonet, Departamento de Matemática Aplicada, E.T.S. Arquitectura
Universidad Politécnica de Valencia, E-46071 Valencia, Spain

Fernando Cobos, Departamento de Matemáticas, Facultad de Ciencias
Universidad Autónoma de Madrid, E-28049 Madrid, Spain

Andreas Defant, FB Mathematik der Universität
D-2900 Oldenburg, F.R. Germany

William R. Derrick, Department of Mathematics, University of Montana
Missoula, Montana 59812, USA

Susanne Dierolf, FB IV Mathematik, Universität Trier
D-5500 Trier, F.R. Germany

Seán Dineen, Department of Mathematics, University College Dublin
Belfield, Dublin 4, Ireland

Lech Drewnowski, Inst. Matematyki, U. A. Mickiewicz, ul. Matejki 48/49
60-769 Poznań, Poland

Miguel Florencio, Departamento de Matemáticas, E.S. Ingenieros Industriales
Avda Reina Mercedes s/n, E-41012 Sevilla, Spain

Klaus Floret, FB Mathematik der Universität
D-2900 Oldenburg, F.R. Germany and
Instituto de Matemática, Universidade Estadual de Campinas
Caixa Postal 6065, 13081 Campinas, SP, Brazil

Pablo Galindo, Departamento de Análisis Matemático, Facultad de Matemáticas Universidad de Valencia, Doctor Moliner 50, E-46100 Burjasot (Valencia), Spain

Domingo García, Departamento de Análisis Matemático, Facultad de Matemáticas Universidad de Valencia, Doctor Moliner 50, E-46100 Burjasot (Valencia), Spain

John Horváth, Department of Mathematics, University of Maryland College Park, MD 20742, USA

Hans Jarchow, Mathematisches Institut, Universität Zürich Rämistr. 74, CH-8001 Zürich, Switzerland

Thomas Kühn, Sektion Mathematik, Universität Leipzig D-O-7010 Leipzig, Germany

Kjeld B. Laursen, Department of Mathematics, University of Copenhagen Universitetsparken 5, DK-2100 Copenhagen Ø, Denmark

José G. Llavona, Departamento de Análisis Matemático, Facultad de Matemáticas Universidad Complutense, E-28040 Madrid, Spain

Manuel Maestre, Departamento de Análisis Matemático, Facultad de Matemáticas Universidad de Valencia, Doctor Moliner 50, E-46100 Burjasot (Valencia), Spain

Reinhold Meise, Mathematisches Institut, Heinrich-Heine-Universität Düsseldorf Universitätsstr. 1, D-4000 Düsseldorf, F.R. Germany

Giorgio Metafune, Dipartimento di Matematica, II Università, v. O. Raimondo I-00173 Roma, Italy

Vicenzo B. Moscatelli, Dipartimento di Matematica, Università C.P. 193, I-73100 Lecce, Italy

Jorge Mujica, Instituto de Matemática, Universidade Estadual de Campinas Caixa Postal 6065, 13081 Campinas, SP, Brazil

Lucimar Nova G., Departamento de Matemáticas, Universidad Nacional Bogotá 1, Colombia

José Orihuela, Departamento de Matemáticas, Universidad de Murcia Campus de Espinardo, E-30100 Espinardo (Murcia), Spain

Pedro J. Paúl, Departamento de Matemáticas, E.S. Ingenieros Industriales Avda Reina Mercedes s/n, E-41012 Sevilla, Spain

Alexander Pełczyński, Institute of Mathematics, Polish Academy of Sciences Śniadeckich 8, 00-950 Warszawa, Poland

Albrecht Pietsch, Department of Mathematics, Friedrich-Schiller-Universität D-O-6900 Jena, Germany

Socorro Ponte, Departamento de Análisis Matemático, Facultad de Matemáticas Universidad Complutense, E-28040 Madrid, Spain

Stephen A. Saxon, Mathematics Department, Lehigh University
Bethlehem, PA 18015, USA and
Mathematics Department, University of Florida
Gainesville, FL 32611, USA

Jari Taskinen, Department of Mathematics, University of Helsinki
Hallituskatu 15, SF-00100 Helsinki, Finland

B.A. Taylor, Department of Mathematics, University of Michigan, Angell Hall
Ann Arbor, Michigan 48109, USA

Tosun Terzioğlu, Department of Mathematics, Middle East Technical University
06531 Ankara, Turkey

Richard M. Timoney, School of Mathematics, 39 Trinity College
Dublin 2, Ireland

Dietmar Vogt, FB Mathematik, Bergische Universität-GH-Wuppertal
Gaußstr. 20, D-5600 Wuppertal, F.R. Germany

M. Wojciechowski, Institute of Mathematics, Polish Academy of Sciences
Śniadeckich 8, 00-950 Warszawa, Poland

TABLE OF CONTENTS

Progress in Functional Analysis
K.D. Bierstedt, J. Bonet, J. Horváth & M. Maestre (Eds.)

The Mathematical Works of Manuel Valdivia

John Horváth

Department of Mathematics, University of Maryland, College Park, MD 20742, USA

Introduction

On January 12, 1968 Ricardo San Juan Llosá, professor of mathematics at the University of Madrid, wrote me the following: "I read cursorily the magistral work of Prof. Köthe on topological vector spaces. A professor ("catedrático") from Valencia, Prof. Valdivia, whose doctoral dissertation I directed, works in this and I think that he will obtain results; he seems to me to be a good researcher."

I hardly know more prophetic words, since the same year appeared the first articles of Valdivia on the closed graph and open mapping theorems, and with these a torrent was released. In the course of the last twenty-two years he published about hundred and twenty papers on topological vector spaces and related subjects, in which he solved difficult open problems, created new concepts, and inspired an enormous quantity of research in Spain as well as in a large number of other countries.

Valdivia's papers contain over a thousand theorems. It would be totally impossible to do justice to all of them in a limited amount of space. I have therefore selected those which are the simplest to state. This has the disadvantage that precisely the most difficult, most general and deepest results will not be mentioned. On the other hand, I will not be able to give an idea of the proofs, which are often constructed with the stupendous ingenuity that characterizes the work of Valdivia.

I shall use mostly the terminology and notation of Bourbaki's treatise {5}, which differ sometimes from those used by Valdivia. Numbers in square brackets refer to the works of Valdivia, those in curly brackets to the works of other authors, listed at the end. To lighten the exposition, I shall occasionally use a loose language. Thus I will speak about a "space" instead of a "locally convex Hausdorff topological vector space"; I shall say "inductive limit" when I mean "the finest locally convex topology for which certain linear maps (which most often will not be specified) are continuous". I will also omit hypotheses which are obvious from the context.

1. SEQUENCES AND FAMILIES OF CONTINUOUS, INTEGRABLE AND HOLOMORPHIC FUNCTIONS

Valdivia's research career did, however, not start with the explosion of 1968. In his doctoral dissertation [1] he considers concepts which are usually defined for maps from a topological space into a metric space or a uniform space. He characterizes uniformly convergent, equicontinuous, Cauchy, and quasi-uniformly convergent sequences of maps from a metric space X into a metric space Y, and then uses the characterizations to define these concepts when X and Y are both topological spaces. Thus let X be a topological space, Y a regular topological space, and (f_n) a sequence of continuous maps from X into Y which converges pointwise to a map $f : X \longrightarrow Y$. Given a neighborhood V of $y = f(x)$, let $n_0 \in \mathbb{N}$ be such that $f_n(x) \in V$ whenever $n \geq n_0$. Valdivia says that (f_n) converges uniformly to f if for every $x \in X$ and every neighborhood V of $f(x)$ the set $\bigcap_{n \geq n_0} f_n^{-1}(V)$ is a neighborhood of x; if X is compact and Y a separated uniform space, then the definition coincides with the usual one. The limit function f is then continuous.

Another example: a family of maps $f_i : X \longrightarrow Y$ $(i \in I)$ is equicontinuous at the point $x \in X$ if given a closed set $F \subset Y$ and an open set $A \supset F$, the subfamily $(f_j)_{j \in J}$, consisting of those f_j for which $f_j(x) \in F$, verifies the condition that $\bigcap_{j \in J} f_j^{-1}(A)$ is a neighborhood of x. Valdivia shows that under mild assumptions concerning X and Y the Arzelá-Ascoli compactness theorem holds.

In the same vein he introduces quasi-uniform convergence, called quasi-approximative in [7], which leads to necessary and to sufficient conditions for the limit of a net of continuous maps to be continuous.

In the last section of [1] Valdivia introduces conditions which ensure that the limit of a sequence of maps will be absolutely continuous. In the same section and in [2] he obtains conditions for

$$\lim \int_X f_n d\mu = \int_X (\lim f_n) d\mu \tag{1}$$

to be satisfied. The basic result he proves is the following: Let $\mu(X) < \infty$ and $f_n(x) \longrightarrow f(x)$ almost everywhere on X. Then (1) holds if and only if the set function $\varphi(E) = \lim \sup |\int_E f_n d\mu|$ is absolutely continuous. If $\mu(X) = \infty$, $X = \bigcup_{j=1}^{\infty} X_j$ with $X_j \subset X_{j+1}$ and $\mu(X_j) < \infty$, then the additional condition that $\lim_{j\to\infty} \int_{X\setminus X_j} f_n d\mu = 0$ uniformly in n will imply (1). One of his theorems is the useful generalization of the Lebesgue dominated convergence theorem called sometimes "Pratt's lemma" {28, Chap.4, Th.17, p.92}.

Article [3] is related to a paper of San Juan {29}. Let D be a connected open subset of $\mathbb{C}$ having 0 on its boundary, and $(m_n)_{n\geq 0}$ a sequence of positive numbers. The problem consists in finding necessary and sufficient conditions that given a sequence (a_n) of complex numbers there exists a function f, holomorphic in D, such that

$$|f(z) - a_0 - a_1 z - \cdots - a_{n-1} z^{n-1}| \leq m_n |z|^n \tag{2}$$

for $z \in D$ and $n \geq 1$. Valdivia gives various criteria, distinct from that of San Juan. For the proofs, he equips the set $A(D)$ of holomorphic functions satisfying (2) with different equivalent metrics for which $A(D)$ is compact.

2. BARRELLED SPACES AND THEIR ANALOGUES

Bourbaki introduced barrelled spaces in order to characterize those spaces for which the uniform boundedness principle is valid. Let me recall that a <u>barrel</u> is an absorbing, balanced, convex, closed set, and that a space E is <u>barrelled</u> if every barrel is a neighborhood of 0 or, equivalently, if every subset of the dual E', which is bounded for the weak topology $\sigma(E',E)$, is equicontinuous. The class of barrelled spaces is trivially stable with respect to inductive limits but a closed subspace of a barrelled space is not necessarily barrelled. Dieudonné proved that every finite-codimensional subspace F of a barrelled space E is barrelled, and Valdivia [17; A, Chapter I, §3, 2(9), p.47] proves the same for subspaces F of countable codimension. The result was proved independently with a different proof by S. Saxon and M. Levin {30} and a simple proof was given by J.H. Webb {36}. To prove the theorem, one considers an increasing sequence $F_0 = F \subset F_1 \subset F_2 \subset \cdots \subset E$ of subspaces, each of codimension one in the next, and starting with a barrel $T = T_0$ in F one constructs a barrel T_n in F_n such that $T_{n+1} \cap F_n = T_n$. If one can show that $\bigcup_{n\geq 0} T_n$ is a barrel in E, the battle is won. Therefore, Valdivia's proof is based on a theorem about increasing sequences of balanced, convex sets, which explains the title of [17], and which is presented with a minor simplification in {35, pp. 54-55}, where on pp. 211-217 a generalization due to M. De Wilde and C. Houet can be found.

Valdivia returns to sequences of convex sets in [50], where he makes use of the topology $\mathcal{T}_{c_0}$ of uniform convergence on sequences converging to 0 in the Mackey sense {19, §28,5}. Let me recall that a sequence (x_n) in the space E <u>converges</u> to $x \in E$ <u>in the Mackey sense</u> if there exists a bounded subset B of E with the following property: for $\varepsilon > 0$ there is an $N \in \mathbb{N}$ such that $x_n - x \in \varepsilon B$ whenever $n \geq N$; it is a <u>Cauchy sequence in the Mackey sense</u> if for $\varepsilon > 0$ there is an $N \in \mathbb{N}$ such that $x_n - x_m \in \varepsilon B$ whenever $n, m \geq N$. The space E is <u>locally complete</u> if every Cauchy sequence in the Mackey sense converges in the Mackey sense to some $x \in E$. If (F,G) is a pairing of two vector spaces, the <u>Mackey topology</u> $\tau(G,F)$ on G is the topology of uniform convergence on $\sigma(F,G)$-compact, convex subsets of F. The space F is the dual of G for a locally convex topology $\mathcal{T}$ on G if and only if $\mathcal{T}$ belongs to the interval $[\sigma,\tau]$ in the ordered set of topologies {5, Chap. IV, §1, no.1, Théorème 1, p.IV.2}. Valdivia says that E is a <u>Mackey space</u> if its topology coincides with $\tau(E,E')$.

In [50] Valdivia gives new proofs of some results of [17], and it is this paper that he follows in Chapter I, §3,1 of [A]. He proves that if E is a Mackey space, E′ is locally complete for $\tau(E',E)$, and the completion $\hat{E}$ of E is barrelled, then E is barrelled [A, Chapter I, §3,1.(7)].

Let E be a barrelled space and (A_n) an increasing sequence of closed, convex subsets of E such that $\bigcup A_n = E$. If $\hat{E}$ is a Baire space, then there is an $N \in \mathbb{N}$ such that A_N has a non-empty interior [17, Th.4; 50, Corol.4.1; A, Chapter I, §3, 1.(4)].

If E is a metrizable barrelled space, and there exists a sequence (A_n) of bounded, balanced, convex, complete subsets such that E is the span of $\cup A_n$, then E is a Banach space.

A balanced subset A of a locally convex space *absorbs* another set B if $\lambda A \supset B$ for some $\lambda > 0$. A set which absorbs all bounded sets is said to be *bornivorous*.

A locally convex space E is:

infrabarrelled (or quasibarrelled) if every bornivorous barrel is a neighborhood of 0,

bornological if every balanced, convex, bornivorous set is a neighborhood of 0,

ultrabornological if every balanced, convex set that absorbs all balanced, convex, compact sets is a neighborhood of 0.

E is a infrabarrelled if and only if every strongly bounded subset of E' is equicontinuous; it is bornological if and only if it is the inductive limit of normed spaces; it is ultrabornological if and only if it is the inductive limit of Banach spaces. One has the implications:

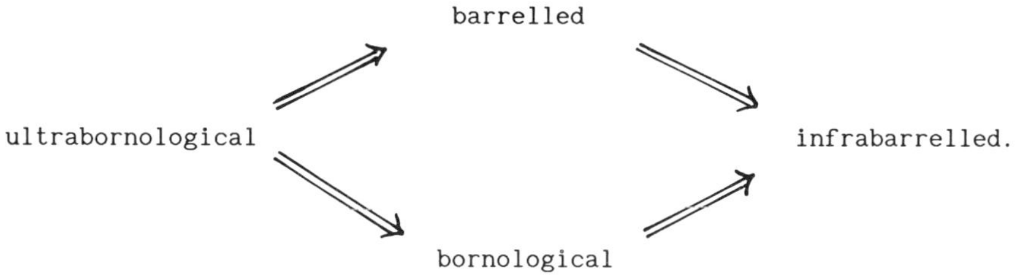

It is easy to give an example of a non-barrelled, bornological space: $\mathbb{R}^{(\mathbb{N})}$ as a subspace of ℓ^2. In 1954 L. Nachbin and T. Shirota considered independently the space $\mathscr{C}(T)$ of continuous functions on the completely regular Hausdorff space T, equipped with the topology of uniform convergence on compact subsets of T. They proved:

1) $\mathscr{C}(T)$ is barrelled if and only if every closed subset of T, on which every $f \in \mathscr{C}(T)$ is bounded, is compact;

2) $\mathscr{C}(T)$ is bornological if and only if T is replete ("realcompact" {14, Chap.8}).

The continuum hypothesis implies that there exists a completely regular space T that satisfies condition 1) but is not replete, and therefore there exist non-bornological barrelled spaces. It was of interest to find such spaces without using the continuum hypothesis. This was first done by Y. Kōmura, and in [22] Valdivia takes a product $E = \prod_{i \in I} E_i$ of an uncountable family of barrelled spaces $\neq \{0\}$ (e.g. lines) and considers the subspace G consisting of those $(x_i)_{i \in I}$ for which $x_i = 0$ with the exception of countably many indices. He proves that every subspace F of E containing G properly, such that the codimension of G in F is finite, is barrelled, dense in E, and not bornological. The same result holds replacing "barrelled" by "infrabarrelled".

Using a similar construction, Valdivia proves in [29] that if E is a separable infinite-dimensional Banach space, then E equipped with $\sigma(E,E')$ can be imbedded as a dense subspace in a barrelled space that is not bornological. In his Silivri lecture of 1973 [27] he gives other imbedding theorems. Thus every metrizable barrelled space F is a closed subspace of some metrizable ultrabornological space E; and every barrelled bornological space F is a closed subspace of some ultrabornological space E. He takes $E = \mathscr{C}(T)$, where T is F' equipped with $\sigma(F',F)$. Since T is replete, an extension due to M. De Wilde and J. Schmets of the Nachbin-Shirota theorem implies that E is ultrabornological. A different construction shows that every bornological space is a closed subspace of some ultrabornological space.

There are in [27] also representations as quotients. Recall that a space is semi-Montel if every bounded closed subset is compact; it is Montel if it is barrelled and semi-Montel {5, Chap. IV, §2, no.5, Définition 4, p. IV.18}. Then every bornological space and every separable space is a quotient of some complete semi-Montel space. Further representations as closed subspaces and as quotient spaces of ultrabornological spaces are contained in [33] for spaces which have families $(U_i)_{i \in I}$ of neighborhoods of 0 such that $\bigcap_{i \in I} U_i = \{0\}$ and the cardinal of I is non-measurable {14, Chapter 12}.

In [51] Valdivia proves that if E is a nuclear space {17, 16.1.4, p.345; 26, 4.1.2, p.70} and F any separable infinite-dimensional Banach space, then there exists a balanced, convex neighborhood V of 0 such that $\widehat{E_V}$ is norm-isomorphic to F; here E_V is the quotient of E modulo the subspace N on which the gauge q_V of V vanishes, equipped with the quotient of q_V, which is a norm on E/N. For special F this result was discovered by several authors {17, 21.2.5, p.483 and 21.11, p.517}. A complete nuclear space E, whose topology is not $\sigma(E,E')$, is the projective limit of an arbitrarily given family of infinite-dimensional Banach spaces.

Answering a question of Dieudonné, Valdivia constructed in [56] a bornological space E whose completion $\hat{E}$ is not bornological. Actually E is a space $\mathscr{C}(T)$ with an appropriate replete T, and it is a hyperplane in $\hat{E}$.

An ultrabornological space is both barrelled and bornological. In the first edition of {5} Bourbaki asserts that it is not known whether the converse is false. In [18] Valdivia shows that if E is the product of an infinite family of ultrabornological spaces $\neq \{0\}$, or the normed space ℓ^1_I, where I is an infinite set of indices, then there exist two sequences (E_n) and (F_n) of dense subspaces of E such that $E_n \supset F_n \supset E_{n+1}$ for all n, each E_n is ultrabornological, and each F_n is barrelled and bornological but not ultrabornological. In [21] he gives examples of barrelled, bornological spaces E that are not even inductive limits of Baire spaces, and in [47] he shows that E can be metrizable.

In his lecture [30] at the Bordeaux conference of 1973 (where I first met Valdivia in person) he produces a barrelled, bornological space that is not the inductive limit of barrelled, normed spaces.

If Ω is an open subset of $\mathbb{R}^n$, and K a compact subset of Ω, denote by $\mathcal{D}(K)$ the vector space of all infinitely differentiable functions on Ω with support in K, equipped with the topology of uniform convergence of $f \in \mathcal{D}(K)$ and all its derivatives. The test function space $\mathcal{D}(\Omega) = \bigcup_K \mathcal{D}(K)$ is equipped with the locally convex inductive limit topology, and its dual $\mathcal{D}'(\Omega)$ is the space of Schwartz distributions which we equip with the strong topology. In [34] Valdivia proves that on $\mathcal{D}'(\Omega)$ there exists a coarser topology which is barrelled and bornological but not ultrabornological. In [62] he shows that on both $\mathcal{D}(\Omega)$ and $\mathcal{D}'(\Omega)$ there exist coarser topologies which are barrelled and bornological but not the inductive limit of Baire spaces. For $\mathcal{D}'(\Omega)$ this is based in part on the following observation: If (G_n) is an increasing sequence of closed subspaces of E' whose union $G = \bigcup G_n$ is $\sigma(E',E)$-dense, then E is bornological for $\tau(E,G)$. De Wilde's closed graph theorem {19, §35, 2.(2)} is used to show that under appropriate hypotheses E is not ultrabornological.

According to [35], if E is the strict inductive limit {5, p.II. 36} of the sequence of spaces E_n which have the topology $\tau(E_n,E'_n)$, and if the E'_n equipped with $\tau(E'_n,E_n)$ are ultrabornological, then E' equipped with $\tau(E',E)$ is ultrabornological.

Dieudonné proved that a finite-codimensional subspace of a bornological space is bornological [A, Chapter 1, §3,2.(3), p.43], and in [14] Valdivia proves that a finite-codimensional subspace of an infrabarrelled space is infrabarrelled [A, Chapter 1, §3,2.(2),p.43]. The situation is completely different for ultrabornological spaces. In his Comptes Rendus note [53] Valdivia constructs in any ultrabornological space, whose topology is not the finest locally convex topology, a hyperplane H that is not ultrabornological. Observe that H is another example of a barrelled, bornological space that is not ultrabornological.

Valdivia gives an example in [21] of a countable-codimensional subspace F of a bornological space E that is not even infrabarrelled, and in [47] he shows that F may be dense in E. However, in [16] he proves that every countable-codimensional subspace of an ultrabornological space is bornological [A, Chapter 1,§3,2.(5),p.45]. To prove this, he uses the following theorem [16, Theorem 3]: If E is the inductive limit of the barrelled spaces E_i and if F is a subspace of E which is countable-codimensional in each $F+E_i$, then F is barrelled and the inductive limit of the spaces $E \cap F_i$. If now E is ultrabornological, then it is the inductive limit of Banach spaces E_i which are a fortiori barrelled (observe that the same conclusion cannot be drawn if E is bornological and the E_i only normed). If F is countable-codimensional in E, it is so in each $F+E_i$. Thus F is the inductive limit of the normed spaces $F \cap E_i$ and so it is bornological.

In [92] Valdivia proves that if F is a dense, non-barrelled subspace of a Fréchet space {5, p.II.26 and 66} E, then there exists an infinite-dimensional closed subspace G of E such that $F \cap G = \{0\}$. It follows that if E is a barrelled space such that E' is complete for $\tau(E',E)$, then every subspace of E with codimension $\leq 2^{\aleph_0}$ is barrelled.

A locally convex space is said to be <u>countably infrabarrelled</u> (resp. <u>countably barrelled</u>) if whenever (V_n) is a decreasing sequence of balanced, convex, closed neighborhoods of 0 such that $V = \bigcap V_n$ is bornivorous (resp. absorbing), then V is a neighborhood of 0. Equivalently, the space E is countably infrabarrelled (resp. countably barrelled) if in E' every $\beta(E',E)$-bounded (resp. $\sigma(E',E)$-bounded) union of a countable family of equicontinuous sets is equicontinuous. A local convex space E is <u>sequentially barrelled</u> (resp. <u>sequentially infrabarrelled</u>) if every $\sigma(E',E)$-bounded (resp. $\beta(E',E)$-bounded) sequence in E' is equicontinuous.

Grothendieck observed that, while the strong dual of a metrizable space is not always barrelled {19, §31,7}, it is countably infrabarrelled, and it obviously has a countable fundamental system of bounded sets. He called spaces having the last two properties $(\mathcal{DF})$-<u>spaces</u> {5,p.IV,58; 19 §29,3).

In [15] Valdivia proves that a finite-codimensional subspace of a $(\mathcal{DF})$-space is a $(\mathcal{DF})$-space. More generally, if F is a subspace of a $(\mathcal{DF})$-space E such that for every bounded subset B of E the codimension of F in the span of $F \cup B$ is finite, then F is a $(\mathcal{DF})$-space. On the other hand, in [21] he gives an example of an infrabarrelled $(\mathcal{DF})$-space which has a subspace of countably infinite codimension that is not a $(\mathcal{DF})$-space.

In [44] Valdivia proves a very general theorem concerning the inheritance of some barrelledness-like properties of finite-codimensional subspaces. Let $\mathcal{B}$ be a <u>cofilter basis</u> on the space E, i.e. a collection of subsets of E such that if B_1 and B_2 belongs to $\mathcal{B}$, then there exists $B_3 \in \mathcal{B}$ containing $B_1 \cup B_2$. Assume that $\mathcal{B}$ consists of bounded, balanced, convex, closed sets which satisfy $\bigcup_{B\in\mathcal{B}} B = E$, and that if $B \in \mathcal{B}$ then $\lambda B \in \mathcal{B}$ for all $\lambda \neq 0$ (i.e. $\mathcal{B}$ is <u>homothecy-invariant</u>). Let $\mathcal{T}_0$ be a topology on E that is compatible with the pairing (E,E'), and let α be an infinite cardinal number. Assume that every subset of E', which is bounded for the topology of uniform convergence on the sets $B \in \mathcal{B}$, and which is the union of at most α sets equicontinuous with respect to $\mathcal{T}_0$ is itself equicontinuous. Let F be a finite-codimensional subspace of E and $\mathcal{B}_F$ the collection of all sets $B \cap F$, where B varies in $\mathcal{B}$. Then every subset of F′, which is bounded for the topology of uniform convergence on the sets of $\mathcal{B}_F$, and which is the union of at most α sets equicontinuous with respect to the topology induced by $\mathcal{T}_0$ on F, is itself equicontinuous.

Various choices of $\mathcal{B}$, $\mathcal{T}_0$ and α yield the inheritance by finite-codimensional subspaces of the property listed in the last column of the table below. We use the following notation: $\mathcal{T}$ is the original topology of E, $\mathcal{B}_1$ is the collection of all bounded, balanced, convex, closed subsets of E, and $\mathcal{B}_2$ is the collection of those $B \in \mathcal{B}_1$ which generate finite-dimensional subspaces.

$\mathcal{B}$	$\mathcal{T}_0$	α	property
$\mathcal{B}_2$	$\sigma(E,E')$	card E′	barrelled
$\mathcal{B}_1$	$\sigma(E,E')$	card E′	infra-barrelled
$\mathcal{B}_2$	$\sigma(E,E')$	$\aleph_0$	sequentially barrelled
$\mathcal{B}_1$	$\sigma(E,E')$	$\aleph_0$	sequentially infrabarrelled
$\mathcal{B}_2$	$\mathcal{T}$	$\aleph_0$	countably barrelled
$\mathcal{B}_1$	$\mathcal{T}$	$\aleph_0$	countably infrabarrelled

From the last row it follows again that a finite-codimensional subspace of a $(\mathcal{DF})$-space is a $(\mathcal{DF})$-space.

In analogy with $(\mathcal{DF})$-spaces, a space is said to be <u>dual metric</u> if it has a countable fundamental system of bounded sets and is sequentially infrabarrelled. Obviously every $(\mathcal{DF})$-space is dual metric, and in [91] Valdivia proves theorems which yield dual metric spaces that are not $(\mathcal{DF})$.

Equip E′ with $\sigma(E',E)$, and let α be an infinite cardinal. Denote by $\mathcal{B}_\alpha$ the collection of all balanced, convex, closed subsets of E′ such that every subset of cardinality at most α is equicontinuous. Paper [26] studies the topology $\mathcal{T}_\alpha$ on E of uniform convergence on elements of $\mathcal{B}_\alpha$.

If $A \subset E$ is the union of a family $(A_i)_{i \in I}$ of precompact sets such that card I is at most α, then the original topology on E coincides with $\mathcal{T}_\alpha$ on A. The case $\alpha = \aleph_0$ yields a theorem of Grothendieck on $(\mathcal{DF})$-spaces {19, §29, 3.(8)}.

E is α-barrelled if every $\sigma(E',E)$-bounded subset of B of E with card $B \leq \alpha$ is equicontinuous. Valdivia proves that if $\alpha < \gamma$, there exist α-barrelled spaces that are not γ-barrelled.

In 1971 C.L. De Vito {8} introduced (ab)-spaces, which sit between bornological and infrabarrelled spaces. In [31] Valdivia proves that every infrabarrelled $(\mathcal{DF})$-space is an (ab)-space, together with some characterizations of (ab)-spaces. In [32] he constructs an example of an infrabarrelled $(\mathcal{DF})$-space that is neither bornological nor barrelled, giving an answer to a question of Grothendieck {15, Question 3, p.120}. Examples of non-bornological, barrelled $(\mathcal{DF})$-spaces can be found in [63], see Section 7.

Grothendieck also asked whether there exist spaces with a countable fundamental system of bounded sets which satisfy the Mackey convergence condition but not the strict Mackey convergence condition {15, III.1, Définition 3, p.105 and p.111}. An example was given by Seth Warner in 1958, and Valdivia proves in [79] that if E is a Banach space such that E′ equipped with $\sigma(E',E)$ is not separable, then there is a topology $\mathcal{T}$ on E, compatible with the pairing (E,E′) such that E equipped with $\mathcal{T}$ is a $(\mathcal{DF})$-space satisfying the requirements of Grothendieck.

Further results on ultrabornological spaces are given in the related papers [40] and [42], which continue a line started in [30] to find conditions under which any infinite-dimensional Banach space is the inductive limit of spaces all isomorphic to the same infinite-dimensional space F. This is the case if F is a Banach space whose dual is separable for $\sigma(F',F)$. This is also the case if F is a sequentially complete space such that: 1) F contains a bounded, countable, total set, 2) in F′ equipped with $\sigma(F',F)$ there is an equicontinuous, countable, total set, 3) if the injective linear map $u : F \longrightarrow F$ has a closed graph, then u is continuous. Since $\mathcal{D}(K)$ as well as its dual satisfy these conditions, every ultrabornological space is the inductive limit of nuclear Fréchet spaces, and also of nuclear $(\mathcal{DF})$-spaces. See {13, §2, pp.210-213}.

In the note [38] Valdivia proves, among more general results, that if

E equipped with $\tau(E,E')$ is sequentially infrabarrelled, then it is infrabarrelled if and only if E' is quasicomplete (i.e. every bounded, closed set is complete) for the topology of uniform convergence on the balanced, convex, bounded, separable subsets of E. The same result holds if we replace "infrabarrelled" by "barrelled".

In [77; A, Chapter I, §5,2; 87] Valdivia investigates the third and the fourth link in the following chain of increasingly larger classes of spaces:

Baire ⇒ unordered Baire-like ⇒ totally barrelled ⇒ suprabarrelled ⇒ Baire-like ⇒ quasi-Baire ⇒ barrelled.

A space E is <u>suprabarrelled</u> if given an increasing sequence (L_n) of subspaces such that $\bigcup L_n = E$, there exists $p \in \mathbb{N}$ such that L_p is dense in E and barrelled. This class was introduced by W.J. Robertson, I Tweddle and F.E. Yeomans {27} under the name of (db)-spaces, and S. Saxon {31} showed that any metrizable $(\mathcal{LF})$-space {19, §19,5} is Baire-like but not suprabarrelled.

A space E is <u>totally barrelled</u> if given a sequence (L_n) of subspaces such that $\bigcup L_n = E$, there exists $p \in \mathbb{N}$ such that L_p is barrelled and its closure $\bar{L}_p$ has finite codimension in E. For unordered Baire-like, Baire-like and quasi-Baire see {25}, Definitions (9.1.19), (9.1.1) and (9.1.9), respectively.

The above quoted references give inheritance results, e.g. to products and to countable codimensional subspaces. A separable, infinite-dimensional Fréchet space always contains a dense subspace that is suprabarrelled but not the inductive limit of unordered Baire-like subspaces. For a similar result see {31, Theorem 4, p.77}.

In Valdivia's joint paper [93] with P. Pérez Carreras it is proved that every non-normable Fréchet space E contains a dense subspace which is a metrizable $(\mathcal{LF})$-space, inductive limit of a sequence (F_n) of Fréchet spaces. If E is reflexive (resp. Montel, Schwartz {15, III.4, Définition 5, p.117; 17, 10.4, p.201}), then the F_n are reflexive (resp. Montel, Schwartz).

Baire-hyperplane spaces [68] are those in which every union of a sequence of closed hyperplanes has a void interior. Every metrizable $(\mathcal{LF})$-space is a Baire-hyperplane space. Let E be either an infinite-dimensional, separable Fréchet space or a locally complete inductive limit of infinite-dimensional, separable Fréchet spaces. A construction similar to one given in [77] shows that there exist two dense subspaces F and G of E such that G is ultrabornological and a hyperplane in F, and F is barrelled and not the inductive limit of Baire-hyperplane spaces.

Answering a question of V. Klee and A. Wilansky, J. Arias de Reyna proved that assuming Martin's axiom, every separable, infinite-dimensional Banach space contains a meager, dense hyperplane. Valdivia proves in [95] that, still assuming Martin's axiom, every separable, infinite-dimensional Baire space contains a dense, meager hyperplane. Assuming the continuum hypothesis, he proves that in every infinite-dimensional Fréchet space E there exists a subspace L such that given a countable set $P \subset E$, there exists a closed, separable subspace G of E with the following properties: G contains P, and $L \cap G$ is a dense, countable-codimensional, meager subspace of G.

It is well known that a product of Baire spaces is not necessarily a Baire space. In [97] Valdivia finds such examples inside the spaces $c_0(I)$ and $\ell^p(I)$, $0 < p < \infty$. More precisely, assume that card $I > \aleph_0$. There exists a family $(X_i)_{i \in I}$ of dense subspaces in any of the listed spaces such that X_i^m is a Baire space for every $m \in \mathbb{N}$, $i \in I$, but $X_i \times X_\kappa$ is not a Baire space whenever $i \neq \kappa$. The case $p = 2$ was considered earlier by J. Arias de Reyna {1}.

In the note [98] published in the proceedings of the 1984 Campinas conference, Valdivia answers a question of L. Nachbin by showing that if the topology of a space E is defined by the collection COS(E) of all seminorms p such that the canonical surjection $E \longrightarrow E/p^{-1}(0)$ is an open mapping, then COS(E) is not necessarily a directed set. He also proves a three-space theorem: If E is a non-barrelled, metrizable space, then there exists a closed subspace F such that neither F nor E/F is barrelled.

Grothendieck {16, Chapter III, Section 3, Exercise 8, p.199} introduced the subspace $\ell_0^\infty(I)$ of $\ell^\infty(I)$ generated by characteristic functions of subsets of the set I, i.e. the space of functions on I which take only a finite number of values. He asserted that it is a non-complete, barrelled, normed space. In [70; A, Chapter I, §7,2, pp.133-136] Valdivia considers a tribe $\mathcal{A}$ of subsets of I and the subspace $\ell_0^\infty(I,\mathcal{A})$ of $\ell_0^\infty(I)$ generated by the characteristic functions of sets belonging to $\mathcal{A}$. He proves that $\ell_0^\infty(I,\mathcal{A})$ is suprabarrelled, from where Grothendieck's assertion follows. The proof is preceded by a long list of interesting propositions, the first two of which are parallel to those in [69], where $I = \mathbb{N}$ and $\mathcal{A}$ consists of all the subsets of $\mathbb{N}$. J. Arias de Reyna {2} proved that $\ell_0^\infty(I,\mathcal{A})$ is not totally barrelled, and more precise results were given by J.C. Ferrando and M. López Pellicer {12}.

A sequence of elements (x_n) in a space E is said to be <u>subseries convergent</u> if for every strictly increasing sequence (n_k) of integers the series $\Sigma_k x_{n_k}$ converges; the concept figures in the statement of the Orlicz-Pettis theorem {10, Chapter IV}. Given a subseries convergent sequence (x_n) in E, let $T : E' \longrightarrow \ell^1$ be the map which associates with the $w \in E'$ the sequence $(<x_n,w>)$; it is continuous if E' and ℓ^1 are equipped with the topologies $\sigma(E',E)$ and $\sigma(\ell^1,\ell_0^\infty)$, respectively. The image of the unit ball of ℓ_0^∞ under the transpose ${}^tT : \ell_0^\infty \longrightarrow E$ spans a barrelled, normed subspace denoted $E[(x_n)]$. D. Bucchioni and A. Goldman {7} introduced the space $G_1' = G_1'(E)$ of all linear forms v on E such that $\Sigma|<x_n,v>| < \infty$ for all subseries convergent sequences (x_n). Valdivia proves in [69] that E equipped with $\tau(E,G_1')$ is the inductive limit of the subspaces $E[(x_n)]$, and answers questions of Bucchioni and Goldman. He also gives conditions which ensure that every subseries convergent sequence is <u>bounded multiplier convergent</u> {10, Chapter IV, Exercise 4, p.29}.

Returning to the tribe $\mathcal{A}$ on I of [70], let $H(\mathcal{A})$ be the space of finitely additive bounded measures on $\mathcal{A}$. Let now T be the topological isomorphism from $H(\mathcal{A})$ onto $(\ell_0^\infty(I,\mathcal{A}))'$ which with each measure μ associates the linear form $\chi_A \longrightarrow \mu(A)$, where χ_A is the characteristic function of $A \in \mathcal{A}$. Let $(\mathcal{A}_n)$ be an increasing sequence of subcollections of $\mathcal{A}$ such that $\cup\mathcal{A}_n = \mathcal{A}$. Using T Valdivia proves the following results: 1) There exists $p \in \mathbb{N}$ such that if $M \subset H(\mathcal{A})$ is bounded on $\mathcal{A}_p$, then M is bounded on $\mathcal{A}$; 2) There exists $p \in \mathbb{N}$ such that if (μ_n) is a sequence in

$H(\mathcal{A})$ and $(\mu_n(A))$ is a Cauchy sequence for every $A \in \mathcal{A}_p$, then (μ_n) is weakly convergent in $H(\mathcal{A})$.

3. BIDUALS AND REFLEXIVITY

In [20] Valdivia proves, generalizing a result of R.C. James, that if E is a real, quasi-complete, non-semi-reflexive {5, Chap.IV, §2, n°2, Définition 2, p.IV.15} space, then there exists a certain $\sigma(E',E)$-closed, convex subset A of E′ not containing the origin, and a balanced, convex, bounded, closed subset B of E such that no $u \in A$ attains its least upper bound on B. In [24] he uses the result of James to show that in any infinite-dimensional Fréchet space E there exists a balanced, convex, bounded, non-closed subset B such that its span E_B, equipped with the gauge of B {5, Chap.II, §2, n°11, p.II.22} as norm, is complete, i.e. a Banach space. Further examples of such sets B are pointed out in the final remark of [42].

Under what conditions is a finite-codimensional subspace F of a Mackey space E itself a Mackey space? Saxon and Levin proved it when E′ is sequentially complete for $\sigma(E',E)$, i.e., E has "property S", and F is closed. In [37] Valdivia proves the analogous result assuming F separable instead of closed. In [44] he shows it under either of the following conditions:

1) E is separable and in E′ equipped with $\sigma(E',E)$ every strongly bounded Cauchy sequence converges;

2) F is dense in E for the topology of uniform convergence on strongly bounded subsets of E′;

3) F is closed and E′ equipped with $\sigma(E',E)$ is locally complete: in this case F may have countable codimension.

In [44] he proves furthermore that if E is the strict inductive limit of a sequence of Fréchet spaces, or if E is an infrabarrelled $(\mathcal{DF})$-space, and F is a dense subspace of E, then F is a Mackey space. If E is a Fréchet space and we equip E′ with $\tau(E',E)$, then the hyperplanes of E′ are Mackey spaces if and only if E is sequentially complete for $\sigma(E,E')$.

Grothendieck {15, Question 8, p.121} asked whether, if E is the strict inductive limit of the sequence (E_n), is E″ the inductive limit of the E''_n? In [66] Valdivia gives a negative answer to the question by showing that if E is a Banach space whose dual E′ is separable for the strong topology, then there exist in E′ equipped with $\tau(E',E)$ two sequences (F_n) and (H_n) of subspaces such that each H_n can be identified with F''_n, the inductive limit of the H_n is a dense subspace of E′ distinct from E′, and the bidual of the inductive limit of the F_n is E′ itself.

R.C. James showed in 1950 that there exist Banach spaces which have finite codimension in their biduals; such spaces are said to be _quasireflexive_. In [80] Valdivia proves that if E is the locally convex direct sum of a sequence of quasireflexive Banach spaces, then E has the _Krein-Shmulian property_, i.e. a convex subset of E′ is $\sigma(E',E)$-closed if its intersection with every $\sigma(E',E)$-closed equicontinuous set is $\sigma(E',E)$-closed. This follows from the fact that the $\sigma(E',E)$-closure of any convex subset of E′ coincides with its sequential closure. The proof involves the subspace of $(E')^*$ formed by the linear forms on E′ which are bounded on equicontinuous sets.

Article [100] is dedicated to prove that if the Banach space E is not quasireflexive, there exists a separable non-quasireflexive closed subspace F such that E/F is not quasireflexive.

A space E is _weakly compactly generated_ {10, p.228} if there exists a balanced, convex $\sigma(E,E')$-compact subset whose span is dense in E. It is _weakly countably determined_ if there exists a metrizable, separable topological space T and a map Φ from T into the collection of all $\sigma(E,E')$-compact subsets of E such that $E = \bigcup_{t\in T} \Phi(t)$ and given a sequence (t_n) in T which converges to $t \in T$ and a neighborhood V of $\Phi(t)$, there exists an $N \in \mathbb{N}$ such that $\Phi(t_n) \subset V$ for $n \geq N$. Every weakly compactly generated space is weakly countably determined.

Let F be a closed subspace of the Banach space E. In [112] it is proved that E is weakly countably determined (resp. weakly compactly generated) under either of the following conditions: 1) E/F is separable and F is weakly countably determined (resp. weakly compactly generated); 2) E″/F is separable and E/F is weakly countably determined (resp. weakly compactly generated).

In what follows I denote by $\tilde{A}$ the closure of $A \subset E$ in E″ equipped with $\sigma(E'',E')$. A decreasing sequence (A_n) of subsets of the space E is quasibounded if for every neighborhood U of 0 there is an $m \in \mathbb{N}$ such that $A_m \subset mU$. In [113] Valdivia proves that a metrizable space E is weakly countably determined if and only if there exists a sequence (U_n) of neighborhoods of 0 such that for every $x \in E$ there is a quasibounded subsequence (U_{n_j}) with $x \in \bigcap_{j=1}^{\infty} \tilde{U}_{n_j} \subset E$. If E is a Fréchet space, the U_n can be chosen balanced, convex. Motivated by this fact, Valdivia says that E is weakly countably convex-determined if there exists a sequence (U_n) of balanced, convex neighborhoods of 0 such that for any $x \in E$ there is a quasibounded subsequence (U_{n_j}) with $x \in \bigcap_{j=1}^{\infty} \tilde{U}_{n_j} \subset E$.

The density character d(E) of the space E is the smallest cardinal c such that there exists a dense subset A of E with cardA = c. Let ω be the first infinite ordinal and μ the first ordinal whose cardinal is d(E). A family $\{P_\alpha : \omega \le \alpha \le \mu\}$ of continuous projectors on the infinite-dimensional space E such that $P_\alpha \circ P_\beta = P_\beta \circ P_\alpha = P_\alpha$ for $\alpha \le \beta$, P_μ is the identity, $d(P_\alpha(E)) \le \text{card}\,\alpha$, and for any limit ordinal α the closure of $\bigcup_{\beta<\alpha} P_\beta(E)$ coincides with $P_\alpha(E)$, is said to be a resolution of the identity in E. The eponymous theorem of [113] asserts that if E is a weakly countably convex-determined, metrizable, infinite-dimensional space, whose topology is defined by an increasing sequence (p_n) of seminorms, then there exists a resolution of the identity on E with $p_n(P_\alpha) = 1$ for all α and n. For weakly compactly generated Banach spaces the result is due to D. Amir and J. Lindenstrauss, for weakly countably determined Banach spaces to L. Vašak. In [108] Valdivia gave a "drastically simplified" proof of the Amir-Lindenstrauss theorem and used the same method to generalize it to weakly compactly generated Fréchet spaces. The proof in [113] employs a similar technique.

In [113] Valdivia also proves that if E is a Fréchet space and the quotient E″/E is separable when E″ is equipped with $\beta(E'',E')$, then E is the topological direct sum of a reflexive subspace and a subspace P such that P'' equipped with $\beta(P'',P')$ is separable. For Banach spaces this was proved in [54] but Valdivia points out that the Fréchet space case is "much more difficult to prove", because the former uses a theorem of Amir-Lindenstrauss on weakly compactly generated Banach spaces, while the latter is based on the following theorem which is new even for Banach spaces: Let F be a closed subspace of the Fréchet space E. If F' equipped with $\beta(F',F)$ is separable and $E'' = E + \tilde{F}$, then there exists a closed subspace G of E such that $G \supset F$, its dual G' equipped with $\beta(G',G)$ is separable, and there is a reflexive subspace L of E'' equipped with $\beta(E'',E')$ verifying $L + \tilde{G} = E''$, $L \cap \tilde{G} = \{0\}$.

A space E is <u>totally reflexive</u> if every Hausdorff quotient of E is reflexive. One of the last questions of Grothendieck to remain open was No.9 {15, p.121}: Is the product of two totally reflexive Fréchet spaces totally reflexive? In [109] Valdivia gives an affirmative answer based on the following characterization: A Fréchet space E is totally reflexive if and only if there exists a sequence (F_n) of reflexive Banach spaces such that E is isomorphic to a closed subspace of $\prod_{n=1}^{\infty} F_n$. The existence of the F_n relies on a 15 pages long construction for a class of separable Fréchet spaces quotients of which have Schauder bases possessing a certain property introduced by Ivan Singer. In parts a method attributed to S. Mazur is used, which occurs also in [104] and [107] (see Section 10).

4. COMPLETENESS

A celebrated theorem of G. Köthe states that the strict inductive limit of a sequence of complete spaces is complete {5, Chap.II, §4, n°6, Proposition 9, p.II. 35; 19, §19,5.(3)}. A very simple proof of this is given in [47, Note 2, p.246].

From the Collins-Pták theorem {19, §21,9.(6)} Valdivia deduces in [6] the following result: Let $\mathcal{F}$ be a collection of complete subspaces of the space E covering E. Then E is complete if and only if every hyperplane H in E', which intersects every equicontinuous subset A of E' in a closed set for the topology induced on A by $\sigma(E',E)$, contains $F^{\perp}$ for some $F \in \mathcal{F}$.

This yields a corollary which, as Valdivia shows by an example, is stronger than Köthe's theorem: Let (E_n) be an increasing sequence of complete subspaces of E such that $E = \bigcup_n E_n$. Denote by $\mathcal{T}_n$ the topology induced on E_n by $\sigma(E,E')$, and by $\mathcal{T}_\infty$ the topology of the inductive limit of the spaces E_n equipped with $\mathcal{T}_n$. If the original topology $\mathcal{T}$ of E is finer than $\mathcal{T}_\infty$, then E is complete for $\mathcal{T}$.

Completeness results connected with strict $(\mathcal{LF})$-spaces are contained in [103]. Thus if E is the strict inductive limit of an increasing sequence of Fréchet spaces and F is a barrelled, closed subspace of E, then E/F is complete.

Let $\mathcal{B}$ be a homothecy-invariant cofilter basis of balanced, convex, bounded, closed subsets of the space E such that $\bigcup_{B \in \mathcal{B}} B = E$. Equip E' with the topology of uniform convergence on the sets $B \in \mathcal{B}$, and let n be a positive integer. In [19] Valdivia proves that E' is complete if and only if every subspace F of E of codimension n, such that $F \cap B$ is closed for all $B \in \mathcal{B}$, is closed. If we assume furthermore that for every $B \in \mathcal{B}$ the span E_B of B, equipped with the gauge of B, is barrelled, then the completeness of E' implies that every countable-codimensional subspace F of E is closed provided that $F \cap B$ is closed for every $B \in \mathcal{B}$.

Let F be a finite-codimensional subspace of E and $\mathcal{B}_F = \{F \cap B : B \in \mathcal{B}\}$. If E' is complete for the topology of uniform convergence on the sets $B \in \mathcal{B}$, then F' is complete for the topology of uniform convergence on the sets $\mathcal{B}_F$. In [44] "complete" is replaced by sequentially complete, quasicomplete and locally complete. If each E_B is barrelled, F may be countable-codimensional.

The subspace F of E is <u>locally dense</u> if for any $x \in E$ there is a sequence in F which converges to x in the Mackey sense. A locally dense subspace of a Mackey space is a Mackey space. Combining the preceding results with {19, §28, 5.(4)} Valdivia obtains a new proof of Dieudonné's theorem, according to which a finite-codimensional subspace of a bornological space is bornological [19, Remark, p.187].

Paper [35] contains the following: Let E be the strict inductive limit of the sequence (E_n), equip E' with $\tau(E',E)$, and let G be a closed subspace of E'. Assume that $G^{\perp} \cap E_n$ is ultrabornological for all n (which happens e.g. if the E_n are Fréchet spaces). If E'/G is sequentially complete, then it is complete. A refinement in [43] of the result involving Fréchet spaces requires E'/G to be locally complete.

In [48] Valdivia considers the bidual E'' of a space E and a subspace G of E'' containing E such that the codimension of E in G is finite. Let me quote two results:

1) Assume that E contains a countable family of balanced, convex, $\sigma(E,E')$-compact subsets whose union is dense in E. Equip G with the topology of uniform convergence on all the subsets of E' that are balanced, convex, $\sigma(E',G)$-compact and equicontinuous on E. If E is complete (resp. quasicomplete), then so is G.

2) Assume that E is separable and quasicomplete for $\tau(E,E')$. Then G equipped with $\tau(G,E')$ is quasicomplete.

The paper also presents a construction of quasireflexive spaces.

The first result of [52] characterizes those spaces E which are topologically isomorphic to $\mathbb{K}^{\mathbb{N}}$. Conditions are given for $\mathbb{K}^{\mathbb{N}}$ or $\mathbb{K}^{(\mathbb{N})}$ to be a complemented subspace of E. Finally the author proves that $(\mathbb{K}^{(\mathbb{N})})^{\mathbb{N}}$ is <u>hypercomplete</u> {19, §34, 4} using the following separation theorem: Let A be a balanced, convex, $\sigma(E,E')$-complete subset of the space E, and G an affine subspace of E such that $G \cap A = \phi$. There exists $u \in E'$ such that $\sup_{a \in A} |\langle u,a \rangle| < 1$ and $\langle u,x \rangle = 1$ for all $x \in G$.

5. CLOSED GRAPH AND OPEN MAPPING THEOREMS

It is remarkable that barrelled spaces, originally introduced to generalize the uniform boundedness principle, also play a role in closed graph theorems. M. Mahowald discovered that if the space is such that every linear map from E into a Banach space with a closed graph is necessarily continuous, then E is barrelled {19,§34,7.(1)}.

A space E is barrelled if and only if any linear map u from E into a space F is almost continuous, i.e. for every neighborhood W of 0 in F the closure of $u^{-1}(W)$ is a neighborhood of 0 in E {19, §34, 6.(1)}. V. Pták introduced under the name of B_r-complete spaces those spaces F for which any almost continuous linear map from a space E into F, whose graph is closed, is necessarily continuous; they are now called mostly infra-Pták spaces {19,§34,6.(8)}.

Let (F,G) be a pairing of two vector spaces and $\mathcal{T}$ a topology on F compatible with the duality, i.e., such that G is the dual of F equipped with $\mathcal{T}$. Then $\mathcal{T}^f$ denotes the finest topology on G which coincides with $\sigma(G,F)$ on every $\mathcal{T}$-equicontinuous subset of G {19,§21.8}. The space E is infra-Pták if and only if every $\mathcal{T}^f$-closed, $\sigma(E',E)$-dense subspace of E′ is $\sigma(E',E)$-closed, i.e. coincides with E′. Analogously E is a Pták space (or B-complete) if every $\mathcal{T}^f$-closed subspace of E′ is $\sigma(E',E)$-closed {19,§34,2}. Obviously every Pták space is infra-Pták but only in [94] did Valdivia settle a problem open since the 1950's by showing that there exist infra-Pták spaces which are not Pták. Take for instance the closure $\tilde{c}_0$ of c_0 in the bidual of ℓ^∞. Since c_0 is infinite-codimensional in $\tilde{c}_0$, there exists a sequence (z_n) of vectors in $\tilde{c}_0$ such that z_{n+1} does not belong to the span L_n of $\ell^\infty \cup \{z_1,\cdots,z_n\}$. If $L = \bigcup L_n$, then $(\ell^\infty)'$ equipped with $\tau((\ell^\infty)',L)$ is infra-Pták but not Pták.

From the dual characterization it follows immediately that every infra-Pták space is complete and that every Fréchet space is a Pták space. Using a result of Smoljanov, Valdivia shows in [34] that $\mathcal{D}'(\Omega)$ is not an infra-Pták space and in [57] that neither is $\mathcal{D}(\Omega)$; a detailed account of the latter result can be found in {24,4.5,pp.77-93}. In [46] he proves that $\mathcal{D}^p(\Omega)$ is not an infra-Pták space for finite p, and in [103] he proves the same thing for Beurling spaces $\mathcal{D}^{(M_p)}(\Omega)$ of ultradifferentiable functions and for spaces of Beurling ultradistributions.

The topology $\mathcal{T}^f$ of E′ associated with a topology $\mathcal{T}$ on E is not necessarily compatible with the vector space structure of E′, and occurs in [36]. If T is a completely regular topological space and $E = \mathcal{C}(T)$, then there is a subset A in E′ equipped with $\mathcal{T}^f$ which is homeomorphic to the k-space associated with T {18, 7.11, p.230}. Using this, Valdivia gives an example of a Banach space E such that E′ equipped with $\mathcal{T}^f$ is not a regular topological space.

Let (E_n) be an increasing sequence of subspaces of the vector space E each equipped with its own topology $\mathcal{T}_n$. In [43] Valdivia compares three topologies on E which induce on each E_n a topology coarser than $\mathcal{T}_n$, namely the finest locally convex such topology $\mathcal{T}$, then finest such vector space topology $\mathcal{T}^i$, and the finest such topology $\mathcal{T}^j$. He proves that $\mathcal{T} = \mathcal{T}^i$, and shows that when $\mathcal{T}^i \neq \mathcal{T}^j$, then for certain topologies $\mathcal{U}$ on E' the topology $\mathcal{U}^f$ is not a vector space topology on E, e.g. if the E_n are Fréchet-Montel spaces, then one can choose $\mathcal{U} = \tau(E',E)$. If there exists in the space E a balanced, convex, bounded, closed subset A such that E_A is infinite-dimensional, then there exists an increasing sequence of subspaces of E such that $\mathcal{T} = \mathcal{T}^j$. If E_A is furthermore complete, there exists a topology $\mathcal{U}$ on E' such that $\mathcal{U}^f$ is not compatible with the vector space structure of E.

Köthe {19, §34,3} and W.H. Summers gave examples of two Pták spaces whose product is not a Pták space. Let F be a Fréchet space. Valdivia proves in [46] that if F is complete for $\sigma(F,F')$, then $F^{\mathbb{N}} \times F^{(\mathbb{N})}$ is a Pták space, and if F is not complete for $\sigma(F,F')$, then $F^{\mathbb{N}} \times F^{(\mathbb{N})}$ is not even an infra-Pták space. If (E_n) and (F_n) are sequences of infinite-dimensional Banach spaces, then according to [47] $(\prod_{n\in\mathbb{N}} E_n) \times (\sum_{n\in\mathbb{N}} F_n)$ is not an infra-Pták space, and one can replace either the E_n or the F_n or both by their duals equipped with their Mackey topology.

In [103] Valdivia considers spaces with nuclear Köthe quotients and also proves that if (E_n) is a sequence of infinite-dimensional Fréchet-Montel spaces, then for their locally convex direct sum $E = \sum_{n\in\mathbb{N}} E_n$ the following assertions are equivalent: a) E is a Pták space, b) every Hausdorff quotient of E is complete, c) every Hausdorff quotient of E is locally complete, d) E is an infra-Pták space, e) any of the preceding four assertions with E replaced by E' equipped with $\tau(E',E)$, f) each E_n is isomorphic to $\mathbb{K}^{\mathbb{N}}$.

In [111] Valdivia shows that if E is the locally convex direct sum of a sequence (E_n) of infinite-dimensional Banach spaces, then the above assertions a) -c) are equivalent and hold if and only if each E_n is quasireflexive .

There exists spaces F which are not infra-Pták and nevertheless every linear map with closed graph from a barrelled space into F is continuous. Therefore Valdivia introduces a class of spaces he calls S-spaces in [4] and Γ_r-spaces in [12] for which the following results hold: a) If E is a barrelled space, F a Γ_r-space, and $u : E \longrightarrow F$ a linear map with closed graph, then u is continuous; 2) If F is not a Γ_r-space, then there exists a barrelled space E, and a linear map $u : E \longrightarrow F$ whose graph is closed but which is not continuous.

Given a space F with topology $\mathcal{T}$, denote by $\mathcal{T}^t$ the coarsest barrelled topology on F that is finer than $\mathcal{T}$ {19,§34,9.(1)}. The Γ_r-spaces F are characterized in [4] by the following property: Denote by $F[\mathcal{T}^t]$ the space F equipped with $\mathcal{T}^t$. Let L be a subspace of $F[\mathcal{T}^t]'$ which is dense for $\sigma(F[\mathcal{T}^t]',F)$. Assume that every bounded subset of L which is closed in F' for $\sigma(F',F)$ is also closed in $F[\mathcal{T}^t]'$ for $\sigma(F[\mathcal{T}^t]',F)$, and that $L \cap F'$ is $\sigma(F',F)$-dense in F'. Then $L = F[\mathcal{T}^t]'$.

The characterization in [12] and characterizations due to Adasch and Kōmura are listed in {35, 5.39, p.71}, see also [A, Chapter 1, §6.2, pp.116-124].

The open mapping theorem or homomorphism theorem is related to the closed graph theorem, and if both spaces are Fréchet then each follows easily from the other {24, p.30}. The barrelled spaces F are characterized also by the property that any linear map u from a locally convex space E onto F is almost open, i.e. for any neighborhood V of 0 in E the closure of u(V) is a neighborhood of 0 in F {19, §34,1.(1)}.

In [5] Valdivia introduces V_r-spaces and V-spaces. For their original description I refer to his paper, and to {35, 5.21, p.65}, where they are denoted $B_r(\mathfrak{T})$-spaces and $B(\mathfrak{T})$-spaces, respectively. In [8] he gives the following characterization:

Call a linear map $u : E \longrightarrow F$ strongly almost open if for any barrel T in E the closure $\overline{u(T)}$ of its image is a neighborhood of 0 in $\overline{u(E)}$. A space E is a V_r-space if and only if every bijective, continuous, linear, strongly almost open map from E onto a space is an isomorphism. A space E is a V-space if and only if every continuous, linear, strongly almost open map into a space is a strict morphism.

Every infra-Pták space is a V_r-space and every Pták space is a V-space. In [13] it is proved that a finite-codimensional subspace of a V-space is also one. Valdivia introduces in [5] also W_r-spaces and W-spaces {35, 5.23, p.67}. Every barrelled space is a W_r-space, and every W_r-space is a W-space. The optimal open mapping theorems are the following: 1) If E is a V_r-space, F a W_r-space, then every bijective, continuous, linear map $E \longrightarrow F$ is an isomorphism. 2) If E is not a V_r-space, then there exists a barrelled space F and a bijective, continuous linear map $E \longrightarrow F$ that is not an isomorphism. 3) If F is not a W_r-space, then there exists a V_r-space E and a bijective, continuous, linear map $E \longrightarrow F$ that is not an isomorphism. 4) If E is a V-space, F a W-space, then every surjective, continuous, linear map is a strict morphism. 5) If E is not a V-space, then there exists a barrelled space F and a surjective, continuous, linear map $E \longrightarrow F$ that is not a strict morphism. 6) If F is not a W-space, then there exists a V-space E and a surjective, continuous, linear map $E \longrightarrow F$ that is not a strict morphism.

In [11] Valdivia proves that if E is a barrelled space which is not Pták, then there exists a barrelled space F and a bijective, continuous, linear map $E \longrightarrow F$ that is not an isomorphism.

Article [12] contains also the following localization theorem: If E is a finite-codimensional subspace of a Baire space, F the inductive limit of an increasing sequence (F_n) of Γ_r-spaces, and $u : E \longrightarrow F$ is a linear map with a closed graph, then there exists an index n such that $u(E) \subset F_n$. From here a closed graph theorem follows.

The papers [12; 16; 17] contain refinements and generalizations of the open mapping theorem for not necessarily continuous linear maps with closed graph, and of the closed range theorem {19, §35,5}.

The Γ_r-spaces figure in [70], where Valdivia considers an increasing sequence of subspaces F_n of a space F such that each F_n is a Γ_r-space and $\bigcup F_n = F$. He proves the following theorems: 1) Let $\mathcal{A}$ be a tribe of subsets of a set I, μ a bounded, finitely additive function defined on $\mathcal{A}$ with values in a space E, and $f : E \longrightarrow F$ a linear map with closed graph. There exists $p \in \mathbb{N}$ such that $f \circ \mu$ takes it values in F_p.

2) If (x_n) is a subseries convergent sequence in F, then there exists $p \in \mathbb{N}$ such that (x_n) is bounded multiplier convergent in F_p.

In [39] Valdivia obtains maximal classes for the closed graph theorem concerning weakly continuous maps. He says that F is a <u>Λ_r-space</u> if every locally complete subspace L of the algebraic dual F^* equipped with $\sigma(F^*,F)$, such that $L \cap F'$ is dense in F' for $\sigma(F',F)$, satisfies $L \supset F'$. If E is a space such that E' is locally complete for $\sigma(E',E)$, and if F is a Λ_r-space, then every linear map $u : E \longrightarrow F$ with closed graph is continuous for $\sigma(E,E')$ and $\sigma(F,F')$. If either E or F does not satisfy the hypothesis, then the conclusion is no longer true. Valdivia also introduces Λ-spaces and proves stability results for the two kinds of spaces.

It is undoubtedly Laurent Schwartz's book on Radon measures, in particular his Borel graph theorem {5, Chap.III, §6, no.1 Théorème 1, p.III.34;34, Part II, Chap. I, no.3, Th.2, p.160}, and the proof given by A. Martineau {22} to the latter, which got Valdivia interested in Polish, Lusin and Suslin spaces [58; 59]. Besides the K-Suslin spaces found already in {22}, he introduces quasi-Suslin, C-Suslin, S-Suslin topological spaces, and semi-Suslin locally convex spaces. I will not recall their definitions, not only because they are somewhat technical, but mainly because Valdivia gives a detailed presentation of the theory in Chapter I, §4 of [A], where also those properties are proved which are listed without proof in {22}. (Warning: the semi-Suslin spaces in [58; 91] and in [A] are different.) Let me, however, tabulate the hypotheses for spaces E and F and the graph Γ of a linear map $f : E \longrightarrow F$ under which f is continuous:

E	F	Γ
Metrizable, Baire	Quasi-Suslin	Closed
Baire	K-Suslin	Closed
Metrizable, Baire	Suslin	Sequentially closed
Ultrabornological	Suslin	Borel
Metrizable, convex-Baire	Semi-Suslin	Sequentially closed
Convex-Baire	Semi-Suslin	Closed
Metrizable, Baire	C-Suslin	$\exists A \subset E$, A meager, graph of $f\|(E\setminus A)$ closed
Metrizable, Baire	S-Suslin	$\exists A \subset E$, A meager, graph of $f\|(E\setminus A)$ sequentially closed.

The result in the fourth row is of course Schwartz's theorem. There are corresponding open mapping theorems and other results in [A; 58; 59]. For instance if G is a K-Suslin non-meager topological group in which there exists a meager subset A such that $G\setminus A$ is metrizable, then G is metrizable, separable and complete for the bilateral uniform structure on G. From here a theorem of De Wilde and Sunyach having the same conclusion follows. One has Suslin ⇒ K-Suslin ⇒ quasi-Suslin; every Fréchet space is semi-Suslin and the latter are related to De Wilde's spaces with webs (see below).

In [91] it is proved that if A is a starshaped, metrizable subset of a topological vector space E, and every closed, bounded subset of E intersects A in a compact set, then A is a K-Suslin space. Choosing $A = E$ one can obtain a result due to C. Bessaga and S. Rolewicz according to which every metrizable Montel space is separable. Answering a question of Grothendieck, Dieudonné proved this for Fréchet-Montel spaces {19, §27, 2.(5)}.

In [60] Valdivia solves the problems stated on p.117 of {34}. He proves that if E is such that $\sigma(E,E') \neq \tau(E,E')$ and E is equipped with $\sigma(E,E')$, then for any F the tensor products $E\hat{\otimes}_\pi F$ and $E\hat{\otimes}_\varepsilon F$ are not Suslin spaces. If E is ultrabornological and its topology is not the finest locally convex topology, then there exists on E' a locally convex topology $\mathcal{T}$ such that if we equip E' with $\mathcal{T}$, the space $\mathcal{L}_c(E',F)$ of continuous linear maps $E' \longrightarrow F$, equipped with the topology of uniform convergence on compact subsets of E', is not a Suslin space.

A theorem of J.W. Tukey states that if A and B are two closed, convex subsets of a Banach space E, A is bounded and $A - B$ is dense in the open unit ball U of E, then $A - B \supset U$. In [64] Valdivia generalizes this as follows: Let $A_1, A_2, \cdots, A_p$ be locally complete, bounded, convex subsets of the metrizable space E, and B a closed, convex subset of E. Setting $A = A_1 + A_2 + \cdots + A_p$ one has $\overset{\circ}{\overline{A+B}} \subset A + B$. The special case $p = 1$, $B = \{0\}$ yields a simple proof of Banach's open mapping theorem. Let E be any space, assume that $A_1, \cdots, A_p$ satisfy the preceding hypotheses, and denote by G the balanced, convex hull of $A_1 \cup \cdots \cup A_p$. Then E_G is a Banach space.

Paper [72], some of whose results can also be found in [A, Chapter I, §6,1, pp.103-116], starts out with the following surprising result: Assume that there exists an infinite-dimensional Banach space E such that every linear map with closed graph from E into the space F is continuous. Then every linear map with closed graph from any ultrabornological space G into F is continuous. The proof uses the theorem from [40] quoted in Section 2, which implies that G is the inductive limit of a family of spaces all isomorphic to a fixed quotient space of E.

Valdivia gives the following conditions on the spaces E and F which ensure that there exist non-continuous linear maps $E \longrightarrow F$ with closed graph:

E	F
E' equipped with $\sigma(E',E)$ is not locally complete	On F' the topology $\tau(F',F)$ is defined by a family of norms, and in F' equipped with $\sigma(F',F)$ there is a countable-dimensional dense suspace $\neq F'$
infrabarrelled, not barrelled	infinite-dimensional separable Banach space
normed, not barrelled	infinite-dimensional Banach space
infrabarrelled, not barrelled	$\mathcal{D}'(\Omega)$

Let $f : E \longrightarrow F$ be a linear map. If there exists in F a fundamental system $(U_i)_{i\in I}$ of balanced, convex, closed neighborhoods of 0 such that $f^{-1}(U_i)$ is closed in E for all $i \in I$, then the graph of f is closed. This yields the following proof of Mahowald's theorem stated at the beginning of this section: Let T be a barrel in E that is not a neighborhood of 0. If φ denotes the canonical map from E onto E_T, let U be the closure of $\varphi(T)$ in the Banach space $\hat{E}_T$. Considering φ as a map $E \longrightarrow \hat{E}_T$ we have $\varphi^{-1}(U) = T$, so φ has a closed graph but is not continuous.

Let $\mathcal{E}$ be a class of spaces stable for inductive limits and containing all finite-dimensional spaces. Valdivia denotes by $\mathcal{E}_r$ the class of those spaces F for which any linear map with closed graph from any $E \in \mathcal{E}$ into F is continuous. The spaces in $\mathcal{E}_r$ are often said to be $\mathcal{E}$-minimal {24, Def.1.3.3, p.31}. In the opposite direction let $\mathcal{F}$ be a class of spaces and denote by $\mathcal{F}_s$ the class of those spaces E for which any linear map with closed graph from E into any $F \in \mathcal{F}$ is continuous. The article [72] contains a number of results concerning $\mathcal{E}_r$ and $\mathcal{F}_s$ of which I will quote one.

Given a locally convex topology $\mathcal{T}$ on F, consider all locally convex topologies $\mathcal{U}$ on F which are finer than $\mathcal{T}$ and are such that F equipped with $\mathcal{U}$ belongs to $\mathcal{E}$. The coarsest $\mathcal{T}^{\mathcal{E}}$ among the topologies $\mathcal{U}$ figures in Kōmura's characterization of $\mathcal{E}$-minimal spaces {24, Theorem 7.1.1, p.121}. Assume that $\mathcal{E}$ has the property that if E has a hyperplane belonging to $\mathcal{E}$, then E belongs to $\mathcal{E}$. Valdivia proves that if F equipped with $\mathcal{T}$ belongs to $\mathcal{E}_r$, then F equipped with $\mathcal{T}^{\mathcal{E}}$ is complete.

He observes that the theorem of Adasch and Eberhardt, according to which if F is barrelled-minimal (i.e. a Γ_r-space), then F equipped with the associated barrelled topology is complete {19, §34, 9.(9)}, can be obtained from this result.

A new chapter opened in the theory of locally convex spaces when in 1967 Marc De Wilde discovered webs. A $\mathcal{C}$-web on a space E associates with each finite sequence $(n_1, n_2, \cdots, n_k)$ of positive integers a subset $C(n_1, \cdots, n_k)$ of E such that $\bigcup_{n\in\mathbb{N}} C(n) = E$, $\bigcup_{n_k\in\mathbb{N}} C(n_1, \cdots, n_{k-1}, n_k) = C(n_1, \cdots, n_{k-1})$, and for every sequence (n_k) of positive integers there exists a sequence (ρ_k) of strictly positive numbers such that $\sum_{k\in\mathbb{N}} \lambda_k x_k$ converges whenever $|\lambda_k| \le \rho_k$ and $x_k \in C(n_1, \cdots, n_k)$. The main theorem states that if f is a linear map from an ultrabornological space into a space with a $\mathcal{C}$-web, and if the graph of f is sequentially closed, then f is continuous {19, §35,2.(2)}. The class of spaces with a $\mathcal{C}$-web has all the stability properties required by a celebrated conjecture of Grothendieck {19, §35,1}.

Paper [89] investigates properties of tensor products. The following table summarizes some of the results. The first two columns state the hypotheses on the spaces E and F which imply that the projective tensor product $E \otimes_\pi F$ {19, §41,2} possesses the property listed in the third column:

E	F	$E \otimes_\pi F$
has $\mathcal{C}$-web	countable-dimensional	has $\mathcal{C}$-web
Suslin	Suslin	Suslin
barrelled, metrizable	Baire-like	Baire-like
unordered Baire-like, metrizable	unordered Baire-like	unordered Baire-like
suprabarrelled, metrizable	suprabarrelled	suprabarrelled
finite-dimensional	$F^{\dim E}$ is a Baire-space	Baire space

In the first and last row the conditions are necessary as well as sufficient. Therefore the projective tensor product of two uncountable-dimensional Suslin spaces will be a Suslin space without a $\mathscr{C}$-web. Consequently, Schwartz's Borel graph theorem is not a special case of De Wilde's theorem, even if the graph is supposed to be sequentially closed. The result in the fifth row is proved in [101].

Let E and F be two Baire spaces with $\mathscr{C}$-webs and $f : E \longrightarrow F$ a linear map with closed graph. The simplest theorem in [88] asserts that there exists a subspace L of F and a topology $\mathscr{T}$ on L, finer than the topology induced by F, such that $f(E) \subset L$, and $f : E \longrightarrow L$ is continuous if L is equipped with $\mathscr{T}$.

Before De Wilde, already W. Słowikowski and D.A. Raikov introduced some rather complicated constructs on certain spaces F whose existence ensures that every linear map with closed graph from an ultrabornological space into F is continuous. In [99] Valdivia investigates the concepts introduced by the three authors, and the relations between them. He also discusses the connection between the closed graph theorem for linear relations and for linear maps.

In [102] Valdivia introduces a new class of target spaces for the closed graph theorem, related to the spaces of Słowikowski and De Wilde. A bounded, balanced, convex subset A of a space F is a <u>Banach disk</u> if F_A is a Banach space. A <u>quasi-($\mathscr{L}\mathscr{B}$)-representation</u> of the space F is a family (A_α) of Banach disks indexed by the sequences $\alpha = (a_1, a_2, \cdots)$ of positive integers such that $F = \bigcup A_\alpha$ and $A_\alpha \subset A_\beta$ whenever $\alpha \leq \beta$, i.e. $a_n \leq b_n$ for all n. If F has a quasi-($\mathscr{L}\mathscr{B}$)-representation, then it is said to be a <u>quasi-($\mathscr{L}\mathscr{B}$)-space</u>. The class of quasi- ($\mathscr{L}\mathscr{B}$)-spaces satisfies the desiderata of Grothendieck. Every locally complete space with a $\mathscr{C}$-web is a quasi-($\mathscr{L}\mathscr{B}$)-space, and every quasi-($\mathscr{L}\mathscr{B}$)-space has a strict web {19, §35, 1}; this answers a question of De Wilde. A space is quasi-($\mathscr{L}\mathscr{B}$) if and only if it has a strict web that is <u>ordered</u>, i.e. $C(n_1, \cdots, n_k) \subset C(m_1, \cdots, m_k)$ whenever $n_j \leq m_j$, $1 \leq j \leq k$.

A space E is said to be <u>strictly barrelled</u> if given any ordered, balanced, convex web $\{C(n_1, \cdots, n_k)\}$ on E, there is a sequence (r_k) of positive integers such that the closure of each set $C(r_1, \cdots, r_k)$ is a neighborhood of the origin. Every totally barrelled space is strictly barrelled. If E is unordered Baire-like and F metrizable, strictly barrelled, then $E \otimes_\pi F$ is strictly barrelled.

Among the open mapping, closed graph and localization theorems proved by Valdivia the simplest one is the following: If E is strictly barrelled and F a quasi-$(\mathcal{LB})$-space, then every linear map $E \longrightarrow F$ with closed graph is continuous.

Closely related to the quasi-$(\mathcal{LB})$-spaces are the semi-$(\mathcal{LB})$-spaces whose definition is more technical and which are introduced in [105]. If the space F has a semi-$(\mathcal{LB})$-representation (resp. if there is a web consisting of Banach disks on F) then there exists a family $(F_\alpha)_{\alpha \in \mathbb{N}^{\mathbb{N}}}$ of Fréchet spaces continuously embedded in F such that if f is a linear map with closed graph from a Baire space (resp. from an unordered Baire-like space) E into F, then $f(E) \subset F_\alpha$ for some $\alpha \in \mathbb{N}^{\mathbb{N}}$ and $f : E \longrightarrow F_\alpha$ is continuous.

In [96] Valdivia considers classes of not necessarily locally convex topological vector spaces, and introduces for these classes the concepts of barrelled, Baire-like, unordered Baire-like, suprabarrelled, totally barrelled and Γ_r-spaces. For these he proves closed graph and localization theorems. He also proves that in every infinite-dimensional Fréchet space there is a dense subspace which is totally barrelled in the class of topological vector spaces but which is not the inductive limit of unordered Baire-like spaces. For separable Fréchet spaces this was proved by P. Dierolf, S. Dierolf and L. Drewnowski. Three space theorems for these concepts are presented in [101]. For instance, if E has a closed subspace F such that F is a Fréchet space and E/F is an $(\mathcal{LF})$-space, then E is an $(\mathcal{LF})$-space. If F and E/F are barrelled (ultrabarrelled, totally barrelled, totally barrelled in the class of topological vector spaces, suprabarrelled, suprabarrelled in the class of topological vector spaces, unordered Baire-like) then the same holds for E.

6. BOUNDED AND COMPACT SETS

In [28] Valdivia brings generalizations of the theorem of Frederick Riesz, according to which if the closed unit ball of a Banach space E is compact, then E is finite-dimensional. Thus if E and F are two vector spaces which form a pairing, $\mathcal{T}$ a vector space topology on E such that there exists a closed $\mathcal{T}$-neighborhood U of 0 in E which contains no line and whose boundary is $\sigma(E,F)$-closed, then E is finite-dimensional. This implies Riesz's theorem, since if the closed, unit ball is compact, so is the unit sphere, which is therefore $\sigma(E,E')$-compact, hence $\sigma(E,E')$-closed. Valdivia gives a similar condition for E to be a product of lines. He calls the following result a topological form of Riesz's theorem: If X is a locally compact topological space with topology $\mathcal{T}$, and $\mathcal{T}'$ is a coarser Hausdorff topology on X such that each point $x \in X$ has a fundamental system of $\mathcal{T}'$- neighborhoods that are connected for $\mathcal{T}$, then $\mathcal{T} = \mathcal{T}'$. If the unit ball of the Banach space E is compact, then E is locally compact. The weak topology $\sigma(E,E')$ is Hausdorff and coarser than the topology $\mathcal{T}$ defined by the norm. The convex $\sigma(E,E')$-neighborhoods of any point form a fundamental system and they are $\mathcal{T}$-connected. Therefore $\mathcal{T} = \sigma(E,E')$ and so E is finite-dimensional.

If F is a closed subspace of the locally convex space E, then $\beta(E',E)$ induces $\beta(F^{\perp},E/F)$ on $F^{\perp} \simeq (E/F)'$ if every bounded set in E/F is contained in the image of a bounded subset of E under the canonical surjection $E \longrightarrow E/F$. In [9] Valdivia gives an example of a space E and a closed subspace F such that some bounded subset of E/F is not contained even in the closure of the image of any bounded set $B \subset E$ (for another example see {32, Chap.IV, Exercise 20(c), p.195}), and gives conditions for the existence of such a set B.

Grothendieck proved that if E is a $(\mathcal{DF})$-space and M a separable subset of E, then the original topology of E and the topology $\beta^{*}(E,E')$ of uniform convergence on strongly bounded subsets of E′ induce the same topology on M {19, §29,3.(8)}. In [10] Valdivia replaces the points of M by compact subsets, i.e. he proves that the same conclusion holds if the union of some countable family of compact sets is dense in M. As a consequence, if the union of a sequence of compact sets is total in the $(\mathcal{DF})$-space E, then E is infrabarrelled, and therefore every semi-Montel

($\mathcal{DF}$)-space is Montel. The last result is interesting because there are semi-reflexive ($\mathcal{DF}$)-spaces that are not reflexive {19, §29,4, final remark}.

The Eberlein-Shmulian type theorems assert the equivalence of compactness in the weak topology with countable compactness or sequential compactness under appropriate hypotheses. In [23] Valdivia gives conditions under which for a cardinal α weak α-compactness implies weak compactness. Recall that a Hausdorff space X is α-compact if for any family $(F_i)_{i\in I}$ of closed subsets, where card $I \leq \alpha$ and $\bigcap_{i\in I} F_i = \phi$, there exists a finite subfamily whose intersection is empty. In the same paper he proves parallel conditions for convex-compact sets {19,§24,3} to be compact. In the countable case he obtains as a corollary of one of his results that if on the locally convex space E there is a coarser metrizable locally convex topology, then every countably $\sigma(E,E')$-compact set is $\sigma(E,E')$-compact, a result proved also by S. Dierolf. Furthermore, if E is bornological, then every countably $\sigma(E',E)$-compact subset of E' is $\sigma(E',E)$-compact.

E.E. Floyd and V.L. Klee proved that in a normed space the closed sets which figure in the Cantor-type characterization of compact sets can be replaced by closed affine subspaces. In [25] Valdivia proves analogues of this for locally convex spaces satisfying conditions concerning dense subsets having a given cardinality. The countable cases are given as corollaries and can be summarized as follows: If A is a bounded, convex, not relatively $\sigma(E,E')$-compact subset of the space E, then there exists in E a sequence (H_n) of closed affine hyperplanes so that $\bigcap_{n=1}^{p} H_n \cap A \neq \phi$ for all p but $\bigcap_{n\in\mathbb{N}} H_n = \phi$ provided one of the following conditions is satisfied: 1) E' is separable for $\sigma(E',E)$; 2) E' equipped with $\tau(E',E)$ is the inductive limit of its closed separable subspaces; 3) If $\mathcal{B}$ is the family of all balanced, convex, bounded, separable subsets of E' equipped with $\sigma(E',E)$, then E equipped with the topology of uniform convergence on the sets of $\mathcal{B}$ is complete; 4) E' equipped with $\tau(E',E)$ is bornological. Similar results hold in which the affine hyperplanes are replaced by closed affine subspaces.

In [41] and [42] Valdivia introduces concepts used also in [40] (see Section 2). A sequence (x_n) in a Banach space E is <u>hyperconvergent</u> to the origin if for every integer $p>0$ the sequence $(2^{pn}\|x_n\|)$ converges to 0. A subset A of E is <u>hyperprecompact</u> if there is a sequence hyperconvergent to the origin whose balanced, convex, closed hull contains A. A closed, hyperprecompact set is <u>hypercompact</u>.

If A is hypercompact, then there exists a hypercompact, balanced, convex set $B\subset E$ such that E_B is a Hilbert space and A is hyperprecompact in E_B, see also {13, p.211}. Using a construction based on this result, Valdivia proves the following theorem:

Assume that the infinite-dimensional Banach space E has a dense subset G with cardinal α. Let $(E_i)_{i\in I}$ be a family of infinite-dimensional Banach spaces such that card I equals the cardinal of the set of all infinite countable subsets of G, and assume that E_i' equipped with $\sigma(E_i',E_i)$ is separable for each $i\in I$. Then for each $i\in I$ there exists a hyperprecompact, balanced, convex set $A_i\subset E$ such that $\bigcup_{i\in I} A_i = E$, each E_i is isomorphic to E_{A_i}, the E_{A_i} form a directed family of subspaces, and E is their inductive limit.

There is a similar result for the case when the E_i' are separable for $\beta(E_i',E_i)$. In that case $E_i''\simeq E_{\bar{A}_i}$, where $\bar{A}_i$ is the closure of A_i in E. As a consequence Valdivia obtains the result that if $(F_i)_{i\in I}$ is a family of infinite-dimensional Banach spaces and card $I\geq 2^{\aleph_0}$, then $\prod_{i\in I} F_i$ is not an infra-Pták space.

Let E be a Fréchet space, A a bounded subset of E and A^σ the closure of A in E'' for $\sigma(E'',E')$. Denote by G the span of $E\cup A^\sigma$ in E'', and assume that E has finite codimension in G. Then [49] contains the following results: 1) The balanced, convex hull of A belongs to G (this generalizes the Krein-Phillips theorem {19,§24,5.(4)}); 2) For every $x\in A^\sigma$ there exists a sequence in A converging to x with respect to $\sigma(E'',E')$. 3) If $G\neq E''$, then there exists a separable, bounded subset M of E such that G is a proper subspace of the span of $E\cup M^\sigma$; 4) If F is a subspace of G containing E, then E' is complete for $\tau(E',F)$.

Valdivia considers furthermore a subspace F of E'' such that $E'' = E+F$, assumes that F is separable for $\beta(E'',E')$ and obtains the following conclusions: 1) E' is barrelled for $\tau(E',E'')$; 2) For any subspace L of E'' containing E the dual E' is complete with respect to $\tau(E',L)$; 3) For every $x \in E''$ there is a sequence in E which converges to x with respect to $\sigma(E'',E')$.

To explain some of the results of the related papers [55] and [78], the best is to use a slight extension of the terminology used by Valdivia in [78]. A subset A of a topological space T is hyperbounded if every continuous function on T is bounded on A. The space T is an $\mathcal{M}$-space if every closed hyperbounded subset of T is compact. Thus one of the two Nachbin-Shirota theorems quoted in Section 2 asserts that $\mathcal{C}(T)$ is barrelled if and only if T is an $\mathcal{M}$-space.

One of the early theorems of [55] states that if the locally convex space E is quasicomplete, and if A is a hyperbounded subset of E equipped with $\sigma(E,E')$, then A is relatively $\sigma(E,E')$-compact. As a consequence, a Banach space E is reflexive if and only if every $\sigma(E,E')$-continuous real-valued function on E is bounded on the closed unit ball. More generally, a quasicomplete space E is semireflexive if and only if every $\sigma(E,E')$-continuous real-valued function on E is bounded on bounded subsets of E.

Denote by A^* the $\sigma(E'^*,E')$-closure of $A \subset E$ in E'^*. The proof of the preceding theorem rests in part on the following remark: If A is a hyperbounded subset of E equipped with $\sigma(E,E')$, then A^* is separably continuous on E' equipped with $\sigma(E',E)$, i.e. each $x'' \in A^*$ is continuous on every separable subspace of E'.

In [55] the following condition is introduced:

(CH) Let A be a convex subset of the space E. For any sequence (H_n) of closed hyperplanes of E, such that $A \cap (\bigcap_{j=1}^{n} H_j) \neq \phi$ for $n \in \mathbb{N}$, one has $\bigcap_{n\in\mathbb{N}} H_n \neq \phi$.

Then: 1) (CH) implies that A^* is separably continuous; 2) Let A be a bounded, convex subset of E; then A^* is separably continuous if and only if (CH) is satisfied; 3) Let A be a bounded, convex subset of E;

then (CH) implies that A is $\beta(E,E')$-bounded; 4) Assume that E is infrabarrelled and that A is a bounded, convex subset of E′ equipped with $\sigma(E',E)$; then condition (CH), where E is replaced by E′, implies that A is relatively $\sigma(E',E)$-compact; 5) Assume that E is quasicomplete for the topology of uniform convergence on $\sigma(E',E)$-compact subsets of E′; let A be a bounded, convex subset of E; then (CH) implies that A is relatively $\sigma(E,E')$-compact.

It follows from 2), 5) and the first remark concerning A^* that if E is quasicomplete for the topology of uniform convergence on $\sigma(E',E)$-compact subsets of E′, then every convex, hyperbounded subset of E equipped with $\sigma(E,E')$ is relatively compact.

In [78] many conditions are given which imply that certain subsets of the space E, in particular hyperbounded, closed subsets, are $\sigma(E,E')$-compact. Several of these (Theorems 1,3,4) involve the smallest subspace S(E′) of E^* equipped with $\sigma(E^*,E)$ which contains E′, and in which every separable, bounded subset is relatively compact. Denote by $\mathcal{A}_1$ the collection of balanced convex, separable, compact subsets of S(E′), by $\mathcal{A}_2$ the collection of all balanced, convex, $\sigma(E',E)$-compact subsets of E′, and let $\mathcal{A} = \{A_1 + A_2;\ A_1 \in \mathcal{A}_1,\ A_2 \in \mathcal{A}_2\}$. Assume that E is complete for the topology of uniform convergence on the sets belonging to $\mathcal{A}$. Then: 1) Every $\sigma(E,E')$-closed, hyperbounded subset of E is $\sigma(E',E)$-compact; 2) Equip E with $\sigma(E,E')$, let A be a closed subset of E, and assume that every bounded, increasing, pointwise convergent sequence of continuous functions on E converges uniformly on A; then A is compact.

Similar results are given in terms of the smallest sequentially complete subspace T(E′) of E^* which contains E′.

7. SEQUENCE SPACES

In this section and the next I can be brief since Chapters 2 and 3 of [A] treat, among others, most of the results obtained by Valdivia in the two areas.

[32] gives a condition on an echelon space λ [A, Chapter 2, §2,2, p.212], {19,§30,8} so that its Köthe α-dual $\lambda^\times$ [A, Chapter 2, §1,1,p.173], {19, §30,1} equipped with $\sigma(\lambda^\times,\lambda)$ has a dense subspace E such that E equipped with the topology induced by $\beta(\lambda^\times,\lambda)$ is infrabarrelled but not

bornological. This is used to construct an example of an infrabarrelled $(\mathcal{DF})$-space that is not bornological (see Section 2).

Let λ be an echelon space, and assume that equipped with $\tau(\lambda,\lambda^{\times})$ it is a Montel space. If λ is not a Schwartz space, then in the paper [63], published in the Proceedings of the 1977 Campinas Meeting, Valdivia constructs in $\lambda^{\times}$ equipped with $\tau(\lambda^{\times},\lambda)$ a dense subspace that is barrelled but not bornological (see Section 2). Conversely, if E is a Fréchet-Schwartz space, then every barrelled subspace of E′ equipped with $\tau(E',E)$ is bornological, cf. [A, Chapter 2, §2,4, p.237].

In [67] Valdivia proves that the echelon space λ equipped with $\tau(\lambda,\lambda^{\times})$ is a Schwartz space if and only if λ has no quotient that is topologically isomorphic to ℓ^1 (cf. [A, Chapter 2, §2,3.(14), p.226]). Also every separable Fréchet space is topologically isomorphic to a quotient of a Fréchet-Montel echelon space.

Most of the results of [73] and [79] can be found in [A, Chapter 2, §2,4, pp.227-237] and are augmented there with (23): Let E be a Fréchet space. Assume that in E′ equipped with $\sigma(E',E)$ for every bounded set A there is a balanced, convex, bounded set $B \supset A$ such that $\sigma(E',E'')$ coincides on A with the weak topology of E'_B. Then every infrabarrelled subspace of E′ equipped with $\beta(E',E)$ is bornological.

If λ is an echelon space which equipped with $\tau(\lambda,\lambda^{\times})$ is a Montel space but not a Schwartz space, then Valdivia proves in [73] that there exists in $\lambda^{\times}$ equipped with $\tau(\lambda^{\times},\lambda)$ a dense subspace which is an $(\mathcal{LB})$-space but is not locally complete, and another dense subspace that has countable dimension, is infrabarrelled but neither barrelled nor bornological. Valdivia also constructs non-bornological, infrabarrelled $(\mathcal{DF})$-spaces (see Section 2), and proves that λ is a Schwartz space if and only if every infrabarrelled subspace of $\lambda^{\times}$ is bornological.

Let λ be an echelon space, equip $\lambda^{\times}$ with $\beta(\lambda^{\times},\lambda)$, and denote by $\varphi^{(1)}$ the subspace of $\lambda^{\times}$ which consists of all limits of sequences in $\mathbb{K}^{(\mathbb{N})}$ that converge in the Mackey sense. Then [79; A. Chapter 2, §2,4.(21), p.235] the following are equivalent: 1) λ equipped with $\tau(\lambda,\lambda^{\times})$ is quasi-normable {15, III.1, Définition 4, p.106; 17, 10.7, p.214}; 2) $\varphi^{(1)}$ is complete; 3) $\lambda^{\times}$ satisfies the Mackey convergence condition. If these conditions are not satisfied, then λ is the strong dual of a non-complete $(\mathcal{LB})$-space.

In [103] Valdivia proves that if the echelon space λ is a Montel space but not a Schwartz space, then λ has a quotient isomorphic to $(\ell^1)^{\mathbb{N}}$; furthermore if H is any separable Fréchet space, then λ has a closed subspace G such that H is isomorphic to a quotient of G. If, however, λ is a Schwartz space not equal to $\mathbb{K}^{\mathbb{N}}$, then there is a sequence (λ_n) of nuclear echelon spaces with continuous norms such that λ has a quotient isomorphic to $\Pi\lambda_n$.

8. FUNCTION SPACES AND THEIR REPRESENTATION AS SEQUENCE SPACES

It is the space $\mathfrak{s}$ of rapidly decreasing sequences $a = (a_m)$ (i.e. such that $\lim_{m\to\infty} m^r a_m = 0$ for any $r \in \mathbb{N}$), equipped with the seminorms $p_r(a) = \max_m r^m |a_m|$, which figures in these representations. For the spaces of distribution theory I use the notation of Laurent Schwartz {33}, for the spaces of ultradifferentiable functions [82] I will refer to Meise's expository article {23}. Here are some of the isomorphisms proved by Valdivia (most of which were found independently by D. Vogt {37}):

[61; A, Chapter 3, §1,12]: $\mathcal{E}(\Omega) \simeq \mathfrak{s}^{\mathbb{N}}$, $\mathcal{E}'(\Omega) \simeq (\mathfrak{s}')^{(\mathbb{N})}$.

[61; A, Chapter 3, §1,13]: $\mathcal{D}(\Omega) \simeq \mathfrak{s}^{(\mathbb{N})}$, $\mathcal{D}'(\Omega) \simeq (\mathfrak{s}')^{\mathbb{N}}$. In the proof that $\mathcal{D}(\Omega)$ is not an infra-Pták space [57] the fact that it has a closed subspace isomorphic to $\mathfrak{s}^{(\mathbb{N})}$ plays an essential role {24, Satz 4.5.12, p.78}.

[65; A, Chapter 3, §1,14.(7),(8),p.388]: Let $\mathcal{D}(n) = \{f \in \mathcal{E}(\mathbb{R});$ $\operatorname{supp} f \subset [-n,\infty)\}$ with the topology induced by $\mathcal{E}(\mathbb{R})$, and $\mathcal{D}_+ = \bigcup_n \mathcal{D}(n)$ with the inductive limit topology. Then $\mathcal{D}_+ \simeq \mathfrak{s}^{\mathbb{N}} \times \mathfrak{s}^{(\mathbb{N})}$, $\mathcal{D}'_+ \simeq (\mathfrak{s}')^{(\mathbb{N})} \times (\mathfrak{s}')^{\mathbb{N}}$.

[75; A, Chapter 3, §1,15.(22), p.398]: $\mathcal{D}(K) \simeq \mathfrak{s}$. Earlier it was only known that $\mathcal{D}(K)$ is isomorphic to a complemented subspace of $\mathfrak{s}$.

[76; A, Chapter 3, §1,16.(8),(9), p.403]: Let Γ be a closed cone in $\mathbb{R}^n$ with nonempty interior and vertex at the origin. For each $z \in \mathbb{R}^n$ let $\mathcal{D}_z = \{f \in \mathcal{E}(\mathbb{R}^n);\ \operatorname{supp} f \subset z + \Gamma\}$ equipped with the topology induced by $\mathcal{E}(\mathbb{R}^n)$, and let $\mathcal{D}_\Gamma = \bigcup_z \mathcal{D}_z$ with the inductive limit topology. Then $\mathcal{D}_\Gamma \simeq (s^{\mathbb{N}})^{(\mathbb{N})}$, $\mathcal{D}'_\Gamma \simeq ((s')^{(\mathbb{N})})^{\mathbb{N}}$.

[82]: Let (M_n) be a sequence of strictly positive numbers such that $M_0 = 1$, $M_n^2 \le M_{n-1} M_{n+1}$ and $\sum(M_{n-1}/M_n) < \infty$, and let I be the closed unit cube of $\mathbb{R}^m$. Then for the spaces of ultradifferentiable functions of Beurling and Roumieu type {23, 4.2 Definition, p.201} we have

$$\mathcal{E}^{(M_n)}(\Omega) \simeq \mathcal{E}^{(M_n)}(I)^{\mathbb{N}} \simeq \mathcal{D}^{(M_n)}(I)^{\mathbb{N}},$$

$$\mathcal{E}^{\{M_n\}}(\Omega) \simeq \mathcal{E}^{\{M_n\}}(I)^{\mathbb{N}} \simeq \mathcal{D}^{\{M_n\}}(I)^{\mathbb{N}},$$

$$\mathcal{D}^{(M_n)}(\Omega) \simeq \mathcal{E}^{(M_n)}(I)^{(\mathbb{N})} \simeq \mathcal{D}^{(M_n)}(I)^{(\mathbb{N})},$$

$$\mathcal{D}^{\{M_n\}}(\Omega) \simeq \mathcal{E}^{\{M_n\}}(I)^{(\mathbb{N})} \simeq \mathcal{D}^{\{M_n\}}(I)^{(\mathbb{N})}.$$

[83]: $\mathcal{O}_M \simeq s \hat{\otimes} s'$. In his thesis Grothendieck proved that $\mathcal{O}_M$ is isomorphic to a complemented subspace of $s \hat{\otimes} s'$ and asked whether the isomorphism proved by Valdivia is true.

[85]: $\mathcal{D}_{L^p} \simeq s \hat{\otimes} \ell^p$, $1 \le p < \infty$.

[86; A, Chapter 3, §1,22, pp.433-439, §2,6,pp.461-465]: Let V be a non-compact differentiable manifold, countable at infinity. Then $\mathcal{E}(V) = \mathcal{C}^\infty(V) \simeq s^{\mathbb{N}}$, $\mathcal{D}(V) \simeq s^{(\mathbb{N})}$, $\mathcal{C}^m(V) \simeq \mathcal{C}^m(I)^{\mathbb{N}} \simeq \mathcal{D}^m(I)^{\mathbb{N}}$, $\mathcal{D}^m(V) \simeq \mathcal{C}^m(I)^{(\mathbb{N})} \simeq \mathcal{D}^m(I)^{(\mathbb{N})}$, $0 \le m \le \infty$. Chapter 3, §2 of [A] contains many results concerning $\mathcal{C}^m$ not published in articles previously.

Denote by $\mathcal{B}(\Omega)$ the vector space of all infinitely differentiable functions f on the open subset Ω of $\mathbb{R}^n$ whose derivatives $\partial^\alpha f$ of all orders $\alpha \in \mathbb{N}^n$ are continuous and bounded, and equip $\mathcal{B}(\Omega)$ with the topology defined by the seminorms $p_\alpha(f) = \sup_{x\in\Omega} |\partial^\alpha f(x)|$. Let $\mathcal{B}_0(\Omega)$ be the subspace of $\mathcal{B}(\Omega)$ consisting of those f which vanish at the boundary of Ω, i.e. given $\alpha \in \mathbb{N}^n$ and $\varepsilon > 0$ there exists a compact subset K of Ω

such that $|\partial^\alpha f(x)| \le \varepsilon$ for $x \in \Omega \cap \complement K$. The space $\mathcal{B}_1(\Omega)$ introduced by P. Dierolf consists of the restrictions to Ω of those $f \in \mathcal{B}(\mathbb{R}^n)$ which vanish outside Ω. Clearly $\mathcal{B}_0(\Omega) \subset \mathcal{B}_1(\Omega) \subset \mathcal{B}(\Omega)$. Properties and representations of these spaces are given in [84] and [A, Chapter 3, §1, 18-21, pp.413-433].

If $\Omega = \mathbb{R}^n$, then $\mathcal{B}_0 \simeq \mathfrak{s}\hat{\otimes} c_0$ and $\mathcal{B}_1 = \mathcal{B} \simeq \mathfrak{s}\hat{\otimes}\ell^\infty$. If $\Omega \ne \mathbb{R}^n$ Valdivia constructs a sequence Q of cubes as follows. Let $a \in \mathbb{Z}^n$ and let Q_1 be the collection of all those cubes $\{x \in \mathbb{R}^n; a_j \le x_j \le a_j+1, 1 \le j \le n\}$ which are contained in Ω and whose distance from $\complement\Omega$ is $\ge\sqrt{n}$. Assume that $Q_1, Q_2, \cdots, Q_k$ has already been defined, and let Q_{k+1} be the collection of those cubes $\left\{x \in \mathbb{R}^n;\ \frac{a_j}{2^k} \le x_j \le \frac{a_j+1}{2^{k+1}},\ 1 \le j \le n\right\}$ which are not contained in any of the cubes of $\bigcup_{\ell=1}^{k} Q_\ell$ and whose distance from $\complement\Omega$ is $\ge \frac{\sqrt{n}}{2^k}$. Then $Q = \bigcup_{k\in\mathbb{N}} Q_k = (Q_r)$ and denote by $\rho_r = 2^{-k(r)}$ the length of the edge of Q_r. Let $\lambda_0(\rho)$ be the vector space of those sequences $a = (a_r)$ which satisfy $\lim_{r\to\infty} \rho_r^{-k} a_r = 0$ for all $k \in \mathbb{N}$, equipped with the topology defined by the seminorms $\sup_r \rho_r^{-k}|a_r|$. With this notation Valdivia proves the isomorphism $\mathcal{B}_0(\Omega) \simeq \mathfrak{s}\hat{\otimes}\lambda_0(\rho)$ and deduces from it conditions for the reflexivity and the nuclearity of $\mathcal{B}_0(\Omega)$ found by a different method by P. Dierolf and J. Voigt {9}. If $\mathcal{B}_0(\Omega)$ is not reflexive, then generalizing the case $\Omega = \mathbb{R}^n$ one has $\mathcal{B}_0(\Omega) \simeq \mathfrak{s}\hat{\otimes} c_0$. If $\mathcal{B}_0(\Omega)$ is nuclear, then $\mathcal{B}_0(\Omega) \simeq \mathfrak{s}$.

$\mathcal{B}_1(\Omega)$ is isomorphic to the bidual of $\mathcal{B}_0(\Omega)$. Let $\lambda_\infty(\rho)$ be the vector space of all sequences $a = (a_r)$ such that $(\rho_r^{-k} a_r)$ is bounded for all $k \in \mathbb{N}$, equipped with the topology defined by the seminorms $\sup_r \rho_r^{-k}|a_r|$. Then $\mathcal{B}_1(\Omega) \cong \mathfrak{s}\hat{\otimes}\lambda_\infty(\rho)$. If $\mathcal{B}_1(\Omega)$ is not reflexive, then $\mathcal{B}_1(\Omega) \cong \mathfrak{s}\hat{\otimes}\ell^\infty$, and if $\mathcal{B}_1(\Omega)$ is nuclear, then obviously $\mathcal{B}_1(\Omega) \simeq \mathcal{B}_0(\Omega) \simeq \mathfrak{s}$.

Let X be a Hausdorff topological space which is the union of an increasing sequence (K_n) of compact subsets such that $K_{n+1} \cap \complement K_n$ is uncountable for each n and every compact subset of X is contained in some K_n. Assume that there exists on X a coarser metrizable topology. If X is a k-space {18,7.11,p.230} then $\mathcal{C}(X)$ is topologically isomorphic to $(\mathcal{C}([0,1]))^{\mathbb{N}}$ for the topology of uniform convergence on compact subsets [71; A, Chapter 3, §3, 6, p.494]. If E is a separable Fréchet space, we can take for X the dual of E equipped with the topology of uniform convergence on compact subsets of E.

Similarly, Valdivia considers in [74; A, Chapter 3, §3,7.(6),p.500] a locally compact space X which is the union of a sequence (K_n) of metrizable compact subsets such that $K_{n+1} \cap \complement K_n$ is uncountable and $K_n \subset \mathring{K}_{n+1}$ for all n. Then the space $\mathcal{K}(X)$ of the continuous functions with compact support on X, equipped with the usual inductive limit topology {6, Chap.III, §1, no.1} is topologically isomorphic to $\mathcal{C}([0,1])^{(\mathbb{N})}$, and its dual, the space $\mathcal{M}(X)$ of Radon measures on X, equipped with the strong topology is topologically isomorphic to $\mathcal{C}([0,1])^{\mathbb{N}}$.

9. HOLOMORPHY

It was during a visit to Dublin in February 1980 that Valdivia got involved in infinite-dimensional holomorphy. Let $\Omega \neq \phi$ be an open connected subset of the space E. Denote by $H_b(\Omega)$ the space of bounded holomorphic functions on Ω. The H_b-hull of a set $A \subset \Omega$ is defined by

$$\hat{A} = \{z \in \Omega;\ |f(z)| \leq \sup_{x \in A}|f(x)|,\ \forall\ f \in H_b(\Omega)\}.$$

If for every A bounded away from the boundary $\partial\Omega$ of Ω also $\hat{A}$ is bounded away from $\partial\Omega$, then we say that Ω is H_b-<u>holomorphically convex</u>. In [81] Valdivia proves among others that if E is the dual of a Fréchet space H, equipped with the topology of uniform convergence on the compact subsets of H, and F is a separable space, then there exists an F-valued holomorphic function f on the H_b-holomorphically convex domain $\Omega \subset E$ such that $f(\Omega)$ is dense in F and Ω is the domain of existence of f.

The proof is based on the following interpolation theorem and on its vector-valued analogue. Let Ω be a set, union of an increasing sequence (A_n) of subsets. Let M be an algebra of scalar-valued functions on Ω such that: 1) Every $f \in M$ is bounded on each A_n, 2) M is a Fréchet space for the topology of uniform convergence on the sets A_n. Denote by

B_n the set $\{z \in \Omega;\ |f(z)| \le \sup_{x \in A_n} |f(x)|,\ \forall f \in M\}$. Let (x_n) be a sequence in Ω such that $\{x_n;\ n \in \mathbb{N}\} \cap B_m$ is finite for every m, and if $n \neq p$, there exists $f \in M$ such that $f(x_n) = 1$, $f(x_p) = 0$. Then given any sequence (α_n) of scalars, there exists an $f \in M$ such that $f(x_n) = \alpha_n$ for all n.

Let E and F be Banach spaces. An m-<u>homogeneous polynomial</u> on E with values in F is a map $P : E \longrightarrow F$ for which there exists an m-linear map $A : E^m \longrightarrow F$ such that $P(x) = A(x,\cdots,x)$ {11, Chapter 1}. In collaration with R. Aron and C. Hervés, Valdivia studies in [90] certain subspaces of the space $\mathcal{P}(^mE;F)$ of all continuous, m-homogeneous polynomials:

$\mathcal{P}_{wu}(^mE;F)$ is the space of all $P \in \mathcal{P}(^mE;F)$ whose restriction to any ball $B \subset E$ is weakly uniformly continuous;

$\mathcal{P}_{w}(^mE;F)$ is the space of all $P \in \mathcal{P}(^mE;F)$ whose restriction to any ball $B \subset E$ is weakly continuous;

$\mathcal{P}_{wc}(^mE;F)$ is the space of all $P \in \mathcal{P}(^mE;F)$ which map all bounded, weak Cauchy sequences in E into Cauchy sequences in F;

$\mathcal{P}_{wsc}(^mE;F)$ is the space of all $P \in \mathcal{P}(^mE;F)$ which map all bounded sequences converging weakly to some element $x \in E$ into sequences in F which converge to $P(x)$.

They prove that the following diagram holds:

$$\begin{array}{ccccc} & & \mathcal{P}_w(^mE;F) & & \\ & = & & \subset & \\ \mathcal{P}_{wu}(^mE;F) & & & & \mathcal{P}_{wsc}(^mE;F) \subset \mathcal{P}(^mE;F) \\ & \subset & & = & \\ & & \mathcal{P}_{wc}(^mE;F) & & \end{array}$$

where the symbol $\subset$ indicates that strict inclusion can occur.

10. THE GEOMETRY OF BANACH SPACES

Let E be an infinite-dimensional Banach space, and G a closed subspace of the bidual E'' containing E.

1) [104] If E'' is separable, then there exists a closed subspace F

of E such that denoting by $\tilde{F}$ the $\sigma(E'',E')$-closure of F in E'', one has $G = E + \tilde{F}$, and both F and E/F are infinite-dimensional.

2) [106] If E'' is separable, there exists a closed subspace F of E such that $G = E + \tilde{F}$, and both F and F' have shrinking bases {17,14.5, p.302; 20, Proposition 1.b.1, p.8; 21,III.3, p.34}.

2) [106] If E' and G are separable, there exists a closed subspace F of E such that $E + \tilde{F} \supset G$ and F has a shrinking basis.

The proofs of 1) and 2) use the Lemma proved in [104] according to which if F is a closed subspace of E, then $E + \tilde{F}$ is a closed subspace of E''. They also use a construction given in [107] which contains results of C. Bessaga, A. Pelczyński, W.B. Johnson and H.P. Rosenthal on basic sequences {17, 14.3, p.296}, and which is similar to the one given in [109] and mentioned at the end of Section 3.

It follows from a result of R.C. James that there exists a separable Banach space L having a topological complement S in L'' such that S is isometric to ℓ^1. This space has the following universal property: Given any separable Banach space E, there exists a closed subspace F of L such that (L/F)''/(L/F) is isometric to E, and L/F has a shrinking basis. This result is contained in [106], as is the following:

Given a separable Banach space E, there exists a Banach space F such that E is isometric to F''/F and both F'' and F have a shrinking basis. This improves a result of J. Lindenstrauss, who proved that E is isomorphic to F''/F.

In [107] it is proved that if E is a non-quasireflexive, separable Banach space, then there exists a closed, non-quasireflexive subspace F of E such that E/F is not quasireflexive and has a Schauder basis (except for the last assertion, this was proved by J.W. Davis and W.B. Johnson). If E is a non-quasireflexive Banach space and E' is sepable, then there exists a closed, non-quasireflexive subspace F of E such that E/F is not quasireflexive and has a shrinking basis.

CONCLUSION

According to the list Professor Manuel Maestre gave me, between March 1970 and April 1988 Valdivia directed thirty doctoral dissertations, and his former students teach now in many Spanish universities. An impressive quantity of articles published in the Revista de la Real Academia de Ciencias bears the mention: "trabajo realizado en el Departamento de Teoria de Funciones y Ecuaciones Funcionales de la Facultad de Matemáticas de Valencia, bajo la dirección del profesor Dr. Don Manuel Valdivia Ureña".

He obtained the chair of Analysis at the Faculty of Science of the University of Valencia in 1965 through that singular Spanish institution called "oposición". In 1969 he obtained a chair, also through "oposición", at the School of Agricultural Engineering of the Polytechnical University of Valencia, where he taught until 1976.

He was elected a corresponding member of the Spanish Academy of Science in 1973 and a full member in 1975. The Academy of Barcelona elected him to membership in 1985, and the Royal Society of Science of Liège to corresponding membership in July of 1990. In 1981 he received the medal of the Order of Alfonso X the Wise for meritorious didactical activity.

In my own name as well as in the name of the participants of the International Functional Analysis Meeting on the occasion of the 60th birthday of Professor M. Valdivia, and of his many friends and admirers all over the world who could not be present at the meeting, I wish him many more years of happy and fruitful research activity.

Publications of Manuel Valdivia

A. Topics in locally convex spaces, North-Holland Mathematics Studies 67; Notas de Matemática 85. 510 pp. North-Holland Publishing Company, Amsterdam-New York-Oxford 1982. MR 84i:46007.

1. Sucesiones de aplicaciones continuas, absolutamente continuas y sumables con diversos tipos de convergencia. Rev. Real Acad. Ci. Exact. Fis. Natur. Madrid 57 (1963), 677-726. MR 33#6201

2. Algunos criterios de convergencia en la teoria de la integración. Rev. Real Acad. Ci. Exact. Fís. Natur. Madrid 59 (1965), 173-210. MR 37#5338.

3. Desarrollos asintóticos y familias compactas de funciones holomorfas. Rev. Real Acad. Ci. Exact. Fís. Natur. Madrid 59 (1965), 339-378.

4. El teorema general de la gráfica cerrada en los espacios vectoriales topológicos localmente convexos. Rev. Real Acad. Ci. Exact. Fís. Natur. Madrid 62 (1968), 545-551. MR 39#1940.

5. El teorema general de la applicación abierta en los espacios vectoriales topológicos localmente convexos. Rev. Real Acad. Ci. Exact. Fís. Natur. Madrid 62 (1968), 553-562. MR 39#3281.

6. Sobre la completitud en los espacios vectoriales topológicos localmente convexos. Collect. Math. 20 (1969), 257-265. MR 46#2384.

7. Convergencia casi-aproximativa. In Honor of Prof. Dr. Iñiguez y Almech on the occasion of his Seventieth Birthday and Academic Retirement, pp.205-210. Consejo Sup. Investigación Ci. Fac. Ci. Zaragoza, Zaragoza, 1969. MR 41#7616.

8. Aplicaciones lineales fuertemente casi abiertas. Rev. Real Acad. Ci. Exact. Fís. Natur. Madrid 63 (1969), 33-37. MR 40#3263

9. Sobre los conjuntos acotados en los espacios vectoriales topológicos cocientes. Rev. Real Acad. Ci. Exact. Fís. Natur. Madrid 63 (1969), 39-44. MR 40#1753.

10. Sobre los conjuntos compactos en espacios $\mathcal{DF}$. Collect.Math 21 (1970), 99-103. MR 44#2020.

11. Sobre ciertos espacios tonelados. Proceedings of the Ninth Annual Conference of Spanish Mathematicians (Granada, 1968), pp.66-69. Inst. "Jorge Juan" Mat., Madrid, 1971. MR 51#11049.

12. Sobre el teorema de la gráfica cerrada. Collect. Math. 22 (1971), 51-72. MR 48#2715.

13. Sobre el teorema de la aplicación abierta. Proceedings of the Tenth Conference of Spanish Mathematicians (La Laguna, 1969), pp.146-150, Inst. "Jorge Juan" Mat. Madrid, 1972. MR 50#2859.

14. A hereditary property in locally convex spaces. Ann. Inst. Fourier (Grenoble) 21 (1971), fasc. 2, 1-2. MR 48#11967.

15. On $\mathcal{DF}$ spaces. Math. Ann. 191 (1971), 38-43. MR 43#6694.

16. On final topologies. J. Reine Angew.Math. 251 (1971), 193-199. MR 45#4108.

17. Absolutely convex sets in barrelled spaces. Ann.Inst.Fourier (Grenoble) 21 (1971), fasc.2, 3-13. MR 48#11968.

18. A class of bornological barrelled spaces which are not ultrabornological. Math. Ann 194 (1971), 43-51. MR 47#2307.

19. On subspaces of countable codimension of a locally convex space. J. Reine Angew. Math. 256 (1972), 185-189. MR 57#7097.

20. On non-semi-reflexive spaces. Manuscripta Math. 7 (1972), 307-313, MR 48#6876.

21. Some examples on quasi-barrelled spaces. Ann. Inst. Fourier (Grenoble) 22 (1972), fasc.2, 21-26. MR 49#1053.

22. On nonbornological barrelled spaces. Ann.Inst.Fourier (Grenoble) 22 (1972), fasc.2, 27-30. MR 49#1050.

23. Some criteria for weak compactness. J. reine angew.Math. 255 (1972), 165-169. MR 45#9095.

24. On bounded sets which generate Banach spaces. Arch.Math. (Basel) 23 (1972), 640-642. MR 48#825.

25. On weak compactness. Studia Math. 49 (1973), 35-40. MR 48#11969.

26. A note on locally convex spaces. Math. Ann 201 (1973), 145-148. MR 49#7725.

27. Some new results on bornological barrelled spaces. Proc.Symposium on Functional Analysis. Istanbul, Silivri, 1973, 85-90. Publ.Math.Res. Inst. - Istanbul, No.1, Math.Res.Inst., Istanbul, 1974. MR 55#6160.

28. Sobre los espacios vectoriales topológicos de dimensiones finitas. Collect. Math. 24 (1973), 3-6. MR 49#1055.

29. Un teorema de inmersión en espacios tonelados que no son bornológicos. Collect.Math. 24 (1973), 7-11. MR 49#1051.

30. Some properties of the bornological spaces. Actes du Deuxième Colloque d'Analyse Fonctionnelle de Bordeaux (Univ. de Bordeaux, 1973), III, pp.51-56. Publ.Dép.Math. (Lyon) 10 (1973), fasc. 4, 329-334. MR 51#1324.

31. On a class of quasi-barrelled spaces. Math.Ann. 202 (1973), 295-300. MR 48#6868.

32. A class of quasi-barrelled (DF)-spaces which are not bornological. Math. Z. 136 (1974), 249-251. MR 50#7979.

33. Quotients of complete locally convex spaces. Manuscripta Math. 14 (1974), 235-240. MR 51#1319.

34. The space of distributions $\mathcal{D}'(\Omega)$ is not B_r-complete. Math. Ann. 211 (1974), 145-149. MR 51#6406.

35. On countable strict inductive limits. Manuscripta Math. 11 (1974), 339-343. MR 49#11195.

36. On certain topologies on a vector space. Manuscripta Math. 14 (1974), 241-247. MR 50#14151.

37. On Mackey spaces. Duke Math. J. 41 (1974) 835-841. MR 50#2860.

38. A note on quasi-barrelled spaces. Arch. Math. Basel 26 (1975), 98-100. MR 51#8777.

39. Mackey convergence and the closed graph theorem. Arch.Math.Basel 25 (1974), 649-656. MR 51#11052, Zbl. 297.46006.

40. Some characterizations of ultrabornological spaces. Ann.Inst.Fourier (Grenoble) 24, 3 (1974), 57-66. MR 51#6344.

41. On inductive limits of Banach spaces. Manuscripta Math. 15 (1975), 153-163. MR 58#23470, Zbl. 327.46005.

42. A class of precompact sets in Banach spaces. J. reine angew. Math. 276 (1975), 130-136. MR 51#13650.

43. Some properties in countable inductive limits. Arch.Math. Basel 26 (1975), 201-208. MR 51#3848.

44. On quasi-completeness and sequential completeness in locally convex spaces. J. reine angew. Math. 276 (1975), 190-199. MR 52#8870.

45. Algunos teoremas de inmersión en espacios topológicos separables. Real Acad. Ci. Exact. Fís. Natur. Madrid. Vol. Homenaje Prof. Lora Tamayo (1975), 71-75.

46. On countable locally convex direct sums. Arch.Math.Basel 26 (1975), 407-413. MR 52#1241.

47. On B_r-completeness. Ann.Inst.Fourier (Grenoble) 25 (1975), fasc.2, 235-248. MR 53#3634.

48. Algunos resultados sobre completitud en espacios localmente convexos. Collect. Math. 26 (1975), No.2, 97-104. MR 54#11019.

49. On metrizable locally convex spaces. Arch. Math. Basel 27 (1976), no.1, 79-85. MR 53#1207.

50. Sucesiones de conjuntos convexos en los espacios vectoriales topológicos. Rev.Mat.Hisp.-Amer. (4) 36 (1976), 92-99. MR 55#10990.

51. Nuclearity and Banach spaces. Proc. Edinburgh Math. Soc. Ser. II, 20 (1977), 205-209. MR 55#8735.

52. On weakly complete sets in a locally convex space. Arch. Math. Basel 28 (1977), 638-643. MR 58#2119.

53. Sur certains hyperplans qui ne sont pas ultra-bornologiques dans les espaces ultra-bornologiques. C.R. Acad. Sci. Paris Ser. A-B 284 (1977), A935-A937. MR 55#6161, Zbl.344.46006.

54. On a class of Banach spaces. Studia Math. 60 (1977), 11-13. MR55#3760.

55. Some new results on weak compactness. J. Functional Analysis 24 (1977), 1-10. MR 58#12266, Zbl.344.46004.

56. On the completion of a bornological space. Arch. Math. Basel 29 (1977), 608-613. MR 57#1064, Zbl.382.46002.

57. The space $\mathcal{D}(\Omega)$ is not B_r-complete. Ann. Inst. Fourier (Grenoble) 27, 4(1977), 29-43. MR 57#17182.

58. Sobre una cierta clase de espacios topológicos. Collect.Math. 28 (1977), No.1, 9-20. MR 58#12955.

59. On the closed graph theorem in topological spaces. Manuscripta Math. 23 (1978), 173-184. MR 58#23477.

60. Sobre espacios localmente convexos de Suslin. Rev. Real. Acad. Ci. Exact. Fis. Natur. Madrid 72 (1978), 215-220. MR82i:46020.

61. Representaciones de los espacios $\mathcal{D}(\Omega)$ y $\mathcal{D}'(\Omega)$. Rev. Real Acad. Ci. Exact. Fis. Natur. Madrid 72 (1978), 385-414. MR 82m:46037, Zbl.404.46021.

62. On the space $\mathcal{D}(\Omega)$. Arch.Math.Basel 30 (1978), 516-522. MR80a:46020.

63. A characterization of echelon Köthe-Schwartz spaces. Approximation Theory and Functional Analysis, Ed. J. B. Prolla, pp.409-419, North-Holland Mathematics Studies No.35, Amsterdam-New York 1979. MR 81d:46009, Zbl.415.46007.

64. Some notes on convex sets. Manuscripta Math. 26 (1979), 381-386. MR80j:46024.

65. Una representación del espacio $\mathcal{D}_+(\Omega)$. Rev.Real Acad.Ci. Exact.Fis. Natur.Madrid, 72 (1978), 559-571. MR 82m:46038.

66. Solution of a problem of Grothendieck. J. reine angew. Math. 305 (1979), 116-121. MR80g:46006.

67. Cocientes de espacios escalonados. Rev.Real Acad. Ci. Exact. Fis. Natur. Madrid 73 (1979), 169-183. MR82i:46017, Zbl.449.46012.

68. On Baire-hyperplane spaces. Proc.Edinburgh Math. Soc.(2) 22 (1979), 247-255. MR81m:46004, Zbl.439.46001.

69. On subseries convergent sequences. Proc.Edinburgh Math. Soc. (2) 22 (1979), 257-262. MR 81a:46008. Zbl.439.46003.

70. On certain barrelled normed spaces. Ann.Inst.Fourier (Grenoble) 29 (1979), fasc.3, 39-56. MR 81d:46006, Zbl.379.46004, 409.46002.

71. Sobre ciertos espacios de funciones continuas. Rev. Real Acad.Ci. Exact. Fis. Natur. Madrid 73 (1979), 485-490. MR82g:46057. Zbl.449.46029.

72. Algunos nuevos resultados sobre el teorema de la gráfica cerrada. Revista Mat. Hispano-Americana (4) 39 (1979), 27-47. MR 83b:46007, Zbl.449.46008.

73. Algunas propiedades de los espacios escalonados. Rev.Real.Acad.Ci. Exact.Fis.Natur.Madrid 73(1979),389-400. MR 82i:46018,Zbl.449.46013.

74. Espacios de medidas de Radon. Rev.Real Acad. Ci. Exact. Fis. Natur. Madrid 74 (1980), 91-98. MR 82e:46040, Zbl.449.46032.

75. A representation of the space $\mathcal{D}(K)$, J. reine angew. Math.320 (1980), 97-98. MR 82j:46052. Zbl.433.56010.

76. On certain infinitely differentiable function spaces. Sémin. P. Lelong-H. Skoda, Analyse, Années 1978/79, Lect. Notes Math. 822, Springer, Berlin (1980), 310-316. MR 82d:46042, Zbl.449.46036.

77. On suprabarrelled spaces. Functional Analysis, Holomorphy and Approximation Theory, Ed. S. Machado, Rio de Janeiro 1978, Lecture Notes in Math. 843, Springer, Berlin (1981), 572-580. MR 82h:46003, Zbl. 469.46001.

78. Sobre ciertas clases de conjuntos en espacios localmente convexos. Collect. Math. 31 (1980), 217-228. MR 83j:46006, Zbl.469.46006.

79. On quasi-normable echelon spaces. Proc. Edinburgh Math.Soc.,II. Ser. 24 (1981), 73-80. MR 82h:46012. Zbl.451.46006, 439.46006.

80. Convex sets in metrizable spaces. Arch. Math. Basel 36 (1981), 244-254. MR 82g:46014, Zbl. 439.46009, 449.46016.

81. Interpolation in certain function spaces. Proc. Roy. Irish Acad. Sect. A 80 (1980), 173-189. MR 82i:46067. Zbl.451.46017, 439.46016.

82. Espacios de funciones ultradiferenciables. Rev. Real Acad.Ci.Exact. Fis. Natur. Madrid 74 (1980), 997-1002. MR 83e:46033.

83. A representation of the space $\mathcal{O}_M$. Math. Z. 177 (1981), 463-478. MR 82i:46057.

84. Sobre el espacio $\mathcal{B}_0(\Omega)$. Rev.Real Acad.Ci.Exact.Fis.Natur.Madrid 74 (1980), 835-863. MR 82j:46053.

85. On the space $\mathcal{D}_{L^p}$. Advances in Math. Supplementary Studies, Vol.7B, Mathematical Analysis and Applications, Essays dedicated to Laurent Schwartz on the occasion of his 65th Birthday, Part B, pp.759-767. Academic Press, New York 1981. MR 83h:46057.

86. Representaciones de los espacios $\mathcal{E}^m(V)$ y $\mathcal{D}^m(V)$. Rev.Real.Acad.Ci. Exact.Fis.Natur.Madrid 75 (1981), 589-596. MR 83e:46032.

87. On totally barrelled spaces (with Pedro Pérez Carreras). Math.Z. 178 (1981), 263-269. MR 82m:46002, Zbl.449.46002.

88. Localization theorems in webbed spaces. Semesterbericht Funktionalanalysis Sommersemester 1982, pp.49-57. Math. Institut, Eberhard-Karls-Universität, Tübingen.

89. A class of locally convex spaces without $\mathcal{C}$-webs. Ann. Inst. Fourier (Grenoble) 32 (1982), fasc.2, 261-269. MR 83j:46015, Zbl.478.46004.

90. Weakly continuous mappings on Banach spaces (with R.M. Aron and C. Hervés.) J. Funct. Anal. 52 (1983), 189-204. MR 84g:46066.

91. On semi-Suslin spaces and dual metric spaces. Functional Analysis, Holomorphy and Approximation Theory, ed. J.A. Barroso, pp.445-459. (Rio de Janeiro, Aug. 4-8, 1980) North-Holland Mathematics Studies No.71. North-Holland, Amsterdam - New York 1982. MR 85d:46005.

92. A property of Fréchet spaces. Functional Analysis, Holomorphy and Approximation Theory, ed. G.I. Zapata, pp. 466-477. (Rio de Janeiro, Aug. 3-7, 1981) North-Holland Mathematics Studies No.86. North-Holland, Amsterdam - New York - Oxford 1984. MR 86h:46008.

93. Sobre espacios $(\mathcal{LF})$ metrizables (with Pedro Pérez Carreras). Collect. Math. 33 (1982), 299-303. MR 85j:46007.

94. B_r-complete spaces which are not B-complete. Math. Z. 185 (1984), 253-259. MR 85b:46009.

95. Subespacios de primera categoria en espacios vectoriales topológicos de Baire. Collect. Math. 34 (1983), 287-296. MR 86f:46005.

96. Classes of barrelled spaces related with the closed graph theorem. Portugal. Math. 40 (1981), 345-366. MR 87g:46009.

97. Products of Baire topological vector spaces. Fund. Math. 125 (1985), 71-80. MR 87d:46007.

98. On certain metrizable locally convex spaces. Complex Analysis, Functional Analysis and Approximation Theory, ed. Jorge Mujica, pp. 287-293, North-Holland Mathematics Studies No.125. Amsterdam - New York - Oxford 1986. MR 88h:46002.

99. On Słowikowski, Raikov and De Wilde closed graph theorems. Aspects of Mathematics and its Applications, ed. J. Barroso. North-Holland Mathematical Library, Vol.34,pp.833-857. Elsevier Science Publishers B.V., Amsterdam, 1986. MR 87m:46015.

100. Una propiedad de los espacios de Banach que no son casirreflexivos. Collect. Math. 36 (1985), 291-299. MR 88a:46012.

101. Three space problems and projective tensor products of locally convex spaces. Proc.Roy.Irish Acad.Sect.A 86(1986), 127-141, MR 88h:46003.

102. Quasi-$(\mathcal{LB})$-spaces. J. London Math. Soc.(2) 35 (1987), 149-168. MR 88b:46012.

103. $(\mathcal{F})$-spaces and strict $(\mathcal{LF})$-spaces. Math. Z. 195 (1987), 345-364. MR 89e:46005.

104. Banach spaces X with X^{**} separable. Israel J. Math. 59 (1987), 107-111. MR 89c:46015.

105. Completing sequences and semi-LB-spaces. Note di Matematica, Lecce 7 (1987), 75-82. MR 90g:46008.

106. Shrinking bases and Banach spaces Z^{**}/Z. Israel J. Math. 62 (1988), 347-354. MR 89k:46020.

107. Bases y casirreflexividad en espacios de Banach. Rev.Real Acad. Ci. Exact. Fis. Natur. Madrid 82 (1988), 45-53. MR 90a:46022.

108. Espacios de Fréchet de generación débilmente compacta. Collect.Math. 38 (1987), 17-25. MR 90a:46002.

109. A characterization of totally reflexive Fréchet spaces. Math.Z. 200 (1989), 327-346. MR 90a:46003.

110. Fréchet spaces with separable strong bidual. Confer.Sem.Mat. Univ. Bari. 1988.

111. Countable direct sums of Banach spaces. Math. Nachr. 141 (1989), 73-79.

112. Some properties of weakly countably determined Banach spaces. Studia Math. 93 (1989), 137-144. MR 90f:46019.

113. Resolution of identity in certain metrizable locally convex spaces. Rev. Real Acad. Ci. Exact Fis. Natur. Madrid 83 (1989), 75-96.

114. Resolution of the identity in certain Banach spaces. Collect.Math. 39 (1988), 127-140. MR91b:46018.

References

1. J. Arias de Reyna, Normed barely Baire spaces, Israel J. Math. 42 (1982), 33-36. Correction, ibid 50 (1985), 264.

2. J. Arias de Reyna, $\ell_0^\infty(\Sigma)$ no es totalmente tonelado. Rev.Real.Acad.Ci. Exact.Fis.Natur. Madrid 79 (1985), 77-78.

3. S. Banach, Théorie des opérations linéaires. Warszawa 1932.

4. N. Bourbaki, Eléments de mathématique, Topologie générale. Hermann, Paris 1958-1961.

5. N. Bourbaki, Eléments de mathématique, Espaces vectoriels topologiques. Masson, Paris 1981.

6. N. Bourbaki, Eléments de mathématique, Intégration. Hermann, Paris 1959-1969.

7. D. Bucchioni and A. Goldman, Sur certains espaces de formes linéaires liés aux mesures vectorielles. Ann. Inst. Fourier (Grenoble) 26, fasc. 3 (1976), 173-209.

8. C.L. De Vito, On Alaoglu's theorem, bornological spaces and the Mackey-Ulam theorem. Math.Ann. 191 (1971), 83-89.

9. P. Dierolf and J. Voigt, Calculation of the bidual for some function spaces, integrable distributions. Math. Ann. 253 (1980), 63-87.

10. J. Diestel, Sequences and series in Banach spaces. Graduate Texts in Mathematics 92. Springer-Verlag, New York-Berlin-Heidelberg-Tokyo 1984.

11. S. Dineen, Complex analysis in locally convex spaces. North-Holland Mathematics Studies 57; Notas de Matemática 83. North-Holland, Amsterdam - New York - Oxford 1981.

12. J.C. Ferrando and M. López Pellicer, Strong barrelledness properties in $\ell_0^\infty(X,\mathcal{A})$ and bounded finite additive measures. Math.Ann. 287 (1990), 727-736.

13. K. Floret, Some aspects of the theory of locally convex inductive limits. Functional analysis: Surveys and recent results II., ed.: K.-D. Bierstedt and B. Fuchssteiner, pp. 205-237. North-Holland Mathematics Studies 38; Notas de Matemática 68. North-Holland, Amsterdam - New York - Oxford 1980.

14. L. Gillman and M. Jerison, Rings of continuous functions. Van Nostrand, Princeton 1960.

15. A. Grothendieck, Sur les espaces (F) et (DF). Summa Brasil. Math. 3 (1954), 57-123.

16. A. Grothendieck, Espaces vectoriels topologiques, 2.a edição. Soc.Mat.São Paulo 1958.

17. H. Jarchow, Locally convex spaces. B.G. Teubner, Stuttgart 1981.

18. J.L. Kelley, General topology. Van Nostrand, Princeton 1955.

19. G. Köthe, Topological vector spaces I, II. Die Grundlehren der mathematischen Wissenschaften in Einzeldarstellungen 159, 237. Springer-Verlag, Berlin - Heidelberg - New York 1969, 1979.

20. J. Lindenstrauss and L. Tzafriri, Classical Banach spaces I. Ergebnisse der Mathematik und ihrer Grenzgebiete 92. Springer-Verlag, Berlin - Heidelberg - New York 1977.

21. J.T. Marti, Introduction to the theory of bases. Springer Tracts in Natural Philosophy 18. Springer-Verlag, New York 1969.

22. A. Martineau, Sur des théorèmes de S. Banach et L.Schwartz concernant le graphe fermé. Studia Math. 30 (1968), 43-51.

23. R. Meise, Representation of distributions and ultradistributions by holomorphic functions. Functional analysis: Surveys and recent results, ed.: K.-D. Bierstedt and B. Fuchssteiner, pp.189-208. North-Holland, Amsterdam - New York - Oxford 1977.

24. M. Oberguggenberger, Der Graphensatz in lokalkonvexen topologischen Vektorräumen. Teubner-Texte zur Mathematik, Band 44. B.G. Teubner, Leipzig 1982.

25. P. Pérez Carreras and J. Bonet, Barrelled locally convex spaces. North-Holland Mathematics Studies 131; Notas de Matemática 113. North-Holland, Amsterdam - New York - Oxford - Tokyo 1987.

26. A. Pietsch, Nuclear locally convex spaces. Ergebnisse der Mathematik und ihrer Grenzgebiete 66. Springer-Verlag, Berlin - Heidelberg - New York 1972.

27. W.J. Robertson, I. Tweddle and F.E. Yeomans, On the stability of barrelled topologies III. Bull. Austral. Math. Soc. 22 (1980), 99-112.

28. H.L. Royden, Real analysis, 3rd edition. Macmillan-Collier, New York-London 1988.

29. R. SanJuan, Sobre la existencia de una función holomorfa que se aproxime asintóticamente a una serie dada con cotas prefijadas, y de una función real indefinidamente derivable en un intervalo con derivadas prefijadas en un punto y acotadas en el intervalo. Collectanea Math. 4 (1951), 3-11.

30. S. Saxon and M. Levin. Every countable-codimensional subspace of a barrelled space is barrelled. Proc.Amer.Math.Soc. 29 (1971), 91-96.

31. S.A. Saxon and P.P. Narayanaswami, Metrizable (LF)-spaces, (db)-spaces, and the separable quotient problem. Bull.Austral.Math.Soc. 23 (1981), 65-80.

32. H.H. Schaefer. Topological vector spaces. Macmillan-Collier, New York - London 1966.

33. L. Schwartz, Théorie des distributions. Hermann, Paris 1966.

34. L. Schwartz. Radon measures on arbitrary topological spaces and cylindrical measures. Tata Institute of Fundamental Research Studies in Mathematics 6. Oxford University Press 1973.

35. L. Waelbroeck, ed., Summer school on topological vector spaces. Lecture Notes in Mathematics 331. Springer-Verlag, Berlin-Heidelberg-New York 1973.

36. J.H. Webb, Subspaces of barrelled spaces. Questiones Math. 4 (1981), 323-324.

37. D. Vogt, Sequence space representations of spaces of test functions and distributions. Functional Analysis, Holomorphy, and Approximation Theory, Ed. G.I. Zapata, pp. 405-443. Lecture notes in pure and applied mathematics, Vol. 83, Marcel Dekker, New York-Basel 1983.

Professor Valdivia kindly supplied the editors with the following list of recent articles, which we include here for the sake of completeness:

Additional list of publications of Manuel Valdivia

115. Projective generators and resolutions of identity in Banach spaces (with J. Orihuela). Rev. Mat. Univ. Complutense Madrid 2, Núm. Supl. (1989), 179 - 199.

116. Some properties of Banach spaces Z**/Z. Geometric essays of Banach spaces. Essays in honour of A. Plans. London Math. Soc. Lect. Note Ser. 140 (1989), 169 - 194.

117. Projective resolution of identity in $C(K)$ spaces. Arch. Math. Basel 54 (1990), 493 - 498.

118. Una propiedad de interpolación en espacios de funciones holomorfas con desarrollos asintóticos. Homenaje Prof. N. Hayek Kalil. Publ. Univ. de la Laguna (Tenerife) (1990), 351 - 360.

119. Topological direct sum decomposition of Banach spaces. Israel J. of Math. 71 (1990), 289 - 296.

120. Every Radon-Nikodym compact space is Eberlein compact (with J. Orihuela and W. Schachermayer). Studia Mathematica 98 (1991), 157 - 174.

121. Resoluciones de la identidad y bases de Markushevich en ciertos espacios de Banach. Rev. Real Acad. Ci. Exact. Fís. Natur. Madrid (to appear).

122. On the extent of the (non) quasi-analytic classes (with J. Schmets). Arch. Math. Basel (to appear).

123. Interpolation in spaces of holomorphic mappings with asymptotic expansions. Proc. Roy. Irish Acad. (to appear).

124. On totally reflexive spaces. Homenaje Prof. Acquaro. Univ. de Bari (to appear).

125. Simultaneous resolutions of the identity operator on normed spaces (to appear).

126. On basis sequence in Banach space. Note Mat. 10 (1990) (volume dedicated to the memory of Prof. G. Köthe).

127. Complemented subspaces of certain Banach spaces. Seminar on Functional Analysis. Universidad de Murcia (to appear).

Progress in Functional Analysis
K.D. Bierstedt, J. Bonet, J. Horváth & M. Maestre (Eds.)

Regularity properties of (LF)–spaces

Dietmar Vogt

Fachbereich Mathematik, Bergische Universität–GH Wuppertal, Gaußstr. 20, D–5600 Wuppertal, Fed. Rep. of Germany.

Dedicated to Professor Manuel Valdivia on the occasion of his 60th birthday.

In the present paper we investigate acyclic and weakly acyclic (LF)–spaces in the sense of Palamodov [14] (or (LF)–spaces of type (M) and (M_0) in the sense of Retakh [18]), and the implications between these properties and completeness, regularity, sequential retractivity. We give complete characterizations in terms of the defining matrices in the case of (LF)–sequence spaces.

In the first section we introduce the classes of acyclic and weakly acyclic (LF)–spaces, following the definition of Palamodov, and derive their basic properties, avoiding all homological tools. In the second section we show Retakh's theorem with a direct proof on the (LF)–side, but also present versions of it which are suitable for applications and evaluation in concrete cases. In the third and fourth section we investigate the connection between the acyclicity properties and completeness properties, regularity and sequential retractivity. The fifth section gives for (LF)–sequence spaces E^p ($1 \leq p \leq +\infty$, $p = 0$) necessary and sufficient conditions for all of these properties in terms of the defining matrix. Here p denotes the ℓ^p–type of the seminorms used in the definition of the sequence space, $p = 0$ corresponds to the c_0–type. The characterizations vary with p. Only the characterization of weak acyclicity for $p = +\infty$ remains open. In the Fréchet–Schwartz case all of these properties coincide.

For inductive limits of Banach spaces all these questions are well investigated. The connection of (M) and (M_0) to various regularity properties in the case of (LB)–spaces has been studied in Palamodov [14],§7, Neus [13], Valdivia [20],I,§9. Conditions for regularity properties in the case of Köthe co–echelon spaces were given in Bierstedt, Meise and Summers [1], [2], Bierstedt and Meise [3], Valdivia [21]. Many of these results are contained as special cases in our theory.

There is also a close connection to the dual problem of studying projective limits of (DF)–spaces. For these questions cf. Grothendieck [10],II,§4, Palamodov [14], Retakh [16], Krone and Vogt [12] and [23], [24], [25].

We use common notation for locally convex spaces (see Köthe [11]). For inductive limits and (LF)-spaces see Floret and Wloka [8], Valdivia [20]. For all undefined terminology we refer to these books.

1. Let E be an (LF)-space, $E_1 \hookrightarrow E_2 \hookrightarrow \ldots$ a defining inductive spectrum. This means that we have an increasing sequence of subspaces of E, each E_k equipped with a Fréchet space topology, with continuous imbeddings and $E = \varinjlim E_k$.

For every k we have a fundamental system $(\| \|_{k,n})_n$ of seminorms in E_k and we assume

$$\begin{aligned} \| \|_{k,n} &\leq \| \|_{k,n+1} \\ \| \|_{k,n} &\geq \| \|_{k+1,n} \end{aligned}$$

for all k and n.

Identifying $\bigoplus_k E_k$ with the set

$$\Big\{x = (x_k)_k \in \prod_k E_k : \ x_k = 0 \text{ up to finitely many } k\Big\}$$

we define a map $q : \bigoplus_k E_k \longrightarrow E$ by

$$qx = \sum_k x_k .$$

The map $\sigma : \bigoplus_k E_k \longrightarrow \bigoplus_k E_k$ defined as

$$\sigma x = (x_k - x_{k-1})_k \ , \ x_{-1} = 0$$

is an isomorphism onto the kernel of q. Hence we have a canonical exact sequence

$$0 \longrightarrow \bigoplus_k E_k \xrightarrow{\ \sigma\ } \bigoplus_k E_k \xrightarrow{\ q\ } E \longrightarrow 0 . \tag{1}$$

The inverse $\sigma^{-1} : \ker q \longrightarrow \bigoplus_k E_k$ is given by

$$\sigma^{-1}(x) = \Big(\sum_{j=1}^{k} x_j\Big)_k .$$

We remark that q is continuous and open, σ continuous but not necessarily open onto its range. This means that σ^{-1} need not to be continuous.

Following Palamodov [14], we call the spectrum *acyclic* if σ^{-1} is continuous; we call it *weakly acyclic* if σ^{-1} is weakly continuous, i.e. continuous with respect to the weak topologies.

Remark: For any increasing sequence $E_1 \hookrightarrow E_2 \hookrightarrow \ldots$ of Fréchet spaces and $E = \bigcup_k E_k$ equipped with the, not necessarily Hausdorff, locally convex inductive topology we can do the same as before. If the spectrum is acyclic then $\ker q \cong \bigoplus_k E_k$ is complete, hence closed. Therefore E is Hausdorff. So, dealing with acyclic spectra, the assumption of E being Hausdorff was no loss of generality.

For an interpretation of weak acyclicity we remark that the definition of σ shows that σ gives for every n an isomorphism

$$\sigma_n : \bigoplus_{k=1}^{n} E_k \longrightarrow \ker q \cap \bigoplus_{k=1}^{n+1} E_k \, .$$

σ_n is clearly a topological isomorphism onto its range. So we obtain the following:

Remark: The spectrum is weakly acyclic if and only if every linear form on $\ker q$ which is continuous on $\ker q \cap \bigoplus_{k=1}^{n} E_k$ for every n, is continuous on $\ker q$, hence can be extended to $\bigoplus_{k=1}^{\infty} E_k$. In other words, $\ker q$ is well located (see Slowikowski [19]) in $\bigoplus_{k=1}^{\infty} E_k$.

To show the significance of these definitions and also, that they depend only on E and not on the particular defining spectrum, we need the following technical lemma (see [24], Lemma 3.9).

Lemma 1.1 *Let the following be a commutative diagram of topological vector spaces and linear maps with algebraically exact rows and columns*

$$\begin{array}{ccccccccc}
 & & 0 & & 0 & & 0 & & \\
 & & \uparrow & & \uparrow & & \uparrow & & \\
0 & \longrightarrow & A & \xrightarrow{i} & B & \xrightarrow{S} & C & \longrightarrow & 0 \\
 & & {\scriptstyle q}\uparrow & & {\scriptstyle Q}\uparrow & & {\scriptstyle T}\uparrow & & \\
0 & \longrightarrow & D & \xrightarrow{h} & E & \xrightarrow{P} & F & \longrightarrow & 0 \\
 & & & & {\scriptstyle k}\uparrow & & {\scriptstyle j}\uparrow & & \\
 & & & & G & \xrightarrow{p} & H & \longrightarrow & 0 \\
 & & & & \uparrow & & \uparrow & & \\
 & & & & 0 & & 0 & &
\end{array}$$

We assume that q, k, P are continuous, that h and j are open onto their range and that p and Q are open. Then i is open onto its range.

PROOF: We may assume that i, h, k, j are identical imbeddings of subspaces. Let U be a neighborhood of zero in A, then there is a neighborhood of zero $W \subset E$ such that $D \cap (W + W) \subset q^{-1}U$. $p(G \cap W)$ is a neighborhood of zero in H, hence there is a neighborhood of zero $V \subset W$ in E such that $H \cap PV \subset p(G \cap W)$. We claim that $A \cap QV \subset U$. In fact, assume $a = Qv$, $a \in A$, $v \in V$. Then $TPv = SQv = 0$, hence $Pv \in H \cap PV \subset p(G \cap W)$. So there exists $w \in G \cap W$ such that $Pv = pw$. We have $v - w \in W + W$ and, because of $P(v - w) = 0$, $v - w \in D$. Therefore $a = Qv = q(v - w) \in U$. Since Q is open, this completes the proof of the lemma.

Proposition 1.2 *Let E be as before with acyclic (weakly acyclic) defining spectrum $E_1 \hookrightarrow E_2 \hookrightarrow \ldots$ and F, G be (LF)-spaces with defining spectra $F_1 \hookrightarrow F_2 \hookrightarrow \ldots$, $G_1 \hookrightarrow G_2 \hookrightarrow \ldots$. Let*

$$0 \longrightarrow G \xrightarrow{\psi} F \xrightarrow{\varphi} E \longrightarrow 0$$

be an algebraically exact sequence with continuous maps ψ and φ such that for every n there is an m with $\psi^{-1}F_n \subset G_m$. Then the map ψ is a topological imbedding (topological imbedding with respect to weak topologies).

Remark: Since φ is, due to the open mapping theorem, always open the sequence is then topologically exact (topologically exact with respect to weak topologies).

PROOF of the theorem: By Grothendieck's factorization theorem (see Grothendieck [10], I,p.16) the inductive spectra $\varphi F_1 \hookrightarrow \varphi F_2 \hookrightarrow \ldots$ and $E_1 \hookrightarrow E_2 \hookrightarrow \ldots$ are equivalent. We choose an increasing sequence $n(k)$ such that $\varphi F_{n(k)} \supset E_k$ for all k and put $\tilde{F}_k = F_{n(k)} \cap \varphi^{-1}E_k$, $\tilde{G}_k = \psi^{-1}\tilde{F}_k$. These are (notice the assumption) defining spectra of Fréchet spaces for F and G respectively. Therefore we may assume without restriction of generality, that for the defining spectra of F and G the exact sequence

$$0 \longrightarrow G_k \xrightarrow{\psi} F_k \xrightarrow{\varphi} E_k \longrightarrow 0 \tag{2}$$

is defined and (topologically!) exact for every k. It is then also topologically exact with respect to weak topologies.

We set up the following commutative diagram (cf.[24], Theorem 3.10) with algebraically exact rows and columns

$$\begin{array}{ccccccccc}
 & & 0 & & 0 & & 0 & & \\
 & & \uparrow & & \uparrow & & \uparrow & & \\
0 & \longrightarrow & G & \xrightarrow{\psi} & F & \xrightarrow{\varphi} & E & \longrightarrow & 0 \\
 & & \uparrow & & \uparrow & & \uparrow & & \\
0 & \longrightarrow & \bigoplus_k G_k & \longrightarrow & \bigoplus_k F_k & \longrightarrow & \bigoplus_k E_k & \longrightarrow & 0 \\
 & & \sigma\uparrow & & \sigma\uparrow & & \sigma\uparrow & & \\
0 & \longrightarrow & \bigoplus_k G_k & \longrightarrow & \bigoplus_k F_k & \longrightarrow & \bigoplus_k E_k & \longrightarrow & 0 \\
 & & \uparrow & & \uparrow & & \uparrow & & \\
 & & 0 & & 0 & & 0 & &
\end{array} \tag{3}$$

The second and third line are derived from (2) by direct summation, hence topologically exact. All maps are continuous. For E σ is a topological imbedding by assumption. Hence Lemma 1.1 yields the result. The same argument applies to the case of weak topologies.

Corollary 1.3 *If some defining spectrum for the (LF)-space E is acyclic (weakly acyclic), then every defining spectrum for E is acyclic (weakly acyclic).*

PROOF: We apply Proposition 1.2 to the canonical exact sequence (1).

This result justifies the following definition.

Definition: An (LF)-space E is called acyclic (weakly acyclic) if it admits an acyclic (weakly acyclic) defining spectrum.

Remark: According to Corollary 1.3 this means that every defining spectrum is acyclic (weakly acyclic).

With this definition Proposition 1.2 turns into

Theorem 1.4 *If E is acyclic (weakly acyclic) and*

$$0 \longrightarrow G \xrightarrow{\psi} F \xrightarrow{\varphi} E \longrightarrow 0$$

an exact sequence of (LF)-spaces such that for every n there is an m with $\psi^{-1}F_n \subset G_m$, then the map ψ is a topological imbedding (topological imbedding with respect to weak topologies).

If on the other hand we use the diagram (3) in a different way, assuming that ψ is a topological imbedding and F is acyclic, which means that σ in the middle column is open onto its range then we obtain from Lemma 1.1 the "acyclic part" of the following theorem. The "weakly acyclic part" follows in the same way.

Theorem 1.5 *If F is acyclic (weakly acyclic) and*

$$0 \longrightarrow G \xrightarrow{\psi} F \xrightarrow{\varphi} E \longrightarrow 0$$

an exact sequence of (LF)-spaces such that ψ is a topological imbedding (topological imbedding with respect to weak topologies), then E is acyclic (weakly acyclic).

2. We shall now derive conditions for acyclicity and weak acyclicity. In particular we shall derive the necessary and sufficient conditions (M) and (M_0) given by Palamodov [14] and Retakh [18]. We shall also prove some useful variants of them.

To treat acyclicity and weak acyclicity at the same time we assume $(|\ |_{k,\alpha})_{\alpha\in A}$ to be a fundamental system for some topology t_k on E_k which is weaker than the original one. We denote by t the corresponding topology on $\bigoplus_k E_k$.

Lemma 2.1 *σ^{-1} is continuous with respect to t if and only if for all sequences $\alpha(k)$ and $C_k \geq 0$ there are sequences $\beta(k)$ and $D_k \geq 0$ such that*

$$\sum_k C_k \,|\sum_{\nu=0}^{k} x_\nu|_{k,\alpha(k)} \leq \sum_k D_k |x_k|_{k,\beta(k)} \tag{4}$$

for all $x = (x_k)_k \in \bigoplus_k E_k$ with $\sum_k x_k = 0$.

Lemma 2.2 *If σ^{-1} is continuous with respect to t then the following holds*

$$\forall\,\mu\;\exists\,n,k\geq\mu\;\forall\,\alpha,K\geq k,R>0\;\exists\,\beta,S>0\;\forall\,x\in E_\mu\;:\;|x|_{k,\alpha}\leq\frac{1}{R}\|x\|_{\mu,n}+S|x|_{K,\beta}\,. \quad (5)$$

PROOF: Assume that (5) does not hold, then there exists μ such that for all k there are $\alpha(k)$, $K(k)\geq k$, $R(k)>0$ such that for every β, $S>0$ there is an $x\in E_\mu$ with

$$|x|_{k,\alpha(k)}>\frac{1}{R(k)}\|x\|_{\mu,k}+S|x|_{K(k),\beta}\,.$$

We apply the previous lemma to the sequences $\alpha(k)$ and $C_k=kR(k)$ and obtain sequences $\beta(k)$ and D_k.

We insert now into (4) the vector $(0,\ldots,0,x,0,\ldots,0,-x,0,\ldots)\in\bigoplus_k E_k$, where x is in the μ–th and $-x$ in the $K(k)$–th place. This gives

$$kR(k)|x|_{k,\alpha(k)}\leq\tilde{D}_\mu\|x\|_{\mu,n(\mu)}+D_{K(k)}|x|_{K(k),\beta(K(k))}\,.$$

Here we estimated $|\;|_{\mu,\beta(\mu)}$ by $C\|\;\|_{\mu,n(\mu)}$ with suitable C, $n(\mu)$ and put $\tilde{D}_\mu=D_\mu\cdot C$. We choose now k so large that $k\geq\tilde{D}_\mu$, $k\geq n(\mu)$ and arrive at a contradiction.

From this we derive the following necessary conditions for acyclicity and weak acyclicity.

Proposition 2.3 *(a) If E is acyclic, then the following condition holds*

$$\forall\,\mu\;\exists\,n,k\geq\mu\;\forall\,m,K\geq k,R>0\;\exists\,N,S>0\;\forall\,x\in E_\mu: \quad (Q)$$

$$\|x\|_{k,m}\leq\frac{1}{R}\|x\|_{\mu,n}+S\|x\|_{K,N}\,.$$

(b) If E is weakly acyclic, then the following condition holds

$$\forall\,\mu\;\exists\,n,k\geq\mu\;\forall\,m,K\geq k\;\exists\,N,S>0\;\forall\,x\in E_\mu: \quad (WQ)$$

$$\|x\|_{k,m}\leq S(\|x\|_{\mu,n}+\|x\|_{K,N})\,.$$

PROOF: (a) is just a reformulation of condition (5) with t_k the given topology on E_k for all k. (b) follows if, for μ,n,k,K chosen in accordance with condition (5), we compare on the linear space E_μ the locally convex metrizable topologies T_1 given by the seminorms $(\|\;\|_{k,m})_m$ and T_2 given by the seminorms $(\|\;\|_{\mu,n}+\|\;\|_{K,N})_N$. Because then (5) implies that $(E,T_1)'\subset(E,T_2)'$ and the result follows from well known theorems.

Lemma 2.4 *If the following holds*

$$\exists n \,\forall \mu \,\exists k \geq \mu \,\forall \alpha, K \geq k, R > 0 \,\exists \beta, S > 0 \,\forall x \in E_\mu : \; |x|_{k,\alpha} \leq \frac{1}{R}\|x\|_{\mu,n} + S|x|_{k,\beta} \qquad (6)$$

then σ^{-1} is continuous with respect to T on $\operatorname{im}\sigma$ and t, where T is the restriction of the strongest locally convex topology on $\bigoplus_k E_k$ which coincides with t on $\{x : \sum_{\nu=0}^{\infty} \|x_\nu\|_{\nu,n} \leq 1\}$, n as in (6).

PROOF: Assuming (6) we may after going to a subsequence of the E_k even assume

$$\forall k \geq 1, \alpha, R > 0 \,\exists \beta, S \,\forall x \in E_{k-1} : \; |x|_{k,\alpha} \leq \frac{1}{R}\|x\|_{k-1,0} + S|x|_{k+1,\beta} \,. \qquad (7)$$

For $\varepsilon > 0$ and given sequences $\alpha(k)$, $0 < C_k \leq C_{k+1}$ we choose inductively sequences $\beta(k)$ and $S_k > 0$. First we choose $\beta(1) \geq \alpha(0), \alpha(1)$ and put $S_1 = 1$. If then $\beta(k)$ is determined, we choose $\beta(k+1) \geq \alpha(k+1), \beta(k)$, $S_{k+1} \geq S_k \geq 1$ such that with $\varepsilon_k = 2^{-k-1}C_k^{-1}\varepsilon$ we have

$$S_k|x|_{k,\beta(k)} \leq \varepsilon_k\|x\|_{k-1,0} + S_{k+1}|x|_{k+1,\beta(k+1)}$$

for all $x \in E_{k-1}$. This can be done because of (7) with $\alpha = \beta(k)$, $R = 2^{k+1}C_k\varepsilon^{-1}S_k$ if we put $\beta = \beta(k+1)$ and $S \cdot S_k = S_{k+1}$.

For $x_\nu \in E_\nu$, $\nu = 0, \ldots, n$, with $\sum_{\nu=0}^{n} x_\nu = 0$ and $1 \leq k \leq n$ we obtain

$$\begin{aligned} S_k|\sum_{\nu=0}^{k-1} x_\nu|_{k,\beta(k)} &\leq \varepsilon_k \sum_{\nu=0}^{k-1} \|x_\nu\|_{\nu,0} + S_{k+1}|\sum_{\nu=0}^{k-1} x_\nu|_{k+1,\beta(k+1)} \\ &\leq \varepsilon_k \sum_{\nu=0}^{k-1} \|x_\nu\|_{\nu,0} + S_{k+1}|x_k|_{k+1,\beta(k+1)} + S_{k+1}|\sum_{\nu=0}^{k} x_\nu|_{k+1,\beta(k+1)} \,. \end{aligned}$$

Repeating this argument inductively we get

$$\begin{aligned} S_k|\sum_{\nu=0}^{k-1} x_\nu|_{k,\beta(k)} &\leq \sum_{l=k}^{n} \varepsilon_l \sum_{\nu=0}^{l-1} \|x_\nu\|_{\nu,0} + \sum_{l=k}^{n} S_{l+1}|x_l|_{l+1,\beta(l+1)} \\ &\leq 2^{-k}C_k^{-1}\varepsilon \sum_{\nu=0}^{\infty} \|x_\nu\|_{\nu,0} + \sum_{l=k}^{\infty} S_{l+1}|x_l|_{l+1,\beta(l+1)} \,, \end{aligned} \qquad (8)$$

hence by use of $S_k \geq 1$

$$|\sum_{\nu=0}^{k} x_\nu|_{k,\beta(k)} \leq |x_k|_{k,\beta(k)} + 2^{-k}C_k^{-1}\varepsilon \sum_{\nu=0}^{\infty} \|x_\nu\|_{\nu,0} + \sum_{l=k}^{\infty} S_{l+1}|x_l|_{l+1,\beta(l+1)}$$

for $k = 1, 2, \ldots$ and therefore, using $\alpha(k) \leq \beta(k)$

$$\begin{aligned}\sum_{k=0}^{\infty} C_k \Big| \sum_{\nu=0}^{k} x_\nu \Big|_{k,\alpha(k)} &\leq C_0 |x_0|_{0,\alpha(0)} + \sum_{k=1}^{\infty} C_k |x_k|_{k,\beta(k)} \\ &\quad + \varepsilon \sum_{\nu=0}^{\infty} \|x_\nu\|_{\nu,0} + \sum_{k=1}^{\infty} C_k \sum_{l=k}^{\infty} S_{l+1} |x_l|_{l,\beta(l+1)} \\ &\leq \varepsilon \sum_{\nu=0}^{\infty} \|x_\nu\|_{\nu,0} + \sum_{l=0}^{\infty} D_l |x_l|_{l,\beta(l+1)}\end{aligned}$$

with $D_l = C_l + \sum_{k=1}^{l} C_k S_{l+1}$. Taking into account our adjustment of seminorms ($n = 0$) this proves the result for an equivalent defining spectrum, which is a subsequence of the given one. The general result now follows easily by use of the triangular inequality.

Remark: T is clearly the inductive limit topology of the topologies T_k on E_k, where T_k is the strongest locally convex topology on E_k which coincides with t_k on $U_{k,n}$.

Lemma 2.5 *If condition (6) of Lemma 2.4 and (a) or (b) or (c) holds, where*

(a) t_k is the given topology on E_k for every k

(b) t_k is the weak topology on E_k for every k

(c) $E_k = F_k'$ for some (LB)-space F_k and t_k is the weak-topology for every k,*

then σ^{-1} is continuous with respect to t.

PROOF: This is an immediate consequence of Lemma 2.4, the Remark and the fact that in case (a) $T_k = t_k$, in cases (b) and (c) E_k equipped with T_k has the same dual as E_k equipped with t_k.

The next lemma is based on a variation of the proof of de Wilde's localization Lemma (cf. Köthe [11],II,§35,6.(1)). For the rest of this section we shall assume in the proofs without loss of generality that

$$U_{k,n} = \{x \in E_k : \|x\|_{k,n} \leq 1\}, \; n = 1, 2, \ldots$$

is a basis of neighborhoods of zero in E_k for all k.

Lemma 2.6 *If E is weakly acyclic then for any sequence $m_1, m_2, \ldots$ there are sequences $n_1, n_2, \ldots$ and $\alpha_1, \alpha_2 \ldots$ such that*

$$\sum_{k=1}^{K} \Big\| \sum_{\nu=1}^{k} x_\nu \Big\|_{k,m_k} \leq \alpha_{K+1} \sum_{k=1}^{K+1} \|x_k\|_{k,n_k}$$

for all $x \in \bigoplus_{k=1}^{K+1} E_k$ with $\sum_{k=1}^{K+1} x_k = 0$.

PROOF: We put

$$B_{k,n} = \{y \in E_k' : |y(x)| \leq 1 \text{ for all } x \in U_{k,n}\} .$$

For a given sequence $m_1, m_2, \dots$ we put

$$B = \prod_{k=1}^{\infty} B_{k,m_k} \subset \prod_{k=1}^{\infty} E_k' .$$

$E_B^* :=$ span B equipped with the Minkowski norm of B is a Banach space. For any integers $n_1, \dots, n_K$ we put

$$C_{n_1,\dots,n_K} = \prod_{k=1}^{K} B_{k,n_k} \times \prod_{k=K+1}^{\infty} E_k' \subset \prod_{k=1}^{\infty} E_k'$$

and

$$D_{n_1,\dots,n_K} = E_B \cap qC_{n_1,\dots,n_K}$$

where $qy = (y_k - y_{k+1}|_{E_k})_k$ for $y = (y_k)_k \in \prod_{k=1}^{\infty} E_k' = (\bigoplus_k E_k)'$. Obviously q is the transpose of σ, and therefore surjective by assumption. It follows that

$$\bigcup_n D_n = E_B, \quad \bigcup_n D_{n_1,\dots,n_K,n} = D_{n_1,\dots,n_K} \tag{9}$$

for $K = 1, 2, \dots$ and arbitrary $n_1, \dots, n_K$.

We construct inductively a sequence $n_1, n_2, \dots$. Since $\bigcup_n D_n = E_B$ there is n_1 such that D_{n_1} is of 2. category (= not meager) in E_B. Assume $n_1, \dots, n_K$ determined and $D_{n_1,\dots,n_K}$ of 2. category in E_B then because of (9) there is n_{K+1} such that $D_{n_1,\dots,n_K,n_{K+1}}$ is of 2. category in E_B. For all K we have now $\varepsilon_K > 0$ such that $\varepsilon_K B \subset \overline{D}_{n_1,\dots,n_K}$ where the closure is taken in E_B. Every $x \in \bigoplus_k E_k$ defines a continuous linear functional on E_B^*. Its norm is $\sum_k \|x_k\|_{k,m(k)}$. For $x \in \bigoplus_{k=1}^{K} E_k$ we obtain therefore

$$\begin{aligned}
\varepsilon_K \sum_{k=1}^{K} \|x_k\|_{k,m_k} &\leq \sup_{\overline{D}_{n_1,\dots,n_K}} |y(x)| = \sup_{D_{n_1,\dots,n_K}} |y(x)| \\
&\leq \sup_{\eta \in C_{n_1,\dots,n_K}} |(q\eta)x| = \sup_{\eta \in C_{n_1,\dots,n_K}} |\eta(\sigma x)| .
\end{aligned}$$

For $(\xi_k)_k \in \bigoplus_k E_k$ with $\sum_{k=1}^{K+1} \xi_k = 0$ we set $x_k = \sum_{\nu=1}^{k} \xi_\nu$ for all k and obtain $x \in \bigoplus_{k=1}^{K+1} E_k$ and therefore

$$\varepsilon_{K+1} \sum_{k=1}^{K} \| \sum_{\nu=1}^{k} \xi_\nu \|_{k,m_k} \leq \sup_{\eta \in C_{n_1,\dots,n_{K+1}}} |\eta(\xi)| = \sum_{k=1}^{K+1} \|\xi_k\|_{k,n_k} .$$

This was to be proved.

Definition: We call E adjustable if the following holds: For every sequence $(\tilde{n}(\nu))_\nu$ there is a sequence $(n(\nu))_\nu$ such that

$$\forall\,\mu,K\,\exists\,\tilde{K},C\,\forall\,\tilde{N},\tilde{S}\,\exists\,N,S : \bigwedge_{\nu=0}^{\mu}(\tilde{S}\|\ \|_{\tilde{K},\tilde{N}}+\|\ \|_{\nu,\tilde{n}(\nu)}) \le S\|\ \|_{K,N}+C\bigwedge_{\nu=0}^{\mu}\|\ \|_{\nu,n(\nu)}\,.$$

Here we define for seminorms $|\ |_\nu$ on E_ν, $\nu=0,\dots,\mu$

$$\Big(\bigwedge_{\nu=0}^{\mu}|\ |_\nu\Big)(x)=\inf\Big\{\sum_{\nu=0}^{\mu}|x_\nu|_\nu : x=\sum_{\nu=0}^{\mu}x_\nu,\ x_\nu\in E_\nu,\ \nu=0,\dots,\mu\Big\}\,.$$

Lemma 2.7 *If E is weakly acyclic, then E is adjustable.*

PROOF: We apply Lemma 2.6 to the sequence $\tilde{n}(1),\tilde{n}(2),\dots$ and obtain sequences $n(1),n(2),\dots$ and $\alpha_1,\alpha_2,\dots$ such that for $x=\sum_{\nu=0}^{\mu}x_\nu$, $x_\nu\in E_\nu$, $\nu=0,\dots,\mu$ we have:

$$\|x\|_{\mu,\tilde{n}(\mu)}\le\alpha_K(\|x\|_{K,n(K)}+\sum_{\nu=0}^{\mu}\|x_\nu\|_{\nu,n(\nu)})\,.$$

This follows by application of the inequality in Lemma 2.6 to $(x_0,\dots,x_\mu,0,\dots,0,-x,\dots)$. It implies obviously the assertion.

Theorem 2.8 *E is acyclic if and only if E is adjustable and satisfies (Q).*

PROOF: Necessity of the condition follows from Proposition 2.3 and Lemma 2.7. To prove sufficiency we write (Q) in the form

$$\exists\,(n(\nu))_\nu\,\forall\,\mu\,\exists\,n,k\ge\mu\,\forall\,m,K\ge k,R>0\,\exists\,N,S>0\,\forall\,x\in E_\mu : \tag{10}$$

$$\|x\|_{k,m}\le\frac{1}{R}\bigwedge_{\nu=0}^{\mu}(\|x\|_{\mu,n(\nu)}+S\|x\|_{K,N})\,.$$

We define on E_k, for every k, the following equivalent fundamental system of seminorms

$$|\ |_{k,m}=\bigwedge_{\nu=0}^{k}\|\ \|_{\nu,n(\nu)+m}\,.$$

Then adjustability and (10) yield

$$\forall\,\mu\,\exists\,k\ge\mu\,\forall\,m,K\ge k,R>0\,\exists\,N,S>0\,\forall\,x\in E_\mu : |x|_{k,m}\le\frac{1}{R}|x|_{\mu,0}+S|x|_{K,N}\,,$$

i.e. (6) in Lemma 2.4 with $n=0$. Lemma 2.5(a) proves the result.

Theorem 2.9 *E is weakly acyclic if and only if E is adjustable and satisfies*

$$\forall\,\mu\,\exists\,n,k\ge\mu\,\forall\,y\in E_k',K\ge k,R>0\,\exists\,N,S>0\,\forall\,x\in E_\mu : \tag{Q_0}$$

$$|y(x)|\le\frac{1}{R}\|x\|_{\mu,n}+S\|x\|_{K,N}\,.$$

PROOF: Necessity of the condition follows from Lemma 2.2 and Lemma 2.7. As in Theorem 2.8 we obtain with an equivalent fundamental system of seminorms

$$\forall \mu \, \exists k \geq \mu \, \forall y \in E_k', K \geq k, R > 0 \, \exists N, S > 0 \, \forall x \in E_\mu : |y(x)| \leq \frac{1}{R}|x|_{\mu,0} + S|x|_{K,N} .$$

An application of the Hahn–Banach–Theorem yields $y = y_1 + y_2$, $|y_1(x)| \leq \frac{1}{R}|x|_{\mu,0}$ for all $x \in E_\mu$, $y_2 \in E_K'$. Hence we have again (6) in Lemma 2.4. Lemma 2.5(b) proves the result.

These theorems are variants of theorems of Retakh [18]. They have the advantage, that in both cases one of the two conditions is in many concrete cases a priori fulfilled. The other one is an inequality in terms of an arbitrary given fundamental system of seminorms.

We shall now give the original result of Retakh [18].

Theorem 2.10 *E is (weakly) acyclic if and only if the following condition is fulfilled: There exists a sequence U_μ of absolutely convex neighborhoods of zero in E_μ, $\mu = 0, 1, 2, \ldots$, such that*

(1) $U_\mu \subset U_{\mu+1}$ for all μ

(2) For every μ there is $k \geq \mu$ such that for all $K \geq k$ the (weak) topology of E_K coincides on U_μ with the (weak) topology on E_k.

Definition: In the notation of Retakh [18] the condition in the acyclic case is called (M), in the weakly acyclic case (M_0).

Condition (M) means that E is a generalized strict inductive limit in the sense of Garling [9] of the sequence $U_1 \subset U_2 \subset \ldots$. For a consideration under this aspect see Valdivia [20], Ch.I,§9.

PROOF of the theorem: Let condition (M) (condition (M_0)) be given and $\| \ \|_\mu$ be the Minkowski seminorm of U_μ. We replace $\| \ \|_{k,n}$ by $\max(\| \ \|_k, \| \ \|_{k,n})$ and obtain a fundamental system of seminorms on each E_k, satisfying condition (6) in Lemma 2.4 for the given (weak) seminorm. Lemma 2.5 proves the assertion.

If E is acyclic (weakly acyclic) we use the proof of Theorem 2.8 (Theorem 2.9) to produce an equivalent fundamental system of seminorms which satisfies condition (6) in Lemma 2.4, which obviously implies condition (M) (condition (M_0)).

The following version of Theorem 2.10 is given in Palamodov [14], Theorems 6.2 and 6.4.

Theorem 2.11 *In Theorem 2.10 (2) may be replaced by*

(2') For every μ there is $k \geq \mu$ such that the (weak) topology of E coincides on U_μ with the (weak) topology of E_k.

PROOF: Using the proof of Lemma 2.4 including the standardization (7) of the fundamental system of seminorms (and the proof of Theorem 2.8) we obtain from inequality (8), applied to $C_k \equiv 1$, $x \in E_\mu = E_{k-1}$, $x = \sum_{l=k}^\infty x_l$, $x_l \in E_l$ for all l:

$$|x|_{k,\alpha} \le \varepsilon \|x\|_{\mu,0} + \sum_{l=k}^{\infty} D_l \|x_l\|_{l,m(l)} \,.$$

We put

$$\|x\| = \inf \{ \sum_{l=k}^{\infty} D_l \|x_l\|_{l,m(l)} : x = \sum_{l=k}^{\infty} x_l,\ x_l \in E_l \text{ for all } l\}$$

and obtain a continuous seminorm on E, hence

$$|x|_{k,\alpha} \le \varepsilon \|x\|_{\mu,0} + \|x\| \,.$$

Putting $U_\mu = \{x : \|x\|_{\mu,0} \le 1\}$, we are ready in the case of (M). For (M_0) we argue as at the end of the proof of Theorem 2.9.

3. In this section we investigate topological properties of acyclic spaces and prove some equivalences under assumptions which will be fulfilled in the case of sequence spaces and spaces of continuous functions. The following is a result of Raikov [15] (for subspaces see Köthe [11], I, §19, 5.(1)). For the sake of completeness we present it with a proof, which is a straightforward adaptation of the proof in Köthe.

Lemma 3.1 *Let E be a locally convex space, $U_1 \subset U_2 \subset \ldots$ an increasing sequence of absolutely convex subsets of E such that $\bigcup_k U_k = E$. We assume that E carries the finest locally convex topology t which coincides on U_k with the restriction of t to U_k for all k. Let $(x_\tau)_{\tau \in T}$ be a Cauchy net in E. Then there exists k such that for every τ_0 and every neighborhood V of zero in E there is $\tau \ge \tau_0$ such that $(x_\tau + V) \cap U_k \ne \emptyset$.*

PROOF: We assume the contrary then

$$\forall\, k\ \exists\, \tau_0, W\ \forall\, \tau \ge \tau_0 \ : \ (x_\tau + w) \cap U_k = \emptyset$$

and therefore

$$\forall\, k\ \exists\, \tau_k, W_k\ \forall\, \sigma, \tau \ge \tau_k \ : \ (x_\tau + W_k) \cap U_k = \emptyset \ , \ x_\sigma - x_\tau \in \frac{1}{2} W_k \,.$$

We may assume $W_{k+1} \subset W_k$, W_k absolutely convex for all k. We put $W_k^{(n)} = W_k \cap U_n$. Then $W_k^{(n)}$ is a neighborhood of zero in U_n. We put

$$V_k = \Gamma\Big(\frac{1}{2} W_1^{(1)} \cup \frac{1}{2} W_2^{(2)} \cup \ldots \cup \frac{1}{2} W_{k-1}^{(k-1)} \cup W_k\Big)$$

and claim that $(x_\tau + V_k) \cap \frac{1}{2} U_k = \emptyset$ for every $\tau \ge \tau_k$.

Assume that this is not the case. Then there is $\tau \ge \tau_k$ and $z \in (x_\tau + V_k) \cap \frac{1}{2} U_k$. We have

$$z = x_\tau + \sum_{i=1}^{k} \alpha_i x_i \in \frac{1}{2} U_k$$

with $x_i \in \frac{1}{2} W_i^{(i)}$, $i = 1, \ldots, k-1$, $x_k \in W_k$, $\sum_{i=1}^{k} |\alpha_i| \leq 1$. Therefore

$$z - \sum_{i=1}^{k-1} \alpha_i x_i = x_\tau + \alpha_k x_k \in U_k \cap (x_\tau + W_k)$$

which contradicts the choice of W_k.

We put $U = \Gamma_{k=1}^{\infty} \frac{1}{2} W_k^{(k)}$, then U is a neighborhood of zero in E and $U \subset V_k$ for all k, since for $m \geq k$ we have $W_m^{(m)} \subset W_k^{(m)} \subset W_k \subset V_k$. We choose τ_0 such that $x_\sigma - x_\tau \in U$ for $\sigma, \tau \geq \tau_0$, and $\sigma_k \geq \tau_0$, $\sigma_k \geq \tau_k$. Then

$$x_{\tau_0} = x_{\sigma_k} + (x_{\tau_0} - x_{\sigma_k}) \in x_{\sigma_k} + U .$$

Hence $2x_{\tau_0} \notin U_k$ for all k, which contradicts the assumptions.

For the following two results, i.e. completeness of acyclic (LF)–spaces and regularity as its consequence, see Palamodov [14], Corollaries 7.1 and 7.2.

Theorem 3.2 *An acyclic (LF)–space is complete.*

PROOF: We assume the U_μ in Theorem 2.10 absolutely convex and open and we may assume $\bigcup_\mu U_\mu = E$. Then according to Lemma 3.1 there is μ and for every τ, V a $\sigma \geq \tau$ and $x_{\tau,V} \in (x_\sigma + V) \cap U_\mu$. Obviously this defines a Cauchy net in U_μ with respect to the topology of E, hence because of Theorem 2.11 with respect to the topology of E_k too.

An (LF)–space E is called *regular* if every bounded set in E is contained in some E_k and bounded there; it is called *sequentially retractive* (see Floret [6]) if every convergent sequence is contained in some E_k and convergent there.

Corollary 3.3 *An acyclic (LF)–space is quasicomplete, regular and sequentially retractive.*

PROOF: Completeness implies quasicompleteness and this, by Grothendieck's factorization theorem, regularity. Because of regularity a convergent sequence is contained in some E_μ and bounded there, hence contained in some λU_μ, U_μ as in 2.10. The result follows from Theorem 2.11.

Definition: The (LF)–space E is said to have closed local neighborhoods if it admits a defining spectrum $E_1 \hookrightarrow E_2 \hookrightarrow \ldots$ such that every E_k has a basis of absolutely convex neighborhoods of zero which are closed in E_k with respect to the topology induced by E.

Remark: (1) If E has closed local neighborhoods then we may assume that $\{x \in E_k : \|x\|_{k,n} \leq 1\}$ is closed in E_k with respect to the topology induced by E, for every k and n.

(2) All (LF)–Köthe spaces, as defined in section 5 have closed local neighborhoods.

Lemma 3.4 *Every weakly acyclic space E has closed local neighborhoods.*

PROOF: This is an immediate consequence of Theorem 2.11.

Our next main result is a modification and an improvement, at least under the assumption of closed local neighborhoods, of Proposition 5.9 in [24]. It needs a preparatory lemma which is a modification of Lemma 5.8 in [24].

Lemma 3.5 *Assume that for numbers $n \leq m \leq N$, $\mu \leq k \leq K$, S and ε the inequality*

$$\|x\|_{k,m} \leq S\|x\|_{K,N} + \varepsilon\|x\|_{\mu,n}$$

holds for all $x \in F \subset E_\mu$ where $F = \{x \in E_\mu : y_\nu(x) = 0,\ \nu = 1, \dots, \nu_0\}$ and $y_1, \dots, y_{\nu_0} \in E'$. Then, with suitable $\tilde{S}$ and $\tilde{N}$ we have

$$\|x\|_{k,m} \leq \tilde{S}\|x\|_{K,\tilde{N}} + \varepsilon\|x\|_{\mu,n}$$

for all $x \in E_\mu$.

PROOF: We may assume that the restrictions of $y_1, \dots, y_{\nu_0}$ to E_μ are linearly independent and that $x_1, \dots, x_{\nu_0} \in E_\mu$ is a dual basis, i.e. $y_k(x_\nu) = \delta_{k,\nu}$ for $k, \nu = 1, \dots, \nu_0$. We put

$$Px = \sum_{\nu=1}^{\nu_0} y_\nu(x)x_\nu, \quad Qx = x - Px$$

for $x \in E_\mu$. Then P and Q are projections in E_μ and $\operatorname{im} Q = F$. We have with suitable C and $\tilde{N} \geq N$

$$\|Px\|_{\mu,N} \leq \sum_{\nu=1}^{\nu_0} |y_\nu(x)|\, \|x_\nu\|_{\mu,N} \leq C\|x\|_{K,\tilde{N}}$$

for all $x \in E_\mu$ and therefore

$$\begin{aligned} \|x\|_{k,m} &\leq \|Px\|_{k,m} + \|Qx\|_{k,m} \leq \|Px\|_{k,m} + S\|Qx\|_{K,N} + \varepsilon\|Qx\|_{\mu,n} \\ &\leq (1 + S + \varepsilon)\|Px\|_{\mu,N} + S\|x\|_{K,N} + \varepsilon\|x\|_{\mu,n} \leq \tilde{S}\|x\|_{K,\tilde{N}} + \varepsilon\|x\|_{\mu,n}\,. \end{aligned}$$

Proposition 3.6 *If E has closed local neighborhoods and does not satisfy*

$$\forall\, \mu\ \exists\, n, k, \varepsilon\ \forall\, K, m, R\ \exists\, N, S\ \forall\, x \in E_\mu : R\|x\|_{k,m} \leq S\|x\|_{K,N} + \varepsilon\|x\|_{\mu,n} \tag{11}$$

then there is μ and a sequence $(x_n)_n$ in E_μ such that $(x_n)_n$ converges to zero in E, but does not converge to zero in any E_k.

PROOF: If E does not satisfy (11) then we have

$$\exists\, \mu\ \forall n, k, \varepsilon\ \exists\, K, m, R\ \forall\, N, S\ \exists\, x \in E_\mu : R\|x\|_{k,m} > S\|x\|_{K,N} + \varepsilon\|x\|_{\mu,n}\,.$$

We assume that $\{x \in E_k : \|x\|_{k,n} \leq 1\}$ is closed in E_k with respect to the topology induced by E, for all k, n. We may replace the sequence $(E_k)_k$ by a subsequence $(E_{n(k)})_k$ and the fundamental systems $(\|\ \|_{n(k),l})_l$ by $(R(l)\,\|\ \|_{n(k),m(l)})_l$, where $m(l)$ is a subsequence and, doing this in an appropriate way, assume that for every k :

$$\forall n, S\ \exists x \in E_1 : \|x\|_{k,k} > S\|x\|_{k+1,n} + 2^k\|x\|_{1,k-1}\,. \tag{12}$$

We choose now for fixed n inductively sequences $(x_{k,n})_{k=2,3,\ldots}$ in E_1. First we choose $x_{2,n}$ according to (12) with $k = 2$, n, $S = 2^k n$ such that $\|x_{2,n}\|_2 = 2$. Then we find $y_{2,n} \in E'$, $\|y_{2,n}\|^*_{2,2} \le 1$ such that $y_{2,n}(x_{2,n}) = 1$. This is possible because of the assumption on closed local neighborhoods.

Assume $x_{2,n}, \ldots, x_{k-1,n}, y_{2,n}, \ldots, y_{k-1,n}$ being chosen. Put

$$X_{k,n} = \{x \in E_1 : y_{\nu,n}(x) = 0 \text{ for } \nu = 2, \ldots, k-1\}\,.$$

The space $X_{k,n}$ fulfills the assumption of Lemma 3.5. Hence (12) is satisfied on $X_{k,n}$. We choose $S = 2^k n$ and find $x = x_{k,n} \in X_{k,n}$ such that

$$\|x\|_{k,k} > 2^k n\|x\|_{k+1,n} + 2^k\|x\|_{1,k-1}$$

and $\|x\|_{k,k} = 2$. Using the assumption on closed local neighborhoods we find $y = y_{k,n} \in E'$, $\|y\|^*_{k,k} \le 1$ such that $y(x) = 1$. We obtain sequences $(x_{k,n})_{k=2,3,\ldots}$ in E_1, $(y_{k,n})_{k=2,3,\ldots}$ in E' such that

$$\begin{array}{lll} \|x_{k,n}\|_{k+1,n} \le \frac{1}{n}2^{-k+1} & \qquad & \|x_{k,n}\|_{1,k-1} \le 2^{-k+1} \\ \|x_{k,n}\|_{k,k} = 2 & & \|y_{k,n}\|^*_{k,k} \le 1 \\ y_{l,n}(x_{k,n}) = 0 \text{ for } k > l & & y_{k,n}(x_{k,n}) = 1\,. \end{array}$$

We define $x_n = \sum_{k=2}^n x_{k,n}$ and have $x_n \in E_1$. We show that $x_n \longrightarrow 0$ in E. Let $\|\ \|$ be any continuous seminorm in E. There exist increasing sequences $m(l)$, C_l such that $\|x\| \le C_l\|x\|_{l,m(l)}$ for all $x \in E_l$. For $l \ge m(1)$ we put $m = m(l+1)$ and obtain for $n \ge m$

$$\begin{aligned} \|x_n\| &\le C_{l+1}\sum_{k=2}^{l}\|x_{k,n}\|_{k+1,n} + C_1\sum_{k=l+1}^{n}\|x_{k,n}\|_{1,k-1} \\ &\le \frac{C_{l+1}}{n} + C_1\cdot 2^{-l+1}\,. \end{aligned}$$

Hence $\|x_n\| \longrightarrow 0$ for $n \longrightarrow \infty$. Next we show that $x_n \not\longrightarrow 0$ in every E_k. For $n \ge k$ we have:

$$\begin{aligned} \|x_n\|_{k,k} &\ge |y_{k,n}(x_n)| = |\sum_{\nu=2}^{k} y_{k,n}(x_{\nu,n})| \ge |y_{k,n}(x_{k,n})| - \sum_{\nu=2}^{k-1}\|x_{\nu,n}\|_{k,k} \\ &\ge 1 - \sum_{\nu=2}^{k-1}\|x_{\nu,n}\|_{\nu+1,n} \ge 1 - \frac{1}{n}\,. \end{aligned}$$

Therefore $(x_n)_n$ does not converge to 0 in E_k.

Obviously condition (11) in Proposition 3.6 and condition (Q) are equivalent. Therefore we have shown

Theorem 3.7 *If E has closed local neighborhoods and is sequentially retractive then it satisfies condition (Q).*

The following theorem we could, due to Lemma 3.4, also formulate as a complete equivalence in the spirit of Theorem 2.8. We prefer the following version.

Theorem 3.8 *If the (LF)–space E is adjustable and has closed local neighborhoods then the following are equivalent:*

(1) E is acyclic

(2) E is sequentially retractive

(3) Condition (Q).

PROOF: This follows from Corollary 3.3, Theorem 3.7 and Theorem 2.8.

4. We are now studying consequences of weak acyclicity on the topological properties of E. First we give a description in terms of the dual spaces. We consider the dual spectrum $\mathcal{E}^* : E_1' \leftarrow E_2' \leftarrow \ldots$ of (DF)–spaces. Setwise by natural identification we have $E' = \lim\limits_{\leftarrow} E_k'$ and the identical map $E_b' \longrightarrow \lim\limits_{\leftarrow} E_k' \longrightarrow E_\sigma'$ is continuous. Since E is barrelled the bounded sets in E_b' and E_σ' coincide, hence also the bounded sets in E_b' and $\lim\limits_{\leftarrow} E_k'$ and these coincide with the equicontinuous sets in E'. We refer to them as bounded sets in E'.

Lemma 4.1 *E is weakly acyclic if and only if* $\mathrm{Proj}^1 \mathcal{E}^* = 0$.

PROOF: This follows from the definition of $\mathrm{Proj}^1 \mathcal{E}^*$ (cf. [25]) and the fact that weak acyclicity is equivalent to σ^t being surjective.

Lemma 4.2 *If E is weakly acyclic then $E_b' = \lim\limits_{\leftarrow} E_k'$ topologically and this space is bornological.*

PROOF: By [24], Theorem 5.6 and our Lemma 4.1 the space $\lim\limits_{\leftarrow} E_k'$ is bornological. This and the remarks before Lemma 4.1 show that then $E_b' = \lim\limits_{\leftarrow} E_k'$.

Lemma 4.3 *If E is weakly acyclic, then for every bounded set $B \subset E$ there is k and an E_k–bounded set $M \subset E_k$, such that B is contained in the E–closure of M.*

PROOF: The polar B^0 is a neighborhood of zero in $\lim\limits_{\leftarrow} E_k'$ by Lemma 4.2. The result follows from the bipolar theorem.

Remark: As Theorems 5.13, 5.14 show, this does not imply that every weakly acyclic E is regular.

Proposition 4.4 *If E is weakly acyclic and every E_k is reflexive then E is regular, reflexive and complete.*

PROOF: To show that E is regular let $B \subset E$ be bounded. We choose an E_k-bounded set M according to Lemma 4.3, which may be assumed absolutely convex and E_k-closed. By reflexivity of E_k it is weakly compact in E_k, hence in E. So M is E-closed and $B \subset M \subset E_k$ is E_k-bounded. The assumption and regularity imply that E is semireflexive and therefore, since barrelled, also reflexive. From Lemma 4.2 then follows that $E = (E_b')_b'$ is complete.

The following will be parallel to Proposition 3.6 and its proof. However there is a characteristic change, cf. also [22], Proposition 4.4 and [24], Proposition 5.9.

Lemma 4.5 *Assume that for all numbers $n \le m \le N$ and $\mu \le k \le K$ the inequality*

$$\|x\|_{k,m} \le S\left(\|x\|_{K,N} + \|x\|_{\mu,n}\right)$$

holds for all $x \in X \subset E_\mu$, where $X = \{x \in E_\mu : y_\nu(x) = 0,\ \nu = 1, \ldots, \nu_o\}$ and $y_1, \ldots, y_{\nu_o} \in E'_{\mu,n}$. Then with suitable $\tilde{S}$ we have

$$\|x\|_{k,m} \le \tilde{S}\left(\|x\|_{K,N} + \|x\|_{\mu,n}\right)$$

for all $x \in E_\mu$.

PROOF: We may assume that $y_1, \ldots, y_{\nu_o}$ are linearly independent and that we have $x_1, \ldots, x_{\nu_o} \in E_\mu$ such that $y_k(x_\nu) = \delta_{k,\nu}$ for $k, \mu = 1, \ldots, \nu_o$. We put

$$Px = \sum_{\nu=1}^{\nu_o} y_\nu(x) x_\nu\,, \quad Qx = x - Px$$

for $x \in E_\mu$. Then P and Q are projections in E_μ and $\operatorname{im} Q = X$. We have

$$\|Px\|_{\mu,N} \le \sum_{\nu=1}^{\nu_o} |y_\nu(x)| \|x_\nu\|_{\mu,N} \le C\|x\|_{\mu,n}$$

for all $x \in E_\mu$ and therefore

$$\begin{aligned} \|x\|_{k,m} &\le \|Px\|_{k,m} + \|Qx\|_{k,m} \le \|Px\|_{k,m} + S\left(\|Qx\|_{K,N} + \|Qx\|_{\mu,n}\right) \\ &\le (1+2S)\|Px\|_{\mu,n} + S\left(\|x\|_{K,N} + \|x\|_{\mu,n}\right) \le \tilde{S}\left(\|x\|_{K,N} + \|x\|_{\mu,n}\right) . \end{aligned}$$

Proposition 4.6 *If the spectrum does not have property (WQ) then there is μ and a sequence $(x_n)_n$ in E_μ such that $(2^n x_n)_n$ converges to 0 in E but $(2^{-n} x_n)_n$ does not converge to 0 in any E_k, $k \ge \mu$.*

PROOF: We have the following

$$\exists \mu\, \forall n, k\, \exists K, m\, \forall N, S\, \exists x \in E_\mu \ :\ \|x\|_{k,m} > S\left(\|x\|_{K,N} + \|x\|_{\mu,n}\right) .$$

We may replace $(E_n)_n$ by a subsequence $(E_{n(k)})_k$ and $\| \|_{n(k),m}$ by a subsequence $\| \|_{n(k),m(l)}$ and so assume that for every k:

$$\forall n, S\, \exists x \in E_1 \;:\; \|x\|_{k,k} > S\,(\|x\|_{k+1,n} + \|x\|_{1,k-1}) \;. \tag{13}$$

For fixed n we choose inductively sequences $x_{k,n} \in E_1$, $y_{k,n} \in E'_{k,k}$, $k = 2, 3, \ldots$.

We choose $x_{2,n}$ according to (13) with $k = 2$, n, $S = 2^{2+2n}$ such that $\|x_{2,n}\|_{2,2} = 2^{n+1}$. Then we choose $y_{2,n} \in E'_{2,2}$, $\|y_{2,n}\|^*_{2,2} \leq 1$ such that $y_{2,n}(x_{2,n}) = 2^n$.

Assume that $x_{2,n}, \ldots, x_{k-1,n}$, $y_{2,n}, \ldots, y_{k-1,n}$ being chosen. Put

$$X_{k,n} = \{x \in E_1 \;:\; y_{\nu,n}(x) = 0 \text{ for } \nu = 2, \ldots, k-1\} \;.$$

Since $y_{\nu,n} \in E'_{\nu,\nu} \subset E'_{1,k-1}$ the space $X_{k,n}$ fulfills the assumption of Lemma 4.5. Hence (13) is satisfied on $X_{k,n}$.

We choose $S = 2^{k+2n}n$ and find $x = x_{k,n} \in X_{k,n}$ such that

$$\|x\|_{k,k} > 2^{k+2n}n\,(\|x\|_{k+1,n} + \|x\|_{1,k-1})$$

and $\|x\|_{k,k} = 2^{n+1}$. We choose $y = y_{k,n} \in E'_{k,k}$, $\|y\|^*_{k,k} \leq 1$ such that $y(x) = 2^n$. We obtain sequences $(x_{k,n})_{k=2,3,\ldots}$ in E_1, $(y_{k,n})_{k=2,3,\ldots}$ in $E'_{k,k}$ such that

$$\begin{array}{lcllcl} \|x_{k,n}\|_{k+1,n} & \leq & \frac{1}{n}2^{-k-n+1} & \qquad \|x_{k,n}\|_{1,k-1} & \leq & \frac{1}{n}2^{-k-n+1} \\ \|x_{k,n}\|_{k,k} & = & 2^{n+1} & \qquad \|y_{k,n}\|^*_{k,k} & \leq & 1 \\ y_{l,n}(x_{k,n}) & = & 0 \text{ for } k > l & \qquad y_{k,n}(x_{k,n}) & = & 2^n. \end{array}$$

We proceed now as in the proof of Proposition 3.6. We define $x_n = \sum_{k=2}^n x_{k,n}$ and have $x_n \in E_1$. Using the same notation as in the proof of Proposition 3.6 the relevant inequalities become

$$\begin{array}{lcll} \|x_n\| & \leq & C_{l+1}\frac{1}{n}2^{-n} & \text{for } n \geq m \\ \|x_n\|_{k,k} & \geq & 2^n - \frac{1}{n}2^{-n} & \text{for } n \geq k. \end{array}$$

This shows that $2^n x_n$ converges to zero in E but $2^{-n}x_n$ does not converge to zero in every E_k.

As an immediate consequence we have

Theorem 4.7 *If E is regular then it satisfies condition (WQ).*

5. We shall now apply the previous results to the case of sequence spaces.

Let $a_{j;k,n}(j, k, n \in \mathbb{N})$ be a matrix with nonnegative extended real valued entries and the following properties

$$a_{j;k,n} \leq a_{j;k,n+1}$$

$$a_{j;k,n} \geq a_{j;k+1,n}$$

for all j, k, n. Moreover we assume that for every j and k there is n with $a_{j;k,n} > 0$ and that for every j there is k such that $a_{j;k,n} < +\infty$ for all n.

For $1 \leq p < +\infty$ we set

$$E_k^p = \left\{ x = (x_1, x_2, \ldots) : \|x\|_{k,n}^p = \sum_j |x_j|^p a_{j;k,n}^p < +\infty \text{ for all } n \right\} .$$

We set

$$E_k^\infty = \left\{ x = (x_1, x_2, \ldots) : \|x\|_{k,n} = \sup_j |x_j| a_{j;k,n} < +\infty \text{ for all } n \right\} ,$$

$$E_k^0 = \left\{ x \in E_k^\infty : \lim_j |x_j| a_{j;k,n} = 0 \text{ for all } n \right\} .$$

All these are Fréchet spaces with fundamental systems of seminorms $(\| \ \|_{k,n})_n$. We set $E^p = \bigcup_k E_k^p$ equipped with the inductive limit topology of the defining spectrum $E_1 \hookrightarrow E_2 \hookrightarrow \ldots$ of Fréchet spaces.

For these spaces one can give a fundamental system of seminorms in a very concrete form (projective description, cf. Bierstedt, Meise and Summers [1]). We denote by $d^\times$ the set of all sequences $a = (a_1, a_2, \ldots)$ which satisfy estimates $a_j \leq C_k a_{j;k,n(k)}$ for all j, k and suitable $C_k, n(k)$. By $D^\times$ we denote the set of all systems $A = (a^{(k)})_k$, $a^{(k)} = (a_1^{(k)}, a_2^{(k)}, \ldots)$ which satisfy $a_j^{(k)} \geq a_j^{(k+1)}$, $a_j^{(k)} \leq \inf_{\nu=1,\ldots,k} C_\nu a_{j;\nu,n(\nu)}$ for all j, k with suitable $C_k, n(k)$.

Proposition 5.1 *A fundamental system of seminorms in E^p is given by the seminorms*

$$\begin{aligned} \|x\|_a &= \Big(\sum_j |x_j|^p a_j^p\Big)^{\frac{1}{p}} , & a \in d^\times , & \quad \text{for } 1 \leq p < +\infty \\ \|x\|_a &= \sup_j |x_j| a_j , & a \in d^\times , & \quad \text{for } p = 0 \\ \|x\|_A &= \inf_k \sup_j |x_j| a_j^{(k)} , & A \in D^\times , & \quad \text{for } p = +\infty . \end{aligned}$$

PROOF: Since these are obviously continuous seminorms on E^p, we have only to show that every continuous seminorm p on E^p can be estimated by some $\| \ \|_a$ or $\| \ \|_A$ respectively.

We have $p(x) \leq C_k \|x\|_{k,n(k)}$ for all k with suitable C_k and $n(k)$. We may choose the sequence C_k increasing so rapidly that $\lim_k C_k a_{j;k,n(k)} = +\infty$ for all j.

For $1 \leq p < \infty$ or $p = 0$ we put

$$a_j = \min_k \{2^k C_k a_{j;k,n(k)}\}$$

and define for $x \in E^p$

$$x_j^{(k)} = \begin{cases} x_j & \text{if } k \text{ is the smallest number with } a_j = 2^k C_k a_{j;k,n(k)} \\ 0 & \text{otherwise.} \end{cases}$$

Then $x = \sum_k x^{(k)}$ and $C_k \|x^{(k)}\|_{k,n(k)} \leq 2^{-k} \|x\|_a$, hence

$$p(x) \leq \sum_k p(x^{(k)}) \leq \sum_k C_k \|x^{(k)}\|_{k,n(k)} \leq \|x\|_a .$$

For $p = +\infty$ we put $A = (a_j^{(k)})$ where

$$a_j^{(k)} = \min_{\nu=1,\ldots,k} \{2^\nu C_\nu a_{j;k,n(\nu)}\}$$

We fix $x \in E^\infty$, $k \in \mathbb{N}$ and define for $1 \le \nu \le k$

$$x_j^{(\nu)} = \begin{cases} x_j & \text{if } \nu \text{ is the smallest number with } a_j^{(k)} = 2^\nu C_\nu a_{j;\nu,n(\nu)} \\ 0 & \text{otherwise.} \end{cases}$$

Then $x = \sum_{\nu=1}^k x^{(\nu)}$ and $C_\nu \|x^{(\nu)}\|_{\nu,n(\nu)} \le 2^{-\nu} \sup_j |x_j| a_j^{(k)}$, hence

$$p(x) \le \sum_{\nu=1}^{k} p(x^{(\nu)}) \le \sum_{\nu=1}^{k} C_\nu \|x^{(\nu)}\|_{\nu,n(\nu)} \le \sup_j |x_j| a_j^{(k)} .$$

Since this holds for every k we have $p(x) \le \|x\|_A$.

The seminorms $\|x\|_a = \sup_j |x_j| a_j$, $a \in d^\times$, define also a locally convex topology on E^∞. By E_0^∞ we denote E^∞ equipped with this topology. Obviously the seminorms

$$\|x\|_{A,0} = \sup_j \inf_k |x_j| a_j^{(k)} , \quad A \in D^\times$$

give an equivalent fundamental system of seminorms for E_0^∞ (and also for E^0). From this it is easy seen that the topology of E^∞ is stronger than the topology of E_0^∞. In fact we obtain, using the idea of the proof of Köthe [11, I, §29, 4(1)], the following description of the topology of E^∞. In the case of (LB)–spaces this corresponds to Bierstedt, Meise, and Summers [1], Corollary 2.8(e). It will be used in the investigation of the completeness of E^∞ in a similar way as there.

Lemma 5.2 *E^∞ carries the strong topology $b(E_0^\infty, E_0^{\infty\prime})$.*

PROOF: Since the topology of E^∞ is stronger than that of E_0^∞ and E^∞ is barrelled, the topology of E^∞ is stronger than $b(E_0^\infty, E_0^{\infty\prime})$.

For $A \in D^\times$ we set $U = \{x \in E^\infty : \|x\|_A < 1\}$. Then $U = \bigcup_k U_k$ where $U_k = \{x \in E^\infty : \sup_j |x_j| a_j^{(k)} < 1\}$. Let $\|x\|_A > 2$, k_0 so that $x \in E_{k_0}$. Then for every $k \ge k_0$ there is j_k and η_k with $|\eta_k| \le 1$ such that $x_{j_k} \eta_k a_{j_k}^{(k)} = 2$. We put $y_k(\xi) = \xi_{j_k} \eta_k a_{j_k}^{(k)}$ for $\xi \in E_0^\infty$. Then $y \in E_0^{\infty\prime}$.

For $k \ge m$ we have $a_{j_k}^{(k)} \le C_m a_{j_k;m,n(m)}$. Hence we have $a_{j_k}^{(k)} \le D_m a_{j_k;m,N(m)}$ for all k with suitable D_m and $N(m)$. We put $a_j = \inf_m D_m a_{j;m,N(m)}$ and obtain $a \in d^\times$. We have

$$|y_k(\xi)| \le |\xi_{j_k}| a_{j_k}^{(k)} \le |\xi_{j_k}| a_{j_k} \le \|\xi\|_a$$

for all $\xi \in E_0^\infty$ and $k \ge k_0$. Therefore $\{y_k : k \ge k_0\}$ is equicontinuous, hence $\sigma(E_0^{\infty\prime}, E_0^\infty)$–relatively compact. Therefore it has a limit point $y \in E_0^{\infty\prime}$.

Since $y_k \in U_m^0$ for $k \ge m$ we have $y \in \bigcap_m U_m^0$ where the polar is taken E_0^∞. Moreover $y(x) = 2$. Therefore $x \notin U^{00} = \bar{U}$ where the closure is taken E_0^∞.

We showed that $\bar{U} \subset \{x : \|x\|_A \le 2\}$. Since $\bar{U}$ is a barrel in E_0^∞, i.e. a $b(E_0^\infty, E_0^{\infty\prime})$–neigbourhood of zero, this shows the result.

Lemma 5.3 *(a) For every $x \in E^0$ and $a \in d^\times$ we have $\lim_j |x_j| a_j = 0$.*

(b) The completion of E^p is

$$\widehat{E}^p = \{x = (x_1, x_2, \ldots) : \sum_j |x_j|^p a_j^p < +\infty \textit{ for all } a \in d^\times\} \textit{ for } 1 \le p < +\infty$$

$$\widehat{E}^0 = \{x = (x_1, x_2, \ldots) : \lim_j |x_j| a_j = 0 \textit{ for all } a \in d^\times\} \textit{ for } p = 0 .$$

In particular for these p the space E^p is complete if and only if it is sequentially complete.

PROOF: (b) is an obvious consequence of (a) and Proposition 5.1. To prove (a) let $x \in E^0$. Then there is k such that for every n we have $\lim_j |x_j| a_{j;k,n} = 0$. There exists a nonnegative sequence $(\lambda_j)_j$ with $\lim_j \lambda_j = +\infty$ such that $\lim_j \lambda_j |x_j| a_{j;k,n} = 0$ for all n. Hence $\lambda x = (\lambda_j x_j)_j \in E^0$ and therefore $\|\lambda x\|_a < +\infty$ for all $a \in d^\times$. This shows (a).

To apply the theory of the previous sections we need the following lemma, which is easy to prove. However its first part contains one of the basic ideas for Krone and Vogt [12], there used in a dual version for spaces of matrices. Here we use it in a more general situation.

Lemma 5.4 *For all p the space E^p is adjustable and has closed local neighborhoods.*

PROOF: While the second part follows from the fact that the $U_{k,n}$ are closed in E_k^p with respect to pointwise convergence, the first part is a consequence of the following observation: Let

$$|x|_\nu = \Big(\sum_j |x_j|^p\, b_{j,\nu}^p\Big)^{\frac{1}{p}}, \quad \nu = 0, \ldots, \mu\,, \qquad |x| = \Big(\sum_j |x_j|^p (\min_{\nu=0,\ldots,\mu} b_{j,\nu})^p\Big)^{\frac{1}{p}}$$

then

$$|x| \le \Big(\bigwedge_{\nu=0}^{\mu} |\;|_\nu\Big)(x) \le \mu |x|\,.$$

The left inequality is clear. To prove the right one we put for $x = (x_1, x_2, \ldots)$

$$x_j^{(\nu)} = \begin{cases} x_j & \text{if } \nu \text{ is the smallest number with } b_{j,\nu} = \min\limits_{n=0,\ldots,\mu} b_{j,n} \\ 0 & \text{otherwise.} \end{cases}$$

Then $x = \sum_{\nu=0}^{\mu} x^{(\nu)}$ and therefore we have for every seminorm $\|\;\|$ with $\|\;\| \le |\;|_\nu, \nu = 0, \ldots, \mu$

$$\|x\| \le \sum_{\nu=0}^{\mu} \|x^{(\nu)}\| \le \sum_{\nu=0}^{\mu} |x^{(\nu)}|_\nu = \sum_{\nu=0}^{\mu} |x^{(\nu)}| \le \mu |x|\,.$$

The same argument applies for sup–norms.

Definition: We say, the matrix $(a_{j;k,n})$ is of type (Q) if it satisfies the following condition:

$$\forall \mu\, \exists n, k\, \forall m, K, R > 0\, \exists N, S\, \forall j : a_{j;k,m} \le \frac{1}{R} a_{j;\mu,n} + S a_{j;K,N}\,.$$

We call it of type (WQ) if it satisfies

$$\forall \mu\, \exists n, k\, \forall m, K\, \exists N, S\, \forall j : a_{j;k,m} \le S(a_{j;\mu,n} + a_{j;K,N})\,.$$

This notation is justified by the following lemma which we admit without proof.

Lemma 5.5 *For every p the space E^p satisfies condition (Q) if and only if the matrix is of type (Q); it satisfies condition (WQ) if and only if the matrix is of type (WQ).*

We are now ready to give the characterizations we are aiming at:

Theorem 5.6 *For $p = 1$ the following are equivalent*

(1) E^1 is acyclic.

(2) E^1 is weakly acyclic.

(3) E^1 is sequentially retractive.

(4) The matrix is of type (Q).

PROOF: The equivalence (1) $\iff$ (3) $\iff$ (4) follows from Theorem 3.8, (1) $\Longrightarrow$ (2) is obvious. (2) $\Longrightarrow$ (4) follows from Lemma 2.2 when applied to the weak seminorms

$$|x|_{k,m} = |\sum_j x_j a_{j;k,m}|$$

on E_k and then to $x = e_1, e_2, \ldots$, the canonical basis vectors.

Theorem 5.7 *For $1 < p \leq +\infty$ the following are equivalent*

(1) E^p is acyclic.

(2) E^p is sequentially retractive.

(3) The matrix is of type (Q).

PROOF: Follows from Theorem 3.8.

Before we prove the analogous theorem for the case $p = 0$, we turn to the characterizations involving matrices of type (WQ). The following is another version of an adjustability lemma.

Lemma 5.8 *If the matrix is of type (WQ) then after an equivalent change of the matrix we may assume*

$$\forall \mu\, \exists k\, \forall m, K\, \exists N, S : a_{j;k,m} \leq S(a_{j;\mu,0} + a_{j;K,N})\,. \qquad (\mathrm{WQ}_1)$$

PROOF: Given a matrix of type (WQ) we choose for every μ an $n(\mu)$ according to the condition and set

$$\tilde{a}_{j;\mu,k} = \inf\{a_{j;\nu,n(\nu)+k};\ \nu = 0, \ldots, \mu\}\,.$$

Lemma 5.9 *For $1 < p < +\infty$ and $p = 0$ condition (WQ_1) implies condition (6) in Lemma 2.4 for E^p with weak seminorms.*

PROOF: For given μ we choose k according to (WQ_1). Let $y \in E_k'$. Then it may be described as

$$y(x) = \sum_j x_j \eta_j a_{j;k,m}$$

with $\eta = (\eta_j) \in l^q$, $\frac{1}{p} + \frac{1}{q} = 1$ for $1 < p < +\infty$, $q = 1$ for $p = 0$, and suitable m.

For given $K, \varepsilon > 0$ we choose first N, S according to (WQ), then j_ν so that

$$\Big(\sum_{j=j_\nu}^{\infty} |\eta_j|^q\Big)^{\frac{1}{q}} \leq \frac{\varepsilon}{S}\,.$$

Then we choose $M \geq N$ so that $a_{j;K,M} > 0$ for $j = 1, \ldots, j_0 - 1$. We obtain

$$y(x) = \sum_{j=j_0}^{\infty} x_j \varepsilon_j \eta_j a_{j;\mu,0} + \sum_{j=1}^{\infty} x_j \tilde{\eta}_j a_{j;K,M}\,.$$

Therefore we have

$$|y(x)| \leq \varepsilon \|x\|_{\mu,0} + |z(x)|$$

where $z \in E'_K$ is given by the second term in the above equation.

Theorem 5.10 *For $1 < p < +\infty$ the following are equivalent*

(1) E^p is weakly acyclic.

(2) E^p is complete.

(3) E^p is quasicomplete (sequentially complete).

(4) E^p is reflexive.

(5) E^p is regular.

(6) The matrix is of type (WQ).

PROOF: (1) $\Longrightarrow$ (2) and (1) $\Longrightarrow$ (4) are Proposition 4.4.
(2) $\Longrightarrow$ (3) $\Longrightarrow$ (5) and (4) $\Longrightarrow$ (5) are clear.
(5) $\Longrightarrow$ (6) is Theorem 4.7.
(6) $\Longrightarrow$ (1) follows from Lemmas 5.8, 5.9 and 2.5(b).

To study the cases of $p = 1, +\infty$ we slightly modify the approach of the beginning of §4. Let (for $p = 1, +\infty$) $\mathcal{E}^p_* : E_{1,*} \longleftarrow E_{2,*} \longleftarrow \ldots$ be the predual projective spectrum, which is defined by

$$E^1_{k,*} = \Big\{x = (x_1, x_2, \ldots) : \|x\|_{k,*} = \sup_j |x_j| a^{-1}_{j;k,n} < \infty\,,\ \lim_j |x_j| a^{-1}_{j;k,n} = 0 \text{ for some } n\Big\} / G^1_k$$

$$E^\infty_{k,*} = \Big\{x = (x_1, x_2, \ldots) : \|x\|_{k,*} = \sum_j |x_j| a^{-1}_{j;k,n} < +\infty \text{ for some } n\Big\} / G^\infty_k$$

where G^p_k is in both cases the subspace of all $x = (x_1, x_2, \ldots)$ such that $x_j = 0$ for all j with $a_{j;k,n} < +\infty$ for all n. Equipped with the decreasing sequence of extended real valued norms $\|\ \|_{k,n,*}$ and their inductive topology this is an (LB)–space. $E^p_* = \lim\limits_{\longleftarrow} E^p_{k,*}$ is a projective limit of (LB)–spaces and $E^p = (E^p_*)'$ setwise by natural identification.

The following lemma (cf. Bierstedt, Meise and Summers [1], Theorem 2.7) will lead us to a description of completeness of E^1, E^∞_0 using results on the projective limit functor (see [24], [25]). By use of Lemma 5.2 we then can handle the case of E^∞.

Lemma 5.11 *E^1 and E_0^∞ are the strong duals of E_*^1 and E_*^∞ respectively.*

PROOF: For $p = 1$ it suffices to observe that all the sets

$$B_a = \{x \in E_*^1 : \sup_j |x_j| a_j^{-1} \leq 1\}\,, \quad a \in d^\times$$

are bounded in E_*^1. Because this implies that the topology generated by the $\| \|_a$, $a \in d^\times$, is weaker than the strong topology $b(E^1, E_*^1)$. Since this is weaker than the topology of E^1, the result follows from Proposition 5.1.

In the case of E_0^∞ we show that the sets

$$B_a = \{x \in E_*^\infty : \sum_j |x_j| a_j^{-1} \leq 1\}\,, \quad a \in d^\times$$

are a fundamental system of bounded sets in E_*^∞. Since these sets are clearly bounded it remains to show that every bounded set B is contained in some B_a.

Because of Bierstedt, Meise and Summers [1], Theorem 2.3, (or of Theorem 5.14, case $p = 1$, which is independent of the present part of the proof) all $E_{k,*}^\infty$ are regular, hence for every k there are $n(k)$ and C_k such that

$$B \subset \{x : \sum_j |x_j| a_{j;k,n(k)}^{-1} \leq C_k\}\,.$$

Put

$$a_j = \min_k \{2^k C_k a_{j;k,n(k)}\}\,.$$

Then we have for $x \in B$

$$\sum_j |x_j| a_j^{-1} \leq \sum_k 2^{-k} C_k^{-1} \sum_j |x_j| a_{j;k,n(k)}^{-1} \leq 1\,.$$

Lemma 5.12 *If* $\mathrm{Proj}^1\, \mathcal{E}_*^1 = 0$ *then E^1 is complete, if* $\mathrm{Proj}^1\, \mathcal{E}_*^\infty = 0$ *then E_0^∞ and E^∞ are complete.*

PROOF: According to [24], §5, under the respective assumption, the space E^p equipped with the strong topology $b(E^p, E_*^p)$ is complete. Therefore, because of Lemma 5.11, E^1 and E_0^∞ respectively are complete. Lemma 5.2 and a standard argument (see Köthe [11, I, §18, 4.(4)b]) prove the assertion.

Lemma 5.13 *If $p = 1, +\infty$ and the matrix is of type (WQ) then* $\mathrm{Proj}^1\, \mathcal{E}_*^p = 0$.

PROOF: An argument strictly parallel to that of the proof of Lemma 5.8 shows condition (6) of Lemma 2.4 for the weak*–seminorms on E_k. Then Lemma 2.5(c) shows that σ is a topological imbedding with respect to the corresponding t. This however is equivalent to $\mathrm{Proj}^1\, \mathcal{E}_*^p = 0$.

Theorem 5.14 *For $p = 1, +\infty$ the following are equivalent*

(1) E^p is complete.

(2) E^p is quasicomplete (sequentially complete).

(3) E^p is regular.

(4) The matrix is of type (WQ).

For $p = +\infty$ the following are also equivalent

(1') E_0^∞ is complete.

(2') E_0^∞ is quasicomplete (sequentially complete).

PROOF: (4) $\Longrightarrow$ (1) and (4) $\Longrightarrow$ (1') are shown in Lemma 5.12 and 5.13. The equivalence of (1),..,(4) then follows as in the proof of Theorem 5.10. (1') $\Longrightarrow$ (2') is clear.

(2') implies by Grothendieck's factorization theorem that every bounded set in E_0^∞, and therefore every E^∞–bounded set, is contained and bounded in some E_k^∞. Hence E^∞ is regular.

Finally we give the characterizations for the case of $p = 0$.

Theorem 5.15 *For $p = 0$ the following are equivalent*

(1) E^0 is weakly acyclic.

(2) The matrix is of type (WQ).

PROOF: (1) $\Longrightarrow$ (2) is Proposition 2.3(b).

(2) $\Longrightarrow$ (1) follows from Lemmas 5.8, 5.9, 2.5(b).

Theorem 5.16 *For $p = 0$ the following are equivalent*

(1) E^0 is acyclic.

(2) E^0 is sequentially retractive.

(3) E^0 is complete.

(4) E^0 is quasicomplete (sequentially complete).

(5) E^0 is regular.

(6) The matrix is of type (Q).

PROOF: (1) $\Longleftrightarrow$ (2) $\Longleftrightarrow$ (6) is Theorem 3.8, (3) $\Longleftrightarrow$ (4) is Lemma 5.3, (1) $\Longrightarrow$ (3) is Theorem 3.2, (3) $\Longrightarrow$ (4) $\Longrightarrow$ (5) are clear.

We complete the proof now by showing (5) $\Longrightarrow$ (6). We use (Q) in the form of condition (11) in Proposition 3.6. Assuming that E^0 does not satisfy (Q) we may, as in the proof of Proposition 3.6, even assume

$\forall\, k, n, S\, \exists\, j : a_{j;k,k} > S a_{j;k+1,n} + 2^k a_{j;0,k-1}$.

We fix k and find for every n a $j = j(k,n)$ such that $j(k,n) < j(k,n+1)$ and for $j = j(k,n)$

$a_{j;k,k} > 2^n a_{j;k+1,n} + 2^k a_{j;0,k-1}$.

We put $J_k = \{j(k,1), j(k,2), \ldots\}$ and define

$$x_j^{(k)} = \begin{cases} \frac{1}{a_{j;k,k}} & \text{for } j \in M_k \\ 0 & \text{otherwise.} \end{cases}$$

Then $x^{(k)} \notin E_k^0$, $x^{(k)} \in E_{k+1}^0 \subset E^0$, $\|x^{(k)}\|_{0,k-1} \leq 2^{-k}$. This implies that the sequence $(x^{(n)})_n$ converges to zero in E^0, but is not contained in any E_k^0.

Many important cases (see e.g. Braun, Meise and Vogt [4]) are covered by the following Lemma.

Lemma 5.17 *If the matrix satisfies*

$$\forall\, k\, \exists\, K\, \forall\, n\, \exists\, N \;:\; \lim_j \frac{a_{j;K,n}}{a_{j;k,N}} = 0 \tag{14}$$

then $E^0 = E^\infty$ and the matrix is of type (Q) if and only if it is of type (WQ).

The previous Theorems then give:

Theorem 5.18 *If the matrix satisfies (14) then for $1 \leq p \leq +\infty$ the following are equivalent*

(1) E^p is acyclic.

(2) E^p is weakly acyclic.

(3) E^p is sequentially retractive.

(4) E^p is complete.

(5) E^p is quasicomplete (sequentially complete).

(6) E^p is regular.

(7) The matrix is of type (WQ) (or of type (Q)).

Remark: Condition (14) is in particular satisfied if for some (every) p the spaces E_k^p, $k = 0, 1, \ldots$, are Fréchet–Schwartz.

Acknowledgement: The author thanks J. Bonet for pointing out an error in the first version of Lemma 5.1.

References

[1] Bierstedt K.D., Meise R.G., Summers W.H., 'Köthe sets and Köthe sequence spaces', in *Functional Analysis, Holomorphy and Approximation Theory, A. Barroso (ed.), North–Holland Math. Studies* **71** (1982), 27-91.

[2] Bierstedt K.D., Meise R., Summers W.H., 'A projective description of weighted inductive limits', *Trans. Amer. Soc.* **272** (1982), 107-160.

[3] Bierstedt K.D., Meise R., 'Distinguished echelon spaces and the projective description of weighted inductive limits of type $\mathcal{V}_d C(X)$', in *Aspects of Mathematics and its Applications, A. Barroso (ed.), North–Holland Math. Library* **34** (1986), 169-226.

[4] Braun R., Meise R., Vogt D., 'Existence of fundamental solutions and surjectivity of convolution operators on classes of ultradifferentiable functions', *Proc. London Math. Soc. (3)* **61** (1990), 344-370.

[5] Dostal M., 'Some recent results on topological vector spaces', *Functional Analysis and Applications, Springer Lecture Notes in Math.* **384** (1974), 20-91.

[6] Floret K., 'Folgenretraktive Sequenzen lokalkonvexer Räume', *J. reine angew. Math.* **259** (1973), 65-85.

[7] Floret K., 'Some aspects of the theory of locally convex inductive limits', in *Functional Analysis: Surveys and Recent Results II, K.D. Bierstedt, B. Fuchssteiner (eds.), North–Holland Math. Studies* **38** (1980), 205-237.

[8] Floret K., Wloka J., 'Einführung in die Theorie lokalkonvexer Räume', *Lecture Notes in Math.* **56,** Springer– Verlag, Berlin–Heidelberg–New York 1968.

[9] Garling D.J.H., 'A generalized form of inductive limit topology for vector spaces', *Proc. London Math. Soc.* **14** (1964), 1-28.

[10] Grothendieck A., 'Produits tensoriels topologiques et espaces nucléaires', *Mem. Amer. Math. Soc.* **16** (1953).

[11] Köthe G., 'Topological vector spaces I,II', *New York–Heidelberg–Berlin,* 1969 and 1979.

[12] Krone J., Vogt D., 'The splitting relation for Köthe spaces', *Math. Z.* **190** (1985), 387-400.

[13] Neus H., 'Über die Regularitätsbegriffe induktiver lokalkonvexer Sequenzen', *Manuscripta Math.* **25** (1978), 135-145

[14] Palamodov V.P., 'Homological methods in the theory of locally convex spaces' (Russian), *Usp. Mat. Nauk* **26** 1 (1971), 3-66, English transl. Russian Math. Surveys **26** 1(1971), 1-64.

[15] Raikov D.A., 'A criterion of completeness of locally convex spaces', *Uspehi Mat. Nauk* **14** (1959), no. 1 (85)8, 223-229. (Russian)

[16] Retakh V.S., 'On the dual of a subspace of a countable inductive limit' (Russian), *Dokl. Akad. Nauk SSSR* **184** (1969), No.1, English transl. *Soviet Math. Dokl.* **10** (1969), 39-41.

[17] Retakh V.S., 'On the conjugate homomorphism in locally convex spaces', *Functional Anal. Appl.* **3** (1969), 307-311.

[18] Retakh V.S., 'Subspaces of a countable inductive limit' (Russian), *Dokl. Akad. Nauk SSSR* **194** (1970), No.6, English transl. *Soviet Math. Dokl.* **11** (1970), 1384-1386.

[19] Slowikowski W., 'Epimorphisms of adjoints to generalized (LF)–spaces', *Lecture notes,* Aarhus 1966.

[20] Valdivia M., 'Topics in Locally Convex Spaces', *North–Holland Math. Studies* **67** Amsterdam (1982).

[21] Valdivia M., 'On quasi–normable echelon spaces', *Proc. Edinburgh Math. Soc.* **24** (1981), 73-80.

[22] Vogt D., 'Some results on continuous linear maps between Fréchet spaces', in *Functional Analysis: Surveys and Recent Results III, K.D. Bierstedt, B. Fuchssteiner (eds.), North–Holland Math. Studies* **90** (1984), 349-381.

[23] Vogt D., 'On the functor $\mathrm{Ext}^1(E,F)$ for Fréchet spaces', *Studia Math.* **85** (1987), 163-197.

[24] Vogt D., 'Lectures on projective spectra of (DF)–spaces', Seminar lectures, AG Funktionalanalysis Düsseldorf/Wuppertal 1987.

[25] Vogt D., 'Topics on projective spectra of (LB)–spaces' in *Advances in the Theory of Fréchet Spaces, T. Terzioğlu (ed.), Kluwer Academic Publishers* (1989), 11-27.

Progress in Functional Analysis
K.D. Bierstedt, J. Bonet, J. Horváth & M. Maestre (Eds.)

Some applications of a decomposition method

A. Aytuna and T. Terzioğlu

Department of Mathematics, Middle East Technical University,
06531 Ankara - Turkey.

Dedicated to Professor Valdivia

Let E be a specific locally convex space. Suppose we have (i) F is isomorphic to a complemented subspace of E (ii) E is isomorphic to a complemented subspace of F. With these assumptions if we can conclude that F is in fact isomorphic to E, we have a so called decomposition method for E. In this general pattern E is at least *stable*, that is $E \simeq E \times E$. Initially the decomposition method was used by Pelczynski [11] in the context of Banach spaces. Valdivia utilized it in [15] to solve a problem posed by Grothendieck. In this article E was $s\widehat{\otimes}_\pi s'$, where s denotes the space of rapidly decreasing sequences. Vogt used this method in [17] for $E = \Lambda_\infty(\alpha)$, a stable nuclear power series space of infinite type. In Valdivia's book [16] and Vogt's articles [17], [18] we find a wealth of results which are obtained by various methods of decomposition.

Since E is a concrete space, for example $E = \Lambda_\infty(\alpha)$ or $E = s\widehat{\otimes}_\pi G$ where G is any locally convex space [16], the problem of imbedding F as a complemented subspace into E can be expected to be easier than imbedding E into F as a complemented subspace. In the case of $E = \Lambda_\infty(\alpha)$, this is quite apparent, because complemented subspaces of a nuclear stable infinite type power series space are characterized by Vogt and Wagner [20]. In a series of articles [4], [5], [6], J. Krone and the present authors have dealt with the problem of imbedding nuclear power series spaces of infinite type and were able to improve the decomposition method of Vogt. In §1 of this article we give a brief description of the results of [4-6] and in §2 indicate some applications. We treat spaces of analytic functions, spaces of C^∞-functions and spaces of functions holomorphic on a open subset of the dual of certain nuclear Fréchet spaces. In §3 we consider the spaces of analytic functions defined on a complete Reinhardt domain. Our notation and terminology is quite standard.

Finally, we would like to thank the Scientific and Technical Research Council of Turkey for partial support.

§1. IMBEDDING SMOOTH SEQUENCE SPACES

Let A be a Köthe set of sequences with the following properties:

(1) each $a \in A$ is a positive non-decreasing sequence

(2) for each $a \in A$ there is $b \in A$ with $(a_n)^2 = O(b_n)$.

Here and throughout $c_n = O(d_n)$ means there is some $\rho > 0$ such that $c_n \leq \rho d_n$ for all $n \in N$. The Köthe space $\lambda(A)$ is then called a smooth sequence space of infinite type or a $G_\infty -$ *space*. Power series spaces of infinite type as well as the strong duals of nuclear power series spaces of finite type are well known examples of G_∞-spaces. Those G_∞-spaces which are metrizable are exactly the (D_1)-spaces of Dragilev. For these facts and for further references about G_∞-spaces we refer to [13]. There are occasions where one encounters in analysis G_∞-spaces which are neither Fréchet nor duals of Fréchet spaces. Here we would like to present a new example. This is a sequence space defined by Meise [10] and appears as the kernel of certain convolution operators.

Let α and β be two sequences of positive real numbers and

$$\Lambda_\beta = \left\{ (y_j) : y_j \geq 0, \quad y_j = 0(e^{\frac{\beta_j}{m}}) \quad \text{for all} \quad m \in N \right\}.$$

$\lambda(\alpha, \beta)$ is the space of all sequences (x_j) such that

$$\pi_{k,y}(x) = \sum |x_j| y_j \, e^{k\alpha_j} < \infty$$

for each $k \in N$ and $y \in \Lambda_\beta$. The seminorms $\pi_{k,y}$ define the topology.

1.1. Proposition. *If α and β are non-decreasing then $\lambda(\alpha, \beta)$ is a G_∞-space.*

Proof: Let $\hat{y}_j = y_j + e^{-\beta_j}$. For $y \in \Lambda_\beta$, $\hat{y} \in \Lambda_\beta$ also and $\hat{y}_j > 0$. Now set $z_j = \max\{\hat{y}_1, \ldots, \hat{y}_j\}$. Then $z \in \Lambda_\beta$ also and $0 < z_j \leq z_{j+1}$. Let Λ_β^+ be the set of all $z \in \Lambda_\beta$ with $0 < z_j \leq z_{j+1}$. We have shown that the norms $\pi_{k,z}$, $k \in N$ and $z \in \Lambda_\beta^+$, define the space $\lambda(\alpha, \beta)$. It is easy to check that $(z_j^2) \in \Lambda_\beta^+$ whenever $z \in \Lambda_\beta^+$ and thus $\lambda(\alpha, \beta)$ is a G_∞-space.

□

Suppose T is a continuous operator from a G_∞-space $\lambda(A)$ into a locally convex space E. Let $\sigma = (\sigma_n)$ satisfy

$$(\ell_1): \qquad 0 < \sigma_n = O(a_n) \quad \text{for some } a \in A.$$
$$(\ell_2): \qquad 1/\sigma_n = O(b_n) \quad \text{for some } b \in A.$$

If there is a continuous seminorm $\| \ \|$ on E such that the inequality

$$|x|_\sigma = \sum |x_n|\sigma_n \leq \|Tx\|$$

holds for each $x \in \lambda(A)$, we call T a *local imbedding* [4]. We note that any $a \in A$ and $e = (1, 1, \ldots)$ satisfy (ℓ_1) and (ℓ_2). For a more complete account of the properties of existence of local imbeddings we refer to [13] and [3].

1.2. Theorem. *Let $\lambda(A)$ be a nuclear stable G_∞-space. Suppose there is a local imbedding of $\lambda(A)$ into E. Each one of the following assumptions implies that E has a complemented subspace which is isomorphic to $\lambda(A)$.*

(i) *$\lambda(A)$ is Fréchet and E is isomorphic to a closed subspace of $\lambda(A)^N$.*

(ii) *E is isomorphic to a closed subspace of $\lambda(A)$.*

(iii) *E is isomorphic to a sequentially complete quotient space of $\lambda(A)$.*

For the proof we refer to [13] which is a minor modification of the proof given in [4] for the case when $\lambda(A)$ is Fréchet. We also would like to note that the barreledness condition on E when (ii) is assumed in [13], was unnecessary.

For the generalization of Vogt's decomposition method [17], we need a short preparation. Let F be a nuclear Fréchet space with a basis and assume $(F, F) \in S_1$, which is a splitting condition [19]. It was proved by Krone in [9] that F is then isomorphic to a G_∞-space $\Lambda_\phi(\alpha)$ which is defined by a sequence of transition functions $\phi = (\phi_k)$. A particular case is of course when $F = \Lambda_\infty(\alpha)$, which gives an improvement of the decomposition method of Vogt.

1.3. Corollary. *Let $\Lambda_\phi(\alpha)$ be a nuclear stable G_∞-space which is defined by transition functions. Let E be a complemented subspace of $\Lambda_\phi(\alpha)$. If there is a local imbedding of $\Lambda_\phi(\alpha)$ into E, then E is isomorphic to $\Lambda_\phi(\alpha)$.*

Proof follows from 1.2. Theorem and 2.4. Proposition of [6].

Suppose E is a nuclear Fréchet space which satisfies the conditions $\underline{DN}$ and Ω of Vogt (cf. eg. [19]). Then there is an exponent sequence α^E, which is unique up to equivalence, such that

$$\Lambda_1(\alpha^E) \subset \Delta(E) \subset \Lambda_\infty(\alpha^E)'$$

holds, where $\Delta(E)$ denotes the diametral dimension of E. We call α^E the *exponent sequence associated to E* [5]. We have

1.4. Theorem. *Let E be a nuclear Fréchet space which has Ω and $\underline{DN}$. Suppose also*

that $\Lambda_1(\alpha^E)$ is nuclear. If F has DN and F is isomorphic to a subspace of $\Lambda_1(\alpha^E)$ then F is isomorphic to a subspace of E.

For the proof and corollaries of this result we refer to [5]. In this imbedding theorem there is no stability assumption which was essential in 1.2. Theorem. However the case when the exponent sequence is stable merits special attention. In this case $\Lambda_\infty(\alpha^E)$ is isomorphic to a subspace of $\Lambda_1(\alpha^E)$ and hence by our theorem isomorphic to a subspace of E. If we also have $\Delta(E) = \Lambda_\infty(\alpha^E)'$, then E is isomorphic to a quotient space of $\Lambda_\infty(\alpha^E)$ [20]. So by 1.2. Theorem we have that E has a complemented subspace which is isomorphic to $\Lambda_\infty(\alpha^E)$.

If E is a nuclear Fréchet space with $\underline{DN}, \Omega$ and $\Delta(E) \subset \lambda(A)'$, then there is a local imbedding of $\lambda(A)$ into E. In particular, there is always a local imbedding of $\Lambda_\infty(\alpha^E)$ into E ([5], [13]).

§2. SOME APPLICATIONS

We start with spaces of analytic functions. We know that the space $\mathcal{O}(\Delta^d)$ of functions analytic in the d-dimensional polydisc Δ^d is isomorphic to $\Lambda_1(n^{1/d})$ and $\mathcal{O}(\mathbb{C}^d)$ to $\Lambda_\infty(n^{1/d})$. If M is a connected Stein manifold of dimension d, the associated exponent sequence of $\mathcal{O}(M)$ is $(n^{1/d})$ [5]. Hence by 1.4. Theorem $\mathcal{O}(\mathbb{C}^d)$ is isomorphic to a subspace of $\mathcal{O}(M)$. Further $\mathcal{O}(M)$ is a complemented subspace of s if and only if $\mathcal{O}(M) \simeq \mathcal{O}(\mathbb{C}^d)$ [4].

Let I be a closed non-trivial ideal of $\mathcal{O}(\mathbb{C}^d)$. We choose $g \in I, g \neq 0$. Then there is a compact subset K of $\mathbb{C}^d$ with non-empty interior and $\rho > 0$ with $|g(z)| \geq \rho$ for all $z \in K$. Hence the map which sends each f to fg gives a local imbedding of $\Lambda_\infty(n^{1/d})$ into I. Thus we get that $I \simeq \mathcal{O}(\mathbb{C}^d)$ by our decomposition method (1.3. Corollary). However there is a better result in this direction. Let X be a Stein space such that $\mathcal{O}(X)$ is isomorphic to some $\Lambda_\infty(\alpha)$. If M is a closed submodule of $\mathcal{O}(X)^p$ for some p, then $M \simeq \mathcal{O}(\mathbb{C}^d)$, $d = \dim X$. We refer to [3] for this and also to [2] for other results in this direction.

Next we consider spaces of C^∞-functions. Let K be a compact subset of R^n with non-empty interior, $D(K)$ the space of C^∞-functions with support in $K, \mathcal{E}(K)$ the space of Whitney-differentiable functions. Our purpose is to show how some results of Valdivia [16] and Tidten [14] can also be obtained as applications of our decomposition results.

Let L be a rectangular region whose interior contains K. Then the sequence

$$0 \longrightarrow J(K,L) \longrightarrow D(L) \longrightarrow \mathcal{E}(K) \longrightarrow 0$$

is exact, where

$$J(K,L) = \{f \in D(L) : D^\alpha f|_K = O \quad \text{for every } \alpha\}.$$

Tidten [14] proved that $J(K,L)$ has property Ω and so it is a complemented subspace of s. For any rectangular region M we have $D(M) \simeq s$ [12]. We fix a rectangular region $L_1 \subset \overset{\circ}{L} \setminus K$, we have immediately an imbedding of $D(L_1) \simeq s$ into $J(K,L)$. Hence by 1.3. Corollary we obtain

$$J(K,L) \simeq s.$$

Since $D(K) \simeq J(L \setminus \overset{\circ}{K}, L)$, we also have $D(K) \simeq s$ (see [14]; Folgerung 2.3 and [16]; Chap.3, §1.15).

We take a rectangular region $L \subset \overset{\circ}{K}$. If we consider elements of $D(L)$ simply as elements of $\mathcal{E}(K)$, we have that s is isomorphic to a subspace of $\mathcal{E}(K)$. Since $\mathcal{E}(K)$ is a quotient space of s, from 1.2. Theorem we have that $\mathcal{E}(K)$ has a complemented subspace which is isomorphic to s ([16]; Chap.3, §1, 11 (6)). The moral of the story is: if you want to imbed s into a space of C^∞-functions look for a rectangle! Here is another example. Let M be an n-dimensional C^∞-manifold and $U \subset M$ open which is diffeomorphic via ϕ to a rectangular region L in R^n. If we define $T : D(L) \to C^\infty(M)$ by

$$T(f)(x) = \begin{cases} f(\phi(x)), & x \in U \\ 0 & \text{otherwise} \end{cases}$$

we obtain an imbedding of $D(L) \simeq s$ into $C^\infty(M)$. Further if M is compact, then it is not difficult to show directly that $C^\infty(M)$ is isomorphic to a complemented subspace of s (cf. [18]; proof of 2.3. Thm.). One can then appeal to our decomposition method to get $C^\infty(M) \simeq s$.This was already proved by Valdivia [16], and Vogt [18] independently.

One can handle the space of global sections of a vector bundle over a compact manifold in a similar fashion. A C^∞-vector bundle ξ over a compact C^∞-manifold M is a direct summand of a trivial bundle. This can be seen by imitating the argument given in Corollary I.4.14 of [1] in the C^∞-category. Using this fact it is quite easy to see that the Fréchet space $C^\infty(\xi)$, of C^∞-global sections of ξ is isomorphic to a complemented subspace of $C^\infty(M)^k$ for some integer k and hence has the properties DN and Ω. For the description of the topology of $C^\infty(\xi)$ we refer the reader to [21] (Chap. IV). Now choose an open set U over which ξ is trivial and fix elements $\sigma_1, \ldots, \sigma_k \in C^\infty(\xi)$ with the property that $\{\sigma_1(x), \ldots, \sigma_k(x)\}$ forms a basis of the fiber ξ_x for each $x \in U$. Furthermore choose a chart (V, Φ) of M such that $\overline{V} \subset U$ and let $K = \Phi^{-1}(L)$ for some compact rectangle $L \subset \Phi(V)$. For a function f in $D(K)(\simeq D(L))$ set $Tf(x) = f(x)\sigma_1(x)$. Then it is easy to see that T is a (local) imbedding of $D(K)$ into $C^\infty(\xi)$. Since $D(K)$ is isomorphic to s, we once again conclude from 1.3. Corollary that $C^\infty(\xi) \simeq s$.

Let E be a nuclear Fréchet and space, O a non-empty open subset of E'_b, $H(O)$ the space of holomorphic functions on O. Börgens, Meise and Vogt [7] have proved that $H(\Lambda_\infty(\alpha)'_b)$ is isomorphic to the power series space $\Lambda_\infty(\beta(\alpha))$. We refer to [7] and [8] for the definition and properties of the sequence $\beta(\alpha)$. Our purpose is to present a modification of a result from [13] (5.5. Proposition).

2.1. Proposition. *Suppose there is a local imbedding of $\Lambda_\infty(\alpha)$ into E and $O \subset E'_b$ is open. Then there is a local imbedding of $\Lambda_\infty(\beta(\alpha))$ into $H(O)$.*

Proof: Let $p_j(f) = \sup\{|f(u)| : u \in U_j^\circ\}$ for $f \in H(E'_b)$ and $U_k \in \mathcal{U}(E)$. For an isomorphism $t : \Lambda_\infty(\beta(\alpha)) \to H(\Lambda_\infty(\alpha)'_b)$ we determine m so that $|x|_m \leq p_m(tx)$ for all $x \in \Lambda_\infty(\beta(\alpha))$. For this m using a result from [13] (3.1. Proposition) we find a local imbedding $\imath : \Lambda_\infty(\alpha) \to E$ with $|y|_m \leq \|\imath y\|_k$ for some k. We have $V_m^o \subset \overline{\imath^{-1}(U_k^o)} = \imath^{-1}(U_k^o)$. Now we define $h : \Lambda_\infty(\beta(\alpha)) \to H(E'_b)$ by

$$h(x)(u) = tx(\imath' u), \quad x \in \Lambda_\infty(\beta(\alpha)), \qquad u \in E'$$

h is continuous and

$$|x|_e \leq p_m(tx) \leq \sup\{|h(x)(u)| : u \in U_k^\circ\} = p_k(h(x)).$$

So h is a local imbedding into $H(E'_b)$.

Since $H(O) \simeq H(u_0 + O)$ we may assume $0 \in O$ and so there is a bounded subset B of E with $B^o \subset O$. Hence $\sigma_k U_k^o \subset O$ for some $\sigma_k > 0$. If $\tilde{h}$ is the composition of h with the restriction map from $H(E'_b)$ into $H(O)$ we have

$$|x|_e \leq \frac{1}{\sigma_k} \sup\{|\tilde{h}(x)(v)| : v \in \sigma_k U_k^o\}$$

and so $\tilde{h} : \Lambda_\infty(\beta(\alpha)) \to H(O)$ is a local imbedding.

If E is a nuclear Fréchet space which satisfies $\underline{DN}$ and Ω and α^E the associated exponent sequence we can assume that α^E is equivalent to $(-\log d_n(U_{k+1}, U_k))$ for each k (cf. [5], §1). From 2.1. Proposition we know that $\Delta(H(O)) \subset \Lambda_\infty(\beta(\alpha^E))'$ for any open subset O of E'_b ([13]; 3.3. Proposition). Let us assume $\Lambda_1(\alpha^E)$ is nuclear; i.e. $(q^{\alpha_n^E}) \in \ell_1$ for every $0 < q < 1$. We find some $R_o > 1$ such that $d_n(U_{k+1}, U_k) = O(n^{-3} R_o^{-\alpha_n})$, where we write α for α^E to simplify the notation. Let Q be a compact subset of O. Then we can find another compact subset $K \subset Q$ such that the associated linking map has diameter $d_n = O(R^{-\beta_n})$ where $1 < R < \sqrt{R_o}$ and $\beta = \beta(\alpha)$. This follows from the estimates obtained by Börgens, Meise and Vogt [8]; 3.3. Lemma and 4.1. Theorem. So we have proved the following which is an improvement of ([13]; §5).

2.2. Proposition. *Let E be a nuclear Fréchet space which satisfies $\underline{DN}$ and Ω, with the associated exponent sequence α. If $\Lambda_1(\alpha)$ is also nuclear and $O \subset E'_b$ is open, we have*

$$\Lambda_1(\beta(\alpha)) \subset \Delta(H(O)) \subset \Lambda_\infty(\beta(\alpha))'.$$

One remarkable property of $\beta(\alpha)$ is that it is always stable [7]. Hence from 1.4. Theorem we get that $H(\Lambda_\infty(\alpha)'_b)$ is isomorphic to a subspace of $H(O)$, provided $H(O)$ has $\underline{DN}$ and Ω. If E is a complemented subspace of s, then $H(E'_b) \simeq H(\Lambda_\infty(\alpha)'_b)$, which means that $H(E'_b)$ has a basis, although we do not know if E has a basis. We refer to [13], for these and some other consequences of our results in the context of infinite-dimensional holomorphy.

§3. SPACES OF FUNCTIONS ANALYTIC IN A REINHARDT DOMAIN

Zaharjuta [22] proved that there exists a continuum of pairwise nonisomorphic spaces in the class of spaces of functions which are analytic in complete Reinhardt domains in $\mathbb{C}^d$, $d \geq 2$. Let $D \subset \mathbb{C}^d$ be a complete Reinhardt domain.

We know that the associated exponent sequence of $\mathcal{O}(D)$ is $(n^{1/d})$. D is completely determined by the function

$$h(\theta) = \sup\left\{\sum_{i=1}^{d} \theta_i \log|z_i| \,:\; z = (z_i) \in D\right\}$$

defined on the simplex $\Sigma = \{\theta \in R^d : \theta_1 + \cdots + \theta_d = 1,\ \theta_i \geq 0\}$. For $n = (n_1 \cdots n_d) \in N^d$, let $|n| = n_1 + \cdots + n_d$ and $\theta(n) = n/|n|$. The set $\pi(D) = \{\theta \in \Sigma : h(\theta) < +\infty\}$ is convex. Let m denote the $(d-1)$-dimensional Lebesgue measure. We have already proved in [5] that $m(\pi(D)) = 0$ if and only if $\mathcal{O}(D)$ is isomorphic to $\mathcal{O}(\mathbb{C}^d) \times F$ for some quotient space F of $\mathcal{O}(\mathbb{C}^d)$. This case corresponds to $\Delta(\mathcal{O}(D)) = \Delta(\mathcal{O}(\mathbb{C}^d))$. Our aim now is to complement this result as follows:

3.1. Theorem. *Let D be a complete Reinhardt domain in $\mathbb{C}^d$. Then one and only one of the following is true:*

a) $\mathcal{O}(D)$ is isomorphic to $\mathcal{O}(\Delta^d) \times G$.

b) $\mathcal{O}(D)$ is isomorphic to $\mathcal{O}(\mathbb{C}^d) \times F$ for some quotient space F of $\mathcal{O}(\mathbb{C}^d)$.

Further a) holds if and only if $m(\pi(D)) > 0$ and b) holds if and only if $m(\pi(D)) = 0$.

Proof: By what we have already proved in [5] (cf. Theorem 1.5. and Lemma 1.4.), we only have to prove that $m(\pi(D)) > 0$ implies (a). To simplify the notation to some extent we take $d = 2$. Let $\theta_{n,m} = n/(m+n)$, $m, n \in N$, $\mathcal{F} = \{(n,m) : \theta_{n,m} \in \pi(D)\}$. We know that $\mathcal{O}(D)$ is a subspace of $\mathcal{O}(\Delta^2)$. The set

$$B = \left\{(\xi_{n,m}) :\ \sum_{(n,m)\in\mathcal{F}} |\xi_{n,m}|^2 e^{2(n+m)h(\theta_{n,m})} \leq 1\right\}$$

is a bounded subset of $\mathcal{O}(D)$. Here we treat $\mathcal{O}(D)$ as the Köthe space $\lambda(A)$ where $A = \{a^p_{n,m}\}$, $a^p_{n,m} = \exp\{(n+m)h_p(\theta_{n,m})\}$ and (h_p) is a sequence of functions on $[0,1]$ increasing to h [22]. The step space $\lambda(A)|_{\mathcal{F}}$ is isomorphic by a diagonal transformation to a step space of $\mathcal{O}(\Delta^2) \simeq \Lambda_1(\sqrt{j})$ and so it is a power series space of finite type. To finish the proof we will show that $\lambda(A)|_{\mathcal{F}}$ has the same diametral dimension as $\Lambda_1(\sqrt{j})$. For this we need a somewhat technical result which we prove first.

Let $\varphi : N \to R$ be a non-decreasing unbounded function. By abuse of conventional notation we set

$$\varphi^{-1}(t) = \max\{n : \varphi(n) \leq t\} = \#\{m : \varphi(m) \leq t\}.$$

Lemma. *Let φ and ψ be as above and there is $c \in N$ such that $\varphi^{-1}(t) \leq c\psi^{-1}(t)$ for all $t \geq t_o$. Then there is $n_o \in N$ such that $\psi(n) \leq \varphi(cn)$ for all $n \geq n_o$.*

Proof: Let $\varphi_1 < \varphi_2 < \cdots$ be the range of φ. Choose any j such that if $\varphi^{-1}(\varphi_{k-1}) < j \leq \varphi^{-1}(\varphi_k)$, then $t_o \leq \varphi_{k-1}$. We have $\varphi(j) = \varphi_k$. Suppose i is such that

$$\psi^{-1}(\psi_{i-1}) < \frac{j}{c} \leq \psi^{-1}(\psi_i).$$

Then $\psi(j/c) = \psi_i$; we assume j/c is an integer. Now choose a with $\psi_a \leq \varphi_k < \psi_{a+1}$. Then $\psi(\varphi_k) = \psi_a$ and $\psi^{-1}(\varphi_k) = \psi^{-1}(\psi_a)$. Letting $t = \varphi_k$ in the assumption we have

$$\varphi^{-1}(\varphi_k) \leq c\psi^{-1}(\varphi_k) = c\psi^{-1}(\psi_a).$$

So

$$\psi^{-1}(\psi_{i-1}) < \frac{j}{c} \leq \frac{\varphi^{-1}(\varphi_k)}{c} \leq \psi^{-1}(\psi_a).$$

Hence $i \leq a$ and so $\psi_i \leq \psi_a \leq \varphi_k$. So we have

$$\psi\left(\frac{j}{c}\right) \leq \varphi(j).$$

Now we let $cn = j$ to get the result. □

Let $R_2 > R_1 > 0$, $\alpha = (\alpha_n)$ an exponent sequence with

$$\sup \frac{\alpha_{n+1}}{\alpha_n} = A < +\infty.$$

Define

$$\varphi(n) = \left(\frac{R_2}{R_1}\right)^{\alpha_n}.$$

Suppose $\psi : N \to R$ is such that

$$\varphi^{-1}(t) \leq c\psi^{-1}(t) \tag{1}$$

where $c \geq 1$ is an integer constant. Then the lemma gives $\psi(n) \leq \varphi(cn)$ for all $n \geq n_o$. By the shift stability of α we determine $a > 1$ with $\alpha_{cn} \leq a\alpha_n$ and so get for $n \geq n_o$

$$\psi(n) \leq \varphi(cn) \leq \left(\frac{R}{R_1^-}\right)^{\alpha_n} \tag{2}$$

where

$$R_1^- = \left(\frac{R_1}{R_2}\right)^{a-1} R_1 < R_1.$$

We now proceed with the proof of the theorem. We first choose a closed interval $I \subset \pi(D)$ with $m(I) > 0$ and arrange things by using Dini's theorem so that $h - h_p \leq p^{-1}$ on I. Let U_k be the Hilbertian ball defined by h_k and set $\psi_k(n) = d_n(B, U_k)^{-1}$, $\varphi_k(n) = \exp(\sqrt{n}/k)$. We first estimate $\psi_k^{-1}(e^t)$ from below. Since

$$\psi_k^{-1}(e^t) = \#\{(n,m) : (n+m)(h - h_k)(\theta_{n,m}) \leq t\}$$

we have

$$\#\{(n,m) :\ \theta_{n,m} \in I,\ n+m \leq tk\} \leq \psi_k^{-1}(e^t).$$

By a result of Zaharjuta [22] (Lemma 2) there is a constant $B > 0$ such that

$$\frac{t^2}{2\sqrt{2}} k^2 m(I) - Btk \leq \psi_k^{-1}(e^t) \tag{3}$$

for all large t.

On the other hand $\varphi_k^{-1}(e^t) = n$ if and only if $\sqrt{n} \leq tk < \sqrt{n+1}$. Hence

$$t^2k^2 - 1 \leq \varphi_k^{-1}(e^t) \leq t^2k^2.$$

Using this and (3) we obtain

$$\liminf_{t\to\infty} \frac{\psi_k^{-1}(e^t)}{\varphi_k^{-1}(e^t)} \geq \frac{m(I)}{2\sqrt{2}} > 0.$$

So we can find an integer $b > 1$ with

$$\varphi_k^{-1}(e^t) \leq b\, \psi_k^{-1}(e^t)$$

which means that (1) is true. Hence by (2) we get

$$\psi_k(n) \leq \exp\left(\frac{a\sqrt{n}}{k}\right) \quad \text{or} \quad e^{-\frac{a\sqrt{n}}{k}} = O\left(d_n(B, U_k)\right).$$

This implies that if $\lim \xi_n d_n(B, U_k) = 0$ for all k, then $(\xi_n) \in \Lambda_1(\sqrt{j}) = \Delta(\Lambda_1(\sqrt{j}))$. Further, if P is the canonical projection of $\lambda(A)$ onto the step space $\lambda(A)|_{\mathcal{F}}$, we have, since $B = P(B)$, $d_n(B, P(U_k)) = d_n(B, U_k)$. Thus we have proved $\Delta(\lambda(A)|_{\mathcal{F}}) \subset \Delta(\Lambda_1(\sqrt{j}))$ but since the other inclusion is trivial, we have reached the conclusion.

REFERENCES

[1] M. Atiyah: *K-Theory*, W.A. Benjamin Inc. New York, 1967.

[2] A. Aytuna: Stein spaces M for which $\mathcal{O}(M)$ is isomorphic to a power series space. Advances in the theory of Fréchet spaces (ed. T. Terzioğlu) 115-149, Kluwer Academic Publishers (1989).

[3] A. Aytuna: The Fréchet space structure of global sections of coherent analytic sheaves. *Math. Balkanica*, **3**, 311-324 (1989).

[4] A. Aytuna, J. Krone and T. Terzioğlu: Complemented infinite type power series subspaces of nuclear Fréchet spaces. *Math. Ann.* **283**, 193-200 (1989).

[5] A. Aytuna, J. Krone and T. Terzioğlu: Imbedding of power series spaces and spaces of analytic functions. *manuscripta math.* **67**, 125-142 (1990).

[6] A. Aytuna, J. Krone and T. Terzioğlu: On complemented subspaces of certain nuclear Köthe spaces. Advances in the theory of Fréchet spaces (ed. T. Terzioğlu) 325-332, Kluwer Academic Publishers (1989).

[7] J. Börgens, R. Meise, D. Vogt: Entire functions on nuclear sequence spaces. *J. reine angew. Math.* **322**, 196-220 (1981).

[8] J. Börgens, R. Meise, D. Vogt: $\Lambda(\alpha)$-nuclearity in infinite-dimensional holomorphy. *Math. Nachr.* **106**, 129-146 (1982).

[9] J. Krone: Zur topologischen Charakterisierung von Unter-und Quotienträumen spezieller nuklearer Ktheräume mit der Splittingmethode. Diplomarbeit, Wuppertal 1984.

[10] R. Meise: Sequence space representations for zero-solutions of convolution equations on ultradifferentiable functions of Roumieu type. *Studia Math.* **92**, 211-230 (1989).

[11] A. Pelczynski: Projections in certain Banach spaces. *Studia Math.* **19**, 209-228 (1960).

[12] A. Pietsch: *Nuclear locally convex spaces.* Springer-Verlag 1972.

[13] T. Terzioğlu: Some invariants of Fréchet spaces and imbeddings of smooth sequence spaces. Advances in the theory of Fréchet spaces (ed. T. Terzioğlu), 305-324, Kluwer Academic Publishers (1989).

[14] M. Tidten: Fortsetzungen von C^∞-Funktionen, welche auf einer abgeschlossenen Menge in R^n definiert ist. *manuscripta math.* **27**, 291-312 (1979).

[15] M. Valdivia: A representation of the space $\mathcal{O}_M$. *Math. Z.* **177**, 463-478 (1981).

[16] M. Valdivia: *Topics in locally convex spaces.* North-Holland Math. Studies, No.67, (1982).

[17] D. Vogt: Ein Isomorphiesatz für Potenzreihenräume. *Arch. Math.* **38**, 540-548 (1982).

[18] D. Vogt: Sequence space representations of spaces of test functions and distributions. Advances in the Functional Analysis, Holomorphy and Approximation Theory. (ed. G.I. Zapata) 405-443, New York-Basel (1983).

[19] D. Vogt: On the functor $Ext^1(E, F)$ for Fréchet spaces. *Studia Math.* **85**, 163-197 (1987).

[20] D. Vogt und M.J. Wagner: Charakterisierung der Unterräume und Quotientenräume der nuklearen stabilen Potenzreihenräume von unendlichem Typ. *Studia Math.* **70**, 65-80 (1981).

[21] R.O. Wells, Jr.: *Differential Analysis on Complex Manifolds*, Prentice-Hall, Englewood Cliffs, N.J. 1973.

[22] V.P. Zaharjuta: Generalized Mityagin invariants and a continuum of pairwise nonisomorphic spaces of analytic functions. *Func. Anal. and App.*, **11**, 182-188 (1978).

Progress in Functional Analysis
K.D. Bierstedt, J. Bonet, J. Horváth & M. Maestre (Eds.)

On the range of the Borel map for classes of non-quasianalytic functions

J. Bonet[a] R. Meise[b] B. A. Taylor[c]

Dedicated to Professor M. Valdivia on the occasion of his 60th birthday

[a] Departamento de Matemática Aplicada, E.T.S. Arquitectura, Universidad Politécnica de Valencia, E–46071 Valencia, Spain.

[b] Mathematisches Institut, Heinrich-Heine-Universität Düsseldorf, Universitätsstraße 1, D–4000 Düsseldorf 1, Federal Republic of Germany.

[c] Department of Mathematics, University of Michigan, Angell Hall, Ann Arbor, Michigan 48109, United States of America.

Abstract

Let $\mathcal{E}_{(\omega)}(\mathbb{R}^N)$ denote the non-quasianalytic class of ω-ultradifferentiable functions of Beurling type on $\mathbb{R}^N$. Using a characterization of the bornological linear subspaces of *(DFS)*-spaces, an idea of Carleson and Hörmander's solution of the $\bar{\partial}$-problem, it is characterized for which weight functions σ a certain Köthe sequence space $\Lambda(\sigma, N)$ is contained in the range of the Borel map $B_N : f \mapsto (f^{(\alpha)}(0))_{\alpha \in \mathbb{N}_0^N}$. The same question is treated also for the class $\mathcal{E}_{\{\omega\}}(\mathbb{R}^N)$ of ω-ultradifferentiable functions of Roumieu type.

Introduction

Let $\omega : [0, \infty[\to [0, \infty[$ be a continuous increasing function which has the following properties:

(α) $\omega(2t) = O(\omega(t))$	(β) $\int_1^\infty \frac{\omega(t)}{t^2} dt < \infty$
(γ) $\log t = o(\omega(t))$	(δ) $\varphi : t \mapsto \omega(e^t)$ is convex.

Define $\varphi^* : [0, \infty[\to \mathbb{R}$ by

$$\varphi^*(x) := \sup_{y>0}\{xy - \varphi(y)\},$$

and let

$$\mathcal{E}_{(\omega)}(\mathbb{R}^N) := \{f \in C^\infty(\mathbb{R}^N) \mid \forall K \subset\subset \mathbb{R}^N \, \forall m \in \mathbb{N} : \sup_{x \in K} \sup_\alpha |f^{(\alpha)}(x)| e^{-m\varphi^*(\frac{|\alpha|}{m})} < \infty\}$$

and

$$\Lambda(\omega, N) := \{x \in \mathbb{C}^{\mathbb{N}_0^N} \mid \forall m \in \mathbb{N} : \|x\|_m := \sup_{\alpha \in \mathbb{N}_0^N} |x_\alpha| e^{-m\varphi^*(\frac{|\alpha|}{m})} < \infty\}.$$

Obviously, the Borel map

$$B_N : \mathcal{E}_{(\omega)}(\mathbb{R}^N) \to \Lambda(\omega, N) \; , \; B_N(f) := (f^{(\alpha)}(0))_{\alpha \in \mathbb{N}_0^N}$$

is a continuous linear map. Meise and Taylor [10] have characterized those functions ω for which B_N is surjective.
In the present note we investigate – in the spirit of Carleson [5] – what can be said about the range of the Borel map B_N when it is not necessarily surjective. Our main result is the following: Assume that ω and σ satisfy $(\alpha) - (\delta)$ and that $\sigma \geq \omega$. Then $B_N(\mathcal{E}_{(\omega)}(\mathbb{R}^N)) \supset \Lambda(\sigma, N)$ is equivalent to

(*) there exist $A, B > 0$ so that $\int_1^\infty \frac{\omega(yt)}{y^2} dy \leq A\sigma(t) + B$ for all $t \geq 0$.

To prove this, we first characterize for Fréchet-Schwartz spaces E, F, G and continuous linear maps $S : G \to F$ and $T : E \to F$ with dense range the relation $S(G) \subset T(E)$ (Cor.2.3) in terms of bounded sets. This characterization is given under the hypothesis that F' is bornological for the initial topology with respect to the map $T^t : F' \to E'_b$. Hence we run into the question which linear subspaces of a *(DFS)*-space are bornological for the induced topology. Using a counterexample of Dubinsky [6] we show that in general even a *(DFN)*-space can contain a linear subspace which is not bornological. This solves a problem of Bierstedt and Meise [2] in the negative (see 2.11). However, for our application, an approximation theorem of Taylor [15] can be used to show that $B_N^t(\mathcal{E}_\omega(\mathbb{R}^N)')$ is bornological under the topology induced by $\Lambda(\omega, N)'_b$. Hence we can apply the characterization mentioned above together with results of Meise and Taylor [10] to get our main result. Similar, but simpler arguments show that the analogous result for ultradifferentiable functions of Roumieu type holds, too (see 3.7).
The authors thank Professor Zobin for an interesting discussion in Oberwolfach in February 1990 which motivated them to start the research for the present paper.

1 Preliminaries

In this section we fix the notation which is used subsequently and we formulate the question which will be answered in section 3.

1.1 Definition. Let $\omega : [0, \infty[\to [0, \infty[$ be a continuous increasing function. Consider the following conditions:

(α) $\omega(2t) = O(\omega(t))$	(β) $\int_1^\infty \frac{\omega(t)}{t^2} dt < \infty$	(β') $\omega = o(t)$
(γ) $\log t = o(\omega(t))$	(δ) $\varphi : t \mapsto \omega(e^t)$ is convex.	

We will call ω a weight if it satisfies $(\alpha), (\beta'), (\gamma)$ and (δ) and we will call ω a non-quasianalytic weight if it satisfies $(\alpha), (\beta), (\gamma)$ and (δ). In both cases we denote by φ^* the Young conjugate of φ, i.e.

$$\varphi^*(y) := \sup_{x>0}\{xy - \varphi(x)\}, \quad y \in [0, \infty[.$$

Also, by abuse of notation we write $\omega(z) := \omega(|z|)$ for $z \in \mathbb{C}^N$.

1.2 Definition. Let ω be a weight and let Ω be an open set in $\mathbb{R}^N$. Then we define

$$\mathcal{E}_{(\omega)}(\Omega) := \{f \in C^\infty(\Omega) \mid \forall K \subset\subset \Omega\, \forall m \in \mathbb{N} : \sup_{x\in K} \sup_{\alpha \in \mathbb{N}_0^N} |f^{(\alpha)}(x)| e^{-m\varphi^*(\frac{|\alpha|}{m})} < \infty\}$$

$$\mathcal{E}_{\{\omega\}}(\Omega) := \{f \in C^\infty(\Omega) \mid \forall K \subset\subset \Omega\, \exists m \in \mathbb{N} : \sup_{x\in K} \sup_{\alpha \in \mathbb{N}_0^N} |f^{(\alpha)}(x)| e^{-\frac{1}{m}\varphi^*(m|\alpha|)} < \infty\}$$

and we endow both spaces with their natural locally convex topologies. The elements of $\mathcal{E}_{(\omega)}(\Omega)$ (resp. $\mathcal{E}_{\{\omega\}}(\Omega)$) are called ω-ultradifferentiable functions of Beurling- (resp. Roumieu-) type.

1.3 Definition. (a) Let $\mathbb{P} = (p_k)_{k\in\mathbb{N}}$ be an increasing sequence of non-negative continuous functions on $\mathbb{C}^N$. Then we define

$$A_{\mathbb{P}}(\mathbb{C}^N) := \operatorname*{ind}_{k\to} A_{p_k},$$

where

$$A_{p_k} := \{f \in A(\mathbb{C}^N) \mid \|f\|_k := \sup_{z\in\mathbb{C}^N} |f(z)| e^{-p_k(z)} < \infty\},$$

and where $A(\mathbb{C}^N)$ denotes the entire functions on $\mathbb{C}^N$. If $\mathbb{P} = (kp)_{k\in\mathbb{N}}$ we write A_p instead of $A_{\mathbb{P}}$. (b) Let $\mathbb{P} = (p_k)_{k\in\mathbb{N}}$ be a decreasing sequence of non-negative continuous functions on $\mathbb{C}^N$. Then we define

$$A^0_{\mathbb{P}} := \operatorname*{proj}_{\leftarrow k} A_{p_k}.$$

If $\mathbb{P} = (\frac{1}{k}p)_{k\in\mathbb{N}}$ then we write A^0_p instead of $A^0_{\mathbb{P}}$.

1.4 Remark. Using a result of Taylor [14] (see also Meyer [11]) one proves that the Fourier-Laplace transform

$$\mathcal{F} : \mathcal{E}_{(\omega)}(\mathbb{R}^N)'_b \to A_p \text{ and } \mathcal{F} : \mathcal{E}_{\{\omega\}}(\mathbb{R}^N)'_b \to \operatorname*{ind}_{k\to} A^0_{\mathbb{P}_k},$$

$$\mathcal{F}(\mu)[z] := \langle \mu_x, e^{-i<x,z>}\rangle,$$

is a linear topological isomorphism, if we let $p(z) := |\operatorname{Im} z| + \omega(z)$ and $\mathbb{P}_k := (k|\operatorname{Im} z| + \frac{1}{m}\omega(z))_{m\in\mathbb{N}}$, $k \in \mathbb{N}$.
From this it follows that $\mathcal{E}_\omega(\mathbb{R}^N)'_b$ is a *(DFN)*-space while $\mathcal{E}_{\{\omega\}}(\mathbb{R}^N)'_b$ is an inductive limit of *(FN)*-spaces.

1.5 Definition. Let ω be a weight. Then we define

$$\Lambda(\omega, N) := \{x \in \mathbb{C}^{\mathbb{N}_0^N} \mid \forall m \in \mathbb{N} : \|x\|_m := \sup_{\alpha \in \mathbb{N}_0^N} |x_\alpha| e^{-m\varphi^*(\frac{|\alpha|}{m})} < \infty\}$$

$$\Lambda^0(\omega, N) := \{x \in \mathbb{C}^{\mathbb{N}_0^N} \mid \exists m \in \mathbb{N} : \|x\|_m := \sup_{\alpha \in \mathbb{N}_0^N} |x_\alpha| e^{-\frac{1}{m}\varphi^*(m|\alpha|)} < \infty\}.$$

1.6 Remark. Well-known nuclearity proofs show that for each weight ω the space $A_\omega(\mathbb{C}^N)$ is a *(DFN)*-space and that $A_\omega^0(\mathbb{C}^N)$ is an *(FN)*-space. Moreover, in these spaces the monomials form an absolute basis and give rise to the isomorphisms

$$A_\omega(\mathbb{C}^N) \cong \Lambda(\omega, N)'_b \quad , \quad A_\omega^0(\mathbb{C}^N) \cong \Lambda^0(\omega, N)'_b.$$

In particular, $\Lambda(\omega, N)$ is an *(FN)*-space, while $\Lambda^0(\omega, N)$ is a *(DFN)*-space. Later we shall use the fact that

$$J : ((a_\alpha)_{\alpha \in \mathbb{N}_0^N}) \mapsto \sum_{\alpha \in \mathbb{N}_0^N} a_\alpha (-i)^{|\alpha|} z^\alpha$$

is an isomorphism from $\Lambda(\omega, N)'_b$ into $A_\omega(\mathbb{C}^N)$ as well as from $\Lambda^0(\omega, N)'_b$ into $A_\omega^0(\mathbb{C}^N)$, where we use the standard identification of a dual of a sequence space.

1.7 The Borel map. For a weight ω, the Borel map $B_N : \mathcal{E}_{(\omega)}(\mathbb{R}^N) \to \Lambda(\omega, N)$ resp. $B_N : \mathcal{E}_{\{\omega\}}(\mathbb{R}^N) \to \Lambda^0(\omega, N)$ is defined as

$$B_N(f) := (f^{(\alpha)}(0))_{\alpha \in \mathbb{N}_0^N}.$$

In Meise and Taylor [10] resp. in Bonet, Meise and Taylor [4] those non-quasianalytic weights ω were characterized for which the Borel map is surjective. The question remained open what could be said about the range of the Borel map in case it is not surjective. The aim of the present article is to answer this question by using the idea of proof of Carleson [5] together with the methods applied in [10] and [4]. The precise question which we want to answer is the following: Let ω be a non-quasianalytic weight and let σ be a weight with $\sigma \geq \omega$. Under which conditions do we have

$$\Lambda(\sigma, N) \subset B_N(\mathcal{E}_{(\omega)}(\mathbb{R}^N)) \text{ resp. } \Lambda^0(\sigma, N) \subset B_N(\mathcal{E}_{\{\omega\}}(\mathbb{R}^N))\ ?$$

To study this, we note that B_N^t coincides with the natural inclusion map $A_\omega \hookrightarrow A_p$ resp. $A_\omega^0 \hookrightarrow \text{ind}_{k\to} A_{\mathbb{P}_k}$ if we identify the corresponding duals by means of 1.4 and 1.6 with spaces of entire functions.

2 Functional analytic lemmas

In order to answer the question which was raised at the end of the previous section, we analyse the situation in an abstract setting. This was done already by Ehrenpreis [8], Lemma 13.14; however, his conditions are hard to evaluate since he uses the weak topology.

2.1 Proposition. *Let E, F, G be (LF)-spaces, E semi-reflexive, and assume that $T \in L(E, F)$ and $S \in L(G, F)$ have dense range and that each bounded subset of $E/\ker T$ is contained in the closure of the image of a bounded set in E. Then $S(G) \subset T(E)$ if and only if for each zero neighborhood U in G'_b there exists a zero neighborhood V in E'_b so that $(T^t)^{-1}(V) \subset (S^t)^{-1}(U)$.*

PROOF. $\Rightarrow$: Denote by $\pi : E \to E/\ker T$ the quotient map and by $\tilde{T} : E/\ker T \to F$ the canonical map induced by T. By hypothesis we have $S(G) \subset T(E)$. Hence we can define $L := \tilde{T}^{-1} \circ S : G \to E/\ker T$. Obviously, L is linear and has closed graph. Hence L is continuous by the closed graph theorem and consequently $L^t : (E/\ker T)'_b \to G'_b$ is continuous, too. Now fix a zero neighborhood U in G'_b arbitrarily. Then $(L^t)^{-1}(U)$ is a zero neighborhood in $(E/\ker T)'_b$. Since the hypothesis implies that π^t is an injective topological homomorphism, we can find a zero neighborhood V in E'_b with $(\pi^t)^{-1}(V) \subset (L^t)^{-1}(U)$. Now we claim that $(T^t)^{-1}(V) \subset (S^t)^{-1}(U)$. To prove this, fix $x \in (T^t)^{-1}(V)$ and note that

$$\pi^t \circ \tilde{T}^t x = T^t x \in V.$$

This implies

$$\tilde{T} x \in (\pi^t)^{-1}(V) \subset (L^t)^{-1}(U).$$

Next note that $S = \tilde{T} \circ L$. Hence we get

$$S^t x = L^t \circ \tilde{T}^t x \in U,$$

which proves our claim.

$\Leftarrow$: Since S^t and T^t are injective by hypothesis, the condition above shows that the initial topology on F' with respect to $T^t : F' \to E'_b$ is finer than the initial topology on F' with respect to $S^t : F' \to G'_b$. Now fix $y \in G$ arbitrarily. Then $y \circ S^t$ is a linear functional on F' which is continuous with respect to the initial topology induced by $T^t : F' \to E'_b$. Hence the theorem of Hahn-Banach implies the existence of $x \in (E'_b)'$ so that $y \circ S^t = x \circ T^t$, i.e. $T^{tt} x = S^{tt} y$. By the semi-reflexivity of E we have $x \in E$ and hence $Tx = Sy$. Since y was chosen arbitrarily this proves $S(G) \subset T(E)$.

2.2 Corollary. *Let E, F and G be (DFS)-spaces and assume that $T \in L(E, F)$ and $S \in L(G, F)$ have dense range. Then the following assertions are equivalent:*

(1) $T(E) \supset S(G)$

(2) for each bounded set B in G there is a bounded set M in E so that $T(M) \supset S(B)$

(3) for each continuous semi-norm p on G'_b there exists a continuous semi-norm q on E'_b so that $p(S^t y) \leq q(T^t y)$ for all $y \in F'$

(4) for each sequence $(f_k)_{k\in \mathbb{N}}$ in F' for which $(T^t f_k)_{k\in\mathbb{N}}$ is a null-sequence in E'_b also $(S^t f_k)_{k\in\mathbb{N}}$ is a null-sequence in G'_b.

PROOF. (1) $\Leftrightarrow$ (2): Because of Proposition 2.1, condition (1) is equivalent to

(*) for each bounded set B in G there is a bounded set M in E so that $(T^t)^{-1}(M^0) \subset (S^t)^{-1}(B^0)$.

Since E is a *(DFS)*-space, we can assume that M is compact and absolutely convex. Hence (*) is seen to be equivalent to (2) by taking polars in F.
The equivalence of (1), (3) and (4) follows easily from Proposition 2.1, since E'_b and F'_b are Fréchet spaces.

2.3 Corollary. *Let E, F and G be (FS)-spaces and assume that $T \in L(E,F)$ and $S \in L(G,F)$ have dense range. Moreover, assume that F', endowed with the initial topology with respect to $T^t : F' \to E'_b$ is bornological. Then the following conditions are equivalent:*

(1) $S(G) \subset T(E)$

(2) for each set B in F' for which $T^t(B)$ is bounded in E'_b, the set $S^t(B)$ is bounded in G'_b.

PROOF. (1) $\Rightarrow$ (2): This follows easily from Proposition 2.1.
(2) $\Rightarrow$ (1): Denote by τ the initial topology on F' with respect to $T^t : F' \to E'_b$. Then (2) implies that $S^t : (F', \tau) \to G'_b$ maps bounded sets into bounded sets. Since (F', τ) is bornological by hypothesis, S^t is continuous. Hence (1) follows from Proposition 2.1.

The following proposition allows us to check when the extra condition in 2.3 is satisfied.

2.4 Proposition. *Let X be a (DFS)-space and let $(B_n)_{n\in\mathbb{N}}$ be a fundamental sequence of bounded sets in X which are closed and absolutely convex. Then for each dense subspace H of X the following conditions are equivalent:*

(1) H is bornological

(2) H is quasibarrelled

(3) H is a (DF)-space

(4) for each bounded set B in X there exists a bounded set M in H so that $B \subset \overline{M}^X$

(5) for each $n \in \mathbb{N}$ there exists $m \in \mathbb{N}$ so that $B_n \subset \overline{B_m \cap H}^{X_{B_m}}$

(6) $X = \bigcup_{n\in\mathbb{N}} \overline{B_n \cap H}^{\sigma(X,X')}$.

PROOF. The implications (1) $\Rightarrow$ (2) $\Rightarrow$ (3) and (5) $\Rightarrow$ (6) are obvious. We shall show (3) $\Rightarrow$ (6) $\Rightarrow$ (4) $\Rightarrow$ (5) and (4) $\Rightarrow$ (2) $\Rightarrow$ (1).
(3) $\Rightarrow$ (6): Since H is a dense *(DF)*-subspace of X and since $(B_n \cap H)_{n \in \mathbb{N}}$ is a fundamental sequence of bounded sets in H, we can apply a result of de Wilde (see e.g. Pérez Carreras and Bonet [13], 8.2.27) to obtain

$$X = \overline{H}^X = \bigcup_{n \in \mathbb{N}} n\overline{(B_n \cap H)}^X = \bigcup_{n \in \mathbb{N}} \overline{B_n \cap H}^{\sigma(X,X')}.$$

(6) $\Rightarrow$ (4): Let B be a closed absolutely convex bounded subset of X and let X_B denote the canonical normed space generated by B. Since X_B is a Banach space, (6) together with Grothendieck's factorization theorem implies (4).
(4) $\Rightarrow$ (5): For fixed $n \in \mathbb{N}$ we use (4) to find $k \in \mathbb{N}$, $k > n$, so that $B_n \subset \overline{B_k \cap H}^X =: D$. Since D is bounded in the *(DFS)*-space X, we can find $m > l > k$ so that D is contained in B_l, so that X_{B_m} and X induce the same topology on B_l and so that B_k is relatively compact in X_{B_l}. Then we have

$$B_n \subset D = \overline{B_k \cap H}^X = \overline{B_k \cap H}^{X_{B_m}}$$

(see e.g. Pérez Carreras and Bonet [13], 6.1.12).
(4) $\Rightarrow$ (2): In the terminology of [13], 8.3.22, condition (4) means that H is a "large" subspace of X. Hence (2) follows directly from [13], 8.3.24.
(2) $\Rightarrow$ (1): This was shown by Valdivia [16], Prop. 2. Here is another argument: Since X is a *(DFS)*-space, H is a quasibarrelled *(DF)*-space satisfying the strict Mackey condition (see e.g. [13], 5.1.29 and 5.1.31). Consequently, H is bornological by Bierstedt and Bonet [1], 1.3(a) and 1.5(6).

2.5 Lemma. *Let $E = \operatorname{ind}_{n\to} E_n$ be the (Hausdorff) inductive limit of an embedding spectrum of locally convex spaces. Let H be a subspace of E for which $H_n := H \cap E_n$ is dense in E_n for each $n \in \mathbb{N}$. Then E and $\operatorname{ind}_{n\to} H_n$ induce the same topology on H.*

This is a particular case of a result of Valdivia to be found e.g. in [13], 6.3.1.

2.6 Proposition. *Let $\{X_n, j_{nm}\}$ be an inductive spectrum of Banach spaces with compact embeddings. Let $X := \operatorname{ind}_{n\to} X_n$, let Y be a subspace of X and put $Y_n := X_n \cap Y, n \in \mathbb{N}$. Then the following conditions are equivalent:*

(1) $\operatorname{ind}_{n\to} \overline{Y}_n^{X_n}$ is a topological subspace of X

(2) $\bigcup_{n \in \mathbb{N}} \overline{Y}_n^{X_n} = \overline{Y}^X$

(3) Y is bornological for the topology induced by X

(4) on Y the topology induced by X coincides with the one of $\operatorname{ind}_{n\to} Y_n$.

PROOF. (2) $\Leftrightarrow$ (3): Since $\overline{Y}^X$ is a *(DFS)*-space, this follows from Proposition 2.4.
(3) $\Leftrightarrow$ (4): This is easy to check.
(1) $\Rightarrow$ (4): Lemma 2.5 implies that $\text{ind}_{n\to} Y_n$ is a topological linear subspace of $\text{ind}_{n\to} \overline{Y}_n^{X_n}$ and hence of X.
(4) $\Rightarrow$ (1): From (4) and Lemma 2.5 we get that on Y the topology of X and of $\text{ind}_{n\to} \overline{Y}_n^{X_n}$ coincide. Since the topology of $\text{ind}_{n\to} \overline{Y}_n^{X_n}$ is finer than the one induced by X, Bierstedt, Meise and Summers [3], 1.2, implies (1).

2.7 Remark. (a) Let F be a Fréchet-Schwartz space and let $(P_k)_{k\in\mathbb{N}}$ be an equicontinuous sequence of continuous linear operators on F such that $\lim_{k\to\infty} P_k x = x$ for all $x \in F$. If H is any linear subspace of F'_b containing $\bigcup_{k\in\mathbb{N}} P_k^t(F')$, then H endowed with the topology induced by F'_b is bornological. In fact, condition (6) of Proposition 2.4 is satisfied, as the following argument shows: For each $u \in F'$ the sequence $(P_k^t u)_{k\in\mathbb{N}}$ converges to u in $(F', \sigma(F', F))$ and is contained in $H \cap B$ for some equicontinuous subset B of F', by the equicontinuity of the sequence $(P_k)_{k\in\mathbb{N}}$.
(b) If A is a Köthe matrix on $\mathbb{N}$ and if $\lambda_1(A)$ is a Fréchet-Schwartz space, then (a) implies that every linear subspace H of $\lambda_1(A)'_b$ which contains $\varphi = \bigoplus_{\mathbb{N}} \mathbb{K}$ is bornological in the topology induced by $\lambda_1(A)'_b$.
(c) If $\lambda_1(A)$ is a Fréchet-Montel space which is not Schwartz, then there exist dense linear subspaces H of $\lambda_1(A)'_b$ containing φ which are quasibarrelled but not bornological, as Valdivia [17] has shown.

2.8 Definition. A Fréchet space F is called countably normable if F can be described as the projective limit of a sequence of Banach spaces with injective connecting maps.

2.9 Proposition. *Let F be a Fréchet-Schwartz space and let $\|\cdot\|$ be a continuous norm on F. Put $U := \{x \in F \mid \|x\| \leq 1\}$ and assume that F'_{U^0} is bornological for the topology induced by F'_b. Then F is countably normable.*

PROOF. We choose a fundamental sequence $(\|\cdot\|_n)_{n\in\mathbb{N}}$ of seminorms on F so that $\|\cdot\|_1 = \|\cdot\|$ and so that the corresponding unit balls $(U_n)_{n\in\mathbb{N}}$ form a neighborhood basis of zero in F. By Dubinsky and Vogt [7], Prop. 5, it suffices to show

(1) for each $n \in \mathbb{N}$ there exists $m > n$ so that every $\|\cdot\|_m$-Cauchy sequence in F which is $\|\cdot\|_1$-null is also $\|\cdot\|_n$-null.

To prove (1) we fix $n \in \mathbb{N}$. Since F'_{U^0} is bornological by hypothesis, Proposition 2.4 implies the existence of $m > n$ so that

(2) $$U_n^0 \subset \overline{U_m^0 \cap F'_{U^0}}^{F'_{U_m^0}}.$$

Furthermore we may assume that m is chosen so that U_n^0 is relatively compact in $F'_{U_m^0}$. Now fix a sequence $(x_k)_{k\in\mathbb{N}}$ which is $\|\cdot\|_m$-Cauchy and $\|\cdot\|_1$-null and define

$$z : F'_{U_m^0} \to \mathbb{K} \quad , \quad z(v) := \lim_{k\to\infty} \langle v, x_k \rangle.$$

Then z is well-defined, linear and continuous on $F'_{U_m^\circ}$, since $(x_k)_{k\in\mathbb{N}}$ is $\|\cdot\|_m$-Cauchy. Moreover, for each $u \in F'_{U^0}$ we have $z(u) = 0$, since $(x_k)_{k\in\mathbb{N}}$ is $\|\cdot\|_1$-null. Consequently z vanishes on $U_m^0 \cap F'_{U^0}$ and by (2) also on U_n^0. This means

(3) $\lim_{k\to\infty}\langle u, x_k\rangle = 0$ for each $u \in U_n^0$.

Now note that U_n^0 is relatively compact in $F'_{U_m^0}$ and that $(x_k)_{k\in\mathbb{N}}$ is a bounded sequence in the dual of this Banach space. From this and (3) it follows $(x_k)_{k\in\mathbb{N}}$ tends to zero uniformly on U_n^0. Hence $(x_k)_{k\in\mathbb{N}}$ is a $\|\cdot\|_n$-null sequence, which proves (1).

Remark. Proposition 2.9 can also be derived from a recent result of Önal and Terzioğlu [12], as we indicate now. Using the notation introduced in the proof of 2.9, the hypothesis of 2.9 implies by Proposition 2.6 that

$$F' = \bigcup_{n\in\mathbb{N}} \overline{F'_{U^0} \cap U_n^0}.$$

This is precisely condition (y) of Önal and Terzioğlu [12]. Now [12], Prop. 2, shows that F is countably normable.

2.10 Corollary. *There exist a (DFN)-space X and a dense linear subspace Y of X for which Y is not bornological under the induced topology.*

PROOF. As Dubinsky [6] (see also Vogt [18]) has shown, there exists a nuclear Fréchet space F which is not countably normable but which admits a continuous norm $\|\cdot\|$. Let $X := F'_b$ and $Y := F'_{U^0}$ where $U = \{x \in F \mid \|x\| \le 1\}$. By Proposition 2.9, Y is not bornological under the induced topology, since F is not countably normable.

2.11 Corollary. *There exists an inductive spectrum $\{X_n, j_{nm}\}$ of Banach spaces with nuclear embeddings so that $X := \operatorname{ind}_{n\to} X_n$ contains a (dense) linear subspace Y for which the topology of $\operatorname{ind}_{n\to} \overline{X_n \cap Y}^{X_n}$ is strictly finer than the topology which is induced by X.*

PROOF. By Proposition 2.6 this follows immediately from Corollary 2.10.

Remark. Corollary 2.11 shows that the open problem which was stated in Bierstedt and Meise [2] in the Remark behind Lemma 13 has a negative solution.

2.12 Remark. In Corollary 2.3 the assertions (1) and (2) are not equivalent if one drops the hypothesis that F' is bornological for the initial topology with respect to $T^t : F' \to E'_b$. To see this, let E be a nuclear Fréchet space which admits a continuous norm but which is not countably normable. The existence of E is guaranteed by Dubinsky [6] and Vogt [18]. Without restriction we can assume that the topology of E is generated by an increasing sequence $(\|\cdot\|_n)_{n\in\mathbb{N}}$ of norms, which are induced by scalar products. For $n \in \mathbb{N}$ we let

$$U_n := \{x \in E : \| x \|_n \le 1\} \quad \text{and} \quad X_n := E'_{U_n^0}.$$

Then $E_b' = \mathrm{ind}_{n\to} X_n =: X$ and X_n is a separable Hilbert space for each $n \in \mathbb{N}$. Therefore we can assume that the *(DFN)*-space φ of all finite sequences is a dense linear subspace of the Hilbert space X_1. Then we have

$$\overline{X_1 \cap X_n}^{X_n} = \overline{X_1}^{X_n} = \overline{\varphi}^{X_n} = \overline{\varphi \cap X_n}^{X_n} \quad \text{for each} \quad n \in \mathbb{N}$$

and hence

$$(*) \qquad Z := \mathrm{ind}_{n\to} \overline{X_1 \cap X_n}^{X_n} = \mathrm{ind}_{n\to} \overline{\varphi \cap X_n}^{X_n}.$$

Note that by Bierstedt and Meise [2], Lemma 13, Z is a *(DFN)*-space.
Next let $j : \varphi \to X$ and $i : \varphi \to Z$ denote the corresponding inclusion maps. Furthermore, define

$$F := \varphi_b', \quad G := Z_b', \quad T := j^t \quad \text{and} \quad S := i^t.$$

Since the spaces E, F and G are reflexive, T and S have dense range.
To see that assertion 2.3(2) holds in the present situation, fix a set B in $F' = \varphi$ for which $T^t(B) = j(B)$ is bounded in X. Since X is a *(DFN)*-space, there exists $n \in \mathbb{N}$ so that $j(B)$ is bounded in X_n. Hence $S^t(B) = i(B)$ is bounded in Z.
Now assume that $S(G) \subset T(E)$, i.e. that assertion 2.3(1) holds. Then denote by τ the topology of $E_b' = X$ and by τ_0 the topology of Z. Next note that by Proposition 2.1 the inclusion map $(\varphi, \tau|_\varphi) \to Z$ is continuous. Hence we have $\tau|_\varphi = \tau_0|_\varphi$. Since τ_0 is finer than $\tau|_Z$ and since φ is dense in Z we get from Bierstedt, Meise and Summers [3], 1.2, that $\tau_0 = \tau|_Z$. Consequently, Proposition 2.6 and (*) imply that $(X_1, \tau|_{X_1})$ is bornological. However, by our choice of E we get from Proposition 2.9 that $(X_1, \tau|_{X_1})$ is not bornological. From this contradiction it follows that the assertion 2.3(1) does not hold.

3 On the range of the Borel map

In this section we use the results of the previous one to investigate which sequence spaces are contained in the range of the Borel map B_N acting on $\mathcal{E}_{(\omega)}(\mathbb{R}^N)$ and $\mathcal{E}_{\{\omega\}}(\mathbb{R}^N)$.

3.1 Definition. Let ω be a non-quasianalytic weight. Then we define

$$\text{(a)} \quad P_\omega(x+iy) := \begin{cases} \omega(x) & y = 0 \\ \dfrac{|y|}{\pi} \displaystyle\int_{-\infty}^{+\infty} \frac{\omega(t)}{(x-t)^2 + y^2} dt & y \neq 0 \end{cases}$$

$$\text{(b)} \quad \kappa_\omega(s) := \int_1^\infty \frac{\omega(st)}{t^2} dt.$$

3.2 Remark. (a) P_ω is continuous and harmonic in the open upper and lower half plane and satisfies $\omega \le P_\omega$.
(b) κ_ω is a weight satisfying $\omega \le \kappa_\omega$. In general κ_ω need not be non-quasianalytic, see Example 3.8 below.

3.3 Lemma. *For a non-quasianalytic weight ω and a weight σ the following conditions are equivalent:*

(1) $P_\omega(z) = O(\sigma(z))$ as z tends to infinity

(2) $\kappa_\omega(z) = O(\sigma(z))$ as z tends to infinity.

PROOF. By the arguments of Carleson [5], p. 198, we have

$$\max_{|z|=y} P_\omega(z) = P_\omega(iy), \quad \text{for all} \quad y > 0.$$

Next note that

$$\begin{aligned} P_\omega(iy) &= \frac{y}{\pi}\int_{-\infty}^{+\infty} \frac{\omega(t)}{t^2+y^2}dt = \frac{2y}{\pi}\int_0^\infty \frac{\omega(t)}{t^2+y^2}dt = \frac{2}{\pi}\int_0^\infty \frac{\omega(ys)}{1+s^2}ds \\ &\leq \frac{2}{\pi}[\omega(y) + \int_1^\infty \frac{\omega(ys)}{s^2}ds] \leq \frac{2}{\pi}(\omega(y)+\kappa_\omega(y)) \leq \frac{4}{\pi}\kappa_\omega(y). \end{aligned}$$

On the other hand

$$P_\omega(iy) \geq \frac{2y}{\pi}\int_{|t|\geq y} \frac{\omega(t)}{t^2+y^2}dt = \frac{2}{\pi}\int_1^\infty \frac{\omega(yt)}{t^2+1}dt \geq \frac{1}{\pi}\int_1^\infty \frac{\omega(yt)}{t^2}dt = \frac{1}{\pi}\kappa_\omega(y).$$

Hence the equivalence of (1) and (2) is shown.

3.4 Lemma. *Let ω be a non-quasianalytic weight and $h : \mathbb{R}^N \to [0,\infty[$ be continuous and positively homogeneous. Put $\mathbb{P} := (h(\operatorname{Im} z) + m\omega(z))_{m\in\mathbb{N}}$, $Q := (h(\operatorname{Im} z) + mP_\omega(z))_{m\in\mathbb{N}}$ and $p(z) := |\operatorname{Im} z| + \omega(z)$. Then we have*

(a) A_Q is closed in A_p.

(b) For each $f \in A_Q$ there exists a sequence $(f_k)_{k\in\mathbb{N}}$ in $A_{\mathbb{P}}$ with $f = \lim_{k\to\infty} f_k$ in A_p.

(c) A_Q is the closure of $A_{\mathbb{P}}$ in A_p.

PROOF. (a) Since A_p is a *(DFS)*-space, it suffices to show that $A_Q \cap A_{p,n}$ is closed for each $n \in \mathbb{N}$ (see Floret and Wloka [9]). Fix $f \in A_{p,n}$ and $(f_j)_{j\in\mathbb{N}}$ in A_Q so that $f_j \to f$ in $A_{p,n}$. Then $(f_j)_{j\in\mathbb{N}}$ is bounded in $A_{p,n}$, hence there exists $A > 0$ such that for each $j \in \mathbb{N}$:

(*) $\qquad |f_j(z)| \leq A\exp(n|\operatorname{Im} z| + n\omega(z)).$

Since f_j is in A_Q we have:

(**) $\qquad |f_j(z)| \leq A_j \exp(h(\operatorname{Im} z) + n_j P_\omega(z)).$

From (*) and (**) it follows as in the proof of [10], 2.2, that there exist $C > 0$ and $k \in \mathbb{N}, k \geq n$ so that for all $j \in \mathbb{N}$

$$|f_j(z)| \leq C \exp(h(\operatorname{Im} z) + kP_\omega(z)).$$

Hence $(f_j)_{j\in\mathbb{N}}$ is bounded in $A_{Q,k}$. Since $f_j \to f$ uniformly on the compact subsets of $\mathbb{C}^N$, we get that $f_j \to f$ in $A_{Q,k+1}$. Hence f is in A_Q.

(b) For $j \in \mathbb{N}$ define

$$\omega_j(t) := \begin{cases} \omega(t) & |t| \leq j \\ A_j \log|t| + B_j & |t| \geq j \end{cases}$$

where A_j and $B_j > 0$ are choosen in such a way that ω_j is continuous, increasing and satisfies 1.1 (δ) and $\omega_j \leq \omega$. Then $\omega_j \to \omega$ and hence $P_{\omega_j} \to P_\omega$. Note also that $P_{\omega_j} = O(\log)$. Using $h(\operatorname{Im} z) + P_{\omega_j}$ as the functions Φ_j in Taylor [15], Thm. 1., it follows that for each $f \in A_{Q,n}$ there exists a sequence $(f_k)_{k\in\mathbb{N}}$ in $A_{\mathbb{P},2n}$ so that $f_k \to f$ in $A_{p,3n}$.

(c) This follows trivially from (a) and (b).

3.5 Corollary. *Let ω, h, $\mathbb{P}$ and p be as in 3.4. Then*

$$A_{\mathbb{P}} = \{f \in A(\mathbb{C}^N) \mid \exists\, m : \sup|f(z)|\exp(-h(\operatorname{Im} z) - m\omega(z)) < \infty\} \subset A_p(\mathbb{C}^N)$$

is bornological in the topology induced by $A_p(\mathbb{C}^N)$.

PROOF. By Lemma 3.4 the closure of $A_{\mathbb{P}}$ in A_p equals the space A_Q, defined in 3.4. Hence Lemma 3.4 (b) implies that for $X = A_p$ and $Y = A_{\mathbb{P}}$, condition (2) of Proposition 2.6 is satisfied. Therefore, 2.6 (3) shows that $A_{\mathbb{P}}$ is bornological for the topology induced by A_p.

3.6 Theorem. *Let ω be a non-quasianalytic weight and let σ be a weight. Then the following assertions are equivalent:*

(1) $B_N(\mathcal{E}_{(\omega)}(\mathbb{R}^N)) \supset \Lambda(\sigma, N)$

(2) $P_\omega(z) = O(\sigma(z))$ as z tends to infinity

(3) $\kappa_\omega(z) = O(\sigma(z))$ as z tends to infinity.

PROOF. Under the identifications mentioned in section 1, we have that $B_N^t : \Lambda(\omega, N)_b' = A_\omega \to A_p = \mathcal{E}_{(\omega)}(\mathbb{R}^N)_b'$ is the natural inclusion map, while the adjoint of the inclusion $S : \Lambda(\sigma, N) \to \Lambda(\omega, N)$ equals the inclusion $A_\omega \hookrightarrow A_\sigma$. Therefore Corollary 3.5 together with Corollary 2.3 shows that (1) is equivalent to:

(*) For each set B in A_ω which is bounded in A_p, B is bounded in A_σ.

In view of Lemma 3.3 it suffices therefore to show that (2) is equivalent to (*).
(2) $\Rightarrow$ (*): Apply the Phragmén-Lindelöf principle as in Meise and Taylor [10].
(*) $\Rightarrow$ (2): To argue by contradiction, assume that there is a sequence $(a_j)_{j\in\mathbb{N}}$ in $\mathbb{C}$ with $\lim_{j\to\infty} \frac{P_\omega(a_j)}{\sigma(a_j)} = \infty$. Without loss of generality we can assume that $\operatorname{Im} a_j > 0$. Then note that by the proof of [10], 2.3. (claim), there exists a sequence $(f_j)_{j\in\mathbb{N}}$ of polynomials, hence in A_ω, so that for suitable $C, D > 0$ and for all $j \in \mathbb{N}$ we have:

$$|f_j(z)| \leq \exp(DP_\omega(z)) \leq \exp(D|\operatorname{Im} z| + DE\,\omega(z) + E) \quad \text{for all} \quad z \in \mathbb{C}$$

$$f_j(a_j) = \exp(P_\omega(a_j)).$$

Hence $(f_j)_{j\in\mathbb{N}}$ is bounded in A_p but unbounded in A_σ, since

$$|f_j(a_j)e^{-\sigma(a_j)}| = e^{P_\omega(a_j)-\sigma(a_j)} = \exp(\sigma(a_j)(\frac{P_\omega(a_j)}{\sigma(a_j)} - 1)) \to \infty.$$

3.7 Theorem. *Let ω be a non-quasianalytic weight and let σ be a weight. Then the following assertions are equivalent:*

(1) for each bounded set K in $\Lambda^0(\sigma, N)$ there exists a bounded set L in $\mathcal{E}_{\{\omega\}}(\mathbb{R}^N)$ with $B_N(L) \supset K$

(2) $B_N(\mathcal{E}_{\{\omega\}}(\mathbb{R}^N)) \supset \Lambda^0(\sigma, N)$

(3) $P_\omega(z) = O(\sigma(z))$ as z tends to infinity

(4) $\kappa_\omega(z) = O(\sigma(z))$ as z tends to infinity.

PROOF. By Corollary 2.2, (1) and (2) are equivalent to each of the following assertions:

(*) For each sequence $(f_k)_{k\in\mathbb{N}}$ in A^0_ω which tends to zero in $A^0_{\|}$, even tends to zero in A^0_σ.

(**) For each $n \in \mathbb{N}$ there exists $m \in \mathbb{N}, C > 0$ so that for all $f \in A^0_\omega$

$$\sup_{z\in\mathbb{C}^N} |f(z)|\exp(-\frac{1}{n}\sigma(z)) \leq C \sup_{z\in\mathbb{C}^N} |f(z)|\exp(-|\operatorname{Im} z| - \frac{1}{m}\omega(z)).$$

In view of Lemma 3.3 the proof will be complete if we show (3) $\Rightarrow$ (**) and (*) $\Rightarrow$ (3).
(3) $\Rightarrow$ (**): Apply the Phragmén-Lindelöf principle as in [4].
(*) $\Rightarrow$ (3): To argue by contradiction, assume that there exists a sequence $(a_j)_{j\in\mathbb{N}}$ in $\mathbb{C}$ with $\lim_{j\to\infty} \frac{P_\omega(a_j)}{\sigma(a_j)} = \infty$. Without loss of generality we can assume that $\operatorname{Im} a_j > 0$. As in [4], 2.4, define

$$f_j(z) := \exp(\sigma(a_j) - \frac{1}{j}P\omega(a_j))h_j(z),$$

where $(h_j)_{j\in\mathbb{N}}$ is the sequence constructed there. Assuming a priori that $6j\sigma(a_j) \leq P_\omega(a_j)$, we see that $(f_j)_{j\in\mathbb{N}}$ tends to zero in $A^0_{\|}$, while $f_j(a_j) = \exp(\sigma(a_j))$ implies that $(f_j)_{j\in\mathbb{N}}$ does not tend to zero in A^0_σ. Since h_j is a polynomial, we have $f_j \in A^0_\omega$.

3.8 Example. For $\alpha > 1$ let ω_α be defined by

$$\omega_\alpha(t) = \max(0, \frac{t}{(\log et)^\alpha}).$$

Then an easy calculation shows that

$$\kappa_{\omega_\alpha}(t) = (\alpha - 1)\, \omega_{\alpha-1}(t).$$

3.9 Remark. Let ω and τ be non-quasianalytic weights, assume that $\kappa_\tau \leq \omega$ and let K be a compact convex subset of $\mathbb{R}^N$ with non-empty interior. Then the results of the present section together with those of Meise and Taylor [10] respectively Bonet, Meise and Taylor [4] imply that the following holds:
Let $(f_\alpha)_{\alpha \in \mathbb{N}_0^N}$ be a family of continuous functions on K which satisfies $f_{(0)}|_{\overset{\circ}{K}} \in C^\infty(\overset{\circ}{K})$ and $(f_{(0)}|_{\overset{\circ}{K}})^{(\alpha)} = f_\alpha|_{\overset{\circ}{K}}$ for all $\alpha \in \mathbb{N}_0^N$.
Then

$$\sup_{\alpha \in \mathbb{N}_0^N} \sup_{x \in K} |f_\alpha(x)| \exp(-m\varphi^*(\frac{|\alpha|}{m})) < \infty \quad \text{for all } m \in \mathbb{N}$$

implies the existence of $g \in \mathcal{E}_{(\tau)}(\mathbb{R}^N)$ satisfying

$$(*) \qquad g^{(\alpha)}|_K = f_\alpha \quad \text{for all } \alpha \in \mathbb{N}_0^N,$$

and

$$\sup_{\alpha \in \mathbb{N}_0^N} \sup_{x \in K} |f_\alpha(x)| \exp(-\frac{1}{m}\varphi^*(m|\alpha|)) < \infty \quad \text{for some } m \in \mathbb{N}$$

implies the existence of $g \in \mathcal{E}_{\{\tau\}}(\mathbb{R}^N)$ satisfying $(*)$.

Acknowledgement: The research of the first named author was partially supported by DGICYT Proyecto no. PS 88–0050.

References

[1] K.-D. Bierstedt, J. Bonet: Dual density conditions for *(DF)*-spaces I, Res. Math. 14 (1988) 242–274

[2] K.-D. Bierstedt, R. Meise: Nuclearity and the Schwartz property in the theory of holomorphic functions on metrizable locally convex spaces, p. 93–129 in "Infinite Dimensional Holomorphy and Applications", M. C. Matos (Ed.), North-Holland Mathematics Studies 12 (1977)

[3] K.-D. Bierstedt, R. Meise, W. H. Summers: A projective description of weighted inductive limits, Trans. AMS 272 (1982), 107–160

[4] J. Bonet, R. Meise, B. A. Taylor: Whitney's extension theorem for ultradifferentiable functions of Roumieu type, Proc. R. Ir. Acad 89A (1989), 53–66

[5] L. Carleson: On universal moment problems, Math. Scand. 9 (1961), 197–206

[6] E. Dubinsky: Nuclear Fréchet spaces without the bounded approximation property, Studia Math. 71 (1981), 85–105

[7] E. Dubinsky, D. Vogt: Fréchet spaces with quotients failing the bounded approximation property, Studia Math. 81 (1985), 71–77

[8] L. Ehrenpreis: Fourier Analysis in Several Complex Variables, Wiley-Interscience Publ. 1970

[9] K. Floret, J. Wloka: Einführung in die Theorie der lokalkonvexen Räume, Springer LNM 56 (1968)

[10] R. Meise, B. A. Taylor: Whitney's extension theorem for ultradifferentiable functions of Beurling type, Ark. Mat. 26 (1988), 265–287

[11] T. Meyer: Die Fourier-Laplace Transformation quasianalytischer Funktionale und ihre Anwendung auf Faltungsoperatoren, Diplomarbeit Düsseldorf 1989

[12] S. Önal, T. Terzioğlu: A normability condition on locally convex spaces, Rev. Matem. Univ. Complutense Madrid (to appear)

[13] P. Pérez Carreras, J. Bonet: Barrelled Locally Convex Spaces, North-Holland Mathematics Studies 131 (1987)

[14] B. A. Taylor: Analytically uniform spaces of infinitely differentiable functions, Comm. Pure Appl. Math. 24 (1971), 39–51

[15] B. A. Taylor: On weighted polynomial approximation of entire functions, Pacific J. Math. 36 (1971), 523–539

[16] M. Valdivia: Algunas propriedades de los espacios escalonados, Rev. Real. Acad. Ci. Madrid 73 (1979), 389–399

[17] M. Valdivia: A class of quasibarrelled *(DF)*-spaces which are not bornological, Math. Z. 136 (1974), 249–251

[18] D. Vogt: An example of a nuclear Fréchet space without the bounded approximation property, Math. Z. 182 (1983), 265–267

Progress in Functional Analysis
K.D. Bierstedt, J. Bonet, J. Horváth & M. Maestre (Eds.)

Biduality in Fréchet and (LB)-spaces

Klaus D. Bierstedt and José Bonet

Dedicated to Professor Manuel Valdivia on the occasion of his 60th birthday

Abstract. Let E be a Fréchet space (resp. an (LB)-space $\operatorname{ind}_n E_n$) and H a topological subspace of E (resp. $H = \operatorname{ind}_n H_n$ for a sequence of norm subspaces H_n of E_n with $H_n \subset H_{n+1}$). We give necessary and sufficient conditions that E is canonically (topologically isomorphic to) the inductive bidual $(H_b')_i'$ or (even) the strong bidual H_b'' of H. This characterization is applied to weighted spaces of holomorphic functions, weighted inductive limits of spaces of holomorphic functions, spaces of differentiable functions with Hölder conditions and other examples in distribution theory, holomorphy and Köthe sequence spaces.

If a bornological locally convex (abbreviated: l.c.) space (E, τ) has a basis of absolutely convex bounded subsets which are compact with respect to a weaker l.c. topology $\tilde{\tau}$ (a condition which is called (BBC) below), it is the inductive dual F_i' of the complete space

$$F = \{u \in E^*;\ u_{|B}\ \tilde{\tau}\text{-continuous for each bounded set } B \subset E\}.$$

This result is due to Mujica (see [11] or [25]). In our (preparatory) Section 1., we are interested in l.c. properties of the predual F and prove (Corollary 2.) that F is barrelled and E the strong dual F_b' of F whenever τ also has a 0-neighborhood base of absolutely convex $\tilde{\tau}$-closed sets (condition (CNC)). While for (LB)-spaces E, F clearly is a Fréchet space, the l.c. properties of the predual F are mainly important in the case that E itself is Fréchet. In fact, a Fréchet space E satisfies both (BBC) and (CNC) if and only if it is (topologically isomorphic to) the strong dual of a complete barrelled (DF)-space F, and a Fréchet space E is quasinormable and satisfies (BBC) and (CNC) if and only if it is the strong dual of a boundedly retractive (LB)-space (Corollary 5.).

Section 2. of the article contains the main results: We characterize (by the conditions (i) and (ii) of Theorem 6.) when for a topological subspace H of a bornological l.c. space E with (BBC), the restriction mapping $R : F \to H_b'$ is a topological isomorphism onto, i.e. E is canonically the inductive bidual $(H_b')_i'$ of H, and hence the strong bidual H_b'' whenever (CNC) holds. In the case of an (LB)-space $E = \operatorname{ind}_n E_n$, the hypotheses have to be modified in order to avoid the topological subspace problem for inductive limits; it turns out that $H = \operatorname{ind}_n H_n$ with (I) and (II) of Theorem 7. must necessarily be a topological subspace of E. In many examples, condition (ii) in Theorem 6. (resp. (II) in Theorem 7.) is already satisfied, and then only condition (i) (resp. (I)), which involves the approximation of bounded sets in E from bounded subsets of H in the topology $\tilde{\tau}$, must be verified to obtain the biduality of H and E.

The results in Section 2. arose from an analysis, and a sweeping generalization, of the functional-analytic part of the arguments in [8]. Hence our first applications again are weighted spaces of holomorphic functions (Subsection 3.A.) and weighted inductive limits of spaces of holomorphic functions (Subsection 3.B), where the arguments of [8] can be utilized to obtain the bidualities $HW(G) = (HW_0(G))''_b$ and $\mathcal{V}H(G) = ((\mathcal{V}_0H(G))'_b)'_i$, mainly for sequences W resp. $\mathcal{V}$ of radial weights on certain balanced domains $G \subset \mathbb{C}^N$. Another example, the canonical biduality $\Lambda^{k,\alpha}(\Omega) = \lambda^{k,\alpha}(\Omega)''_b$ of spaces of k-times continuously differentiable functions on an open set $\Omega \subset \mathbb{R}^N$ which satisfy Hölder conditions with exponent $\alpha \in (0,1)$ on each compact subset of Ω is worked out in Subsection 3.C in some detail. At the end, in Subsection 3.D, we include other examples from distribution theory (the result $\check{\mathcal{B}}(\Omega) = \mathcal{B}_0(\Omega)''_b$ of Dierolf, Voigt [10]), infinite dimensional holomorphy (the space $H_b(U)$ of holomorphic functions of bounded type and results of Galindo, García, Maestre [14], [15]) and from Köthe echelon and co-echelon spaces, where our methods yield simple proofs for the existence of a predual with certain l.c. properties or for known bidualities.

Acknowledgement. *Part of this project was undertaken in the framework of an "Acción Integrada" between the authors and their institutions, and we gratefully acknowledge the support of "Deutscher Akademischer Austauschdienst (DAAD)" and "Ministerio de Educación y Ciencia (MEC)".*

Notation. Our notation on locally convex spaces is standard. (E.g. see the books of Horváth [18], Jarchow [20], Köthe [22] and Pérez Carreras-Bonet [27].) For a locally convex space E, E^* denotes the space of all linear functionals on E while E' is the topological dual. Concerning locally convex inductive limits, see [2]. If $E = \mathrm{ind}_n E_n$ is the inductive limit of an increasing sequence $(E_n)_{n\in\mathbb{N}}$ of linear subspaces of E (with continuous inclusions), $\mathrm{ind}_n E_n$ is said to be *boundedly retractive* if for each bounded set $B \subset E$ there is $n \in \mathbb{N}$ such that B is contained and bounded in E_n and that the topologies of E and E_n coincide on B. For weighted spaces and weighted inductive limits, cf. [6], and for our notation on Köthe echelon and co-echelon spaces, cf. [7].

1. Preduals of bornological spaces

In the sequel, $E = (E, \tau)$ will be a bornological locally convex (l.c.) space on which

(BBC) *there exists a (necessarily weaker) l.c. Hausdorff topology $\tilde{\tau}$ such that each τ-bounded subset of E is contained in an absolutely convex τ-bounded and $\tilde{\tau}$-compact set.*

In fact, since (E, τ) is bornological, (BBC) implies $\tilde{\tau} \leq \tau$. Moreover, (BBC) means precisely that τ has a basis $\mathcal{B}_0$ of bounded sets which are absolutely convex and $\tilde{\tau}$-compact (whence Banach disks). It follows that E is ultrabornological (i.e., an inductive limit of Banach spaces).

As Mujica (see Dineen [11, p. 417] and Mujica, Nachbin [25, Theorem 1.1]) proved, such a space E can always be represented as the inductive dual $F'_i = \text{ind}_U F'_{U^\circ}$ (where U runs through the system $\mathcal{U}$ of all absolutely convex 0-neighborhoods of F) of a "canonical" l.c. space F, viz.

$$\begin{aligned} F &= \{u \in E^*;\ u_{|B} \text{ is } \tilde{\tau}\text{-continuous for each bounded set } B \text{ in } (E,\tau)\} \\ &= \{u \in E^*;\ u_{|B} \text{ is } \tilde{\tau}\text{-continuous for each } B \in \mathcal{B}_0\}, \end{aligned}$$

endowed with the topology of uniform convergence on the sets $B \in \mathcal{B}_0$. We certainly have $(E,\tilde{\tau})' \subset F \subset E' = (E,\tau)'$ since $(E,\tilde{\tau})' \subset F$ holds by definition and (E,τ) bornological as well as $\tilde{\tau} \leq \tau$ imply $F \subset E'$. Clearly, F carries the topology $\beta(E',E)_{|F}$ of uniform convergence on the bounded subsets of E.

1. Theorem (Mujica). *F is complete, and the evaluation mapping $J : E \to F'$, given by $[J(x)](u) = u(x)$ for all $x \in E$ and $u \in F$, is a topological isomorphism of E onto the inductive dual F'_i of F.*

The (short, elementary) proof of this theorem is given here since neither [11] nor [25] includes a proof and since it will also serve to deduce the following new corollary on the locally convex structure of the predual F.

2. Corollary. (a) *If one identifies E with F'_i (via J), then each bounded subset of E is equicontinuous on F, and F'_i is a regular inductive limit.*

(b) *If (E,τ) also satisfies*

(CNC) *τ has a 0-neighborhood base $\mathcal{U}_0$ of absolutely convex $\tilde{\tau}$-closed sets,*

then F is (quasi-) barrelled, F'_i equals the strong dual F'_b, and E is topologically isomorphic to F'_b (via J).

PROOF. 1. Obviously, F is closed in the complete space $E'_b = (E',\beta(E',E))$, whence it is complete with respect to the topology $\beta(E',E)_{|F}$. J is well-defined and linear; since $(E,\tilde{\tau})' \subset F$ and $\tilde{\tau}$ is a Hausdorff l.c. topology, J must already be injective.

Next note that the injection $(B,\tilde{\tau}) \to (E,\sigma(E,F))$ is continuous for every bounded set $B \subset E$ since $u_{|B}$ is $\tilde{\tau}$-continuous for any $u \in F$. Then each $B \in \mathcal{B}_0$ must be $\sigma(E,F)$-compact, and hence

$$(+) \qquad B = \overline{B}^{\sigma(E,F)} = B^{\bullet\circ}$$

by the bipolar theorem, where $B^\bullet = B^\circ \cap F$ is the polar of B in F (as compared with the polar B° in $< E, E' >$).

But any $l \in F'$ belongs to $(U \cap F)^\circ$ for some 0-neighborhood U in E'_b. By (BBC), there exists $B \in \mathcal{B}_0$ with $l \in (B^\circ \cap F)^\circ = B^{\bullet\circ} = B$, which serves to show that $J : E \to F'$ is surjective. From now on, we will identify E with F' algebraically (via J).

As B varies in $\mathcal{B}_0$, the corresponding sets $B^\bullet$ and $B^{\bullet\circ}$ describe bases of 0-neighborhoods in the space F and bases of the equicontinuous sets in F', respectively. In view of this, it clearly follows from (+) that J is a topological isomorphism of E (= $\operatorname{ind}_B E_B$ = the bornological space associated with E) onto $F'_i = \operatorname{ind}_U F'_{U^\circ}$. This finishes the proof of Theorem 1.

2. By (BBC) each bounded subset C of $E = F'_i$ is contained in an some $B \in \mathcal{B}_0$ and hence, by equation (+), in the polar of some 0-neighborhood $B^\bullet = B^\circ \cap F$ in F. Thus C is equicontinuous on F, and $F'_i = \operatorname{ind}_U F'_{U^\circ}$ must be a regular inductive limit.

For the rest of the proof of Corollary 2., put $G := (E, \tilde{\tau})' \subset F$ and let $\mathcal{U}_0$ be as in (CNC). Each $U \in \mathcal{U}_0$ is $\sigma(E, G)$-closed and a fortiori $\sigma(E, F)$-closed, whereby $U = U^{\bullet\circ}$ (the first polar again taken in F). As the equicontinuous set $U^\bullet$ is $\beta(F, E)$-bounded, U must already be a 0-neighborhood in $F'_b = (F', \beta(F', F))$, and we have proved $F'_i = F'_b$. At this point, the last assertion of (a) implies F quasibarrelled. □

Clearly, a space E satisfying both (BBC) and (CNC) must be quasi-complete. (This follows from the closed neighborhood condition (CNC), together with the completeness consequence of (BBC), cf. Köthe [22, §18, 4.(4)b)], but is also implied by 2.(b) since E turns out to be isomorphic to the strong dual of a barrelled space.) On the other hand, if a l.c. space E is the strong dual of some (quasi-) barrelled l.c. space F, then (BBC) and (CNC) hold for $\tilde{\tau} = \sigma(E, F)$ (but, of course, E need not be bornological).

We will see below (after Remark 4.) that condition (BBC) for some l.c. topology $\tilde{\tau} \leq \tau$ does not imply that (CNC) holds for this topology. On the other hand, it already follows from (CNC) that each τ-bounded subset of E is contained in an absolutely convex τ-bounded and $\tilde{\tau}$-*closed* set. (In fact, if $\mathcal{U}_0$ is a 0-neighborhood base of absolutely convex $\tilde{\tau}$-closed subsets, the sets

$$B((M_U)_U) := \bigcap_{U \in \mathcal{U}_0} M_U U,$$

where $M_U > 0$ for each $U \in \mathcal{U}_0$, form a basis of the τ-bounded sets in E, and each $B((M_U)_U)$ is absolutely convex and $\tilde{\tau}$-closed.) Thus, if (CNC) holds and if $(E, \tilde{\tau})$ is a semi-Montel space, then we must also have (BBC).

The first consequence at this point is Mujica's completeness theorem for (LB)-spaces (see [23]) which was inspired by a theorem of Banach-Dixmier-Waelbroeck-Ng on dual Banach spaces (cf. Waelbroeck [30, Prop. 1] and Ng [26]). It is stated here for the sake of completeness.

3. Corollary (Mujica). *Let $E = (E, \tau)$ be an (LB)-space, $E = \operatorname{ind}_n E_n$ for an increasing sequence $(E_n)_{n \in \mathbb{N}}$ of Banach spaces with continuous injections, and suppose:*

(*) *There exists a l.c. Hausdorff topology $\tilde{\tau} \leq \tau$ on E such that the closed unit ball B_n of each E_n is $\tilde{\tau}$-compact, $n = 1, 2, \ldots$.*

(a) *Then $F := \{u \in E^*;\ u_{|B_n}$ is $\tilde{\tau}$-continuous for each $n \in \mathbb{N}\}$, endowed with the topology of uniform convergence on the sets B_n ($n \in \mathbb{N}$), is a Fréchet space (in fact, a closed subspace of E'_b) such that the evaluation mapping J yields a topological isomorphism of E onto F'_i, and hence E must be complete.*

(b) *If, in addition,* (CNC) *holds, then F is distinguished, and E is topologically isomor-*

phic to the strong dual F_b'.

PROOF. Grothendieck's result on the "almost regularity" of countable inductive limits of (DF)-spaces (cf. [22, §29, 5.(4)]) and (*) certainly imply that $\text{ind}_n E_n$ is regular, and hence $(nB_n)_{n\in\mathbb{N}}$ is a fundamental sequence of bounded sets in E. Therefore, condition (*) implies (BBC) (and the definition of F agrees with the previous one).

In view of 1., the proof of (a) is finished by noting that F is metrizable in the present case and that the inductive dual of a metrizable l.c. space must be complete (cf. [22, §29, 4.(2)]). (b) is a simple consequence of 2.(b). □

We add the following

4. Remark. *If, under the conditions of 3.(a),* $\text{ind}_n E_n$ *is boundedly retractive, then F is quasinormable, and the converse holds as well.*

Clearly, bounded retractivity of the regular inductive limit $E = \text{ind}_n E_n$ means precisely that E satisfies the strict Mackey convergence condition. Thus, 4. follows from Bonet [9] (a Fréchet space F whose inductive dual F_i' is sequentially retractive must be quasinormable) and [27, 8.3.35] (if F is quasinormable, $F_i' = F_b'$ satisfies the strict Mackey convergence condition).

For a non-distinguished Fréchet space G, $E := G_i'$ is a complete (LB)-space which satisfies (*) for $\tilde{\tau} := \sigma(G', G)$, but does not satisfy (CNC). (If G_i' had a basis of 0-neighborhoods which are absolutely convex and $\sigma(G', G)$-closed, then G would be distinguished in view of the bipolar theorem.) For an (LB)-space E with (*), it may actually turn out to be rather hard to check whether (CNC) holds. Mujica [23] notes that (CNC) is satisfied if for each 0-neighborhood U in E, there exists a sequence $(V_n)_{n\in\mathbb{N}}$ of absolutely convex $\tilde{\tau}$-closed 0-neighborhoods V_n in E with $V_n \cap B_n \subset U$ for $n = 1, 2, \ldots$ (and it is sometimes easier to verify this condition).

We now turn to the (important) case that E is a Fréchet space. Here, in contrast to the preceding case, the locally convex structure of the predual F (and hence our Corollary 2.) becomes rather important. — We recall that a metrizable l.c. space satisfies Heinrich's density condition (DC) (cf. [3]) if and only if the bounded subsets of its strong dual are metrizable. Similarly, a (DF)-space F has the dual density condition (DDC) (cf. [4]) if its bounded subsets are metrizable; then F must be quasibarrelled. (The original definitions of (DC) and (DDC) are more complicated and will not be repeated here.)

5. Corollary. *Let $E = (E, \tau)$ be a Fréchet space.*

(a) *If E satisfies* (BBC), *then F (as defined at the beginning of this section) has a fundamental sequence of bounded sets, and hence F_b' is metrizable. If, in addition, the bounded subsets of E_b' are metrizable (i.e., E has* (DC)*), then the bounded subsets of F are metrizable, too.*

(b) *E satisfies both* (BBC) *and* (CNC) *if and only if E is (topologically isomorphic to) the strong dual of a complete barrelled (DF)-space F.*

(c) *E satisfies* (BBC), (CNC) *and* (DC) *if and only if it is the strong dual of a complete barrelled (DF)-space F with* (DDC).

(d) *E is quasinormable and satisfies* (BBC) *and* (CNC) *if and only if it is the strong dual of a (complete) boundedly retractive (LB)-space F.*

PROOF. (a) As a (closed) topological subspace of the (complete) (DF)-space E'_b, F has a fundamental sequence $\mathcal{B}_0$ of bounded sets (viz., $(U_n^\bullet)_{n\in\mathbb{N}}$ for a decreasing fundamental sequence $(U_n)_n$ of 0-neighborhoods in E). Similarly, the metrizability of the bounded subsets is inherited by topological subspaces.

(b) If F is a (quasi-) barrelled (DF)-space and $E = F'_b$, then E is a Fréchet space with (BBC) and (CNC) for $\tilde{\tau} = \sigma(E, F)$. The converse can easily be deduced from (a) and Proposition 2.(b).

One direction of (c) follows from (a) and (b). Conversely, it certainly suffices to prove:

If F is a (DF)-space with (DDC), *then its strong dual F'_b satisfies* (DC) *(i.e., if all the bounded sets in the (DF)-space F are metrizable, then the bounded subsets of its strong bidual $(F'_b)'_b$ must be metrizable, too).*

However, taking polars in the original definition of the dual density condition for a (DF)-space F ([4, Def. 1]), one clearly ends up with the original definition of the density condition for the Fréchet space F'_b ([3, Def. 1]).

(d) Suppose first that E is the strong dual of a boundedly retractive (LB)-space F, and let $(B_n)_{n\in\mathbb{N}}$ denote a fundamental sequence of absolutely convex bounded sets in F. It only remains to verify the quasinormability of E.

Since F satisfies the strict Mackey convergence condition, for each $n \in \mathbb{N}$ there is $m > n$ such that for each $\alpha > 0$, there exists a 0-neighborhood U in F with $B_n \cap U \subset \alpha B_m$. By taking polars, we see that for each $n \in \mathbb{N}$ there is $m > n$ such that for each $\varepsilon > 0$ there exists a 0-neighborhood V in F with $B_m^\circ \subset V^\circ + \varepsilon B_n^\circ$. As $(B_n^\circ)_{n\in\mathbb{N}}$ is a basis of 0-neighborhoods for $E = F'_b$ and V° is bounded in E, we have obtained that E is quasinormable.

Suppose now that E is quasinormable with (BBC) and (CNC). We have to show that F is a boundedly retractive (LB)-space. (b) already implies that F is a (DF)-space which, as a topological subspace of E'_b, inherits the strict Mackey convergence condition (see [27, 8.3.35 and 5.1.31.(i)]). By [4, 1.3.(a) and 1.4.(c)] (but this can also be proved in a direct way), F must now be bornological, whence $F = \text{ind}_n\, F_{(U_n^\circ \cap F)'}$, where $(U_n)_{n\in\mathbb{N}}$ denotes a decreasing fundamental sequence of 0-neighborhoods in E (and U_n° is the polar in E'). It is easy to see that $F_{(U_n^\circ \cap F)}$ equals $E'_{U_n^\circ} \cap F$, endowed with the norm of $E'_{U_n^\circ}$, and is a Banach space. At this point, we know that F is an (LB)-space which, because of the strict Mackey convergence condition, must be boundedly retractive. □

2. Biduality

In the first part of Section 2., we assume that E is a bornological l.c. space with property

(BBC) of Section 1. (but we do not suppose that (CNC) holds as well). $F \subset E'_b$ is defined as before; we identify E with the inductive dual F'_i (via the evaluation mapping J).

For a (given) topological subspace H of E, we will now characterize under which conditions restriction $R : F \to H'_b$ is a topological isomorphism onto; i.e., when "E is canonically the inductive bidual $(H'_b)'_i$ of H, and hence equals the strong bidual $(H'_b)'_b$ if (CNC) holds".

6. Theorem. *Let H be a topological subspace of E. Then the restriction map $R : F \to H'_b$, $R(u) := u_{|H}$ for $u \in F$, is a topological isomorphism of F onto H'_b if and only if the following two conditions hold:*

(i) *For each bounded $B \subset E$ there is an absolutely convex bounded subset C of H such that B is contained in the closure $\overline{C} := \overline{C}^{\tilde{\tau}}$ of C in $(E, \tilde{\tau})$.*

(ii) *The restriction of each $v \in H'$ to any bounded subset C of H is $\tilde{\tau}$-continuous.*

PROOF. In the sequel, polars in E' will be denoted by $^\circ$ and polars in H' by $^\bullet$. All closures are taken in $(E, \tilde{\tau})$. Since F is a topological subspace of E'_b, R is well-defined, linear and continuous. We *claim* that (i) holds if and only if R is injective and open onto its image.

Assume first that R is injective and open onto $R(F)$, and fix a bounded set $B \subset E$. Then $B^\circ \cap F$ is a 0-neighborhood of F, and since R is open onto $R(F) \subset H'_b$, we can find an absolutely convex bounded subset C of H with

$$C^\bullet \cap R(F) \subset R(B^\circ \cap F).$$

We must now have $B \subset \overline{C}$, and hence (i): Otherwise, there would exist $b \in B \setminus \overline{C}$. In this case, the Hahn-Banach theorem yields $u \in (E, \tilde{\tau})' \subset F$ with $|u(c)| \leq 1$ for all $c \in C$, but $|u(b)| > 1$. $v := R(u) = u_{|H}$ belongs to $C^\bullet \cap R(F) \subset R(B^\circ \cap F)$, whence there exists $\tilde{u} \in B^\circ \cap F$ with $R(u) = v = R(\tilde{u})$. Since R is injective, it follows that $u = \tilde{u} \in B^\circ$, which contradicts $|u(b)| > 1$.

Now assume that (i) holds, let $u \in F$ satisfy $0 = R(u) = u_{|H}$ and take $x \in E$ arbitrary. (i) yields a bounded set $C \subset H$ with $x \in \overline{C}$. By (BBC), we can then find an absolutely convex bounded and $\tilde{\tau}$-compact set $B \subset E$ with $C \subset B$ and thus, a fortiori, $\overline{C} \subset B$. By definition, $u \in F$ is $\tilde{\tau}$-continuous on B, and since u vanishes on $C \subset H$, we also get $u(x) = 0$. Thus $u = 0$ (in F), whereby R is injective. To show that R is open onto $R(F)$, fix a bounded set $B \subset E$ (or, equivalently, a 0-neighborhood $B^\circ \cap F$ in F). By (i), there is $C \subset H$ bounded with $B \subset \overline{C}$. If $v \in C^\bullet \cap R(F)$, there exists $u \in F$ with $v = R(u) = u_{|H}$ and $1 \geq |v(c)| = |u(c)|$ for all $c \in C$. By (BBC), there is an absolutely convex bounded and $\tilde{\tau}$-compact subset D of E containing C, and thus also $\overline{C}$. Now $|u| \leq 1$ on C, $B \subset \overline{C}$ and $u \in F$ is $\tilde{\tau}$-continuous on D; hence $|u| \leq 1$ on B, $u \in B^\circ \cap F$ and $v = R(u) \in R(B^\circ \cap F)$. We have shown $C^\bullet \cap R(F) \subset R(B^\circ \cap F)$, which clearly implies that R is open onto $R(F)$.

Finally, supposing condition (i), we will show that (ii) is true if and only if R is surjective. One direction of this is trivial: If $v \in H'$ and R is surjective, we can find $u \in F$ with $v = R(u) = u_{|H}$. Then by the definition of F, the restriction of v to each bounded subset of H must be $\tilde{\tau}$-continuous; i.e., (ii) holds.

Conversely, assume (ii) and take $v \in H'$ arbitrary. We must extend v to an element $u \in F$.

Given $x \in E$, we know by (i) that there is an absolutely convex bounded set $C \subset H$ with $x \in \overline{C}$. By (ii), v is actually $\tilde{\tau}$-uniformly continuous on (the absolutely convex set) C, and since it takes its values in the complete space $\mathbb{R}$ or $\mathbb{C}$, v has a uniquely determined $\tilde{\tau}$-uniformly continuous extension $\overline{v}_C$ to $\overline{C}$. Put $u(x) := \overline{v}_C(x)$. It is a matter of routine to check that $u(x)$ is indeed well-defined (i.e., independent of C). Thus, we have defined $u \in E^*$ with $u_{|H} = v$. It remains to verify that the restriction of u to each bounded subset B of E is $\tilde{\tau}$-continuous (so that $u \in F$). But for any such B, (i) yields an absolutely convex bounded set $C \subset H$ with $B \subset \overline{C}$, and since v was extended to $\overline{v}_C$ on $\overline{C}$ in a $\tilde{\tau}$-uniformly continuous way, $u_{|B}$ must again be $\tilde{\tau}$-continuous. □

Remark. *Let* (E, τ) *be a bornological l.c. space which has property* (BBC) *(with respect to a l.c. topology* $\tilde{\tau} \leq \tau$*) and* H *a topological subspace of* E *which satisfies conditions* (i) *and* (ii) *of Theorem 6.*

(a) *Then* H *must be quasibarrelled.*

(b) *If, in addition,* (CNC) *holds,* H *is also distinguished.*

PROOF. (a) Let $\overline{\tau}$ denote the finest l.c. topology on E which coincides with $\tilde{\tau}$ on all bounded sets. Now (i) also means that for each bounded subset B of E, there is an absolutely convex bounded set $C \subset H$ with $B \subset \overline{C} :=$ the closure of C in $(E, \overline{\tau})$. Since (E, τ) is bornological, (BBC) clearly implies $\overline{\tau} \leq \tau$; on the other hand, for the canonical predual F of E, $F = (E, \overline{\tau})'$ holds, whence $\sigma(E, F) \leq \overline{\tau}$.

To show that H is quasibarrelled, fix an absolutely convex, bornivorous and closed subset T of H and put $S := \overline{T}$, where the closure is again taken in $(E, \overline{\tau})$. Then S is absolutely convex, but also *bornivorous* in (E, τ): By (i), each τ-bounded set $B \subset E$ is contained in the $(E, \overline{\tau})$-closure of an absolutely convex bounded set $C \subset H$; since T is bornivorous in H, we can find $\lambda > 0$ with $\lambda C \subset T$, whereby $\lambda B \subset S$. At this point, S must be a 0-neighborhood in the bornological space (E, τ), and hence $S \cap H$ is a 0-neighborhood in H. However,

$$S \cap H = \overline{T}^{(H, \overline{\tau}_{|H})} = \overline{T}^{(H, \tau_{|H})} = T$$

since $\sigma(H, F) \leq \overline{\tau}_{|H} \leq \tau_{|H}$ holds, $H' = F$ by 6., and since the closed absolutely convex subsets are the same for all compatible topologies.

(b) By 6., the strong dual H'_b of H is (topologically isomorphic to) F, which must now be barrelled in view of Corollary 2.(b). □

While Theorem 6. is sufficient (say) for the treatment of Fréchet spaces E, the case of (LB)-spaces E (i.e., the setting of Corollary 3.) requires some additional considerations (since in this case, an application of 6. would presuppose a positive solution to a topological subspace problem in (LB)-spaces).

For the rest of this section, $E = (E, \tau)$ will be an (LB)-space, $(E, \tau) = \text{ind}_n\ (E_n, \tau_n)$ for an increasing sequence of Banach spaces E_n with injections $E_n \to E_{n+1}$ of norm ≤ 1, $n = 1, 2, \ldots$ We assume that (*) there exists a l.c. Hausdorff topology $\tilde{\tau} \leq \tau$ on E such that the closed unit ball B_n of each E_n is $\tilde{\tau}$-compact, $n = 1, 2, \ldots$, define the Fréchet space F as in Corollary 3.(a) and identify E with the inductive dual F'_i (via J). For each

$n \in \mathbb{N}$, let H_n be a closed norm subspace of E_n and $C_n := B_n \cap H_n$ the closed unit ball of H_n. Supposing that $H_n \subset H_{n+1}$ for each $n \in \mathbb{N}$, we put $H = (H, \tau') := \operatorname{ind}_n (H_n, \tau_{n|H_n})$. Clearly, H is an (LB)-space, Hausdorff, and continuously injected in E.

We would like to know when restriction $R : F \to H'_b$ yields a topological isomorphism onto. Then, again, E is canonically the inductive bidual $(H'_b)'_i$ of H.

7. Theorem. (a) *Under our assumptions, the restriction mapping $R : F \to H'_b$, given by $R(u) = u_{|H}$ for $u \in F$, is always well-defined, linear and continuous.*

(b) *R is a topological isomorphism of F onto H'_b if and only if the following two conditions are satisfied:*

(I) *For each $n \in \mathbb{N}$ there are $m \geq n$ and $M \geq 1$ with $B_n \subset \overline{MC_m}$ (where the closure on the right hand side is taken in $(E, \tilde{\tau})$).*

(II) *The restriction of each $v \in H' = (H, \tau')'$ to any C_n is $\tilde{\tau}$-continuous, $n = 1, 2, \ldots$.*

Whenever R is a topological isomorphism onto, $H = (H, \tau')$ is a topological subspace of $E = (E, \tau)$.

Proof. The proof is similar to the one of Theorem 6. Note, however, that it is not clear a priori that (H, τ') is a topological subspace of (E, τ), and while $\operatorname{ind}_n\ E_n$ is regular in view of (*), $\operatorname{ind}_n\ H_n$ definitely need not be regular.

First note that the topology of H'_b is given by the increasing sequence $(q_n)_{n \in \mathbb{N}}$ of seminorms,

$$q_n(v) = \sup_{x \in C_n} |v(x)| \quad \text{for } v \in H'_b \ (n \in \mathbb{N}),$$

since by Grothendieck's result on the "almost regularity" of (LB)-spaces (cf. [22, §29, 5.(4)]), each bounded subset of H is contained in the τ'-closure of some nC_n. On the other hand, the topology of F is trivially given by the increasing sequence $(p_n)_{n \in \mathbb{N}}$ of seminorms,

$$p_n(u) = \sup_{x \in B_n} |u(x)| \quad \text{for } u \in F \ (n \in \mathbb{N}).$$

Since for each $u \in F$ and $n \in \mathbb{N}$, $q_n(R(u)) = \sup_{x \in C_n} |u(x)| \leq \sup_{x \in B_n} |u(x)| \leq p_n(u)$, R is clearly continuous.

We next *claim* that (I) is equivalent to the fact that R is injective and open onto $R(F)$. If (I) holds, then the injectivity of R follows from (*) and the definition of F just as in the proof of Theorem 6. To show that R is open onto $R(F)$, take $n \in \mathbb{N}$ and $\varepsilon > 0$. (I) yields $m \geq n$ and $M \geq 1$ with $B_n \subset \overline{MC_m}$; put $\delta := \varepsilon / M$. If $v \in R(F)$ with $q_m(v) \leq \delta$, choose $u \in F$ with $v = R(u) = u_{|H}$ and note that, by $\overline{MC_m} \subset MB_m$ (in view of (*)) and by the $\tilde{\tau}$-continuity of $u \in F$ on B_m,

$$p_n(u) = \sup_{x \in B_n} |u(x)| \leq M \sup_{x \in C_m} |u(x)| = M q_m(v) \leq \varepsilon.$$

Thus, $\{v \in R(F);\ q_m(v) \leq \delta\} \subset R(\{u \in F;\ p_n(u) \leq \varepsilon\})$, and R is open onto $R(F)$.

Conversely, let R be injective and open onto $R(F)$. Then for each $n \in \mathbb{N}$ there are numbers $N = N(n) \in I\!N$, $N \geq n$, and $\varepsilon = \varepsilon(n) > 0$ such that

$$\{v \in R(F);\ q_N(v) \leq \varepsilon\} \subset R(\{u \in F;\ p_n(u) \leq 1\}).$$

Suppose that (I) fails; there exists $n \in I\!N$ with $B_n \not\subset \overline{mC_m}$ for each $m \geq n$. Choose $m > \max(N(n), \varepsilon(n)^{-1})$ and $b \in B_n \setminus \overline{mC_m}$. Hahn-Banach yields $u \in (E, \tilde{\tau})' \subset F$ with $|u(x)| \leq 1/m$ for all $x \in C_m$, but $|u(b)| > 1$. Now $v := R(u) = u_{|H}$ belongs to $R(F)$ and satisfies

$$q_{N(n)}(v) \leq q_m(v) = \sup_{x \in C_m} |u(x)| \leq 1/m < \varepsilon(n).$$

Hence there is some $\tilde{u} \in F$ with $p_n(\tilde{u}) \leq 1$ and $v = R(\tilde{u})$. The injectivity of R implies $u = \tilde{u}$, but this clearly contradicts $|\tilde{u}(b)| \leq 1$.

In the rest of the proof, we will assume condition (I) and show that then (II) holds if and only if R is surjective. Again, the implication R surjective $\Rightarrow$ (II) is trivial, and we assume now that (II) holds. Then an arbitrary $v \in H'$ must be extended to some $u \in F$. This can be done exactly as in the proof of 6.: By (II), $v_{|nC_n}$ is $\tilde{\tau}$-uniformly continuous ($n \in I\!N$) and has a uniquely determined $\tilde{\tau}$-uniformly continuous extension $\overline{v_n}$ to $\overline{nC_n}$. In view of (I), one can easily check that these extensions $\overline{v_n}$ are coherent and combine to yield a well-defined $u \in E^*$ whose restriction to each B_n, $n = 1, 2, \ldots$, is $\tilde{\tau}$-continuous.

Alternatively, it is also possible (and not too hard, using, say, [27, 8.1.12]) to prove the following

8. Lemma. *Let $\overline{\tau}$ (resp. $\hat{\tau}$) denote the finest l.c. topology on E (resp. H) whose restriction to each B_n (resp. C_n), $n = 1, 2, \ldots$, coincides with $\tilde{\tau}$. Then* (I) *implies $\overline{\tau}_{|H} = \hat{\tau}$.*

By (II) any $v \in H'$ is actually $\hat{\tau}$-continuous on H; hence Lemma 8. implies $v \in (H, \overline{\tau}_{|H})'$. By use of the Hahn-Banach theorem, one can find $u \in (E, \overline{\tau})'$ with $u_{|H} = v$. But then $u_{|B_n}$ is $\tilde{\tau}$-continuous for each n, and thus $u \in F$ with $R(u) = v$.

We will now assume (I) and (II) and identify H'_b with F (via R) and F'_i with E (via J). Then $(H'_b)'_i = E$ holds canonically. To conclude that in this case (H, τ') is a topological subspace of (E, τ), it is clearly enough to observe that any (LB)-space H is a topological subspace of $(H'_b)'_i$: Certainly H is a topological subspace of its strong bidual $H''_b := (H'_b)'_b$; hence $(H'_b)'_i$ already induces a stronger topology on H. On the other hand, in view of the continuity of the embedding $H \to H''_b$, the canonical mapping $H \to (H'_b)'_i$ has closed graph, and thus must be continuous by Grothendieck's closed graph theorem (see [17, Théorème B, 2.]); i.e., $(H'_b)'_i$ also induces a weaker topology on H. □

9. Corollary. *If, in the situation of 7.,* (I) *and* (II) *hold, then* $\mathrm{ind}_n H_n$ *boundedly retractive implies F quasinormable. In this case, E is (canonically topologically isomorphic to) the strong dual of F and the strong bidual of H, and* $\mathrm{ind}_n E_n$ *is boundedly retractive as well.*

This is a simple consequence of 7.(b), 5.(d) and 4.

3. Examples

A. Weighted Fréchet spaces of holomorphic functions

In the sequel, G denotes an open subset of $\mathbb{C}^N$ ($N \geq 1$); the space $H(G)$ of all holomorphic functions on G will be endowed with the topology of uniform convergence on the compact subsets of G. For an increasing sequence $W = (w_n)_{n\in\mathbb{N}}$ of strictly positive continuous functions w_n on G, the weighted spaces of holomorphic functions are defined by

$$HW(G) := \{f \in H(G);\ \text{for each } n \in \mathbb{N},\ p_n(f) := \sup_{z\in G} w_n(z)|f(z)| < \infty\} \quad \text{and}$$
$$HW_0(G) := \{f \in H(G);\ \text{for each } n \in \mathbb{N},\ w_n f \text{ vanishes at } \infty \text{ on } G\}.$$

Under the topology τ given by the sequence $(p_n)_{n\in\mathbb{N}}$ of norms, $E := HW(G)$ is a Fréchet space; its topological subspace $H := HW_0(G)$ is closed, hence itself a Fréchet space.

Let $\tilde{\tau} \leq \tau$ denote the restriction of the compact-open topology to E. Then E has a 0-neighborhood base $\mathcal{U}_0 = (U_n)_{n\in\mathbb{N}}$ of absolutely convex $\tilde{\tau}$-closed sets, viz.

$$U_n = \{f \in HW(G);\ p_n(f) \leq 1/n\} = \{f \in E;\ |f(z)| \leq 1/(nw_n(z)) \text{ for all } z \in G\},$$

n = 1,2,... Put

$$\bar{V} := \{\bar{v} > 0 \text{ continuous on } G;\ \text{for each } n \in \mathbb{N},\ \sup_{z\in G} w_n(z)\bar{v}(z) < \infty\}$$

and for arbitrary $\bar{v} \in \bar{V}$,

$$B_{\bar{v}} := \{f \in HW(G);\ |f(z)| \leq \bar{v}(z) \text{ for all } z \in G\},\quad C_{\bar{v}} := B_{\bar{v}} \cap HW_0(G).$$

Clearly each $B_{\bar{v}}$ is absolutely convex, τ-bounded and $\tilde{\tau}$-compact by Montel's theorem. Since a set $B \subset E$ is bounded if and only if B is contained in some $B_{\bar{v}}$ (cf. [7, Proposition 2.5], together with [6, 0.2., Proposition] and [5, 4.2]), $E = HW(G)$ satisfies (BBC) and (CNC). By Corollary 5.(b), E is (topologically isomorphic to) the strong dual of the complete barrelled (DF)-space

$$F = \{u \in E^*;\ u_{|B} \text{ is } \tilde{\tau}\text{-continuous for each bounded set } B \text{ in } E\},$$

endowed with the topology $\beta(F, E)$.

Next we show that condition (ii) of Theorem 6. is always satisfied for $H = HW_0(G)$. In fact, by the Hahn-Banach and Riesz representation theorems (cf. [28]), for each $l \in H'$ there are $n \in \mathbb{N}$ and a bounded Radon measure μ on G with

$$l(f) = \int_G f w_n \, d\mu \quad \text{for all } f \in H.$$

Now the restriction of l to an arbitrary bounded subset C of H must clearly be $\tilde{\tau}$-continuous by the inner regularity of μ. Thus we have proved:

10. Proposition. *$((HW_0(G))'_b)'_b = HW(G)$ holds canonically if and only if for each* *bounded set $B \subset HW(G)$ there is an absolutely convex bounded set C in $HW_0(G)$ such that B is contained in the closure of C in $HW(G)$ with respect to the compact-open topology $\tilde{\tau}$. (In this case, $HW_0(G)$ must be distinguished.)*

Proposition 10. generalizes (the isomorphic version of) [8, Theorem 1.1 and Corollary 1.2] to weighted Fréchet spaces of holomorphic functions. In [8, Section 2.] the case of radial weights on the unit disk $D \subset \mathbb{C}$, the complex plane $\mathbb{C}$ or on some related domains $G \subset \mathbb{C}^N$ was discussed. To a large part, this discussion also applies in the present (more general) setting. We will now give some details for entire functions of one variable.

For the time being, take $G = \mathbb{C}$ and assume that all the weights in $W = (w_n)_{n \in \mathbb{N}}$ are *radial*, i.e. $w_n(z) = w_n(|z|)$ holds for all $z \in \mathbb{C}$, $n = 1, 2, \ldots$ By the argument of [6, 0.2, Prop.], it will then be no loss of generality to suppose that also all weights $\bar{v} \in \bar{V}$ are radial. Finally, p_k denotes the function $p_k(z) = z^k$ for $z \in \mathbb{C}$, $k = 0, 1, \ldots$, and $\mathcal{P}_k := \operatorname{span}\{p_0, ..., p_k\}$ for $k \in \mathbb{N}_0$, $\mathcal{P}_{-1} := \{0\}$.

For fixed $k \in \mathbb{N}_0$, $p_k \in HW(\mathbb{C})$ (resp. $p_k \in HW_0(\mathbb{C})$) trivially implies that $HW(\mathbb{C})$ (resp. $HW_0(\mathbb{C})$) contains $\mathcal{P}_k$. Since all the weights are radial, one can easily check (e.g., see the indirect proof in [8, final part of Section 2.], which involves Taylor series expansion about 0 and the maximum modulus theorem) that

$$p_k \notin HW(\mathbb{C}) \text{ (resp. } p_k \notin HW_0(\mathbb{C})) \implies HW(\mathbb{C}) \subset \mathcal{P}_{k-1} \text{ (resp. } HW_0(\mathbb{C}) \subset \mathcal{P}_{k-1}).$$

11. Proposition. *Under the present assumptions, the following properties are equivalent:*

(i) *For each bounded $B \subset HW(\mathbb{C})$ there is an absolutely convex bounded $C \subset HW_0(\mathbb{C})$ such that B is contained in $\overline{C}$ (closure in $(HW(\mathbb{C}), \tilde{\tau})$).*

(ii) *For each $\bar{v} \in \bar{V}$, $B_{\bar{v}}$ is contained in the $\tilde{\tau}$-closure of $C_{\bar{v}}$.*

(iii) *For arbitrary $k \in \mathbb{N}_0$, $p_k \in HW(\mathbb{C})$ implies $p_k \in HW_0(\mathbb{C})$.*

PROOF. Trivially (ii) $\Rightarrow$ (i). And in view of the preceding remarks, (i) $\Rightarrow$ (iii) because $p_k \in HW(\mathbb{C}) \setminus HW_0(\mathbb{C})$ would imply $HW_0(\mathbb{C}) \subset \mathcal{P}_{k-1}$, and since the last space is $\tilde{\tau}$-closed, we would get a contradiction to (i).

Finally, (iii) takes care of the case $HW(\mathbb{C}) \subset \mathcal{P}_k$ for some k, and thus in order to show that (iii) $\Rightarrow$ (ii), it suffices to prove that (ii) holds whenever all weights w_n are (radial and) *rapidly decreasing* (i.e., $p_k \in HW(\mathbb{C})$ or, equivalently, $p_k \in HW_0(\mathbb{C})$ for all $k \in \mathbb{N}_0$). For this purpose fix $f \in B_{\bar{v}}$ and let f_r denote the function $f_r(z) = f(rz)$ for $z \in \mathbb{C}$, $0 < r < 1$. We have $f_r \to f$ uniformly on compact subsets of $\mathbb{C}$ as $r \to 1_-$. Since $\bar{v}$ is radial, the maximum modulus theorem implies that for any $z \in \mathbb{C}$ there is $\alpha \in \mathbb{C}$, $|\alpha| = 1$, with

$$|f_r(z)| \leq |f(\alpha z)| \leq \bar{v}(\alpha z) = \bar{v}(z).$$

At this point, it remains to verify that $f_r \in HW_0(\mathbb{C})$ for $0 < r < 1$ (and hence $(f_r)_r \subset C_{\bar{v}}$). For this one can follow the proof of [8, Example 2.2] almost literally. □

It is obvious from 10. and 11.(iii) that for a sequence W of radial weights on $\mathbb{C}$,

$$HW(\mathbb{C}) = HW_0(\mathbb{C}) \implies HW(\mathbb{C}) = ((HW(\mathbb{C}))'_b)'_b,$$

i.e. $HW(\mathbb{C})$ is reflexive. On the other hand, we have:

12. Corollary. *If W is an increasing sequence of radial rapidly decreasing weights on $\mathbb{C}$, then $((HW_0(\mathbb{C}))'_b)'_b = HW(\mathbb{C})$ holds canonically.*

In fact, the argument after [8, 2.2] demonstrates that this corollary remains true for an increasing sequence of rapidly decreasing weights w on $\mathbb{C}^N$ which are *radial* in the sense that

$$w(\lambda z) = w(z) \text{ for all } z \in \mathbb{C}^N \text{ and every } \lambda \in \mathbb{C} \text{ with } |\lambda| = 1.$$

Similarly, following the same method of proof (cf. [8, Example 2.1]), we also get (in an even simpler way)

$$((HW_0(D))'_b)'_b = HW(D)$$

for any increasing sequence $W = (w_n)_n$ of radial weights on the unit disk D which satisfies $\lim_{|z|\to 1_-} w_n(z) = 0$ for $n = 1, 2, \ldots$ And this result carries over to related types of balanced (bounded) domains $G \subset \mathbb{C}^N$ and radial weights which vanish at ∂G, cf. the remarks after [8, 2.1].

B. Weighted (LB)-spaces of holomorphic functions

In part B., $\mathcal{V} = (v_n)_{n\in\mathbb{N}}$ denotes a decreasing sequence of strictly positive continuous functions v_n on an open subset G of $\mathbb{C}^N$, and

$$E_n := Hv_n(G) = \{f \in H(G);\ \|f\|_n = \sup_{z\in G} v_n(z)|f(z)| < \infty\},$$
$$H_n := H(v_n)_0(G) = \{f \in H(G);\ v_n f \text{ vanishes at } \infty \text{ on } G\}$$

are the weighted Banach spaces (endowed with $\|.\|_n$) of holomorphic functions on G which are associated with $\mathcal{V}$, $n = 1, 2, \ldots$ The closed unit ball of E_n (resp. H_n) will be denoted by B_n (resp. C_n). We are interested in the (LB)-spaces

$$(E, \tau) = \mathcal{V}H(G) := \operatorname*{ind}_n Hv_n(G) \quad \text{and} \quad (H, \tau') = \mathcal{V}_0H(G) := \operatorname*{ind}_n H(v_n)_0(G).$$

Again, $\tilde{\tau} \le \tau$ is the restriction of the compact-open topology to E. By Montel's theorem each B_n is $\tilde{\tau}$-compact, and hence condition (*) in Corollary 3. is satisfied. It follows that $\mathcal{V}H(G)$ is the inductive dual F'_i of the Fréchet space

$$F = \{u \in E^*;\ u_{|B_n} \text{ is } \tilde{\tau}\text{-continuous for each } n \in \mathbb{N}\}.$$

From Remark 4. and [6, different proof of 2.3], we get that F is quasinormable if $\mathcal{V} = (v_n)_n$ is *regularly decreasing*, i.e.

for each $n \in \mathbb{N}$ there is $m \ge n$ such that for each subset Y of X with $\inf_{y\in Y} v_m(y)/v_n(y) > 0$, we also have $\inf_{y\in Y} v_k(y)/v_n(y) > 0$ for $k = m+1, m+2, \ldots$

Then F must be distinguished, whence $\mathcal{V}H(G) = F_b'$.

By the same argument as in A. above, one can see that condition (II) of Theorem 7. is always satisfied for $H = \mathcal{V}H_0(G)$, and thus we obtain:

13. Proposition. *$((\mathcal{V}_0H(G))_b')_i' = \mathcal{V}H(G)$ holds canonically if and only if for each $n \in I\!N$ there are $m \geq n$ and $M \geq 1$ with $B_n \subset \overline{MC_m}$ (closure in $(\mathcal{V}H(G), \tilde{\tau})$). In this case, we have $((\mathcal{V}_0H(G))_b')_b' = \mathcal{V}H(G)$ whenever $\mathcal{V}$ is regularly decreasing.*

As we did in A., we now turn to the case that $\mathcal{V} = (V_n)_n$ is a sequence of radial weights on $G = \mathbb{C}$. The following proposition (in which closures are always taken in $(\mathcal{V}H(\mathbb{C}), \tilde{\tau})$) is parallel to 11. and can be proved in exactly the same way.

14. Proposition. *For a decreasing sequence $\mathcal{V}$ of radial weights on $\mathbb{C}$, the following conditions are equivalent:*

(i) *For each $n \in I\!N$ there are $m \geq n$ and $M \geq 1$ with $B_n \subset \overline{MC_m}$.*

(ii) *For $n \in I\!N$ there is $m \geq n$ such that for all $k \in I\!N_0$, $p_k \in Hv_n(\mathbb{C})$ implies $p_k \in H(v_m)_0(\mathbb{C})$.*

Under the assumptions of 14., it is clear that

$$\mathcal{V}H(\mathbb{C}) = \mathcal{V}_0H(\mathbb{C}) \implies \mathcal{V}H(\mathbb{C}) = ((\mathcal{V}H(\mathbb{C}))_b')_i' = ((\mathcal{V}H(\mathbb{C}))_b')_b'$$

(for the last equality see the end of the proof of 7.), i.e. $\mathcal{V}H(\mathbb{C})$ is reflexive. On the other hand, if all v_n are radial and rapidly decreasing on $\mathbb{C}$, [8, 2.2] directly shows that (even) $B_n \subset \overline{C_n}$ holds for every $n \in I\!N$, whence:

15. Corollary. *If $\mathcal{V}$ is a decreasing sequence of radial rapidly decreasing weights on $\mathbb{C}$, then $((\mathcal{V}_0H(\mathbb{C}))_b')_i' = \mathcal{V}H(\mathbb{C})$ holds canonically and $\mathcal{V}_0H(\mathbb{C})$ is a topological subspace of $\mathcal{V}H(\mathbb{C})$. If $\mathcal{V}$ is, in addition, regularly decreasing, we also have $((\mathcal{V}_0H(\mathbb{C}))_b')_b' = \mathcal{V}H(\mathbb{C})$.*

Corollary 15. remains true for a decreasing sequence of rapidly decreasing weights on $\mathbb{C}^N$ which are radial in the sense explained after 12. Actually, the examples of [8] again allow to deduce biduality results for weighted inductive limits of spaces of holomorphic functions on D (resp. on certain balanced bounded domains $G \subset \mathbb{C}^N$) if the weights v_n in the decreasing sequence $\mathcal{V}$ are radial and vanish at ∂D (resp. radial in the sense after 12. and vanish at ∂G).

C. Spaces of differentiable functions with Hölder conditions

Let $k \in I\!N_0$, $0 < \alpha < 1$, and let Ω denote an open subset of $I\!R^N$; fix a basis $(K_n)_{n \in I\!N}$ of compact subsets of Ω with $K_n \subset \overset{\circ}{K}_{n+1}$, $n = 1, 2, \ldots$ In this section we investigate the following algebras of C^k-functions on Ω whose partial derivatives of order k satisfy a Hölder condition with exponent α on each compact subset of Ω:

$$\Lambda^{k,\alpha}(\Omega) := \{f \in C^k(\Omega);\ \text{for each compact } K \subset \Omega,$$
$$q_K(f) = \sup\left\{\frac{|D^\beta f(x) - D^\beta f(y)|}{|x-y|^\alpha};\ |\beta| = k,\ x,y \in K, x \neq y\right\} < \infty\},$$
$$\lambda^{k,\alpha}(\Omega) := \{f \in \Lambda^{k,\alpha}(\Omega);\ \text{for each compact } K \subset \Omega \text{ and each } \beta \text{ with } |\beta| = k,$$
$$\lim \frac{D^\beta f(x) - D^\beta f(y)}{|x-y|^\alpha} = 0 \text{ as } |x-y| \to 0 \text{ with } x,y \in K\}.$$

Both spaces are Fréchet under the topology given by the sequence of seminorms

$$p_n(f) := \max\left(\sum_{0 \le |\beta| \le k} \sup_{x \in K_n} |D^\beta f(x)|,\ q_{K_n}(f)\right),\quad n \in I\!N.$$

There are continuous inclusions $C^{k+1}(\Omega) \hookrightarrow \lambda^{k,\alpha}(\Omega) \hookrightarrow \Lambda^{k,\alpha}(\Omega) \hookrightarrow C^k(\Omega)$.

The Fréchet spaces of differentiable functions with Hölder conditions were utilized by Kaballo [21]. Galindo [13, Theorem 3] proved that $\Lambda^{k,\alpha}(\Omega)$ (resp. $\lambda^{k,\alpha}(\Omega)$) is (topologically) isomorphic to $(l_\infty)^{I\!N}$ (resp. $(c_0)^{I\!N}$). The main result of this subsection is parallel to a classical theorem of Glaeser [16, p. 96, Prop. VI] for compact Ω (cf. also Frampton, Tromba [12, Theorem B] for compact C^∞-Riemannian manifolds).

16. Proposition. *The strong bidual of $\lambda^{k,\alpha}(\Omega)$ is canonically isomorphic to $\Lambda^{k,\alpha}(\Omega)$.*

PROOF. Obviously $(E,\tau) := \Lambda^{k,\alpha}(\Omega)$ satisfies condition (CNC) with respect to the restriction $\tilde{\tau}$ of the (classical) topology of $\mathcal{E}^k(\Omega) = C^k(\Omega)$. As $(M_n)_{n \in I\!N}$ runs through all sequences of positive numbers, the following sets form a basis of bounded subsets of E:

$$B = B((M_n)_n) := \{f \in \Lambda^{k,\alpha}(\Omega);\ \text{for each } n \in I\!N,\ p_n(f) \le M_n\}.$$

Hence, in order to show condition (BBC), it suffices to prove that each set B of this type is $\tilde{\tau}$-compact. By the same obvious argument as for (CNC), B must already be $\tilde{\tau}$-closed. But an easy application of Ascoli's theorem (see Horváth [18, p. 241, Prop. 11]) yields that B is also $\tilde{\tau}$-relatively compact.

We must now check (i) and (ii) in Theorem 6. for $H := \lambda^{k,\alpha}(\Omega)$; as usual, (i) is the main point. Without loss of generality, we can take the bounded set B in (i) of the form $B((M_n)_n)$ for a sequence of numbers $M_n > 0$. For each $l \in I\!N$ choose a C^∞-function h_l on $I\!R^N$ which is identically 1 on K_l and has its support inside $\mathring{K}_{l+1}$, and take a sequence $(\varepsilon_l)_{l \in I\!N}$ with

$$\text{supp}(h_l) + \overline{B}(0, 2\varepsilon_l) \subset \mathring{K}_{l+1}\ (l \in I\!N) \quad \text{and} \quad \lim_{l \to \infty} \varepsilon_l = 0.$$

Next, for a nonnegative C^∞-function ρ on $I\!R^N$ with $\text{supp}(\rho) \subset B(0,1)$ and $\int_{I\!R^N} \rho(x)dx = 1$, put $\rho_l(x) := \varepsilon_l^{-N} \rho(x/\varepsilon_l)$ ($x \in I\!R^N$ and $l \in I\!N$). Finally, for arbitrary $f \in \Lambda^{k,\alpha}(\Omega)$ define

$$f_l := (f h_l) \star \rho_l,\quad l = 1, 2, \ldots$$

Thus $f_l \in C^\infty(\Omega) \subset \lambda^{k,\alpha}(\Omega) = H$ with $\text{supp}(f_l) \subset \text{supp}(h_l) + B(0, \varepsilon_l) \subset K_{l+1}$ for each $l \in I\!N$, and it is well-known that $f_l \to f$ in $C^k(\Omega)$. Thus, the closure of the set

$$C := \{f_l;\ f \in B,\ l \in I\!N\} \subset H$$

in $(E, \tilde{\tau})$ contains B, and it remains to prove that C is bounded in $\lambda^{k,\alpha}(\Omega)$.

For that purpose, fix the seminorm p_n. Since B is compact in $C^k(\Omega)$ and $h_l \to 1$ in this space, it is easy to verify (e.g., using Horváth [18, 4, §7, Exercise 6 and 4, §10, Proposition 2]) that C must already be bounded in $C^k(\Omega)$, i.e.

$$\sum_{|\beta| \leq k} \sup_{x \in K_n} |D^\beta f_l(x)| \leq M < \infty$$

independent of $f \in B$ and $l \in \mathbb{N}$. To show that also $\sup_{g \in C} q_{K_n}(g) < \infty$, take $f \in B$, $l \in \mathbb{N}$, β with $|\beta| = k$ and $x, y \in K_n$ with $x \neq y$; then

$$\begin{aligned} \frac{|D^\beta f_l(x) - D^\beta f_l(y)|}{|x-y|^\alpha} &\leq \int_{\mathbb{R}^N} \frac{|D^\beta(fh_l)(x-z) - D^\beta(fh_l)(y-z)|}{|(x-z)-(y-z)|^\alpha} \rho_l(z) dz \\ &\leq \sup \left\{ \frac{|D^\beta(fh_l)(a) - D^\beta(fh_l)(b)|}{|a-b|^\alpha};\ a, b \in K_n + \operatorname{supp}(\rho_l),\ a \neq b \right\}. \end{aligned}$$

Now $K_n + \operatorname{supp}(\rho_l) \subset K_{n+1}$ for $l \geq n$, while for $l < n$, by an easy argument, the sup has only to be extended over all

$$a, b \in (K_n + \operatorname{supp}(\rho_l)) \cap \operatorname{supp}(h_l) \subset \operatorname{supp}(h_l) + B(0, 2\varepsilon_l) \subset K_{l+1} \subset K_n.$$

Hence the last term above is

$$\begin{aligned} &\leq q_{K_{n+1}}(fh_l) \ \leq p_{n+1}(fh_l) \\ &\leq (2^{k+1}(\operatorname{diam} K_{n+1})^{1-\alpha} + 2)\ p_{n+1}(h_l)\ p_{n+1}(f) \\ &\leq (2^{k+1}(\operatorname{diam} K_{n+1})^{1-\alpha} + 2) \max\left(\max_{l=1,\ldots,n} p_{n+1}(h_l), 1 \right) \sup_{f \in B} p_{n+1}(f), \end{aligned}$$

where an inequality of Glaeser [16, p. 92] is invoked in the estimate from the first to the second line and where the last estimate comes from the fact that h_l is identically 1 on K_{n+1} for $l \geq n+1$. Since the right hand side is independent of $f \in B$, $l \in \mathbb{N}$ and $|\beta| = k$, C is bounded in $\lambda^{k,\alpha}(\Omega)$.

Condition (ii) will follow after a *preliminary remark*: For $f \in \lambda^{k,\alpha}(\Omega)$ and β with $|\beta| = k$, define $l_\beta(f) : \Omega \times \Omega \longrightarrow \mathbb{R}$ by

$$l_\beta(f)(x, y) := \frac{D^\beta f(x) - D^\beta f(y)}{|x-y|^\alpha} \quad \text{for } x, y \in \Omega,\ x \neq y$$

and $l_\beta(f)(x, x) = 0$ for $x \in \Omega$. By definition of the space $\lambda^{k,\alpha}(\Omega)$ and of its topology, each $l_\beta(f)$ is continuous on $\Omega \times \Omega$ and if we endow $C(\Omega \times \Omega)$ with the compact-open topology, the linear map

$$\Psi : \lambda^{k,\alpha}(\Omega) \longrightarrow C^k(\Omega) \times \prod_{|\beta|=k} C(\Omega \times \Omega), \quad \Psi(f) = (f, \{l_\beta(f);\ |\beta| = k\}),$$

is an injective topological homomorphism.

Now fix $v \in H' = \lambda^{k,\alpha}(\Omega)'$ and an absolutely convex bounded subset C of $\lambda^{k,\alpha}(\Omega)$, without loss of generality $C = B((M_n)_n) \cap \lambda^{k,\alpha}(\Omega)$ for a sequence $(M_n)_n$ of positive numbers. By the preceding remark and the Hahn-Banach theorem, there are $u \in C^k(\Omega)'$ and for each β with $|\beta| = k$ a measure μ_β with compact support in $\Omega \times \Omega$ such that

$$v(f) = u(f) + \sum_{|\beta|=k} \int_{\Omega\times\Omega} l_\beta(f) d\mu_\beta, \quad f \in \lambda^{k,\alpha}(\Omega).$$

To verify that $v_{|C}$ is $\tilde{\tau}$-continuous (at 0), take a sequence $(f_j)_{j\in\mathbb{N}} \subset C$ with $\tilde{\tau}$-$\lim_j f_j = 0$. Clearly $\lim_j u(f_j) = 0$ and for arbitrary $|\beta| = k$, there is $n \in \mathbb{N}$ with $\operatorname{supp}\mu_\beta \subset K_n \times K_n$. We have

$$\sup_{j\in\mathbb{N}} \sup_{(x,y)\in K_n\times K_n} |l_\beta(f_j)(x,y)| \leq \sup_{j\in\mathbb{N}} q_{K_n}(f_j) \leq \sup_{j\in\mathbb{N}} p_n(f_j) \leq M_n$$

and $\lim_j l_\beta(f_j) = 0$ pointwise on $K_n \times K_n$. Applying Lebesgue's dominated convergence theorem, we conclude

$$\lim_{j\to\infty} \int_{\Omega\times\Omega} l_\beta(f_j) d\mu_\beta = 0, \quad \text{hence } \lim_{j\to\infty} v(f_j) = 0,$$

as desired. □

D. Other examples in distribution theory, infinite dimensional holomorphy and Köthe sequence spaces

a) $\check{\mathcal{B}}(\Omega) = \mathcal{B}_0(\Omega)''_b$

$\mathcal{B} = \mathcal{B}(\mathbb{R}^N)$ denotes the space of the C^∞-functions on $\mathbb{R}^N$ which, together with all their derivatives, are uniformly bounded on $\mathbb{R}^N$, endowed with the Fréchet topology given by the seminorms

$$p_n(f) := \sup_{|\alpha|\leq n} \sup_{x\in\mathbb{R}^N} |D^\alpha f(x)|, \quad f \in \mathcal{B},$$

$n = 1, 2, \ldots$ For an open set $\Omega \subset \mathbb{R}^N$, we consider

$$\begin{aligned}\check{\mathcal{B}}(\Omega) &:= \{f \in \mathcal{B}(\mathbb{R}^N);\ D^\alpha f(x) = 0 \text{ for all } x \in \mathbb{R}^N \setminus \Omega \text{ and all multi-indices } \alpha\},\\ \mathcal{B}_0(\Omega) &:= \{f \in C^\infty(\Omega);\ \text{for each } \alpha \text{ and each } \varepsilon > 0 \text{ there is a compact set } K \subset \Omega \text{ with } \\ &\qquad |D^\alpha f(x)| < \varepsilon \text{ for all } x \in \Omega \setminus K\}.\end{aligned}$$

Each $f \in \mathcal{B}_0(\Omega)$ can be extended as a C^∞-function to $\mathbb{R}^N$ by putting $D^\alpha f \equiv 0$ on $\mathbb{R}^N \setminus \Omega$ for all α; then $\mathcal{B}_0(\Omega)$ is a closed subspace of the Fréchet space $\check{\mathcal{B}}(\Omega)$, with the topology τ induced by $\mathcal{B}$. If $\tilde{\tau}$ is the restriction of the topology of $C^\infty(\mathbb{R}^N)$ (i.e. uniform convergence of the functions and all their derivatives on compact subsets of $\mathbb{R}^N$), then $\check{\mathcal{B}}(\Omega)$ clearly satisfies (CNC) and since $\tilde{\tau}$ is semi-Montel, (BBC) holds, too. (See our remark after the proof of 1./2.)

According to Valdivia [29, p. 428, (1)], for each bounded $B \subset \check{\mathcal{B}}(\Omega)$ there is an absolutely convex bounded subset C of $\mathcal{B}_0(\Omega)$ such that B is contained in the $\tilde{\tau}$-closure of C.

Moreover, if $T \in \mathcal{B}_0(\Omega)'$, we can find $n \in I\!N$ and a family $(\mu_\alpha)_{|\alpha|\leq n}$ of bounded Radon measures on Ω such that

$$T(f) = \sum_{|\alpha|\leq n} (-1)^{|\alpha|} \int_\Omega D^\alpha f \, d\mu_\alpha, \quad f \in \mathcal{B}_0(\Omega);$$

see Horváth [18, Corollary on p. 347]. Hence the restriction of T to each bounded set in $\mathcal{B}_0(\Omega)$ is $\tilde{\tau}$-continuous.

It now follows from Theorem 6. that $\check{\mathcal{B}}(\Omega)$ *is canonically the strong bidual of* $\mathcal{B}_0(\Omega)$. This result is known and for $\Omega \neq I\!R^N$ due to P. Dierolf, J. Voigt [10, Theorem 4.8]; another approach can be found in Valdivia's book [29, Ch. III, §1, 21.].

b) Holomorphic functions of bounded type

Let E be a complex Banach space and U a balanced open subset of E. For $U \neq E$, $B \subset U$ is called *U-bounded* if it is bounded and has positive distance from $E \setminus U$; for $U = E$ the U-bounded sets are exactly the bounded subsets. $H_b(U)$ denotes the space of the holomorphic functions on U which are bounded on all U-bounded sets, equipped with the topology τ of uniform convergence on these sets; the elements of $H_b(U)$ are called *holomorphic functions of bounded type*.

By Isidro [19] and Ansemil, Ponte [1], the Fréchet space $H_b(U)$ is quasinormable. If $\tilde{\tau}$ denotes the compact-open topology, $\tilde{\tau}$ is semi-Montel and $H_b(U)$ clearly satisfies (CNC) and (BBC). It follows directly from Corollary 5.(d) that $H_b(U)$ is the strong dual of a boundedly retractive (LB)-space. This was first established by Galindo, García, Maestre [14, Prop. 2] in a more involved way. Another (interesting) representation of the (LB)-predual of $H_b(U)$ was then obtained by Mujica [24, Prop. 7.3].

Recently Galindo, García, Maestre [15, Prop. 16] have utilized our Corollary 2.(b) to prove the following: *If E is a bornological (or just quasibarrelled) (DF)-space and $H_b(E)$ the space of the holomorphic functions on E which are bounded on the bounded subsets of E, endowed with the topology of uniform convergence on the bounded sets, then $H_b(E)$ is a Fréchet space which is the strong dual of a complete barrelled (DF)-space.*

c) Köthe echelon and co-echelon spaces

Corollary 5.(b) can be applied to show that for any Köthe matrix A on an index set I, the echelon space $\lambda_1 = \lambda_1(I, A)$ (of order 1) is the strong dual of a complete barrelled (DF)-space. This should be compared with Valdivia [29, Ch. II, §2, 1.] and with the representation of λ_1 as the strong dual of the co-echelon space $k_0 = k_0(I, V)$, $V = (1/a_n)$, of order 0. (See [7, 2.8.(d)]; but note that the (LB)-space k_0 need not be complete in general.)

Theorem 6. can be used to obtain the biduality $\lambda_\infty = (\lambda_0)''_b$, and it also follows from Theorem 7. that $((k_0)'_b)'_i = k_\infty$. (In all these cases, $\tilde{\tau}$ is the topology of pointwise convergence on I. See [3] for a necessary and sufficient condition for $(k_0)''_b = k_\infty$.)

References

[1] J. M. Ansemil, S. Ponte, An example of a quasi-normable Fréchet function space which is not a Schwartz space, pp. 1 – 8 in *"Advances in Functional Analysis, Holomorphy and Approximation Theory"*, Lecture Notes in Math. **843**, Springer-Verlag (1981).

[2] K. D. Bierstedt, An introduction to locally convex inductive limits, pp. 35 – 133 in *"Functional Analysis and its Applications"*, World Scientific (1988).

[3] K. D. Bierstedt, J. Bonet, Stefan Heinrich's density condition for Fréchet spaces and the characterization of the distinguished Köthe echelon spaces, *Math. Nachr.* **135** (1988), 149 – 180.

[4] K. D. Bierstedt, J. Bonet, Dual density conditions in (DF)-spaces, I., *Results in Math.* **14** (1988), 242 – 274.

[5] K. D. Bierstedt, R. Meise, Distinguished echelon spaces and the projective description of weighted inductive limits of type $\mathcal{V}_d\mathcal{C}(X)$, pp. 169 – 229 in *"Aspects of Mathematics and its Applications"*, North-Holland Math. Library (1986).

[6] K. D. Bierstedt, R. Meise, W. H. Summers, A projective description of weighted inductive limits, *Transact. Amer. Math. Soc.* **272** (1982), 107 – 160.

[7] K. D. Bierstedt, R. G. Meise, W. H. Summers, Köthe sets and Köthe sequence spaces, pp. 27 – 91 in *"Functional Analysis, Holomorphy and Approximation Theory"*, North-Holland Math. Studies **71** (1982).

[8] K. D. Bierstedt, W. H. Summers, Biduals of weighted Banach spaces of analytic functions, to appear.

[9] J. Bonet, A question of Valdivia on quasinormable Fréchet spaces, to appear in *Canadian Math. Bull.*

[10] P. Dierolf, J. Voigt, Calculation of the bidual for some function spaces. Integrable distributions, *Math. Ann.* **253** (1980), 63 – 87.

[11] S. Dineen, *Complex Analysis in Locally Convex Spaces*, North-Holland Math. Studies **57** (1981).

[12] J. Frampton, A. J. Tromba, On the classification of spaces of Hölder continuous functions, *J. Funct. Anal.* **10**, 336 – 345 (1972).

[13] P. Galindo, On vector-valued Lipschitz function spaces, *Canadian Math. Bull.* **30** (1987), 43 – 48.

[14] P. Galindo, D. García, M. Maestre, Holomorphic mappings of bounded type, to appear in *J. Math. Anal. Appl.*

[15] P. Galindo, D. García, M. Maestre, Holomorphic mappings of bounded type on (DF)-spaces, these Proceedings.

[16] G. Glaeser, Etude de quelques algèbres tayloriennes, *J. Analyse Math.* **6** (1958), 1 - 125.

[17] A. Grothendieck, Produits tensoriels topologiques et espaces nucléaires, *Mem. Amer. Math. Soc.* **16** (1955).

[18] J. Horváth, *Topological Vector Spaces and Distributions I*, Addison-Wesley Publ. Co. (1966).

[19] J. M. Isidro, On the distinguished character of the function spaces of holomorphic functions of bounded type, *J. Funct. Anal.* **38** (1980), 139 - 145.

[20] H. Jarchow, *Locally Convex Spaces*, Teubner-Verlag, Stuttgart (1981).

[21] W. Kaballo, Lifting theorems for vector valued functions and the ε-tensor product, pp. 149 - 165 in *"Functional Analysis: Surveys and Recent Results" (I)*, North-Holland Math. Studies **27** (1977).

[22] G. Köthe, *Topological Vector Spaces I*, Springer-Verlag (1969).

[23] J. Mujica, A completeness criterion for inductive limits of Banach spaces, pp. 319 - 329 in *"Functional Analysis, Holomorphy and Approximation Theory II"*, North-Holland Math. Studies **86** (1984).

[24] J. Mujica, Linearization of holomorphic mappings of bounded type, these Proceedings.

[25] J. Mujica, L. Nachbin, Linearization of holomorphic mappings on locally convex spaces, to appear in *J. Math. Pures Appl.*

[26] K.-F. Ng, On a theorem of Dixmier, *Math. Scand.* **29** (1971), 279 - 280.

[27] P. Pérez Carreras, J. Bonet, *Barrelled Locally Convex Spaces*, North-Holland Math. Studies **131** (1987).

[28] W. H. Summers, Dual spaces of weighted spaces, *Transact. Amer. Math. Soc.* **151** (1970), 323 - 333.

[29] M. Valdivia, *Topics in Locally Convex Spaces*, North-Holland Math. Studies **67** (1982).

[30] L. Waelbroeck, Duality and the injective tensor product, *Math. Ann.* **163** (1966), 122 - 126.

Authors' addresses:

Klaus D. Bierstedt
FB 17, Mathematik
Universität-GH-Paderborn
Warburger Str. 100
Postfach 1621
D-4790 Paderborn, Germany

José Bonet
Departamento de Matemática Aplicada
E. T. S. Arquitectura
Universidad Politécnica de Valencia
Camino de Vera
E-46071 Valencia, Spain

Progress in Functional Analysis
K.D. Bierstedt, J. Bonet, J. Horváth & M. Maestre (Eds.)

Holomorphic Mappings of Bounded Type on (DF)-Spaces

Pablo Galindo, Domingo García and Manuel Maestre *

Departamento de Análisis Matemático, Facultad de Matemáticas, Universidad de Valencia, Doctor Moliner 50, 46100 Burjasot (Valencia), Spain.

Dedicated to our Teacher, Professor Manuel Valdivia, on the occasion of his sixtieth birthday

Abstract

We study the holomorphic functions of bounded type defined on (DF)-spaces. We prove that they are of uniformly bounded type. The space of all these functions is a Fréchet space with its natural topology. Some consequences and related results are obtained.

In this paper we prove that the classical result of Grothendieck [10] which states that on a (DF)-space any linear mapping whose restriction to each bounded subset is (uniformly) continuous is necessarily continuous remains valid for G-holomorphic mappings. From this starting point we show that the space $\mathcal{H}_b(E)$ of all holomorphic functions defined on a (DF)-space E which are bounded on bounded sets endowed with the topology of the uniform convergence on bounded sets is a Fréchet space.

We also prove that every $f \in \mathcal{H}_b(E)$ (E being a (DF)-space) is, as in the trivial case of normed spaces, of uniformly bounded type, that is, there is a convex balanced 0-neighbourhood V in E such that $f \in \mathcal{H}_b(E_V)$. Furthermore $\mathcal{H}_b(E)$ coincides with the sequentially retractive inductive limit $\lim_V \mathcal{H}_b(E_V)$. As an application we prove that every $f \in \mathcal{H}_b(E)$ can be extended to an holomorphic function of bounded type on E'', where E'' is the bidual of E. This generalizes a result of Aron-Berner [2].

*Supported in part by DGICYT pr. no. P.S.88/0144. The third author is also supported in part by DGICYT pr. no. P.S.88/0050.

In the special case of a countable inductive limit E of normed spaces the existence of a predual of $\mathcal{H}_b(E)$ is shown.

The following properties a) and b) are two properties of the strong dual of a metrizable locally convex space; these were used by Grothendieck to define the class of (DF)-spaces:

A locally convex space E is said to be a (DF)-space if a) it has a fundamental sequence of bounded sets and b) if every subset of E that is the intersection of countably many closed convex balanced 0-neighbourhoods and that absorbs every bounded set is a 0-neighbourhood.

We refer to [13] for the general theory of locally convex spaces and to [7] for the general theory of infinite dimensional holomorphy. If E and F are locally convex spaces, A is a subset of E and $\mathcal{F}$ is a family of F-valued mappings defined on A, we will call (in a natural way) $\mathcal{F}$ *equiuniformly continuous on* A if for each 0-neighbourhood W in F, there exists a 0-neighbourhood V in E such that if $x, y \in A$ and $x - y \in V$ then $f(x) - f(y) \in W$ for every $f \in \mathcal{F}$. As usual, if $f : A \subseteq E \rightarrow F$, we will denote $\| f \|_A := \sup\{\| f(x) \| : x \in A\}$.

We will denote by $\mathcal{H}_b(E; F)$ the space of all holomorphic mappings from E into F which are bounded on the bounded subsets of E, endowed with the topology τ_b of the uniform convergence on bounded subsets of E. Properties about these spaces when E is a Banach space have been studied in [1], [3], [9], [11] and [12].

We begin with an auxiliary result.

1.Lemma.- *Let E be a locally convex space and let C, D, U and W be convex balanced subsets of E such that $2W \subset U$. Then*

$$(C + \frac{1}{2} U) \cap (D + W) \subset (C + U) \cap D + W.$$

Proof:

Let x belong to $(C + \frac{1}{2} U) \cap (D + W)$ then there are $c \in C$, $d \in D$, $u \in U$ and $w \in W$ such that $x = c + \frac{1}{2}u$ and $x = d + w$. Since $x = (x - w) + w$ and $x - w = d \in D$, we only need to check that $x - w \in C + U$, but this is obvious because

$$x - w = c + \frac{1}{2}u - w \in C + \frac{1}{2} U - W \subset C + \frac{1}{2} U + \frac{1}{2} U = C + U. \blacksquare$$

2.Proposition.- *Let E be a (DF)-space and let $(A_n)_{n\in\mathbf{N}}$ be an increasing fundamental sequence of bounded convex balanced sets in E. If F is a Banach space and $\mathcal{F}$ is a family of mappings defined on E with values in F such that for each bounded subset B of E the family $\{f_{|B} : f \in \mathcal{F}\}$ is equiuniformly continuous, then there is a 0-neighbourhood Z in E such that $A_1 \subset Z$ and*

$$\| f \|_Z \leq \| f \|_{A_n} + \frac{1}{2^n} \quad \text{for } n = 1,2,.. \text{ and } f \in \mathcal{F}.$$

In particular $\mathcal{F} \subset \mathcal{H}_b(E;F)$, if all mappings in $\mathcal{F}$ are G-holomorphic.
Proof:

By assumption $\{f_{|A_n} : f \in \mathcal{F}\}$ is an equiuniformly continuous family on A_n, so there exists a convex balanced 0-neighbourhood V_{n-1} in E such that if $x, y \in A_n$ and $x - y \in V_{n-1}$ then $\| f(x) - f(y) \| < 1/2^n$ for $n = 1,2,..$ and $f \in \mathcal{F}$. Obviously we can suppose $V_{n+1} \subset V_n$, $n = 1,2,..$.

We define $B_1 := A_1$ and $B_n := (B_{n-1} + V_{n-1}) \cap A_n$ for $n \geq 2$. $(B_n)_{n\in\mathbf{N}}$ is an increasing sequence of subsets of E. Since every f is uniformly continuous on the convex balanced set A_n we have that f is bounded on A_n for each n.

First of all, we prove that

$$\| f \|_{B_n} \leq \| f \|_{B_{n-1}} + \frac{1}{2^n} \text{ for } n = 2,3,.. \text{ and } f \in \mathcal{F}. \tag{1}$$

Indeed, given $x \in B_n$ there exist $b_{n-1} \in B_{n-1}$ and $v_{n-1} \in V_{n-1}$ such that $x = b_{n-1} + v_{n-1}$ then $x - b_{n-1} = v_{n-1} \in V_{n-1}$, $x \in A_n$ and $b_{n-1} \in B_{n-1} \subset A_{n-1} \subset A_n$. Thus

$$\| f(x) - f(b_{n-1}) \| < \frac{1}{2^n}$$

hence

$$\| f(x) \| \leq \| f(b_{n-1}) \| + \| f(x) - f(b_{n-1}) \| \leq \| f \|_{B_{n-1}} + \frac{1}{2^n}.$$

Now we put

$$Z := \bigcap_{n=1}^{\infty} (A_n + \frac{1}{2} V_n).$$

It is clear from the definition of (DF)-space that Z is a 0-neighbourhood in E. We claim that $Z \subset \bigcup_{n=p}^{\infty} B_n$ for every $p \in \mathbf{N}$.

Let us show by induction that

$$\bigcap_{k=1}^{n} (A_k + \frac{1}{2} V_k) \subset B_n + \frac{1}{2} V_n \quad \text{for } n = 1,2,.. \tag{2}$$

If $n = 1$, then $A_1 + \frac{1}{2} V_1 = B_1 + \frac{1}{2} V_1$.
Given $n \geq 2$, if we suppose that (2) is true for $n - 1$, then we have

$$\bigcap_{k=1}^{n}(A_k + \frac{1}{2} V_k) = (\bigcap_{k=1}^{n-1}(A_k + \frac{1}{2} V_k)) \cap (A_n + \frac{1}{2} V_n) \subset$$

$$(B_{n-1} + \frac{1}{2} V_{n-1}) \cap (A_n + \frac{1}{2} V_n).$$

Now, if we apply Lemma 1 by taking $C = B_{n-1}$, $U = V_{n-1}$, $D = A_n$ and $W = \frac{1}{2} V_n$, we have

$$\bigcap_{k=1}^{n}(A_k + \frac{1}{2} V_k) \subset ((B_{n-1} + V_{n-1}) \cap A_n) + \frac{1}{2} V_n = B_n + \frac{1}{2} V_n.$$

Therefore (2) is true.

Now the claim follows from

$$Z = Z \cap E = Z \cap (\bigcup_{n=p}^{\infty} A_n) \subset \bigcup_{n=p}^{\infty}(A_n \cap (\bigcap_{k=1}^{n-1}(A_k + \frac{1}{2} V_k))) \subset$$

$$\bigcup_{n=p}^{\infty}(A_n \cap (B_{n-1} + \frac{1}{2} V_{n-1})) \subset \bigcup_{n=p}^{\infty} B_n.$$

For each $z \in Z$ there exists $m \in \mathbf{N}$ such that $z \in B_m$, hence according to (1),

$$\| f(z) \| \leq \| f \|_{B_{m-1}} + \frac{1}{2^m},$$

and applying a trivial induction

$$\| f(z) \| \leq \| f \|_{B_p} + \frac{1}{2^{p+1}} + \ldots + \frac{1}{2^m} < \| f \|_{B_p} + \frac{1}{2^p}.$$

Thus

$$\| f \|_Z \leq \| f \|_{A_p} + \frac{1}{2^p} \quad \text{for } p = 1, 2, \ldots \text{ and } f \in \mathcal{F},$$

and the conclusion follows. ∎

This proposition extends the classical result of Grothendieck mentioned at the very beginning in the following way: "If f is a mapping from a (DF)-space E into a Banach space F whose restriction to each bounded subset of E is uniformly continuous then f is continuous on E." (Given $\epsilon > 0$, choose $m \in \mathbf{N}$ such that $1/2^m < \epsilon$ and a fundamental sequence $(A_n)_{n \in \mathbf{N}}$ of bounded convex balanced subsets of E with $A_1 = \ldots = A_m = \{0\}$.

Now, apply Proposition 2 to the family $\mathcal{F} := \{g\}$ where the mapping g is defined by $g(x) := f(x + x_o) - f(x_o)$ for all $x \in E$ and $n = m$.) This points out the central role of the uniform continuity. As a consequence we obtain the following:

3.Theorem.- *Let E be a (DF)-space and let F be a Banach space. A G-holomorphic mapping $f : E \rightarrow F$ belongs to $\mathcal{H}_b(E; F)$ if and only if $f_{|B}$ is uniformly continuous on B for each bounded subset B of E .*

Proof:

The sufficient condition is a particular case of Proposition 2. Conversely, given $f \in \mathcal{H}_b(E; F)$, its Taylor series expansion at the origin $\sum_{m=0}^{\infty} \frac{\hat{d}^m f(0)}{m!}$ converges absolutely and uniformly on bounded sets, from which the conclusion follows bearing in mind the well-known fact that continuous homogeneous polynomials are uniformly continuous on bounded sets (see the proof of Prop. 1.14 in [7]). ∎

4.Theorem.- *Let E be a (DF)-space and let F be a Banach space. Then $\mathcal{H}_b(E; F)$ is a Fréchet space.*

Proof:

Consider a Cauchy sequence $(f_n)_{n\in\mathbf{N}}$ in $\mathcal{H}_b(E; F)$. Since for any convex balanced subset B of E, $(f_{n|B})_{n\in\mathbf{N}}$ is a Cauchy sequence in the Banach space $(C^*(B, F), \| \cdot \|_B)$ of all bounded continuous mappings from B into F, the mapping f defined by $f(z):= \lim_n f_n(z)$ for all $z \in E$ is well defined and $(f_{n|B})_{n\in\mathbf{N}}$ converges uniformly on B to $f_{|B}$. Thus $f_{|B} \in C^*(B, F)$.

By applying Theorem 3 we have that $f_{n|B}$ is uniformly continuous on B for every $n \in \mathbf{N}$ and since $f_{|B}$ is the uniform limit on B of $(f_{n|B})_{n\in\mathbf{N}}$, then $f_{|B}$ is G-holomorphic and the conclusion follows from Proposition 2. ∎

5.Remark.- A locally convex space E is said to be a (gDF)-space if (i) there exists a fundamental sequence of bounded sets in E, and (ii) given a fundamental sequence $(A_n)_{n\in\mathbf{N}}$ of bounded convex balanced sets in E and an arbitrary sequence $(V_n)_{n\in\mathbf{N}}$ of convex balanced 0-neighbourhoods in E, then the set

$$\bigcap_{n=1}^{+\infty} (A_n + V_n)$$

is a 0-neighbourhood in E (see [13]).

It is clear that every (DF)-space is a (gDF)-space.

The reader may observe that only properties (i) and (ii) are used in the proofs of

Proposition 2, Theorem 3 and Theorem 4. Therefore all these results hold for (gDF)-spaces as well. For instance we have the following:

6.Theorem.- *Let E be a (gDF)-space and let F be a Banach space. Then $\mathcal{H}_b(E;F)$ is a Fréchet space.*

We recall that a mapping $f \in \mathcal{H}(E;F)$ (E, F locally convex spaces) is said to be of uniformly bounded type if there exist a convex balanced 0-neighbourhood V in E, a bounded convex balanced set B in F and a mapping $\hat{f} \in \mathcal{H}_b(E_V;F_B)$ such that the following diagram is commutative:

$$\begin{array}{ccc} E & \xrightarrow{f} & F \\ \downarrow{\scriptstyle \pi_V} & & \uparrow \\ E_V & \xrightarrow{\hat{f}} & F_B \end{array}$$

where $E_V := (E/Kerp_V,\ p_V)$ and $F_B := (span(B),\ p_B)$, p_V and p_B being the Minkowski gauges of V and B, respectively, and π_V is the canonical projection from E into E_V.

The space $\mathcal{H}_{ub}(E;F)$ of all holomorphic mappings of uniformly bounded type will be endowed with the locally convex inductive topology given by

$$\mathcal{H}_{ub}(E;F) := \lim_{V,B\rightarrow} \mathcal{H}_b(E_V;F_B).$$

7.Proposition.- *If E is a* (DF)*-space and F is a Banach space, then $\mathcal{H}_b(E;F) = \mathcal{H}_{ub}(E;F)$ algebraically and topologically.*

Proof:

The inclusion mapping $\mathcal{H}_{ub}(E;F) \rightarrow \mathcal{H}_b(E;F)$ is obviously continuous. By our assumptions on E and F, the space $\mathcal{H}_b(E;F)$ is metrizable, hence bornological and so to complete the proof it is enough to show that each convergent sequence $(f_j)_{j\in\mathbf{N}}$ in $\mathcal{H}_b(E;F)$ is contained and bounded in $\mathcal{H}_{ub}(E;F)$.

Consider a fundamental sequence $(A_n)_{n\in\mathbf{N}}$ of bounded convex balanced closed subsets of E such that $A_n \subseteq \frac{1}{n+1} A_{n+1}$ for all $n \in \mathbf{N}$.

Since $(f_j)_{j\in\mathbf{N}}$ is uniformly Cauchy on bounded sets, we have that it is equiuniformly continuous on bounded sets. So Proposition 2 can be applied for every $n \in \mathbf{N}$ to obtain a convex balanced closed 0-neighbourhood Z_n in E such that

$$\sup_{j\in\mathbf{N}} \| f_j \|_{Z_n} \leq \sup_{j\in\mathbf{N}} \| f_j \|_{A_{n+1}} + \frac{1}{2} \quad \text{and } 2A_n \subset Z_n.$$

In addition, $(f_j)_{j\in\mathbf{N}}$ is bounded on bounded sets, so we have

$$\sup_{j\in\mathbf{N}} \| f_j \|_{Z_n} < \infty$$

for all $n \in \mathbf{N}$. Set

$$V := \bigcap_{n=1}^{\infty} \frac{1}{n}(A_n + \frac{1}{2} Z_n).$$

Since E is a (DF)-space, V is a 0-neighbourhood. Moreover, for every $n \in \mathbf{N}$ we have $nV \subseteq A_n + \frac{1}{2} Z_n \subseteq Z_n$, which implies that $(f_j)_{j\in\mathbf{N}}$ is contained and bounded in $\mathcal{H}_b(E_V; F)$. ∎

The algebraic identity in the above proposition leads to the following result about the extension of holomorphic mappings of bounded type defined on linear topological subspaces of entire mappings of bounded type. This result is due to Meise and Vogt [15] for (DFM)-spaces. Their proof also works in our case.

We recall that a locally convex space E is hilbertizable if E admits a fundamental system of seminorms which are induced by semi-inner products.

8.Proposition.- *Let E be a hilbertizable locally convex space. Let G be a Banach space and let F be a subspace of E which is a* (DF)*-space. Then every $f \in \mathcal{H}_b(F;G)$ has an extension $g \in \mathcal{H}_b(E;G)$.*

9.Remark.- If E is a (DF)-space and $f \in \mathcal{H}_b(E)$ then there exists an extension $\hat{f} \in \mathcal{H}_b(\hat{E})$ where $\hat{E}$ is the completion of E. Indeed, let $\sum_{m=0}^{\infty} \frac{\hat{d}^m f(0)}{m!}$ be the Taylor series expansion at 0 of f. Recalling that each bounded subset of $\hat{E}$ is contained in the $\hat{E}$-closure of some bounded subset of E ([19] 8.3.18) and that $\frac{\hat{d}^m f(0)}{m!}$ is uniformly continuous on the bounded subsets of E for all $m \in \mathbf{N}$, one can observe that every $\frac{\hat{d}^m f(0)}{m!}$ can be extended to an m-homogeneous polynomial P_m on $\hat{E}$ which is uniformly continuous on the bounded subsets of $\hat{E}$. Now, Theorem 3 implies that $P_m \in \mathcal{P}(^m\hat{E})$ for all $m \in \mathbf{N}$. Furthermore, for a bounded set B in $\hat{E}$ we have

$$2^m \| P_m \|_B = \| P_m \|_{2B} \le \| \frac{\hat{d}^m f(0)}{m!} \|_A \le \| f \|_A,$$

where A is a bounded balanced subset of E with $2B \subset \overline{A}^{\hat{E}}$. This implies, due to the completeness of $\mathcal{H}_b(\hat{E})$ (Thm. 6) , that the series $\sum_{m=0}^{\infty} P_m$ converges in $\mathcal{H}_b(\hat{E})$ to a function $\hat{f} \in \mathcal{H}_b(\hat{E})$ which is an extension of f. ∎

10.Proposition.- *If E is a* (DF)*-space and E'' is its bidual then for each $f \in \mathcal{H}_b(E)$ there exists $j \in \mathcal{H}_b(E'')$ such that the restriction of j to E coincides with f.*

Proof:

We know by Proposition 7 that $\mathcal{H}_b(E) = \mathcal{H}_{ub}(E)$, i.e. there exists a convex balanced 0-neighbourhood V in E such that f is bounded on nV for $n = 1, 2, \ldots$ Thus f factors through E_V, that is, there exists $g \in \mathcal{H}_b(E_V)$ such that $f = g \circ \pi_V$. Since E_V is a normed space, by the above remark we can extend g to a holomorphic function $\hat{g}$ of bounded type on the completion $\hat{E}_V$ of E_V. By applying the Aron-Berner Theorem [2] there exists $h \in \mathcal{H}_b(\hat{E}_V'')$ such that $h_{|\hat{E}_V} = \hat{g}$. If we denote by i the natural injection of E_V into its completion $\hat{E}_V$ we have $i \circ \pi_V : E \rightarrow \hat{E}_V$. Thus the second adjoint of $i \circ \pi_V$, $(i \circ \pi_V)^{tt}$, is a continuous linear mapping from E'' into the Banach space E_V''. Thus

$$h \circ (i \circ \pi_V)^{tt} : \quad E'' \rightarrow \mathbf{C}$$

is a holomorphic function, j, of bounded type on E'' and it is clear that its restriction to E coincides with f. ∎

11.Proposition.- *If E is a* (DF)*-space and F is a Banach space, then the inductive limit $\mathcal{H}_b(E;F) = \lim_V \mathcal{H}_b(E_V;F)$ is sequentially retractive.*

Proof:

Let $(f_j)_{j \in \mathbf{N}}$ be a sequence in $\mathcal{H}_b(E;F)$ converging to 0. Consider a fundamental sequence $(A_n)_{n \in \mathbf{N}}$ of closed bounded convex balanced subsets of E such that $A_n \subset \frac{1}{n+1} A_{n+1}$ for every $n \in \mathbf{N}$.

By applying Proposition 2, we get for every $n \in \mathbf{N}$ a closed convex balanced 0-neighbourhood Z_n in E with $2A_n \subseteq Z_n$ such that

$$\sup_{j \in \mathbf{N}} \| \; f_j \; \|_{Z_n} \leq \sup_{j \in \mathbf{N}} \| \; f_j \; \|_{A_m} + \frac{1}{2^{m-n}} \quad \text{for all } m \geq n+1.$$

Therefore, $(\| \; f_j \; \|_{Z_n})_{j \in \mathbf{N}}$ converges to 0 for each $n \in \mathbf{N}$.

Since E is a (DF)-space the set $V := \bigcap_{n=1}^{\infty} \frac{1}{n}(A_n + \frac{1}{n} Z_n)$ is a 0-neighbourhood, and moreover for every $n \in \mathbf{N}$ we have $nV \subset A_n + \frac{1}{2} Z_n \subseteq Z_n$, which implies that $(f_j)_{j \in \mathbf{N}}$ is contained and converges to 0 in $\mathcal{H}_b(E_V;F)$. ∎

12.Remark.- The reader may observe again that only properties (i) and (ii) of the definition of (gDF)-space (see Remark 5) are used in the proofs of Remark 9 and Propositions 7, 10 and 11. Therefore all these results remain true for (gDF)-spaces.

Since every (gDF)-space satisfies the countable neighbourhood property ([19] 8.3.5), Propositions 7 and 11 also hold when F is a metrizable space and Proposition 8 also holds when G is a Fréchet space (see [15] 3.3 and [9] Lemma 2).

The dual of a Fréchet space endowed with the topology of the uniform convergence on the compact sets is called a (DFC)-space. The (gDF)-spaces which are semi-Montel are precisely the (DFC)-spaces ([19] 8.3.49).

The equality $\mathcal{H}(E) = \mathcal{H}_{ub}(E)$ has been studied by several authors. Colombeau and Mujica [5] proved it for (DFM)-spaces. In a previous paper [9] we extended their result to (DFC)-spaces. Meise and Vogt [15] found necessary and sufficient conditions for it in the setting of Fréchet nuclear spaces. Recently, Lindström [14] has shown that for a quasinormable space E the mentioned equality implies that E is a Schwartz space.

13.Remark.- *If E is a complete* (gDF)*-space, then $\mathcal{H}(E) = \mathcal{H}_b(E)$ if and only if E is a* (DFC)*-space. In this case $\mathcal{H}(E) = \mathcal{H}_{ub}(E)$.*

Proof:

Suppose $\mathcal{H}(E) = \mathcal{H}_b(E)$ ($= \mathcal{H}_{ub}(E)$ by Proposition 7). Since every (gDF)-space is quasinormable ([13] 12.4.7), by ([14] Cor. 6) E has to be a Schwartz space. Consequently ([13] 10.4) E is a semi-Montel space, hence a (DFC)-space ([19] 8.3.49).

Conversely, we suppose E is a (DFC)-space. Since E is a semi-Montel space, then $\mathcal{H}(E) = \mathcal{H}_b(E)$. The conclusion follows from Proposition 7.

For the case of a (DF)-space the above result is essentially due to Lindström [14].

14.Remark.- The equality $\mathcal{H}(E) = \mathcal{H}_{ub}(E)$ can also be characterized within the class of spaces of complex continuous functions. In fact, if X is a completely regular Hausdorff topological space and $E = C(X)$ (which we suppose endowed with the compact open topology), we have that $\mathcal{H}(E) = \mathcal{H}_{ub}(E)$ if and only if $C(X)$ is a Schwartz space (i.e., the compact sets in X are finite ([13] 10.8.1)). In order to check this, we first observe that $\mathcal{H}_b(C(X)) = \mathcal{H}_{ub}(C(X))$ for any X. Indeed, if $f \in \mathcal{H}(C(X))$, there is a compact subset K of X such that f factors through $C(K)$, i.e., there is $g \in \mathcal{H}(C(K))$ such that $f = g \circ \pi_K$ where π_K denotes the restriction to K (see [7] 2.26). Since by the Tietze-Urysohn Theorem for each bounded subset B of $C(K)$ there is a bounded set $A \subset C(X)$ such that $\pi_K(A) = B$, it follows that $g \in \mathcal{H}_b(C(K))$ whenever $f \in \mathcal{H}_b(C(X))$ and therefore $f \in \mathcal{H}_{ub}(C(X))$.

Now, since every $C(X)$ is a quasinormable space ([13] 10.8.2), the equality $\mathcal{H}(C(X))$

$= \mathcal{H}_{ub}(C(X))$ implies that $C(X)$ is Schwartz space by [14] Cor. 6. The reverse implication follows immediately from the fact that every $f \in \mathcal{H}(C(X))$ factors through $C(K)$ for some compact $K \subset X$ and that $C(K)$ is finite dimensional.

Now we are going to consider a special class inside the (DF)-spaces, the spaces (LN), i.e., the countable inductive limits of normed spaces [19]. It is well-known that a locally convex space is an (LN)-space if and only if it is a bornological (DF)-space.

15.Proposition.- *Let E be an (LN)-space. Let $\mathcal{F}$ be a family of G-holomorphic mappings from E into a Banach space F. If the family $\mathcal{F}$ is bounded on the bounded subsets of E, then $\mathcal{F}$ is an equicontinuous subset of $\mathcal{H}_b(E;F)$.*

Proof:

Let $(A_n)_{n\in\mathbf{N}}$ be a fundamental sequence of bounded convex balanced subsets of E. Given $\epsilon > 0$ and $x_o \in E$ we shall prove that there exists a sequence $(\lambda_n)_{n\in\mathbf{N}}$ of positive real numbers such that

$$\| f(x) - f(x_o) \| \leq \epsilon \quad \text{for every } x \in x_o + \sum_{n=1}^{\infty} \lambda_n B_n \text{ and every } f \in \mathcal{F} .$$

We may suppose without loss of generality that $x_o = 0$ and $f(0) = 0$ for every $f \in \mathcal{F}$. (Otherwise we would consider $\mathcal{G} := \{g : g(x) = f(x + x_o) - f(x_o) , x \in E, f \in \mathcal{F} \}$.)

By assumption there exists $M > 0$ such that

$$\| f \|_{A_1} \leq M \text{ for every } f \in \mathcal{F}.$$

Applying the Cauchy inequalities, for any $0 < \lambda < 1$ we have that

$$\| f \|_{\lambda A_1} \leq \sum_{m=1}^{\infty} \| \frac{\hat{d}^m f(0)}{m!} \|_{\lambda A_1} = \sum_{m=1}^{\infty} \lambda^m \| \frac{\hat{d}^m f(0)}{m!} \|_{A_1}$$

$$\leq \frac{\lambda}{1-\lambda} \| f \|_{A_1} \leq \frac{\lambda M}{1-\lambda},$$

hence there exists $0 < \lambda_1 < 1$ such that

$$\| f \|_{\lambda_1 A_1} \leq \frac{\epsilon}{2} \quad \text{for every } f \in \mathcal{F}.$$

Suppose we have obtained $\lambda_1, \ldots, \lambda_k > 0$ such that

$$\| f \|_{C_k} \leq \epsilon \sum_{n=1}^{k} \frac{1}{2^n} \quad \text{for every } f \in \mathcal{F}, \tag{3}$$

where $C_k := \lambda_1 A_1 + \ldots + \lambda_k A_k$. By using the Cauchy inequalities it is easy to check that

$$\sum_{m=q+1}^{\infty} \| \frac{\hat{d}^m f(0)}{m!} \|_{C_k+\lambda A_{k+1}} \leq \frac{1}{2^{q-1}} \| f \|_{2C_k+2A_{k+1}}$$

for every $f \in \mathcal{F}$, $q \in \mathbf{N}$ and $0 < \lambda < 1$, so there exists $p \in \mathbf{N}$ such that

$$\sum_{m=p+1}^{\infty} \| \frac{\hat{d}^m f(0)}{m!} \|_{C_k+\lambda A_{k+1}} \leq \frac{\epsilon}{2^{k+3}} \quad \text{for every } f \in \mathcal{F} \text{ and every } 0 < \lambda \leq 1. \tag{4}$$

Moreover, a routine calculation leads to

$$\| \sum_{m=1}^{p} \frac{\hat{d}^m f(0)}{m!} \|_{C_k+\lambda A_{k+1}} \leq$$

$$\| \sum_{m=1}^{p} \frac{\hat{d}^m f(0)}{m!} \|_{C_k} + \lambda \sup_{x \in C_k,\ y \in A_{k+1}} \| \sum_{m=1}^{p} \sum_{j=1}^{m} \binom{m}{j} \frac{d^m f(0)}{m!} (x)^{m-j} (\lambda y)^{j-1} (y) \|,$$

where $\frac{d^m f(0)}{m!}$ is the symmetric m-linear mapping associated to $\frac{\hat{d}^m f(0)}{m!}$. By the polarization formula, $\frac{d^m f(0)}{m!}$ is bounded on the bounded subsets of E, hence there exists $Q > 0$ such that

$$\| \sum_{m=1}^{p} \frac{\hat{d}^m f(0)}{m!} \|_{C_k+\lambda A_{k+1}} \leq \| \sum_{m=1}^{p} \frac{\hat{d}^m f(0)}{m!} \|_{C_k} + \lambda\, Q$$

for every $f \in \mathcal{F}$ and $0 < \lambda < 1$.
Applying (3) and (4) we have that

$$\| \sum_{m=1}^{p} \frac{\hat{d}^m f(0)}{m!} \|_{C_k+\lambda A_{k+1}} \leq \epsilon \sum_{n=1}^{k} \frac{1}{2^n} + \frac{\epsilon}{2^{k+3}} + \lambda Q \quad \text{for every } f \in \mathcal{F}.$$

If we let $\lambda_{k+1} := \frac{\epsilon}{2^{k+3} Q}$ it follows that

$$\| f \|_{C_k+\lambda_{k+1} A_{k+1}} \leq \| f - \sum_{m=1}^{p} \frac{\hat{d}^m f(0)}{m!} \|_{C_k+\lambda_{k+1} A_{k+1}} +$$

$$\| \sum_{m=1}^{p} \frac{\hat{d}^m f(0)}{m!} \|_{C_k+\lambda_{k+1} A_{k+1}} \leq \epsilon \sum_{n=1}^{k+1} \frac{1}{2^n}$$

for every $f \in \mathcal{F}$. Therefore $Z := \sum_{n=1}^{\infty} \lambda_n A_n$ is a 0-neighbourhood of E and $\| f \|_Z \leq \epsilon$ for every $f \in \mathcal{F}$. This completes the proof. ∎

The above proposition extends to (LN)-spaces the corresponding result for (DFM)-spaces obtained by Dineen ([6] and [7] Example 2.47) from where the pattern of proof has been taken.

The classical theorem of Dixmier [8] about the existence of a predual of a Banach space has been extended by Mujica [16] who has successfully applied this sort of theorem to obtain preduals of some spaces of holomorphic functions [16], [17] (see also Mujica-Nachbin [18]). Very recently, Bierstedt and Bonet [4] have proved the following improvement of Mujica's Theorem: *"Let (X, τ) be a bornological locally convex space. Assume that there exists a locally convex topology s on X such that (i) there is a fundamental system of convex balanced bounded sets which are s-compact and (ii) there is a basis of s-closed 0-neighbourhoods in (X, τ). Then there exists a complete barrelled locally convex space G such that the strong dual of G coincides with (X, τ)."*

Proposition 15 together with the above result allows us to obtain a predual of $\mathcal{H}_b(E)$ for an (LN)-space E.

16.Proposition.- *If E is an (LN)-space, then there exists a complete barrelled space G such that the strong dual of G is $\mathcal{H}_b(E)$.*
Proof:
We consider $s = \tau_o$, the topology of uniform convergence on the compact subsets of E. Let $(A_n)_{n\in\mathbf{N}}$ be a fundamental sequence of bounded sets in E. It will be enough to prove that the set $\mathcal{F} := \{f \in \mathcal{H}_b(E) : \| f \|_{A_n} \leq \mu_n, \ n = 1, 2, \ldots\}$ is τ_o-compact for any sequence $(\mu_n)_{n\in\mathbf{N}}$ of positive real numbers. By Proposition 15, $\mathcal{F}$ is equicontinuous. If we denote by $\overline{\mathcal{F}}^p$ the pointwise closure of $\mathcal{F}$ in $\mathbf{C}^E$, then $\overline{\mathcal{F}}^p$ is an equicontinuous set (see [20] Th. 13.3.3) . Moreover, by Tychonov's theorem $\overline{\mathcal{F}}^p$ is a compact subset of $\mathbf{C}^E$, and by [20] Th. 13.3.2 the compact open topology τ_o coincides with the topology of pointwise convergence in $\overline{\mathcal{F}}^p$ .Thus the τ_o-closure of $\mathcal{F}$, $\overline{\mathcal{F}}^{\tau_o}$ ($= \overline{\mathcal{F}}^p$), is a τ_o-closed subset of $\overline{\mathcal{F}}^p$ and so $\overline{\mathcal{F}}^{\tau_o}$ is τ_o-compact. Now it is easy to check that $\overline{\mathcal{F}}^{\tau_o} = \mathcal{F}$ and the conclusion follows.
■

17.Remark.- Proposition 16 can be extended to the case of a quasibarrelled (DF)-space E because the proof of Proposition 15 shows that every bounded subset of $\mathcal{H}_b(E)$ is equicontinuous .

In the case of normed spaces E the authors have proved [9] that there is an (LB)-space, $\mathcal{P}_b(E)$, whose strong dual coincides with $\mathcal{H}_b(E)$(see also [4]).

18.Remark.- If E is an (LN)-space and F is a Banach space then the inductive limit $\mathcal{H}_b(E;F) = \lim_V \mathcal{H}_b(E_V;F)$ is regular (this improves Proposition 11). Indeed, if $\mathcal{F}$ is a bounded subset of $\mathcal{H}_b(E;F)$ and A is a bounded subset of E then the family $\mathcal{G} := \{g : g(x) = f(x+x_o) - f(x_o) , x \in E, f \in \mathcal{F} , x_o \in A\}$ is again a bounded subset of $\mathcal{H}_b(E;F)$, so applying Proposition 15 to $\mathcal{G}$ we obtain that the restriction of $\mathcal{F}$ to each bounded set A is equiuniformly continuous. Now the conclusion follows from the proof of Proposition 7.

Acknowledgments: Thanks are given to Prof. Bonet for suggesting us to deal with holomorphic mappings on (DF)-spaces.

References

[1] J.M. Ansemil and S. Ponte, *An example of a quasi-normable Fréchet function space which is not a Schwartz space.* Adv. in Funct. Anal., Hol. and Approx. Theory (ed.: S. Machado). Lecture Notes in Math. 843. Springer-Verlag. Berlin (1981) 1-8.

[2] R.M. Aron, P.D. Berner, *A Hahn-Banach extension theorem for analytic mappings.* Bull. Soc. Math. France 106 (1978) 3-24.

[3] R.M. Aron, B. Cole and T.W. Gamelin, *Spectra of algebras of analytic functions on a Banach space.* J. Reine Angew. Math. 415 (1991) 51-93.

[4] K.D. Bierstedt and J. Bonet, Biduality in Fréchet and (LB)-spaces. Progress in Functional Analysis (ed.: K. Bierstedt, J. Bonet, J. Horvath and M. Maestre). Math. Studies. North-Holland. Amsterdam. (To appear).

[5] J.F. Colombeau and J. Mujica, *Holomorphic and differentiable mappings of uniform bounded type.* Funct. Anal., Hol. and Approx. Theory (ed.: J. Barroso). Math. Studies 71. North-Holland. Amsterdam (1982) 189-200.

[6] S. Dineen, *Holomorphic functions on strong duals of Fréchet-Montel spaces.* Inf. Dim. Hol. and Appl. (ed.: M.C. Matos). Math. Studies 12. North-Holland. Amsterdam (1977) 147-176.

[7] S. Dineen, *Complex Analysis in Locally Convex Spaces.* Math. Studies 57. North-Holland. Amsterdam. 1981.

[8] J. Dixmier, *Sur un théorème de Banach.* Duke Math. J. 151 (1948) 1057-71.

[9] P. Galindo, D. García and M. Maestre, *Holomorphic mappings of bounded type.* J. Math. Ann. Appl. (To appear).

[10] A. Grothendieck, *Sur les espaces (F) et (DF).* Summa Brasil. Math. 3 (1954) 57-123.

[11] J.M. Isidro, *Topological duality on the function space* $(\mathcal{H}_b(U;F),\tau_b)$. Proc. Irish Acad. 79A12 (1979) 115-130.

[12] J.M. Isidro, *On the distinguished character of the function spaces of holomorphic mappings of bounded type.* J. Funct. Anal. 38 (1980) 139-145.

[13] H. Jarchow, *Locally Convex Spaces.* Teubner. Stuttgart. 1981.

[14] M. Lindström, *A characterization of Schwartz spaces.* Math. Z. 198 (1988) 423-430.

[15] R. Meise and D. Vogt, *Holomorphic functions on nuclear sequence spaces.* Departamento de Teoría de Funciones. Universidad Complutense. Madrid. 1986.

[16] J. Mujica, *A completeness criterion for inductive limits of Banach spaces.* Funct. Anal., Hol. and Approx. Theory II. Math. Studies 86. North-Holland. Amsterdam (1984) 319-329.

[17] J. Mujica, *Linearization of bounded holomorphic mappings on Banach spaces.* Trans. Amer. Math. Soc. (To appear).

[18] J. Mujica and L. Nachbin, *Linearization of holomorphic mappings in locally convex spaces.* J. Math. Pures Appl. (To appear).

[19] P. Pérez and J. Bonet, *Barrelled Locally Convex Spaces.* Math. Studies 131. North-Holland. Amsterdam. 1987.

[20] A. Wilansky, *Topology for Analysis.* Xerox College Publishing. Lexington, Massachussets. Toronto. 1970.

Progress in Functional Analysis
K.D. Bierstedt, J. Bonet, J. Horváth & M. Maestre (Eds.)

LINEARIZATION OF HOLOMORPHIC MAPPINGS OF BOUNDED TYPE

Jorge Mujica

Instituto de Matemática
Universidade Estadual de Campinas
Caixa Postal 6065
13 081 Campinas, SP, Brazil

Dedicated to Manuel Valdivia on the occasion of his sixtieth birthday

We study the relationships between preduals of spaces of holomorphic functions of bounded type and preduals of spaces of bounded holomorphic functions. In this manner we obtain information about the former from the available information about the latter.

INTRODUCTION

Several authors have obtained linearization theorems for various classes of holomorphic mappings. It seems that the first general result of this kind is due to Mazet [13], who constructed, for each open subset U of a locally convex space E, a complete locally convex space $G(U)$ and a holomorphic mapping $\delta_U : U \to G(U)$ such that, for each complete locally convex space F and each holomorphic mapping $f : U \to F$, there is a unique continuous linear mapping $T_f : G(U) \to F$ such that $T_f \circ \delta_U = f$. Related results of Schottenloher [19] and Ryan [18] were not so satisfactory. In a recent paper Nachbin and the author [16] gave a new proof and several applications of the Mazet linearization theorem.

In another paper the author [15] obtained a linearization theorem for bounded holomorphic mappings. He constructed, for each open subset U of a Banach space E, a Banach space $G^\infty(U)$ and a bounded holomorphic mapping $\delta_U : U \to G^\infty(U)$ such that, for each Banach space F and each bounded holomorphic mapping $f : U \to F$, there is a unique continuous linear mapping $T_f : G^\infty(U) \to F$ such that $T_f \circ \delta_U = f$.

In a more recent paper Galindo et al. [7] obtained a linearization theorem for holomorphic mappings of bounded type. They constructed, for each balanced, open subset U of a Banach space E, an (LB)-space $G_b(U)$ and a holomorphic mapping of bounded type

$\delta_U : U \to G_b(U)$ such that, for each Banach space F and each holomorphic mapping of bounded type $f : U \to F$, there is a unique continuous linear mapping $T_f : G_b(U) \to F$ such that $T_f \circ \delta_U = f$.

The Banach spaces which appear in the (LB)-representation of $G_b(U)$ obtained by Galindo et al. [7] are rather involved. In this lecture we improve their result by showing that if (U_n) is a suitable sequence of open subsets of U, then $G_b(U)$ can be represented as the inductive limit of the Banach spaces $G^\infty(U_n)$. This representation theorem is used to derive several properties of the space $G_b(U)$ from the corresponding properties of the spaces $G^\infty(U_n)$.

In this lecture we actually deal with a class of holomorphic mappings more general than the class of holomorphic mappings of bounded type. This more general class appears in a natural way in the study of $G(U)$, and the results obtained here will be used in a subsequent paper to derive further properties of $G(U)$.

I am grateful to Klaus Floret for several helpful comments when this paper was being prepared.

I wish to thank the organizers of the meeting honoring Manuel Valdivia for their kind invitation and financial support to attend and deliver this lecture.

1. NOTATION AND TERMINOLOGY

We follow the standard terminology from topological vector spaces, as found for instance in the books of Grothendieck [10] or Horváth [11].

The letters E and F represent locally convex spaces, always assumed complex and Hausdorff. If A is a convex, balanced, bounded subset of E, then E_A denotes the vector subspace of E generated by A and normed by the Minkowski functional of A.

$\mathcal{L}(E;\ F)$ denotes the vector space of all continuous linear mappings from E into F. If $F = \mathbb{C}$ then we write E' instead of $\mathcal{L}(E;\ \mathbb{C})$. E'_b denotes the strong dual of E, whereas E'_i denotes the *inductive dual* of E. We recall that $E'_i = \text{ind } E'_{V^0}$, where V varies among the convex, balanced 0-neighborhoods in E, and V^0 denotes the polar of V in E'. We refer to Bierstedt [3] for the properties of the inductive dual.

We follow the standard terminology from infinite dimensional complex analysis, as found for instance in the books of Dineen [6] or the author [14].

If $m \in \mathbb{N}_0 = \mathbb{N} \cup \{0\}$, then $\mathcal{P}(^mE;\ F)$ denotes the vector space of all continuous m-homogeneous polynomials from E into F. If $F = \mathbb{C}$, then we write $\mathcal{P}(^mE)$ instead of $\mathcal{P}(^mE;\ \mathbb{C})$.

If U is an open subset of E, then $\mathcal{H}(U;\ F)$ denotes the vector space of all holomorphic mappings from U into F, whereas $\mathcal{H}^\infty(U;\ F)$ denotes the vector subspace of all $f \in \mathcal{H}(U;\ F)$ such that $f(U)$ is bounded in F. If F is a Banach space, then $\mathcal{H}^\infty(U;\ F)$ is a Banach space for the sup norm. If $F = \mathbb{C}$, then we write $\mathcal{H}(U)$ instead of $\mathcal{H}(U;\ \mathbb{C})$, and $\mathcal{H}^\infty(U)$ instead of $\mathcal{H}^\infty(U;\ \mathbb{C})$. τ_c denotes the compact open topology on $\mathcal{H}(U;\ F)$ or on any subset of $\mathcal{H}(U;\ F)$.

If $f \in \mathcal{H}(U;\ F)$, $a \in U$ and $m \in \mathbb{N}_0$, then $P^m f(a) \in \mathcal{P}(^mE;\ F)$ denotes the mth homogeneous polynomial in the Taylor series expansion of f at a.

If $f \in \mathcal{H}(U; F)$, where F is a Banach space, then for each set $A \subset U$ we write $||f||_A = \sup_{x \in A} ||f(x)||$.

2. THE SPACES $\mathcal{H}^\infty(\mathcal{U}; F)$ AND $G^\infty(\mathcal{U})$

Let E and F be locally convex spaces, let U be an open subset of E, and let $\mathcal{U} = (U_n)$ be a countable, open cover of U. Let $\mathcal{H}^\infty(\mathcal{U}; F)$ denote the locally convex space

$$\mathcal{H}^\infty(\mathcal{U}; F) = \{f \in \mathcal{H}(U; F) : f(U_n) \text{ is bounded in } F \text{ for every } n\},$$

equipped with the topology of uniform convergence on all the sets U_n. If F is a Banach space, then $\mathcal{H}^\infty(\mathcal{U}; F)$ is a Fréchet space. If $F = \mathbb{C}$, then we write $\mathcal{H}^\infty(\mathcal{U})$ instead of $\mathcal{H}^\infty(\mathcal{U}; \mathbb{C})$.

2.1. THEOREM. *Let U be an open subset of a locally convex space E, and let $\mathcal{U} = (U_n)$ be a countable, open cover of U. Then there are a complete, barrelled (DF)-space $G^\infty(\mathcal{U})$ and a mapping $\delta_U \in \mathcal{H}^\infty(\mathcal{U}; G^\infty(\mathcal{U}))$ with the following universal property: For each complete locally convex space F and each mapping $f \in \mathcal{H}^\infty(\mathcal{U}; F)$, there is a unique mapping $T_f \in \mathcal{L}(G^\infty(\mathcal{U}); F)$ such that $T_f \circ \delta_U = f$. This property characterizes $G^\infty(\mathcal{U})$ uniquely up to a topological isomorphism.*

Proof. For each sequence $\alpha = (\alpha_n)$ of strictly positive numbers, let $B_{\mathcal{U}}^\alpha$ denote the set

$$B_{\mathcal{U}}^\alpha = \{f \in \mathcal{H}^\infty(\mathcal{U}) : ||f||_{U_n} \le \alpha_n \text{ for every } n\},$$

and let $G^\infty(\mathcal{U})$ denote the locally convex space

$$G^\infty(\mathcal{U}) = \{u \in \mathcal{H}^\infty(\mathcal{U})' : u|B_{\mathcal{U}}^\alpha \text{ is } \tau_c\text{-continuous for every } \alpha\},$$

equipped with the topology of uniform convergence on all the sets $B_{\mathcal{U}}^\alpha$. It is clear that

$$\mathcal{H}^\infty(\mathcal{U}) = \text{ind } \mathcal{H}^\infty(\mathcal{U})_{B_{\mathcal{U}}^\alpha}$$

and each of the sets $B_{\mathcal{U}}^\alpha$ is τ_c-compact, by the Ascoli theorem. Furthermore, each of the sets $\{f \in \mathcal{H}^\infty(\mathcal{U}) : ||f||_{U_n} \le \varepsilon\}$ is τ_c-closed. Thus an application of [16, Theorem 1.1] shows that the evaluation mapping

$$J : \mathcal{H}^\infty(\mathcal{U}) \to G^\infty(\mathcal{U})'_i = G^\infty(\mathcal{U})'_b$$

is a topological isomorphism.

Clearly the space $G^\infty(\mathcal{U})$ is complete. That $G^\infty(\mathcal{U})$ is barrelled follows from a general result of Bierstedt and Bonet [4], or directly from the fact that $J(B_{\mathcal{U}}^\alpha)$ is an equicontinuous subset of $G^\infty(\mathcal{U})'$ for every α. Now since $\mathcal{H}^\infty(\mathcal{U})$ is a Fréchet space, its strong dual

$\mathcal{H}^\infty(\mathcal{U})'_b$ is a (DF)-space. As a subspace of $\mathcal{H}^\infty(\mathcal{U})'_b$, the space $G^\infty(\mathcal{U})$ has a fundamental sequence of bounded sets, and is therefore a (DF)-space.

From this point on the proof of Theorem 2.1 is similar to the proof of [16, Theorem 2.1], and is therefore omitted.

2.2. PROPOSITION. *Let U be an open subset of a locally convex space E, and let $\mathcal{U} = (U_n)$ be a countable, open cover of U. Then:*

(a) For each bounded set $B \subset G^\infty(\mathcal{U})$, there are $n \in \mathbb{N}$ and $\rho > 0$ such that $B \subset \rho\overline{\Gamma}\delta_U(U_n)$, where $\overline{\Gamma}A$ denotes the closed, convex, balanced hull of A.

(b) For each complete locally convex space F, the mapping

$$f \in \mathcal{H}^\infty(\mathcal{U};\ F) \to T_f \in \mathcal{L}(G^\infty(\mathcal{U});\ F)$$

is a topological isomorphism, when $\mathcal{L}(G^\infty(\mathcal{U});\ F)$ is endowed with the topology of uniform convergence on the bounded subsets of $G^\infty(\mathcal{U})$.

Proof. Since the evaluation mapping

$$J : \mathcal{H}^\infty(\mathcal{U}) \to G^\infty(\mathcal{U})'_b$$

is a topological isomorphism, we can prove (a) by imitating the proof of [16, Lemma 4.2]. And (b) is clearly a direct consequence of (a).

2.3. PROPOSITION. *Let E and F be locally convex spaces, with F complete. Let U be an open subset of E, let $\mathcal{U} = (U_n)$ be a countable, open cover of U, and let $(f_i) \subset \mathcal{H}^\infty(\mathcal{U};\ F)$. Then the set $\cup_i f_i(U_n)$ is bounded in F for every $n \in \mathbb{N}$ if and only if the family $(T_{f_i}) \subset \mathcal{L}(G^\infty(\mathcal{U});\ F)$ is equicontinuous.*

The proof of this proposition is similar to the proof of [16, Proposition 2.5], and is therefore omitted.

2.4. PROPOSITION. *Let U be an open subset of a Banach space E, and let $\mathcal{U} = (U_n)$ be a countable, open cover of U, with each U_n bounded. Then E is topologically isomorphic to a complemented subspace of $G^\infty(\mathcal{U})$.*

Proof. Under the hypotheses of the proposition, the inclusion mapping $U \hookrightarrow E$ belongs to $\mathcal{H}^\infty(\mathcal{U};\ E)$. By Theorem 2.1 there exists $T \in \mathcal{L}(G^\infty(\mathcal{U});\ E)$ such that $T(\delta_x) = x$ for every $x \in U$. Then the proof of [15, Proposition 2.3] or [16, Proposition 2.6] applies.

3. AN (LB)-REPRESENTATION OF $G^\infty(\mathcal{U})$

Let U be an open subset of a locally convex space E, and let $\mathcal{U} = (U_n)$ be a countable open cover of U. Let B_n denote the closed unit ball of $\mathcal{H}^\infty(U_n)$, and let $G^\infty(U_n)$ denote the Banach space

$$G^\infty(U_n) = \{u \in \mathcal{H}^\infty(U_n)' : u|B_n \text{ is } \tau_c\text{-continuous }\},$$

with the induced norm. As pointed out in the proof of [15, Theorem 2.1], a theorem of Ng [17] guarantees that the evaluation mapping

$$J_n : \mathcal{H}^\infty(U_n) \to G^\infty(U_n)'$$

is an isometric isomorphism.

Clearly the Fréchet space $\mathcal{H}^\infty(\mathcal{U})$ can be represented as the projective limit of the Banach spaces $\mathcal{H}^\infty(U_n)$ by means of the restriction mappings

$$R_n : \mathcal{H}^\infty(\mathcal{U}) \to \mathcal{H}^\infty(U_n).$$

Consider the dual mappings

$$R_n' : \mathcal{H}^\infty(U_n)' \to \mathcal{H}^\infty(\mathcal{U})'.$$

One can readily verify that $R_n' u \in G^\infty(\mathcal{U})$ for every $u \in G^\infty(U_n)$, that is, the mapping R_n is continuous for the topologies $\sigma(\mathcal{H}^\infty(\mathcal{U}), G^\infty(\mathcal{U}))$ and $\sigma(\mathcal{H}^\infty(U_n), G^\infty(U_n))$. Thus if

$$S_n : G^\infty(U_n) \to G^\infty(\mathcal{U})$$

denotes the restriction of R_n' to $G^\infty(U_n)$, then the mappings R_n and S_n are dual to each other with respect to the dual pairs $\langle \mathcal{H}^\infty(\mathcal{U}), G^\infty(\mathcal{U})\rangle$ and $\langle \mathcal{H}^\infty(U_n), G^\infty(U_n)\rangle$. We will see that, under suitable hypotheses, the (DF)-space $G^\infty(\mathcal{U})$ can be represented as the inductive limit of the Banach spaces $G^\infty(U_n)$ by means of the mappings S_n. We begin with the following proposition.

3.1. PROPOSITION. *Let U be a balanced, open subset of a Banach space E, and let $\mathcal{U} = (U_n)$ be a sequence of balanced, bounded, open subsets of U such that $U = \cup_{n=1}^\infty U_n$. Then:*

(a) The mapping $R_n : \mathcal{H}^\infty(\mathcal{U}) \to H^\infty(U_n)$ has a $\sigma(\mathcal{H}^\infty(U_n), G^\infty(U_n))$-dense range.

(b) The mapping $S_n : G^\infty(U_n) \to G^\infty(\mathcal{U})$ is injective.

Proof. (a) Let $f \in \mathcal{H}^\infty(U_n)$. Then, with the notation of [15, Proposition 5.2], the sequence of Cesàro means $(\sigma_m f)$ converges to f in $(\mathcal{H}^\infty(U_n), \tau_\gamma)$. Since, by [15, Propositions 4.7 and 4.9], $(\mathcal{H}^\infty(U_n), \tau_\gamma)' = G^\infty(U_n)$, we conclude that $(\sigma_m f)$ converges to f for the topology $\sigma(\mathcal{H}^\infty(U_n), G^\infty(U_n))$. Since each $\sigma_m f$ is a polynomial, and since each U_k is bounded, we conclude that $\sigma_m f \in \mathcal{H}^\infty(\mathcal{U})$ for every $m \in \mathbb{N}$, thus proving (a).

Since (b) is a direct consequence of (a), the proof of the proposition is complete.

Thus, under the hypotheses of Proposition 3.1, we may regard each $G^\infty(U_n)$ as a vector subspace of $G^\infty(\mathcal{U})$.

3.2. THEOREM. *Let U be a balanced, open subset of a Banach space E, and let $\mathcal{U} = (U_n)$ be a sequence of balanced, bounded, open subsets of U such that $U = \cup_{n=1}^\infty U_n$ and $\rho_n U_n \subset U_{n+1}$, with $\rho_n > 1$, for every $n \in \mathbb{N}$. Then*

$$G^\infty(\mathcal{U}) = ind\, G^\infty(U_n),$$

and this inductive limit is boundedly retractive.

The key to the proof of Theorem 3.2 is the following lemma.

3.3. LEMMA. *Let U be a balanced, open subset of a Banach space E, and let $\mathcal{U} = (U_n)$ be a sequence of balanced, bounded, open subsets of U such that $U = \cup_{n=1}^\infty U_n$ and $\rho_n U_n \subset U_{n+1}$, with $\rho_n > 1$, for every $n \in \mathbb{N}$. Then $G^\infty(\mathcal{U})$ and $G^\infty(U_{n+1})$ induce the same topology and the same uniform structure on the closed unit ball of $G^\infty(U_n)$.*

Proof. Let A_k and B_k denote the closed unit balls of $G^\infty(U_k)$ and $\mathcal{H}^\infty(U_k)$, respectively. By a lemma of Grothendieck [10, p. 98, Lemma], to prove Lemma 3.3 it suffices to show that for each $\varepsilon > 0$ there exists $\alpha = (\alpha_k)$, with $\alpha_k > 0$, such that

$$A_n \cap (B_{\mathcal{U}}^\alpha)^0 \subset \varepsilon A_{n+1}. \tag{3.1}$$

Let $f \in B_{n+1}$. Since $\rho_n U_n \subset U_{n+1}$, the Cauchy integral formulas yield the inequalities

$$\sum_{m=N}^\infty \|P^m f(0)\|_{U_n} \le \sum_{m=N}^\infty \rho_n^{-m} \|f\|_{U_{n+1}} \le \sum_{m=N}^\infty \rho_n^{-m},$$

and this is less than $\varepsilon/2$ for N sufficiently large. On the other hand, since $U_k \subset c_k\, U_{n+1}$, with $c_k > 0$, for every $k \in \mathbb{N}$, we have that

$$\sum_{m=0}^{N-1} \|P^m f(0)\|_{U_k} \le \sum_{m=0}^{N-1} c_k^m \|f\|_{U_{n+1}} \le \sum_{m=0}^{N-1} c_k^m$$

for every $k \in \mathbb{N}$. Set $\alpha = (\alpha_k)$, where $\alpha_k = \frac{2}{\varepsilon} \sum_{m=0}^{N-1} c_k^m$, for every $k \in \mathbb{N}$. Then the preceding inequalities show that

$$B_{n+1} \subset \frac{\varepsilon}{2}\, B_{\mathcal{U}}^\alpha + \frac{\varepsilon}{2}\, B_n, \tag{3.2}$$

and (3.1) follows directly from (3.2).

Proof of Theorem 3.2. Clearly $\cup_{n=1}^\infty G^\infty(U_n) \subset G^\infty(\mathcal{U})$, and the inclusion mapping ind $G^\infty(U_n) \hookrightarrow G^\infty(\mathcal{U})$ is continuous.

On the other hand, by Theorem 2.1 and by [15, Theorem 2.1], we have the canonical topological isomorphisms

$$G^\infty(\mathcal{U})'_b = \mathcal{H}^\infty(\mathcal{U}) = \text{proj } \mathcal{H}^\infty(U_n) = \text{proj } G^\infty(U_n)'.$$

Thus, by polarity, each bounded subset of $G^\infty(\mathcal{U})$ is included in the closure in $G^\infty(\mathcal{U})$ of a bounded subset of some $G^\infty(U_n)$. Since $G^\infty(U_{n+1})$ is complete, it follows from Lemma 3.3 that

$$\overline{A_n}^{G^\infty(\mathcal{U})} = \overline{A_n}^{G^\infty(U_{n+1})}$$

and thus each bounded subset of $G^\infty(\mathcal{U})$ is included in some $G^\infty(U_n)$ and is bounded there. Thus $G^\infty(\mathcal{U}) = \cup_{n=1}^\infty G^\infty(U_n)$, and the spaces $G^\infty(\mathcal{U})$ and ind $G^\infty(U_n)$ induce the same topology on each bounded subset of $G^\infty(\mathcal{U})$. Since we already know from Theorem 2.1 that $G^\infty(\mathcal{U})$ is a (DF)-space, a theorem of Grothendieck [8, p. 68, Théorème 3] guarantees that the identity mapping $G^\infty(\mathcal{U}) \to$ind $G^\infty(U_n)$ is continuous. This completes the proof.

An examination of the proof of Lemma 3.3 yields the following result.

3.4. PROPOSITION. *Let E and F be Banach spaces, let U be a balanced, open subset of E, and let $\mathcal{U} = (U_n)$ be a sequence of balanced, bounded, open subsets of U such that $U = \cup_{n=1}^\infty U_n$ and $\rho_n U_n \subset U_{n+1}$, with $\rho_n > 1$, for every $n \in \mathbb{N}$. Then $\mathcal{H}^\infty(\mathcal{U}; F)$ is a quasi-normable Fréchet space.*

4. THE APPROXIMATION PROPERTY

To begin with we state the following consequence of [15, Theorem 5.4].

4.1. THEOREM. *Let U be a balanced, bounded, open subset of a Banach space E. Then $G^\infty(U)$ has the approximation property if and only if E has the approximation property.*

With the aid of Theorem 3.2 we can easily extend this result to $G^\infty(\mathcal{U})$.

4.2. THEOREM. *Let U be a balanced, open subset of a Banach space E, and let $\mathcal{U} = (U_n)$ be a sequence of balanced, bounded, open subsets of U such that $U = \cup_{n=1}^\infty U_n$ and $\rho_n U_n \subset U_{n+1}$, with $\rho_n > 1$, for every $n \in \mathbb{N}$. Then $G^\infty(\mathcal{U})$ has the approximation property if and only if E has the approximation property.*

Proof. By Proposition 2.4, E is topologically isomorphic to a complemented subspace of $G^\infty(\mathcal{U})$. Hence E has the approximation property if so does $G^\infty(\mathcal{U})$.

If, conversely, E has the approximation property, then each $G^\infty(U_n)$ has the approximation property, by Theorem 4.1. By Theorem 3.2 we may apply a result of Bierstedt and Meise [5, Satz 1.2] to conclude that $G^\infty(\mathcal{U}) =$ ind $G^\infty(U_n)$ has the approximation property.

5. HOLOMORPHIC FUNCTIONS ON QUOTIENT SPACES

In [15, Corollary 4.12] we gave an explicit description of $G^\infty(U)$. By examining the proofs of [15, Lemma 4.6 and Theorems 4.5 and 4.11], we can sharpen the conclusion of [15, Corollary 4.12] as follows.

5.1. THEOREM. *Let U be an open subset of a Banach space E. Then $G^\infty(U)$ consists of all linear functionals $u \in G^\infty(U)'$ of the form*

$$u = \sum_{j=1}^{\infty} \alpha_j \, \delta_{x_j},$$

with $(\alpha_j) \in l^1$ and $(x_j) \subset U$. Moreover,

$$||u|| = inf \sum_{j=1}^{\infty} |\alpha_j|,$$

where the infimum is taken over all such representations of u.

5.2. THEOREM. *Let E and F be Banach spaces, and let $\pi \in \mathcal{L}(E;\ F)$ be surjective. Let U be an open subset of E, and let $V = \pi(U)$. Let $S : G^\infty(U) \to G^\infty(V)$ be the unique continuous linear mapping such that the following diagram is commutative:*

$$\begin{array}{ccc} U & \xrightarrow{\pi} & V \\ \delta_U \downarrow & & \downarrow \delta_V \\ G^\infty(U) & \xrightarrow{S} & G^\infty(V) \end{array}$$

Then S maps the ball $\{u \in G^\infty(U) : ||u|| < 1\}$ onto the ball $\{v \in G^\infty(V) : ||v|| < 1\}$.

Proof. Since $\delta_V \circ \pi \in \mathcal{H}^\infty(U;\ G^\infty(U))$, [15, Theorem 2.1] guarantees that $||S|| = ||\delta_V \circ \pi|| = 1$. Hence $||Su|| < 1$ for every $u \in G^\infty(U)$ with $||u|| < 1$. If, conversely, $v \in G^\infty(V)$, with $||v|| < 1$, then by Theorem 5.1 we can find $(y_j) \subset V$ and $(\alpha_j) \in l^1$, with $\sum_{j=1}^{\infty} |\alpha_j| < 1$, such that

$$v = \sum_{j=1}^{\infty} \alpha_j \, \delta_{y_j}.$$

Write $y_j = \pi(x_j)$, with $x_j \in U$, for every $j \in I\!N$.

If we define

$$u = \sum_{j=1}^{\infty} \alpha_j \, \delta_{x_j},$$

then $u \in G^\infty(U)$, $||u|| < 1$ and

$$Su = \sum_{j=1}^{\infty} \alpha_j \, S \, \delta_{x_j} = \sum_{j=1}^{\infty} \alpha_j \delta_{y_j} = v.$$

5.3. THEOREM. *Let E and F be Banach spaces, and let $\pi \in \mathcal{L}(E; F)$ be surjective. Let U be a balanced, open subset of E, and let $\mathcal{U} = (U_n)$ be a sequence of balanced, bounded, open subsets of U such that $U = \cup_{n=1}^{\infty} U_n$ and $\rho_n U_n \subset U_{n+1}$, with $\rho_n > 1$, for every $n \in I\!N$. Let $V = \pi(U)$, and let $\mathcal{V} = (V_n)$, where $V_n = \pi(U_n)$ for every $n \in I\!N$. Let $S : G^\infty(\mathcal{U}) \to G^\infty(\mathcal{V})$ be the unique continuous linear mapping such that the following diagram is commutative:*

$$\begin{array}{ccc} U & \xrightarrow{\pi} & V \\ \delta_U \downarrow & & \downarrow \delta_V \\ G^\infty(\mathcal{U}) & \xrightarrow{S} & G^\infty(\mathcal{V}) \end{array}$$

Then S is surjective and open. Each bounded subset of $G^\infty(\mathcal{V})$ is the image under S of some bounded subset of $G^\infty(\mathcal{U})$.

Proof. For each $n \in I\!N$ we have a commutative diagram

$$\begin{array}{ccc} G^\infty(U_n) & \hookrightarrow & G^\infty(\mathcal{U}) \\ S_n \downarrow & & \downarrow S \\ G^\infty(V_n) & \hookrightarrow & G^\infty(\mathcal{V}) \end{array}$$

Let B be a bounded subset of $G^\infty(\mathcal{V})$. By Theorem 3.2, B is included in some $G^\infty(V_n)$ and is bounded there. By Theorem 5.2, there is a bounded set A in $G^\infty(U_n)$ such that $B = S_n(A)$. Thus $B = S(A)$, with A bounded in $G^\infty(\mathcal{U})$. In particular we have shown that S is surjective, and therefore open, by an open mapping theorem due to Grothendieck (see [9, Introduction, p. 17, Théorème B]) or [10, p. 200, Théorème 2]).

6. HOLOMORPHIC FUNCTIONS ON PRODUCT SPACES

6.1. THEOREM. *Let E and F be Banach spaces, and let U and V be open sets in E and F, respectively. Then:*

(a) $\mathcal{H}^\infty(U \times V)$ is isometrically isomorphic to $\mathcal{H}^\infty(U; \mathcal{H}^\infty(V))$.

(b) $G^\infty(U \times V)$ is isometrically isomorphic to $G^\infty(U)\tilde{\otimes}_\pi G^\infty(V)$.

Proof. (a) Clearly the mapping

$$f \in \mathcal{H}^\infty(U \times V) \to \tilde{f} \in \mathcal{H}^\infty(U; \mathcal{H}^\infty(V))$$

defined by $\tilde{f}(x)(y) = f(x, y)$ for all $x \in U$ and $y \in V$, is an isometric isomorphism.

(b) On one hand the mapping

$$(x, y) \in U \times V \to \delta_x \otimes \delta_y \in G^\infty(U)\tilde{\otimes}_\pi G^\infty(V)$$

is holomorphic and $||\delta_x \otimes \delta_y|| = ||\delta_x||\,||\delta_y|| = 1$ for all $x \in U$ and $y \in V$. By the universal property of $G^\infty(U \times V)$ (see [15, Theorem 2.1]), there is a continuous linear mapping

$$S : G^\infty(U \times V) \to G^\infty(U)\tilde{\otimes}_\pi G^\infty(V),$$

with $||S|| = 1$, and such that $S\delta_{(x,\,y)} = \delta_x \otimes \delta_y$ for all $x \in U$ and $y \in V$. On the other hand the mapping

$$(u,\, v) \in G^\infty(U) \times G^\infty(V) \to u \times v \in G^\infty(U \times V)$$

defined by $(u \times v)(f) = u(v \circ \tilde{f})$ for every $u \in G^\infty(U)$, $v \in G^\infty(V)$ and $f \in \mathcal{H}^\infty(U \times V)$, is bilinear and $||u \times v|| \leq ||u||\,||v||$ for all $u \in G^\infty(U)$ and $v \in G^\infty(V)$. By the universal property of tensor products, there is a continuous linear mapping

$$T : G^\infty(U)\tilde{\otimes}_\pi G^\infty(V) \to G^\infty(U \times V),$$

with $||T|| \leq 1$, and such that $T(u \otimes v) = u \times v$ for all $u \in G^\infty(U)$ and $v \in G^\infty(V)$. Hence $T \circ S(\delta_{(x,\,y)}) = \delta_{(x,\,y)}$ and $S \circ T(\delta_x \otimes \delta_y) = \delta_x \otimes \delta_y$ for all $x \in U$ and $y \in V$. It follows that S is an isometric isomorphism, with inverse T.

6.2. THEOREM. *Let E and F be Banach spaces, and let U and V be balanced, open sets in E and F, respectively. Let $\mathcal{U} = (U_n)$ be a sequence of balanced, bounded, open subsets of U such that $U = \cup_{n=1}^\infty U_n$ and $\rho_n U_n \subset U_{n+1}$, with $\rho_n > 1$, for every $n \in \mathbb{N}$. Likewise, let $\mathcal{V} = (V_n)$ be a sequence of balanced, bounded, open subsets of V such that $V = \cup_{n=1}^\infty V_n$ and $\sigma_n V_n \subset V_{n+1}$, with $\sigma_n > 1$, for every $n \in \mathbb{N}$. If $\mathcal{U} \times \mathcal{V}$ denotes the sequence $(U_n \times V_n)$, then:*

(a) $\mathcal{H}^\infty(\mathcal{U} \times \mathcal{V})$ is topologically isomorphic to $\mathcal{H}^\infty(\mathcal{U};\, \mathcal{H}^\infty(\mathcal{V}))$.

(b) $G^\infty(\mathcal{U} \times \mathcal{V})$ is topologically isomorphic to $G^\infty(\mathcal{U})\tilde{\otimes}_\pi G^\infty(\mathcal{V})$.

Proof. (a) Clearly the mapping

$$f \in \mathcal{H}^\infty(\mathcal{U} \times \mathcal{V}) \to \tilde{f} \in \mathcal{H}^\infty(\mathcal{U};\, \mathcal{H}^\infty(\mathcal{V}))$$

defined by $\tilde{f}(x)(y) = f(x,\, y)$ for all $x \in U$ and $y \in V$, is a topological isomorphism.

(b) A glance at the diagram

$$\begin{array}{ccc} U_n \times V_n & \overset{\delta_{U_n} \otimes \delta_{V_n}}{\longrightarrow} & G^\infty(U_n)\tilde{\otimes}_\pi G^\infty(V_n) \\ \downarrow & & \downarrow \\ U \times V & \overset{\delta_{\mathcal{U}} \otimes \delta_{\mathcal{V}}}{\longrightarrow} & G^\infty(\mathcal{U})\tilde{\otimes}_\pi G^\infty(\mathcal{V}) \end{array}$$

shows that the mapping

$$(x,\, y) \in U \times V \to \delta_x \otimes \delta_y \in G^\infty(\mathcal{U})\tilde{\otimes}_\pi G^\infty(\mathcal{V})$$

belongs to $\mathcal{H}^\infty(\mathcal{U} \times \mathcal{V};\, G^\infty(\mathcal{U})\tilde{\otimes}_\pi G^\infty(\mathcal{V}))$. By Theorem 2.1 there is a continuous linear mapping

$$S : G^\infty(\mathcal{U} \times \mathcal{V}) \to G^\infty(\mathcal{U})\tilde{\otimes}_\pi G^\infty(\mathcal{V})$$

such that $S\delta_{(x,\,y)} = \delta_x \otimes \delta_y$ for all $x \in U$ and $y \in V$. On the other hand, since $G^\infty(\mathcal{U}) =$ ind $G^\infty(U_n)$ and $G^\infty(\mathcal{V}) =$ ind $G^\infty(V_n)$, the bilinear mapping

$$(u,\, v) \in G^\infty(\mathcal{U}) \times G^\infty(\mathcal{V}) \to u \times v \in G^\infty(\mathcal{U} \times \mathcal{V})$$

defined by $(u \times v)(f) = u(v \circ \tilde{f})$ for every $u \in G^\infty(\mathcal{U})$, $v \in G^\infty(\mathcal{V})$ and $f \in \mathcal{H}^\infty(\mathcal{U} \times \mathcal{V})$, is separately continuous. By a theorem of Grothendieck on (DF)-spaces (see [8, p. 66, Corollaire] or [10, p. 226, Corollaire 1]), this bilinear mapping is actually continuous. From now on the proof proceeds exactly as the proof of Theorem 6.1.

7. HOLOMORPHIC MAPPINGS OF BOUNDED TYPE

Let U be an open subset of a Banach space E. If $U \neq E$ and $x \in U$, then $d_U(x)$ denotes the distance from x to $E \backslash U$. Since $|d_U(x) - d_U(y)| \leq ||x - y||$ for all $x,\, y \in U$, we see that d_U is a continuous function on U. If $U = E$, then for convenience we define $d_U(x) = \infty$ for every $x \in U$. A set $A \subset U$ is said to be *U-bounded* if A is bounded and $\inf_{x \in A} d_U(x) > 0$.

If F is a locally convex space, then $\mathcal{H}_b(U;\, F)$ denotes the locally convex space

$$\mathcal{H}_b(U;\, F) = \{f \in \mathcal{H}(U;\, F) : f(A) \text{ is bounded in } F \text{ for each } U\text{-bounded set } A\},$$

equipped with the topology of uniform convergence on all U-bounded sets. If $F = \mathbb{C}$, then we write $\mathcal{H}_b(U)$ instead of $\mathcal{H}_b(U;\, \mathbb{C})$. The members of $\mathcal{H}_b(U;\, F)$ are called *holomorphic mappings of bounded type.*

It is clear that $\mathcal{H}_b(U;\, F) = \mathcal{H}^\infty(\mathcal{U};\, F)$ if $\mathcal{U} = (U_n)$ is any fundamental sequence of open U-bounded sets. In the next proposition we give a fundamental sequence $\mathcal{U} = (U_n)$ of open U-bounded sets which satisfies the hypotheses of all the results in the preceding sections.

7.1. PROPOSITION. *Let U be an open subset of a Banach space E, and let (U_n) be defined by*

$$U_n = \{x \in U : ||x|| < n \quad \text{and} \quad d_U(x) > 2^{-n}\}.$$

Then:

(a) (U_n) is a fundamental sequence of open, U-bounded sets.

(b) If U is balanced, then each U_n is balanced as well and, furthermore, there exists (ρ_n), with $\rho_n > 1$, such that $\rho_n U_n \subset U_{n+1}$ for every $n \in \mathbb{N}$.

The proof of this proposition is straightforward, and is left to the reader. For each $n \in \mathbb{N}$ we may take $\rho_n = 1 + 1/n2^n$.

Theorem 2.1 yields in particular the following result, which improves a result of Galindo et al. [7].

7.2. PROPOSITION. *Let U be an open subset of a Banach space E. Then there are a complete, barrelled (DF)-space $G_b(U)$ and a mapping $\delta_U \in \mathcal{H}_b(U; G_b(U))$ with the following universal property: For each complete locally convex space F and each mapping $f \in \mathcal{H}_b(U; F)$, there is a continuous linear mapping $T_f \in \mathcal{L}(G_b(U); F)$ such that $T_f \circ \delta_U = f$. This property characterizes $G_b(U)$ uniquely up to a topological isomorphism.*

Theorem 3.2 yields in particular the following result, which improves a result of Galindo et al [7].

7.3. PROPOSITION. *Let U be a balanced, open subset of a Banach space E, and let (U_n) be a fundamental sequence of open U-bounded sets such that each U_n is balanced and $\rho_n U_n \subset U_{n+1}$, with $\rho_n > 1$, for every $n \in I\!N$. Then $G_b(U) = ind\, G^\infty(U_n)$ and this inductive limit of boundedly retractive.*

It is clear too that Proposition 3.4 yields in particular a result of Ansemil and Ponte [2] and Isidro [12].

Theorem 4.2 yields in particular the following result.

7.4. PROPOSITION. *Let U be a balanced, open subset of a Banach space E. Then $G_b(U)$ has the approximation property if and only if E has the approximation property.*

Theorem 5.3 yields in particular the following result.

7.5. PROPOSITION. *Let E and F be Banach spaces, and let $\pi \in \mathcal{L}(E; F)$ be surjective. Let V be a balanced, open subset of F, let $U = \pi^{-1}(V)$, and let $S : G_b(U) \to G_b(V)$ be the unique continuous linear mapping such that the following diagram is commutative:*

$$\begin{array}{ccc} U & \xrightarrow{\pi} & V \\ \delta_U \downarrow & & \downarrow \delta_V \\ G_b(U) & \xrightarrow{S} & G_b(V) \end{array}$$

Then S is surjective and open. Each bounded subset of $G_b(V)$ is the image under S of some bounded subset of $G_b(U)$.

Proof. Let (U_n) be a fundamental sequence of open U-bounded sets such that each U_n is balanced and $\rho_n U_n \subset U_{n+1}$, with $\rho_n > 1$, for every $n \in I\!N$. Let $V_n = \pi(U_n)$ for every $n \in I\!N$. Since $U = \pi^{-1}(V)$, it follows that each V-bounded set is the image under π of some U-bounded set. Whence it follows that (V_n) is a fundamental sequence of open V-bounded sets. Thus Theorem 5.3 applies and yields the desired conclusion.

7.6. Remark. There are Banach spaces E and F, there is a surjective mapping $\pi \in \mathcal{L}(E; F)$, and there is a convex, balanced, open set U in E such that, if we set $V = \pi(U)$, then the canonical mapping $S \in \mathcal{L}(G_b(U); G_b(V))$ is not surjective. Indeed, if S were surjective, then S would be open, by an open mapping theorem due to Grothendieck (see [9, Introduction, p. 17, Théorème B] or [10, p. 200, Théorème 2]). Then, by a theorem of Grothendieck on (DF)-spaces (see [8, p. 76, Proposition 5] or [10, p. 228, Proposition 4]), each bounded subset of $G_b(V)$ would be included in the closure of the image under S of some bounded subset of $G_b(U)$. Then the mapping $g \in \mathcal{H}_b(V) \to g \circ \pi \in \mathcal{H}_b(U)$ would be an embedding, that is, a topological isomorphism between $\mathcal{H}_b(V)$ and a subspace of $\mathcal{H}_b(U)$. But this is not always true, as a counterexample of Ansemil et al. [1] shows.

Finally Theorem 6.2 yields in particular the following result.

7.7. PROPOSITION. *Let E and F be Banach spaces, and let U and V be balanced, open sets in E and F, respectively. Then:*

(a) $\mathcal{H}_b(U \times V)$ is topologically isomorphic to $\mathcal{H}_b(U; \mathcal{H}_b(V))$.

(b) $G_b(U \times V)$ is topologically isomorphic to $G_b(U) \tilde{\otimes}_\pi G_b(V)$.

REFERENCES

[1] **J. M. Ansemil, R. M. Aron and S. Ponte**, *Embeddings of spaces of holomorphic functions of bounded type.* Preprint, 1990.

[2] **J. M. Ansemil and S. Ponte**, *An example of a quasi-normable Fréchet function space which is not a Schwartz space.* In: Functional Analysis, Holomorphy and Approximation Theory, edited by S. Machado, pp. 1-8. Lecture Notes in Mathematics, vol. 843. Springer, Berlin, 1981.

[3] **K. D. Bierstedt**, *An introduction to locally convex inductive limits.* In: Functional Analysis and its Applications, edited by H. Hogbe-Nlend, pp. 35-133. World Scientific, Singapore, 1988.

[4] **K. D. Bierstedt and J. Bonet**, *Biduality in Fréchet and (LB)-spaces.* In: Progress in Functional Analysis, edited by J. Bonet et al. North-Holland Mathematics Studies. North-Holland, Amsterdam, to appear.

[5] **K. D. Bierstedt and R. Meise**, *Bemerkungen über die Approximationseigenschaft lokalkonvexer Funktionenräume.* Math. Ann. 209 (1974), 99-107.

[6] **S. Dineen**, *Complex Analysis in Locally Convex Spaces.* North-Holland Mathematics Studies, vol. 57. North-Holland, Amsterdam, 1981.

[7] **P. Galindo, D. Garcia and M. Maestre**, *Holomorphic mappings of bounded type.* J. Math. Anal. Appl., to appear.

[8] **A. Grothendieck**, *Sur les espaces (F) et (DF)*. Summa Brasil. Math. 3 (1954), 57-123.

[9] **A. Grothendieck**, *Produits tensoriels topologiques et espaces nucléaires*. Mem. Amer. Math. Soc. 16 (1955).

[10] **A. Grothendieck**, *Espaces Vectoriels Topologiques,* 3ª edição. Universidade de São Paulo, 1964.

[11] **J. Horvath**, *Topological Vector Spaces and Distributions*, vol. I. Addison-Wesley, Reading, Massachusetts, 1966.

[12] **J. M. Isidro**, *On the distinguished character of the function spaces of holomorphic mappings of bounded type*. J. Funct. Anal. 38 (1980), 139-145.

[13] **P. Mazet**, *Analytic Sets in Locally Convex Spaces*. North-Holland Mathematics Studies, vol. 89. North-Holland, Amsterdam, 1984.

[14] **J. Mujica**, *Complex Analysis in Banach Spaces*. North-Holland Mathematics Studies, vol. 120. North-Holland, Amsterdam, 1986.

[15] **J. Mujica**, *Linearization of bounded holomorphic mappings on Banach spaces*. Trans. Amer. Math. Soc. 324 (1991), 867-887.

[16] **J. Mujica and L. Nachbin**, *Linearization of holomorphic mappings on locally convex spaces*. J. Math. Pures Appl., to appear.

[17] **K. F. Ng**, *On a theorem of Dixmier*. Math. Scand. 29 (1971), 279-280.

[18] **R. A. Ryan**, *Applications of topological tensor products to infinite dimensional holomorphy*. Ph. D. thesis, Trinity College Dublin, 1980.

[19] **M. Schottenloher**, *ε-products and continuation of analytic mappings*. In : Analyse Fonctionnelle et Applications, edité par L. Nachbin, pp. 261-270. Hermann, Paris, 1975.

Progress in Functional Analysis
K.D. Bierstedt, J. Bonet, J. Horváth & M. Maestre (Eds.)

Spaces of Holomorphic Functions and Germs on Quotients

J. M. Ansemil[a,1], R. M. Aron[b] and S. Ponte[a,1]

[a]Departamento de Análisis Matemático, Facultad de Matemáticas.
Universidad Complutense, 28040-Madrid. Spain

[b]Department of Mathematical Sciences
Kent State University, Kent, Ohio-44242. USA

Dedicated to Prof. M. Valdivia on the occasion of his 60th birthday

Abstract

Let E be a complex locally convex space, F a closed subspace of E and let Π be the canonical quotient mapping from E onto E/F. If U is an open subset of E and K is a compact subset of E, Π induces a linear injective mapping Π^* from $\mathcal{H}[\Pi(U)]$ (resp. $\mathcal{H}[\Pi(K)]$) into $\mathcal{H}(U)$ (resp. $\mathcal{H}(K)$), where $\mathcal{H}(V)$ (resp. $\mathcal{H}(L)$) stands for the space of holomorphic functions (resp. germs) on the open subset V (resp. the compact subset L) of a complex locally convex space. This mapping Π^* is defined by $\Pi^*(g) = g \circ \Pi$; $g \in \mathcal{H}[\Pi(U)]$ (resp. $\mathcal{H}[\Pi(K)]$).

When we consider the natural topologies on the spaces of holomorphic functions and holomorphic germs, Π^* is always continuous. In some cases Π^* is also a topological isomorphism onto its image (embedding) but this is not always true. The aim of this paper is to give a survey of the results concerning when Π^* is, and is not, an embedding. We also include some new results with their proofs.

1. INTRODUCTION

If E is a locally convex space and F is a closed subspace of E, the canonical quotient mapping $\Pi : E \to E/F$ induces an algebraic isomorphism Π^* from $(E/F)'$ onto the subspace $F^\perp$ of E' which consists of the $\varphi's$ in E' such that $\varphi \mid F = 0$, by defining $\Pi^*(\psi) = \psi \circ \Pi$; $\psi \in (E/F)'$.

[1] Partially supported by the D.G.I.C.Y.T. of Spain. Project PB87/1031.

This algebraic isomorphism is also a topological isomorphism when we consider $(E/F)'$ and E' endowed with the corresponding weak-star topologies. If E is a Fréchet space, then Π^* is also a topological isomorphism for the corresponding topologies of uniform convergence on compact subsets [9, p. 279] and, if E is a DF space, then Π^* is a topological isomorphism for the corresponding strong topologies [9, p. 401].

Let us assume from now on that E is a complex locally convex space.

Given a non void open subset U of E, we will denote by $\mathcal{H}(U)$ the vector space of all holomorphic functions on U. We recall that a function $f : U \to \mathbf{C}$ is holomorphic in U if it is continuous and for every a in U and b in E, the mapping $\lambda \to f(a+\lambda b)$ is holomorphic in an open neighbourhood of 0 in the complex plane.

On the vector space $\mathcal{H}(U)$ the following natural locally convex topologies are often considered:

-The compact open topology τ_o: the locally convex topology on $\mathcal{H}(U)$ generated by the family of all seminorms $f \in \mathcal{H}(U) \to |f|_K = \sup_{z\in K} |f(z)|$ when K ranges over the family of all compact subsets of U.

-The Nachbin ported topology τ_ω: the locally convex topology on $\mathcal{H}(U)$ generated by the family of all seminorms ported by the compact subsets of U. A seminorm p on $\mathcal{H}(U)$ is ported by the compact subset K of U if for every open subset V of U, $V \supset K$, there exists a positive constant C such that

$$p(f) \leq C|f|_V$$

for every $f \in \mathcal{H}(U)$.

-The τ_δ topology: the locally convex topology on $\mathcal{H}(U)$ generated by the family of all seminorms p on $\mathcal{H}(U)$ which have the following property: for every increasing countable open cover (U_n) of U, there exist $C > 0$ and $n_o \in \mathbf{N}$ such that

$$p(f) \leq C\,|f|_{U_{n_o}}$$

for every $f \in \mathcal{H}(U)$.

These topologies have been studied by many authors. See [8] as a general reference for properties of τ_o, τ_ω, and τ_δ. In particular we mention that $(\mathcal{H}(U), \tau_\delta)$ is an ultrabornological space [8, Prop. 2.38].

Given a compact subset K of E, we will denote by $\mathcal{H}(K)$ the vector space of all holomorphic germs on K. A holomorphic germ on K is a equivalence class determined in $\bigcup\{\mathcal{H}(U) : U \supset K,\ U \text{ open}\}$ by the relation $f \cong g$ if f agrees with g on some open neighbourhood of K.

On the vector space $\mathcal{H}(K)$ the two following locally convex topologies are often considered:

-The τ_o topology: the locally convex topology defined by

$$(\mathcal{H}(K), \tau_o) = \operatorname{ind}_{U \supset K}(\mathcal{H}(U), \tau_o(\mathcal{H}(U))).$$

-The τ_ω topology: the locally convex topology defined by

$$(\mathcal{H}(K), \tau_\omega) = \operatorname{ind}_{U \supset K}(\mathcal{H}(U), \tau_\omega(\mathcal{H}(U))).$$

The τ_o topology on $\mathcal{H}(K)$ has been extensively studied by Mujica in [12] when E is a Fréchet space. The τ_ω topology is a very natural topology on $\mathcal{H}(K)$. $(\mathcal{H}(K), \tau_\omega)$ can be obtained as an inductive limit of Banach spaces and therefore it is an ultrabornological space. We mention [8] as a general reference for the properties of $(\mathcal{H}(K), \tau_\omega)$.

In what follows, E will denote a complex Fréchet space, unless otherwise stated.

2. SPACES OF HOLOMORPHIC FUNCTIONS ON QUOTIENTS

Given a closed subspace F of E, let Π denote the canonical mapping from E onto E/F and let Π^* be the mapping from $\mathcal{H}[\Pi(U)]$ into $\mathcal{H}(U)$ defined by $\Pi^*(f) = f \circ \Pi$, $f \in \mathcal{H}[\Pi(U)]$. This mapping is linear, injective and continuous when we consider on $\mathcal{H}[\Pi(U)]$ and $\mathcal{H}(U)$ the corresponding τ_o, τ_ω or τ_δ topologies (the same type of topology on $\mathcal{H}[\Pi(U)]$ and $\mathcal{H}(U)$).

2.1. Remark.- Π^* is not onto unless $F = \{0\}$. Indeed, if Π^* were onto there would exist, for every $\varphi \in E'$, a $g \in \mathcal{H}[\Pi(U)]$ (resp. $\mathcal{H}[\Pi(K)]$) such that $g \circ \Pi = \varphi$. Then we could define, for every φ in E', a function ψ in $(E/F)'$ by the expression $\psi([\Pi(z)] = \varphi(z)$ and this would imply that every φ in E' vanishes on F, which gives that $F = \{0\}$.

For Banach spaces, Aron, Moraes and Ryan proved in [5, Th. 3.4 (i)] the following.

2.2. Theorem.- If E is a Banach space, F is a closed subspace of E and U is an arbitrary open subset of E, then the mapping

$$\Pi_o^* : g \in (\mathcal{H}[\Pi(U)],\ \tau_o) \longrightarrow g \circ \Pi \in (\mathcal{H}(U),\ \tau_o)$$

is an isomorphism of $(\mathcal{H}[\Pi(U)],\ \tau_o)$ onto a closed subspace of $(\mathcal{H}(U),\ \tau_o)$.

The key to the proof of this result is the fact that for every compact subset L in $\Pi(U)$ there is a compact subset J in U such that $L = \Pi(J)$. This result was generalized by Ponte in [13, Th. 1] to the more general setting of the Fréchet spaces. It is also true for Riemann domains, see [10].

The situation with respect to the τ_ω topology in the Banach space case was also studied in [5, Th. 3.4 (ii)] where the following theorem is obtained.

2.3 Theorem.- If U is a balanced open subset of a Banach space E and F is a closed subspace of E, then

$$\Pi_\omega^* : g \in (\mathcal{H}[\Pi(U)],\ \tau_\omega) \longrightarrow g \circ \Pi \in (\mathcal{H}(U),\ \tau_\omega)$$

is an isomorphism of $(\mathcal{H}[\Pi(U)],\ \tau_\omega)$ onto a closed subspace of $(\mathcal{H}(U),\ \tau_\omega)$.

To obtain the proof of this result an explicit description of a fundamental system of the τ_ω continuous seminorms is used (see [4] or [8, Prop. 4.39]).

In the case of Fréchet spaces the situation with respect to the τ_ω topology can be different, as Ponte proved in [13] where the following two results are obtained:

2.4. Theorem.- If E is a Fréchet Schwartz space and F is a closed subspace of E, then

$$\Pi_\omega^* : g \in (\mathcal{H}[\Pi(U)],\ \tau_\omega) \longrightarrow g \circ \Pi \in (\mathcal{H}(U),\ \tau_\omega)$$

is a topological isomorphism from $(\mathcal{H}[\Pi(U)],\ \tau_\omega)$ onto a closed subspace of $(\mathcal{H}(U),\ \tau_\omega)$ for every balanced open subset U of E.

The key to the proof of this theorem is the fact that $\tau_\omega = \tau_o$ on spaces of holomorphic functions on balanced open subsets of Fréchet Schwartz spaces [12].

2.5. Theorem.- Let $E = \lambda(A)$ be a Fréchet Montel Köthe echelon space which is not a Schwartz space and let F be a closed subspace of E such that E/F is not a Montel space. Then

$$\Pi_\omega^* : g \in (\mathcal{H}[\Pi(U)], \ \tau_\omega) \longrightarrow g \circ \Pi \in (\mathcal{H}(U), \ \tau_\omega)$$

is not an embedding for every balanced open subset U of E.

This theorem can be generalized to all Fréchet Montel non-Schwartz spaces as follows:

2.6. Theorem.- Let E be a Fréchet Montel non-Schwartz space and let F be a closed subspace of E such that E/F is not a Montel space. Then

$$\Pi_\omega^* : g \in (\mathcal{H}[\Pi(U)], \ \tau_\omega) \longrightarrow g \circ \Pi \in (\mathcal{H}(U), \ \tau_\omega)$$

is not an embedding for every open subset U of E.

Proof.- Let us assume that for a fixed open subset U of E, Π_ω^* is an embedding and let p be a τ_ω continuous seminorm on $(E/F)'$. Then, for fixed ξ in U,

$$\hat{p}(g) = p(dg(\xi)), \ g \in \mathcal{H}[\Pi(U)]$$

defines a τ_ω continuous seminorm on $\mathcal{H}[\Pi(U)]$. Since we are assuming that Π_ω^* is an embedding, there is a τ_ω continuous seminorm q on $\mathcal{H}(U)$ such that

$$\hat{p}(g) \leq C.q(g \circ \Pi) \text{ for all } g \in \mathcal{H}[\Pi(U)]$$

with some constant $C > 0$. In particular

$$p(\psi) = \hat{p}(\psi) \leq \ \mathrm{C.q}(\psi \circ \Pi) \text{ for all } \psi \in (E/F)'.$$

This implies that

$$\psi \in ((E/F)', \tau_\omega) \longrightarrow \psi \circ \Pi \in (E', \tau_\omega)$$

is an embedding. But since E is Montel $\tau_\omega = \tau_o$ on E' [8, p. 25] and then $\tau_\omega = \tau_o$ on $(E/F)'$ and this implies E/F is a Montel space [8, p. 23] which contradicts the hypothesis. Q.E.D.

We note that Bellenot proved in [7] that for every Fréchet Montel non-Schwartz space E there exists a closed subspace F such that E/F is not a Montel space.

The situation with respect to the τ_δ topology is still unclear. It was noted in [13] that there exist nuclear Fréchet spaces E and closed subspaces F of E such that

$$\Pi_\delta^* : g \in (\mathcal{H}[\Pi(U)],\ \tau_\delta) \longrightarrow g \circ \Pi \in (\mathcal{H}(U),\ \tau_\delta)$$

is not an embedding for every balanced open subset U of E.

2.7. Theorem.- Let U be an open subset of E and let F be a closed subspace of E. Then

a) If Π_δ^* is an embedding, $\Pi^*(\mathcal{H}[\Pi(U)])$ is a bornological and barrelled subspace of $(\mathcal{H}(U), \tau_\delta)$.

b) If $\Pi^*(\mathcal{H}[\Pi(U)])$ is a bornological subspace of $(\mathcal{H}(U), \tau_\delta)$, then Π_δ^* is an embedding.

c) If U is balanced and $\Pi^*(\mathcal{H}[\Pi(U)])$ is a barrelled subspace of $(\mathcal{H}(U), \tau_\delta)$, then Π_δ^* is an embedding.

Proof.- a) If Π_δ^* is an embedding, then $\Pi^*(\mathcal{H}[\Pi(U)])$ is a bornological and barrelled subspace of $(\mathcal{H}(U), \tau_\delta)$ because $(\mathcal{H}[\Pi(U)], \tau_\delta)$ is ultrabornological.

b) Assume that $\Pi^*(\mathcal{H}[\Pi(U)])$ is bornological. Since the mapping

$$f \circ \Pi \in (\Pi^*(\mathcal{H}[\Pi(U)]), \tau_o(\mathcal{H}(U))) \longrightarrow f \in (\mathcal{H}[\Pi(U)], \tau_o)$$

is continuous, it takes bounded subsets of $(\Pi^*(\mathcal{H}[\Pi(U)]), \tau_o(\mathcal{H}(U)))$ into bounded subsets of $(\mathcal{H}[\Pi(U)], \tau_o)$. The result follows from the fact that τ_o and τ_δ define the same bounded sets (E is Fréchet, see [8, Ex. 2.46]).

c) Assume that $\Pi^*(\mathcal{H}[\Pi(U)])$ is barrelled. Since the mapping

$$f \circ \Pi \in (\Pi^*(\mathcal{H}[\Pi(U)]), \tau_o(\mathcal{H}(U))) \longrightarrow f \in (\mathcal{H}[\Pi(U)], \tau_o)$$

is continuous, the mapping

$$f \circ \Pi \in (\Pi^*(\mathcal{H}[\Pi(U)]), \tau_\delta(\mathcal{H}(U))) \longrightarrow f \in (\mathcal{H}[\Pi(U)], \tau_o).$$

is also continuous. Recalling that τ_δ is the barrelled topology associated with τ_o on $\mathcal{H}(U)$ [2, Cor. 3] we have the continuity of the mapping

$$f \circ \Pi \in (\Pi^*(\mathcal{H}[\Pi(U)]), \tau_\delta(\mathcal{H}(U))) \longrightarrow f \in (\mathcal{H}[\Pi(U)], \tau_\delta).$$

Q.E.D.

In the case of Banach spaces there are no results concerning when

$$\Pi^*_\delta : g \in (\mathcal{H}[\Pi(U)],\ \tau_\delta) \longrightarrow g \circ \Pi \in (\mathcal{H}(U),\ \tau_\delta)$$

is or is not an embedding, except for the cases when $\tau_\delta = \tau_\omega$ on $\mathcal{H}[\Pi(U)]$. As is mentioned in [5, p. 626], an affirmative result for $E = l^1$ would solve an old problem in infinite dimensional holomorphy concerning the equality between τ_ω and τ_δ on $\mathcal{H}(U)$ for open subsets of separable Banach spaces (see [6, p. 69]).

Let us denote by $\mathcal{P}(^mE)$ the vector space of all m-homogeneous continuous polynomials on E. If F is a closed subspace of E, the canonical quotient mapping Π also induces a linear injective mapping from $\mathcal{P}(^m(E/F))$ into $\mathcal{P}(^mE)$ given by $P \to P \circ \Pi$. This mapping is also an embedding when the compact open topologies are considered (we are only considering Fréchet spaces). It is also an embedding for τ_ω if Π^*_ω is an embedding. The proof of this is similar to that in Theorem 2.6 where the result is obtained for $m = 1$.

2.8. Theorem.- If U is a balanced open subset of E and $\Pi^*(\mathcal{H}[\Pi(U)])$ is a DF subspace of $(\mathcal{H}(U), \tau_\omega)$ (resp. $(\mathcal{H}(U), \tau_\delta)$) then Π^*_ω (resp. Π^*_δ) is an embedding if

$$P \in (\mathcal{P}(^m(E/F)), \tau_\omega) \longrightarrow P \circ \Pi \in (\mathcal{P}(^m E), \tau_\omega)$$

is an embedding for every $m \in \mathbf{N}$.

Proof.- Since $\Pi^*(\mathcal{H}[\Pi(U)])$ is a DF subspace of the space $(\mathcal{H}(U), \tau_\omega)$ (respectively $(\mathcal{H}(U), \tau_\delta)$), to prove that

$$f \circ \Pi \in (\Pi^*(\mathcal{H}[\Pi(U)]), \tau_\omega(\mathcal{H}(U))) \longrightarrow f \in (\mathcal{H}[\Pi(U)], \tau_\omega)$$

(respectively

$$f \circ \Pi \in (\Pi^*(\mathcal{H}[\Pi(U)]), \tau_\delta(\mathcal{H}(U))) \longrightarrow f \in (\mathcal{H}[\Pi(U)], \tau_\delta))$$

is continuous it is enough to show that if $(f_\alpha \circ \Pi)_\alpha \subset \Pi^*(\mathcal{H}[\Pi(U)])$ is a bounded net with respect to $\tau_\omega(\mathcal{H}(U))$ which converges to $f \circ \Pi$ in $(\Pi^*(\mathcal{H}[\Pi(U)]), \tau_\omega(\mathcal{H}(U)))$ (resp. in $(\Pi^*(\mathcal{H}[\Pi(U)]), \tau_\delta(\mathcal{H}(U))))$ we have that (f_α) converges to f in $(\mathcal{H}(U), \tau_\omega)$ (resp. $(\mathcal{H}(U), \tau_\delta)$ (we recall that τ_ω and τ_δ define the same bounded sets since E is a Fréchet space).

By [8, Lemma 3.28] we have that

$$\frac{1}{m!}\hat{d}^m(f_\alpha \circ \Pi(0)) \to \frac{1}{m!}\hat{d}^m(f \circ \Pi(0)) \text{ for every } m \in \mathbf{N} \text{ when } \alpha \to \infty.$$

This implies that

$$\frac{1}{m!}\hat{d}^m(f_\alpha(0)) \circ \Pi \to \frac{1}{m!}\hat{d}^m(f(0)) \circ \Pi \text{ for every } m \in \mathbf{N} \text{ when } \alpha \to \infty$$

and our hypothesis gives that

$$\frac{1}{m!}\hat{d}^m(f_\alpha(0)) \to \frac{1}{m!}\hat{d}^m(f(0)) \text{ for every } m \in \mathbf{N} \text{ when } \alpha \to \infty.$$

By using again [8, Lemma 3.28] we get that $f_\alpha \to f$ in $(\mathcal{H}(U), \tau_\omega)$ (resp. $(\mathcal{H}(U), \tau_\delta)$). Q.E.D.

2.9. Remark.- We note that if E is a nuclear Fréchet space then

$$P \in (\mathcal{P}(^m(E/F)), \tau_\omega) \longrightarrow P \circ \Pi \in (\mathcal{P}(^m E), \tau_\omega)$$

is an embedding for every $m \in \mathbf{N}$ (because with this hypothesis $\tau_\omega = \tau_o$ on $\mathcal{P}(^m(E/F))$ and $\mathcal{P}(^m E)$) but, as we have said before, Π_δ^* is not an embedding in general.

3. SPACES OF HOLOMORPHIC GERMS ON QUOTIENTS

Using a deep result of Mujica [12] which states that when K is a compact subset of a Fréchet space then $(\mathcal{H}(K), \tau_o)$ is the inductive limit, in the category of all topological spaces and continuous mappings, of a sequence of compact spaces, the following theorem is obtained in [3]:

3.1. Theorem.- Let E be a Fréchet space and let F be a closed subspace of E. Then

$$\Pi_o^* : g \in (\mathcal{H}[\Pi(K)],\ \tau_o) \longrightarrow g \circ \Pi \in (\mathcal{H}(K),\ \tau_o)$$

is an embedding for every compact subset K of E.

Concerning the τ_ω topology in the Banach spaces setting the following result was obtained in [5]:

3.2. Theorem.- If E is a Banach space and F is a closed subspace of E, then

$$\Pi_\omega^* : g \in (\mathcal{H}[\Pi(K)],\ \tau_\omega) \longrightarrow g \circ \Pi \in (\mathcal{H}(K),\ \tau_\omega)$$

is an embedding for every balanced compact subset K of E.

The key to the proof of the above theorem is the use of a special family of τ_ω continuous seminorms which define the τ_ω topology when K is balanced.

In the general setting of Fréchet spaces the following result has been obtained in [3].

3.3. Theorem.- *a*) If E is a Fréchet Schwartz space and F is a closed subspace of E, then Π_ω^* is an embedding for every compact subset K of E.

b) For every Fréchet Montel non-Schwartz space there is a closed subspace F of E such that Π_ω^* is not an embedding for every compact subset K of E.

Part (a) is a consequence of the fact that $\tau_\omega = \tau_o$ on spaces of holomorphic germs on compact subsets of Fréchet Schwartz spaces [12], and the proof of part (a) is analogous to that in Theorem 2.6.

3.4. Remark.- An analogue of Theorem 2.7 is also true for germs and then, if we take E and F as in Theorem 2.7 (b), we have that for every compact subset K of E, $\Pi^*(\mathcal{H}[\Pi(K)])$ is a closed subspace of the ultrabornological space $(\mathcal{H}(K), \tau_\omega)$ which is neither bornological nor barrelled. ($\Pi^*(\mathcal{H}[\Pi(K)])$ is closed in $(\mathcal{H}(K), \tau_\omega)$ because, as a consequence of Theorem 3.1, it is closed in $(\mathcal{H}(K), \tau_o)$.)

For germs the analogue of Theorem 2.8 can be improved in the following way.

3.5. Theorem.- If K is a balanced compact subset of E, then Π_ω^* is an embedding if and only if

$$P \in (\mathcal{P}(^m(E/F)), \tau_\omega) \longrightarrow P \circ \Pi \in (\mathcal{P}(^m E), \tau_\omega)$$

is an embedding for every $m \in \mathbf{N}$.

The proof of this theorem follows as the proof of Theorem 2.8 using the fact that $(\mathcal{H}(K), \tau_\omega)$ is a DF space [11, Prop. 2.4].

3.6. Remark.- Most of the above results for germs are actually true in the more general setting of Riemann domains over Fréchet spaces (see [3]).

4. SPACES OF HOLOMORPHIC FUNCTIONS OF BOUNDED TYPE ON QUOTIENTS

Let U be an open subset of a complex locally convex space E. A subset $V \subset U$ is said to be U-bounded if it is bounded and there is a neighbourhood W of 0 in E such that $V + W \subset$ U. We will denote by $\mathcal{H}_b(U)$ the space of all holomorphic functions on U which are bounded on every U-bounded set. We will always consider $\mathcal{H}_b(U)$ endowed with the locally convex topology defined by the family of seminorms

$$f \in \mathcal{H}_b(U) \longrightarrow |\, f\,|_V$$

when V ranges over the family of all U-bounded sets. If F is a closed subspace of E and Π is the canonical quotient mapping from E onto E/F, Π induces an injective, linear continuous mapping Π_b^* from $\mathcal{H}_b[\Pi(U)]$ into $\mathcal{H}_b(U)$ by $\Pi_b^*(g) = g \circ \Pi$, where $g \in \mathcal{H}_b[\Pi(U)]$.

With the aim of studying when Π_b^* is an embedding the following result was proved in [1].

4.1. Proposition.- Given an open subset U of E let us consider the following conditions:

a) Π_b^* is an embedding.

b) For every $\Pi(U)$-bounded subset B there is a U-bounded subset A such that B is contained in the closed convex hull of $\Pi(A)$.

c) For every $\Pi(U)$-bounded subset B there is a U-bounded subset A such that B is contained in the closure of $\Pi(A)$.

Then (a)$\Rightarrow$ (*b*) and (c)$\Rightarrow$ (*a*). Moreover, if U is convex then (a)$\Leftrightarrow$ (*b*) $\Leftrightarrow$ (*c*).

From this Proposition the following two theorems are obtained in [1].

4.2. Theorem.- If E is either a DF or a DFC space, then

$$\Pi_b^* : g \in \mathcal{H}_b(E/F) \longrightarrow g \circ \Pi \in \mathcal{H}_b(E)$$

is an embedding for every closed subspace F of E.

4.3. Theorem.- *a*) Let F be a closed subspace of E such that E/F is a Fréchet Montel space, then

$$\Pi_b^* : g \in \mathcal{H}_b(\Pi(U)) \longrightarrow g \circ \Pi \in \mathcal{H}_b(U)$$

is an embedding for every open subset U of E.

b) If E is a Fréchet Montel space and F is a closed subspace of E such that E/F is not a Montel space, then Π_b^* is not an embedding for every open subset U of E.

In part (a) we do not need to assume that E is Fréchet Schwartz. If E/F is a Montel space, then the bounded sets are relatively compact and in Fréchet spaces the compact subsets of a quotient can be lifted.

4.4. Remark.- The conditions in Theorem 4.3 (a) are satisfied for every Fréchet Schwartz space E and every closed subspace F of E. The conditions in Theorem 14 (b) are satisfied for every Fréchet Montel non-Schwartz space and some special closed subspaces of E ([7]).

In the above Theorem 4.2 only the case $U = E$ is considered. When E is a Banach space the same is true for open balls. But this is not true for arbitrary open subsets as the following example shows.

4.5. Example.- Let E be c_o or $l^p, 1 \leq p \leq \infty$, let Π be the quotient mapping defined by $\Pi[(z_n)] = (z_{2n})$ and let U be the open subset of E defined by

$$U = \bigcup_{n=1}^{\infty} B(e_{2n} + ne_{2n+1}, \frac{1}{2})$$

where (e_n) denotes the canonical basis of E. Then Π_b^* is not an embedding.

Proof.- Let $B = (e_n)_{n=1}^{\infty}$. B is a $\Pi(U)$-bounded set because $B(e_n, \frac{1}{2})$ is contained in $\Pi(U)$ for every $n \in \mathbf{N}$. Let A be a bounded subset of U. Then there is $m \in \mathbf{N}$ such that A is contained in the union

$$\bigcup_{n=1}^{m} B(e_{2n} + ne_{2n+1}, \frac{1}{2}).$$

Then e_{2m} does not belong to the closed convex hull of $\Pi(A)$, which implies that B is not contained in this hull and condition (b) in Proposition 4.1 is not satisfied. Q.E.D.

The same result is even true if we consider as U the absolutely convex hull of the above U [1, Ex. 1].

The next example, given in [1], shows that the hypothesis of U to be convex is necessary to have the equivalence between conditions (a) and (c) in the above Proposition 4.1.

4.7. Example.- Let E be c_o or $l_p, 1 \leq p \leq \infty$, let Π be the quotient mapping on E given by $\Pi[(z_n)] = (z_{2n})$, and let U be the open subset of E given by $U = \bigcup_{n=3}^{\infty} U_n$ where each U_n is the union of the sets

$$V_n = \bigcup_{\|y\|<\frac{1}{4},\ y_{2n}=0} B(e_{2n}+y, \frac{1}{2n})$$

and

$$W_n = B(e_{2n}+e_{2n+1}, \frac{1}{4}) \backslash \{z : | z_{2n}-1 | \leq \frac{1}{4n}\}.$$

Then $B = (e_n)_{n=1}^{\infty}$ is a $\Pi(U)$-bounded set, B is not contained in the closure of $\Pi(A)$ for any U-bounded set A, and the mapping Π_b^* is an embedding.

Indeed, let A be a U-bounded set and let $r > 0$ such that $A + B(O, r) \subset U$. Fix n such that $\frac{1}{2n} < r$ and let $z \in U$ such that $\| \Pi(z) - e_n \| < \frac{1}{4n}$. Then $z \notin A$, which implies that B is not contained in the closure of $\Pi(A)$.

To prove that Π_b^* is an embedding the following two results are first proved:

a) The family $B^r = \bigcup_{n=3}^{\infty} B_n^r$, with $r \in (0, \frac{1}{8})$, where

$$B_n^r = \left[\bigcup_{\|y\|<\frac{1}{4},\ y_{2n}=0} B(e_{2n}+\Pi(y), \frac{1}{2n} - r) \right] \bigcup B(e_n, \frac{1}{4} - r)$$

is a fundamental family of $\Pi(U)$-bounded sets.

b) Let $n \geq 3$, $r \in (0, \frac{1}{8})$ and B_n^r be as in (a). Define

$$A_n^r = \left[\bigcup_{\|y\|<\frac{1}{4},\ y_{2n}=0} B(e_{2n}+y, \frac{1}{2n}-\frac{r}{2}) \right] \bigcup$$

$$\bigcup \left[B(e_{2n}+e_{2n+1}, \frac{1}{4}-\frac{r}{5}) \backslash \{z : |z_{2n}-1| \leq \frac{1}{4n}+\frac{r}{4}\} \right].$$

Then $A^r = \bigcup_{n=3}^{\infty} A_n^r$ is a U-bounded set, and for any $b \in B_n^r$, either $b \in \Pi(A_n^r)$ or there is $s > 0$ such that

i) $b_n \in D(1, s)$

ii) $b + (\lambda - b_n)e_n \in \Pi(U_n)$ for every $\lambda \in \mathbf{C}$ with $|\lambda - 1| \leq s$, and

iii) $b + (\lambda - b_n)e_n \in \Pi(A_n^r)$ for every $\lambda \in \mathbf{C}$ with $|\lambda - 1| = s$.

From this, the proof that Π_b^* is an embedding follows as a consequence of the maximum modulus theorem for holomorphic functions of a complex variable.

Finally we mention that it is possible that Π_b^* is not an embedding even when U is bounded as the following example (also given in [1]) shows.

4.8. Example.- Let E be c_o or $l_p, 1 \leq p \leq \infty$, let Π be the quotient mapping in E given by $\Pi[(z_n)] = (z_{2n})$, and let U be the open subset in E given by $U = \bigcup_{n=5}^{\infty} U_n$, where

$$U_n = \left[\bigcup_{\|y\|<\frac{1}{4}, y_{2n}=0} B(e_{2n}+y, \frac{1}{2n}) \right] \bigcup$$

$$\bigcup \left[B(e_{2n}+e_{2n+1}, \frac{1}{4}) \backslash \left[\{z : z_{2n}=1\} \bigcup \{z : |z_{2n+1}-1| \geq \frac{1}{n}\} \right] \right].$$

Then U is an open bounded subset of E and $f \in \mathcal{H}_b(\Pi(U)) \longrightarrow f \circ \Pi \in \mathcal{H}_b(U)$ is not an embedding.

To see this consider, for each $j \in \mathbf{N}$, the function $f_j \in \mathcal{H}_b(\Pi(U))$ defined by $f_j(z) = z_j^j$. Then (f_j) does not converge uniformly to 0 on $B = (e_n)_{n=1}^{\infty}$ and $(f_j \circ \Pi)$ converges uniformly to 0 on every U-bounded subset because for every U-bounded set A there is $m \in \mathbf{N}$ such that $A \subset \bigcup_{n=1}^{m} U_n$ and for $j > m$ we have that $| f_j \circ \Pi |_{U_m} < (\frac{1}{2})^j$.

REFERENCES

[1] Ansemil, J.M., Aron, R.M., Ponte, P.: Embeddings of Spaces of Holomorphic Functions of Bounded Type. *To appear in J. London Math. Soc.*

[2] Ansemil, J.M., Ponte, S.: Topologies Associated with the Compact Open Topology on $\mathcal{H}(U)$. *Proc. Roy. Irish Acad.* **82A**, 1 (1982) 121-128.

[3] Ansemil, J.M., Ponte, S.: Spaces of Holomorphic Germs on Quotients. *To appear in J. Math. Anal. Appl.*.

[4] Aron, R.M.: Holomorphic Functions on Balanced Open Subsets of Banach Spaces. *Bull. Amer. Math. Soc.* **78**, 4 (1972) 624-627.

[5] Aron, R.M., Moraes, L.A., Ryan, R.A.: Factorization of Holomorphic Mappings in Infinite Dimensions, *Math. Ann.* **277** (1987) 617-628.

[6] Barroso, J.A., Nachbin, L.: Some Topological Properties of Holomorphic Mappings in Infinitely Many Variables. *Advances in Holomorphy. Barroso, J.A. (Ed.). North-Holland Math. Stud.* **34** (1979) 67-91.

[7] Bellenot, S. F.: Basic Sequences in Non-Schwartz-Fréchet Spaces. *Trans. Amer. Math. Soc.* **258**, 1 (1980) 198-216.

[8] Dineen, S.: Complex Analysis in Locally Convex Spaces. *North-Holland Math. Stud.* **57**, 1981.

[9] Köthe, G.: Topological Vector Spaces I, *Springer*, 1969.

[10] Moraes, L. A., Paques, O. W., Zaine, M. C.: F-Quotients and Envelopes of F-Holomorphy. *To appear in J. Math. Anal. Appl.*.

[11] Mujica, J.: Spaces of Germs of Holomorphic Functions. *Adv. in Math. Suppl. Stud.* **4** (1979) 1-41.

[12] Mujica, J.: A Banach-Dieudonné Theorem for Germs of Holomorphic Functions. *J. Funct. Anal.* **57**, 1 (1984) 31-48.

[13] Ponte, S.: A Remark about the Embedding $(H(E/F), \tau) \to (H(E), \tau)$ with $\tau = \tau_o, \tau_\omega$ in Fréchet Spaces. *Note Math.* **9**, 2 (1989).

Progress in Functional Analysis
K.D. Bierstedt, J. Bonet, J. Horváth & M. Maestre (Eds.)

Automatic Continuity of Intertwiners in Topological Vector Spaces

K. B. Laursen
Department of Mathematics
University of Copenhagen
Universitetsparken 5
2100 Copenhagen Ø, Denmark

Abstract.
We discuss some recent developments in the automatic continuity theory of intertwining operators in the category of (F)-spaces and -algebras and illustrate the situation with special reference to multipliers.

0. INTRODUCTION.

The classical issue which has motivated much of the development of automatic continuity theory is the uniqueness of norm problem. One famous instance of this, going back to a question Kaplansky posed around 1947, asks whether the space $C(X)$ of continuous functions on a compact Hausdorff space X has another (i.e. inequivalent) algebra norm. This was answered (in the affirmative, modulo the continuum hypothesis) in 1976, by Dales and by Esterle. Obviously, other similar topological questions arise; in 1972, for instance, Graham Allan showed that on the algebra of all formal power series (in one indeterminate) there is only one (F)-algebra topology, namely that of coordinate-wise convergence.

The uniqueness of topology question extends naturally to questions of continuity of homomorphisms. Such maps are special cases of intertwining maps and many of the techniques developed handle this extra generality, with only a relatively modest increase in the level of complication; as this gain in generality captures additional interesting classes of maps, e.g. module derivations, the extra effort pays off.

The study of intertwining maps was initiated by Johnson and Sinclair, cf. [J S], and their methods have remained influential. Here we sketch how to extend some of these ideas to several categories of topological vector spaces. What is done is certainly not all new; and some of the presentation has benefited from access to parts of the forthcoming monograph [D 2]. Of course, other research deserves notice as well; suffice it here to mention as examples the several interesting papers by Albrecht and Neumann, e.g. [A N].

1. MITTAG-LEFFLER.

MITTAG-LEFFLER THEOREM. *Let $X_k, k = 0, 1, ..$ be a sequence of complete metrizable spaces and $f_k : X_k \to X_{k-1}, k = 1, 2, ..$ be a sequence of continuous maps, each with dense image. Let $g_k := f_1 \circ .. \circ f_k, k = 1, 2, ..$ Then $\bigcap_{k=1}^{\infty} g_k(X_k)$ is dense in X_0.*

A remarkable result (quoted here from [D 1], where a proof may be found), but a *countable* one - and no significant extensions seem possible. As a consequence, most recent developments of automatic continuity theory have attempted to apply the Mittag-Leffler theorem by using some technique of chopping up the associated topological spaces, to localize the possible discontinuity of the map under investigation, thus reducing to a countable situation.

2. BASICS OF AUTOMATIC CONTINUITY: SEPARATING SPACE.

Let X and Y denote two Hausdorff topological vector spaces. For this class of spaces we shall use the abbreviation TVS. Let $\theta : X \to Y$ be a linear map. We define the *separating space* $\mathfrak{S}(\theta)$ as follows:

$$\mathfrak{S}(\theta) := \{y \in Y \mid \text{there is a net } x_\alpha \to 0 \text{ in } X \text{ for which } \theta x_\alpha \to y\}.$$

Elementary facts about the separating space are: $\mathfrak{S}(\theta)$ is a closed linear subspace of Y and $\mathfrak{S}(\theta) = \{0\}$ if and only if θ has a closed graph. If R is a continuous linear map of Y into a TVS Z, then $R\mathfrak{S}(\theta) \subseteq \mathfrak{S}(R\theta)$. We shall need the following strengthening of this last remark, valid with further assumptions on X and Y. We use the term (F)-space for a complete metrizable TVS where the topology is given by a translation invariant metric. A TVS is *countably boundedly generated* if it is a countable union of bounded sets. If X is an (F)-space, so that the Baire category theorem is valid, then X is easily seen to be countably boundedly generated if and only if X is locally bounded.

PROPOSITION 2. *If X and Y are (F)-spaces and if R is a continuous linear map from Y to a TVS Z, while $\theta : X \to Y$ is linear, then $R\mathfrak{S}(\theta)^- = \mathfrak{S}(R\theta)$. In particular, $R\theta$ is continuous if and only if $R\mathfrak{S}(\theta) = \{0\}$.*

PROOF: The inclusion $R\mathfrak{S}(\theta)^- \subseteq \mathfrak{S}(R\theta)$ is valid in general. For the reverse inclusion let $S := R\mathfrak{S}(\theta)^-$ and consider the quotient maps $Q : Y \to Y/\mathfrak{S}(\theta)$ and $Q_1 : Z \to Z/S$. Since $\ker Q \subseteq \ker(Q_1 R)$, it follows that there is a linear map $R_1 : Y/\mathfrak{S}(\theta) \to Z/S$ for which $R_1 Q = Q_1$R. Moreover, by openness of Q, R_1 is continuous. Since X and $Y/\mathfrak{S}(\theta)$ are (F)-spaces, $Q\theta$ is continuous, by the closed graph theorem, hence $R_1 Q\theta = Q_1 R\theta$ is continuous and so $\mathfrak{S}(Q_1 R\theta) = \{0\}$, from which we conclude that $\mathfrak{S}(R\theta) \subseteq R\mathfrak{S}(\theta)^-$.

3. AUTOMATIC CONTINUITY, CONTINUED: STABILITY LEMMA.

Arguably, the most basic tool in automatic continuity is the *stability lemma*. It has several forms, but all have proofs based on a gliding hump. This one dates back to [L]; the present phrasing is taken from [D 2, 4.2.9], more or less.

STABILITY LEMMA. *If $\theta : X_0 \to Y$ is a linear map from the (F)-space X_0 to the countably boundedly generated (F)-space Y with separating space $\mathfrak{S}(\theta)$, and if $T_i : X_i \to X_{i-1}$ are continuous linear maps between (F)-spaces, for i = 1,2,.., then there is an integer n_0 for which*

$$\mathfrak{S}(\theta T_1 T_2 \ldots T_n) = \mathfrak{S}(\theta T_1 T_2 \ldots T_{n_0}) \quad \textit{for all } n > n_0.$$

A proof of this may be found in [A N]. It would appear that this result, like the Mittag-Leffler theorem, is an inherently countable result. But this is not the case; uncountable versions, involving semi-groups, exist, cf. e.g. [S]; we will not pursue this, though, as these results are not needed here. It is worth pointing out about the stability lemma, that the assumption of countable bounded generation of the space Y is not just a feature of the proof; as pointed out in [D 4.2] there are fairly simple consequences of it which are not true if Y is assumed only to be a (F)-space.

LEMMA 4. *If X and Y are (F)-spaces, if $T_1,..,T_n$ are continuous linear maps from (F)-spaces $X_1,..,X_n$ into X for which $X = T_1X_1 + .. + T_nX_n$, and if $\theta : X \to Y$ is a linear map, then $\mathfrak{S}(\theta) = [\mathfrak{S}(\theta T_1) + \ldots + \mathfrak{S}(\theta T_n)]^-$.*

PROOF: If $S := [\mathfrak{S}(\theta T_1) + \ldots + \mathfrak{S}(\theta T_n)]^-$, then the inclusion $S \subseteq \mathfrak{S}(\theta)$ is straightforward. Conversely, if $Q : X \to X/S$ then, for $j = 1,..,n, Q\mathfrak{S}(\theta T_j) = \{0\}$, so by proposition 2, each $Q\theta T_j$ is continuous. The map $(x_1,..,x_n) \to (T_1x_1,..,T_nx_n)$ is open and so $Q\theta$ is continuous. It follows, again by proposition 2, that $\mathfrak{S}(\theta) \subseteq S$.

4. LOCALIZING THE SINGULARITIES.

We consider now a regular Hausdorff space Ω and the family $\mathcal{F}(\Omega)$ of all closed subsets of Ω as well as a TVS X and its collection $\mathcal{S}(X)$ of closed linear subspaces. A map $\mathcal{E} : \mathcal{F}(\Omega) \to \mathcal{S}(X)$ is *decomposable* if $X = \mathcal{E}(U^-) + \mathcal{E}(V^-)$, whenever $\{U, V\}$ is an open cover of Ω, and is *stable* if $\mathcal{E}(\emptyset) = \{0\}, \mathcal{E}(\Omega) = X$ and $\mathcal{E}(\bigcap F_\alpha) = \bigcap \mathcal{E}(F_\alpha)$, for any family $\{F_\alpha\} \subseteq \mathcal{F}(\Omega)$. Note that a stable map preserves the inclusion order. To describe this latter fact we use the term *order-preserving*.

LEMMA 5 (cf. [L], [D 2, 4.3.23]). *Let Ω be a regular Hausdorff space, let X be an (F)-space and let Y be a countably boundedly generated (F)-space. Suppose $\mathcal{E}_X : \mathcal{F}(\Omega) \to \mathcal{S}(X)$ is order-preserving and decomposable, while $\mathcal{E}_Y : \mathcal{F}(\Omega) \to \mathcal{S}(Y)$ is stable. Let $\theta : X \to Y$ be a linear map for which*

$$\mathfrak{S}(\theta \mid \mathcal{E}_X(A)) \subseteq \mathcal{E}_Y(A), \quad \text{for every} \quad A \in \mathcal{F}(\Omega) \qquad (*).$$

Then there is a finite set $F \subseteq \Omega$ for which $\mathfrak{S}(\theta) \subseteq \mathcal{E}_Y(F)$.

PROOF: First we show that if $\{U_i\}$ is a sequence of open sets for which $U_i \cup U_j = \Omega$ whenever $i \neq j$, then there is an integer i so that for all $j > i, \quad \mathfrak{S}(\theta) = \mathfrak{S}(\theta \mid \mathcal{E}_X(U_j^-))$.

To see this, let $V_n := \bigcap_{j<n} U_j$; then $V_n \cup U_n = \Omega$ so that $X = \mathcal{E}_X(U_n^-) + \mathcal{E}_X(V_n^-)$. Hence by lemma 4, $\mathfrak{S}(\theta) = [\mathfrak{S}(\theta \mid \mathcal{E}_X(U_n^-)) + \mathfrak{S}(\theta \mid \mathcal{E}_X(V_n^-))]^-$. On the other hand,

by the stability lemma, applied to the injections of $\mathcal{E}_X(V_n^-)$ into $\mathcal{E}_X(V_{n-1}^-)$ or into X, the sequence $\{\mathfrak{S}(\theta \mid \mathcal{E}_X(V_n^-))\}$ stabilizes. Moreover, since $\mathfrak{S}(\theta \mid \mathcal{E}_X(V_n^-)) \subseteq \mathfrak{S}(\theta \mid \mathcal{E}_X(U_{n-1}^-))$, the displayed claim follows.

To prove the existence of the finite set, set $F := \bigcap U^-$, where U ranges over all open sets for which $\mathfrak{S}(\theta) \subseteq \mathcal{E}_Y(U^-)$. Since $\mathcal{E}_Y$ is stable, $\mathfrak{S}(\theta) \subseteq \bigcap \mathcal{E}_Y(U^-) = \mathcal{E}_Y(F)$. If F is not finite then, by regularity, we can find a countable collection of pairwise disjoint closed sets $F_i \subseteq \Omega$ for which $F_i \cap F \neq \varnothing$, and hence, refining this slightly and taking complements, a sequence of pairwise covering open sets U_i for which F is not a subset of U_i^-. By the first part of this proof, eventually $\mathfrak{S}(\theta) = \mathfrak{S}(\theta \mid \mathcal{E}_X(U_i^-))$, and hence, eventually $\mathfrak{S}(\theta) \subseteq \mathcal{E}_Y(U_i^-)$, by assumption. But then $F \subseteq U_i$. This contradicts the definition of F, and completes the proof.

Definition. With $\theta, \mathcal{E}_X$ and $\mathcal{E}_Y$ as above, we shall say that θ *has* $(\mathcal{E}_X, \mathcal{E}_Y)$*- localizable discontinuities* if $(*)$ holds.

5. DECOMPOSITION OF SPECTRA.

If X is a TVS and T is a continuous linear map on X then we shall use the symbol $\sigma(T)$ to denote the spectrum of T, i.e. the complement in the complex plane of the set of points $\lambda \in \mathbb{C}$ for which the operator $T - \lambda$ has a continuous inverse on X.

If T and S are given continuous linear maps on X and Y, respectively, then $\lambda \in \mathbb{C}$ is a *critical eigenvalue for* (S, T) if λ is an eigenvalue for S and $(T - \lambda)X$ is of infinite codimension in X.

For any linear map S on a vector space X we may introduce the *algebraic spectral subspaces* $E_S(A)$: for given $A \subseteq \mathbb{C}$ we define $E_S(A)$ to be the largest S-invariant subspace of X on which all restrictions of $S - \lambda$, as λ ranges over $\mathbb{C} \backslash A$, are surjective. It is clear from the definition that $E_S(A) \subseteq \bigcap_{\lambda \notin A, n \in \mathbb{N}} (S - \lambda)^n X$, and sometimes equality holds. As an example of this, suppose S is injective and take $A := \mathbb{C} \setminus \{0\}$. Then $E_S(A) = \bigcap S^n X$.

We shall call a continuous linear map T on a TVS X *admissible* if the map $A \to E_T(A)$ is a stable map from $\mathcal{F}(\mathbb{C})$ into $\mathcal{S}(X)$. The restrictive part of this definition is that the subspaces $E_T(A)$ must be closed.

The map T is said to be *semiadmissible* if there is linear continuous injection i into another TVS $X\hat{}$ and an admissible map $T\hat{}$ on $X\hat{}$ for which $T\hat{\imath} = iT$.

Finally, we shall call T *decomposable* if there is a map $\mathcal{E}_T : \mathcal{F}(\mathbb{C}) \to \mathcal{S}(X)$, whose range consists of T-invariant subspaces, and which is stable, decomposable and satisfies the condition that $\sigma(T \mid \mathcal{E}_T(F)) \subseteq F$ for every $F \in \mathcal{F}(\mathbb{C})$.

Examples. a) Super-decomposable operators with no non-trivial divisible subspaces on Banach spaces are admissible [L N 1].

b) Multipliers on commutative semi-simple Banach algebras are semiadmissible [L N 2]. This class of examples will be explained and extended later.

Suppose that X and Y are topological vector spaces and that T and S are continuous linear maps on X and on Y, respectively. Suppose $\theta : X \to Y$ is linear, but not necessarily continuous. We shall say that θ is an *intertwiner* (of S and T) if $S\theta = \theta$T. More general concepts of intertwining may be and have been developed, but to preserve the flow of the main ideas we shall consider only the form defined here.

Examples. If X and Y are topological algebras, i.e. topological vector spaces equipped with a multiplication which is continuous in each variable separately, and if $\theta : X \to Y$ is an algebra homomorphism, then θ intertwines multiplication by a fixed element x on X and multiplication by $\theta(x)$ on Y.

LEMMA 6. *Suppose that X and Y are topological vector spaces, that T is decomposable on X, and that S is admissible on Y. Suppose $\theta : X \to Y$ is an intertwiner of S and T. Then $\theta\mathcal{E}_T(A) \subseteq E_S(A)$ for every $A \in \mathcal{F}(\mathbb{C})$ and hence θ has $(\mathcal{E}_T, E_S)$-localizable discontinuities.*

PROOF: Let $A \subseteq \mathbb{C}$ be given and take $\lambda \notin A$: since $\sigma(T \mid \mathcal{E}_T(A)) \subseteq A, (T-\lambda)\mathcal{E}_T(A) = \mathcal{E}_T(A)$, hence $(S-\lambda)\theta\mathcal{E}_T(A) = \theta(T-\lambda)\mathcal{E}_T(A) = \theta\mathcal{E}_T(A)$, from which it follows, by maximality of $E_S(A)$, that $\theta\mathcal{E}_T(A) \subseteq E_S(A)$. Since $E_S(A)$ is closed, the second claim follows.

THEOREM 7. *Let X be an (F)-space and Y be a countably boundedly generated (F)-space. Suppose T is decomposable on X and that S is admissible on Y. Suppose $\theta : X \to Y$ is an intertwiner of S and T. Then there is a non-trivial polynomial p such that $p(S)\theta$ is continuous. We may assume that all roots of p are critical eigenvalues of (S, T).*

PROOF: By lemma 6, lemma 5 applies and hence there is a finite set $F \subseteq \mathbb{C}$ for which $\mathfrak{S}(\theta) \subseteq E_S(F)$. Apply the stability lemma to the sequence $T - \lambda$, where $\lambda \in F$. This yields a polynomial p for which $\mathfrak{S}(\theta p(T)) = \mathfrak{S}(\theta p(T)(T - \lambda))$ for every $\lambda \in F$. Since θ intertwines S and T this means that $(S - \lambda)p(S)\mathfrak{S}(\theta)^- = (p(S)\mathfrak{S}(\theta))^-$ for every $\lambda \in F$, by proposition 2. By Mittag-Leffler, there is a dense subspace $W \subseteq p(S)\mathfrak{S}(\theta)^-$ for which $(S - \lambda)W = W$ for every $\lambda \in F$. This means that $W \subseteq E_S(\mathbb{C} \setminus F)$, by the definition of the algebraic spectral subspaces. Since $W \subseteq \mathfrak{S}(\theta) \subseteq E_S(F)$, we obtain that $W \subseteq E_S(F) \cap E_S(\mathbb{C} \setminus F) = E_S(\emptyset)$, hence that $W = \{0\}$, because S is assumed admissible. Consequently, $p(S)\mathfrak{S}(\theta) = \{0\}$. All injective factors $S - \lambda$ may be removed from $p(S)$; what is left will still annihilate $\mathfrak{S}(\theta)$. Thus we have obtained a polynomial p, all of whose roots are eigenvalues of S, for which $p(S)\theta$ is continuous. Moreover, since $p(S)\theta = \theta p(T)$, all non-critical eigenvalues may also be removed from the factorization of p. In fact, if λ is non-critical, then $\dim(X/(T - \lambda)X) < \infty$ and since X is an (F)-space, finite codimension of $(T - \lambda)X$ is enough to ensure that $(T - \lambda)X$ is closed; moreover, $T - \lambda$ is an open map of X onto $(T - \lambda X)$. Thus, if $p(T) = r(T)(T - \lambda)$ then $\theta p(T)$ is continuous on X if and only if $\theta r(T)$ is continuous on $(T - \lambda)X$ and this is the case if and only if $\theta r(T)$ is continuous on X. Thus only the critical eigenvalues are left.

THEOREM 8 (cf. [L N 2]). *If T is a continuous linear operator on an (F)-space and T is a quotient of a decomposable operator on an (F)-space, if S is a continuous linear operator on an (F)-space and S is semiadmissible with respect to an admissible operator on a countably boundedly generated (F)-space, then every intertwiner for the pair (S, T) is continuous if and only if (S, T) has no critical eigenvalues.*

PROOF: Suppose that (S, T) has no critical eigenvalues. Assume first that S is admissible and that Y is locally bounded. Let q be a quotient map and $T\hat{}$ be decomposable such that $qT\hat{} = Tq$. Then $S(\theta q) = (\theta q)T\hat{}$. By theorem 7 there is a polynomial p, all of whose roots are critical eigenvalues of $(S, T\hat{})$, for which $p(S)\mathfrak{S}(\theta q) = \{0\}$. Thus $p(S)\theta q = \theta p(T)q$ is continuous. Since q is an open map, we conclude that $\theta p(T)$ is continuous. By construction of p, all its roots are eigenvalues of S. Let λ be one such root. If $\dim(X/(T-\lambda)X) < \infty$ and if $p(T) = r(T)(T-\lambda)$ then, just as in the proof of theorem 7, $\theta p(T)$ is continuous on X if and only if $\theta r(T)$ is continuous on $(T-\lambda)X$ if and only if $\theta r(T)$ is continuous on X. Thus we may remove from the polynomial $p(T)$ all factors $T - \lambda$ for which λ is not a critical eigenvalue of (S, T), without altering the continuity of its product with θ. Since we assume that (S, T) has no critical eigenvalues this means removing all linear factors of $p(T)$ and completes the proof when S is admissible.

We shall use this to handle the case when S is semiadmissible. Let i be a continuous linear injection and $S\hat{}$ a corresponding admissible map on a locally bounded space for which $S\hat{}i = iS$. Since $S\hat{}i\theta = i\theta T$, we may apply what has been proved so far, and obtain a polynomial p, all of whose roots are critical eigenvalues of $(S\hat{}, T)$, for which $i\theta p(T) = p(S\hat{})i\theta = ip(S)\theta$ is continuous, hence for which $p(S)\theta$ is continuous, by proposition 2. Let $q(S)$ be the product of all the factors $S - \lambda$ of $p(S)$ for which λ is an eigenvalue of S, i.e. for which $S - \lambda$ is not injective. Then continuity of $p(S)\theta$ is equivalent to continuity of $q(S)\theta$, by proposition 2. It follows that $\theta q(T)$ is continuous. Moreover, all the roots of q are critical eigenvalues of (S, T). This proves the theorem: if (S, T) has no critical eigenvalues, then θ is continuous.

The necessity is classical and easy: a one-dimensional discontinuous intertwiner can be constructed if (S, T) has a critical eigenvalue, cf. e.g. [S]: suppose $\lambda \in \mathbb{C}$ is a critical eigenvalue of (S, T) and let $y \in Y$ be an eigenvector. Use the infinite codimension of $(T-\lambda)X$ to select a discontinuous linear functional ϕ on X vanishing on $(T-\lambda)X$. Define $\theta(x) := \phi(x)y$. Then θ is discontinuous and $(S-\lambda)\theta = \theta(T-\lambda) = 0$, so that $S\theta = \theta T$.

Remark. While the passage from theorem 7 to theorem 8 has not enlarged the class of spaces related to the operator T, it has improved the situation around S slightly, since the local boundedness condition may be removed (at the cost of assuming the existence of the injection i and the operator $S\hat{}$). Note that the existence of the injection i does not force Y to be locally bounded, as an elaboration of an example of Bourgin [B] will show: the space of sequences (x_i) for which $\sum \mid x_i \mid^{p_i} < \infty$, where p_i is a sequence of positive reals converging monotonically to 0, is not locally bounded, but imbeds continuously into ℓ^2, say. However, the biggest gain in theorem 8 probably lies in the fact that it has generalized the kinds of operators T and S to which the theory applies. There are results of Albrecht and Eschmeier to show this; in still unpublished work they have identified the class of operators which are quotients of decomposable operators (on Banach space), and thereby shown it to be quite a bit larger than the class of decomposable maps. Corollary 10 below illustrates the increased scope as well: there are no explicit assumptions on the range operator there, other than it being a multiplier. This is enough to ensure semiadmissibility, but not admissibility.

6. APPLICATIONS.

We mention only a few possibilities.

For $a \in \mathbb{R} \setminus \{0\}$ and a function $f : \mathbb{R} \to \mathbb{C}$, the *shift operator* S_a is defined as usual: $(S_a f)(t) := f(t-a)$. We then have

PROPOSITION 9. *If p and $q \in]0, \infty[$ and if $\theta : L^p(\mathbb{R}) \to L^q(\mathbb{R})$ is a linear operator which commutes with a non-trivial shift operator, then θ is continuous.*

SKETCH OF PROOF: Let $p \in]0, \infty[$ and let $X := L^p(\mathbb{R})$. The shift S_a has no eigenvalues on X. For any $f \in C^\infty(\mathbb{C})$ let $\{c_n(f)\}$ denote the Fourier coefficients of the restriction of f to the unit circle $\mathbb{T}$. Since $c_n(f) = o(n^{-k})$, as $|n| \to \infty$, for any $k \in \mathbb{N}$, and since S_a is an invertible isometric map of X, it is not difficult to show that the map

$$\phi : f \to \sum_{n \in \mathbb{Z}} c_n(f) S_a^n$$

is an algebra homomorphism from $C^\infty(\mathbb{C})$ into the algebra of continuous linear operators on X, for which $\phi(1) = I$, the identity on X, and $\phi(z) = S_a$.

Now define, for a given closed subset $A \subseteq \mathbb{C}$,

$$\mathcal{E}(A) := \{x \in X \mid \phi(u)x = 0 \text{ for every } u \in C^\infty(\mathbb{C}) \text{ with } \operatorname{supp}(u) \cap A = \emptyset\}.$$

The map $A \to \mathcal{E}(A)$ has all the desired properties: it is stable and decomposable, each closed subspace $\mathcal{E}(A) \subseteq X$ is S_a-invariant and $(S_a - \lambda) \mid \mathcal{E}(A)$ is invertible on $\mathcal{E}(A)$. Moreover, $\mathcal{E}(A) = E_{S_a}(A)$. Now use theorem 8.

We conclude with a brief look at multipliers. If A is an algebra, then a map $S : A \to A$ is called a *multiplier* if $S(a)b = S(b)a$ for all $a, b \in$ A. Specific examples will appear in corollary 10. The following remarks are to be taken as mere suggestions for further investigation. An (F)-algebra A is an algebra (over the complex field $\mathbb{C}$) which is an (F)-space in which multiplication $(a, b) \to ab : A \times A \to A$ is separately (and hence jointly) continuous. Incidentally, by a result of Pełczyński [Z, Theorem 2.3], if a unital (F)-algebra A is locally bounded, then the topology on A is given by a submultiplicative, p-homogeneous norm (p-homogeneous means there is a real number $p \in]0,1]$ such that for any $t \in \mathbb{C}$ and any $x \in A$, $\| tx \| = |t|^p \| x \|$). However, for multiplier considerations, the unital case is perhaps not so interesting since on a unital algebra, every multiplier is given by an element of the algebra itself.

Suppose A is a commutative, semi-simple (F)-algebra with completely regular space $\Delta(A)$ of continuous linear multiplicative functionals. The Gelfand transform $a \to \hat{a} : A \to C(\Delta(A))$ is continuous and injective. Suppose S is a multiplier on A. A reference for the Banach algebra theory of multipliers is [La]; the arguments from there may be adapted to show that S is necessarily linear. Continuity follows readily from the closed graph theorem, combined with the semi-simplicity: if $a_\alpha \to 0$ and $Sa_\alpha \to x_0$, then, for any $x \in A$, $xSa_\alpha = a_\alpha Sx \to 0 = xx_0$, and hence $x_0 = 0$. Also, as in [La], there is a continuous function $\hat{S} : \Delta(A) \to \mathbb{C}$ for which $\widehat{(Sa)} = \hat{S}\hat{a}$, for any $a \in A$. Since $C(\Delta(A))$ is a regular algebra, it seems likely that multiplication by $\hat{S}$ is an admissible operator, cf. [L N 1]. From this it would follow that the multiplier S is subadmissible and theorem 8 would yield (cf. [L N 2]) that if X is an (F)-space and A a commutative semi-simple locally bounded (F)-algebra, while T is a decomposable operator on X and

S a multiplier on A, then every $\theta : X \to A$ which intertwines T and S is continuous if and only if (S, T) has no critical eigenvalues.

A specific instance of this is the following. Decomposable convolution operators are discussed at length in [L N 3] and [L N 4].

COROLLARY 10 [L N 2]. *Let G be a locally compact abelian group and let $\mu, \nu \in M(G)$ be measures on G. Let the convolution operators*

$$T_\mu, \quad \text{resp.} \quad T_\nu : L^1(G) \to L^1(G),$$

be defined by $T_\mu(f) := f * \mu$, resp. $T_\nu(f) := f * \nu$, for $f \in L^1(G)$. Suppose T_μ is decomposable. Then every intertwining linear transformation $\theta : L^1(G) \to L^1(G)$ for (T_ν, T_μ) is continuous if and only if (T_ν, T_μ) has no critical eigenvalues.

Remarks. a) If $\hat{\nu}$ is non-constant on every open subset of the dual group Γ, then T_ν has no eigenvalues, hence (T_ν, T_μ) has no critical eigenvalues.

b) The apparent gain in generality in assuming T_μ of corollary 10 to be just a quotient of a decomposable operator is indeed only apparent. In [L N 3] it is shown that such a T_μ is itself decomposable.

7. REFERENCES.

[A N] Albrecht, E., Neumann, M. M.: Automatische Stetigkeitseigenschaften einiger Klassen linearen Operatoren, Math. Ann. 240 (1979), 251-280.

[A] Allan, G. R.: Embedding the algebra of formal power series in a Banach algebra, Proc. London Math. Soc. (3) 25 (1972), 329-340.

[B] Bourgin, D. G.: Linear topological spaces, Amer. J. Math. 65 (1943), 637-659.

[D 1] Dales, H. G.: Automatic continuity, a survey, Bull. London Math. Soc. 10 (1978), 129-183.

[D 2] Dales, H. G.: *Banach algebras and automatic continuity* (forthcoming monograph).

[J S] Johnson, B. E., Sinclair, A. M.: Continuity of linear operators commuting with continuous linear operators II, Trans. Amer. Math. Soc. 146 (1969), 533-540.

[La] Larsen, R.: *Theory of Multipliers*, Springer-Verlag, Berlin 1971.

[L] Laursen, K. B.: Some remarks on automatic continuity, Springer Lecture Notes 512.

[L N 1] Laursen, K. B., Neumann, M. M.: Decomposable operators and automatic continuity, J. Operator Theory 15 (1986), 33-51.

[L N 2] Laursen, K. B., Neumann, M. M.: Automatic continuity of intertwining linear operators on Banach spaces, preprint February 1991 (18 pp.) (submitted).

[L N 3] Laursen, K.B., Neumann, M.M.: Local spectral properties of multipliers of Banach algebras, preprint September 1990 (10 pp.), (submitted).
[L N 4] Laursen, K. B., Neumann, M.M.: *Decomposable multipliers and applications to harmonic analysis*, preprint November 1990 (38 pp.), (submitted).
[S] Sinclair A. M.: *Automatic continuity of linear operators*, London Math Society Lecture Notes Series 21, Cambridge University Press.
[Z] Zelasko, W.: Metric generalizations of Banach algebras, Rozprawy Mathematyczne XLVII.

Progress in Functional Analysis
K.D. Bierstedt, J. Bonet, J. Horváth & M. Maestre (Eds.)
1992 Elsevier Science Publishers B.V.

Barrelled function spaces

Lech Drewnowski[a], Miguel Florencio[b] and Pedro J. Paúl[b]*

[a]Inst. Matematyki, U.A. Mickiewicz, ul. Matejki 48/49, 60-769 Poznań, Poland

[b]E.S. Ingenieros Industriales, Avda. Reina Mercedes s/n, 41012 Sevilla, Spain

This paper is dedicated to Don Manuel Valdivia on the occasion of his 60th birthday.

Abstract

Let (X, Σ, μ) be a finite measure space, Λ a complete, solid lattice of scalar functions defined on X, and E a normed space. We study barrelledness properties of the space $\Lambda(E)$ of strongly measurable functions $f : X \to E$ such that $\|f(\cdot)\| \in \Lambda$.

1991 Mathematics Subject Classification: 46A08, 46E40, (46G10)
Keywords: Barrelled spaces. Lattices of integrable functions. Bochner integral.

In what follows, (X, Σ, μ) stands for a measure space, where μ is a finite, positive, countably additive measure defined on a σ–algebra Σ of subsets of X.

Let Λ be a complete, solid, locally convex lattice of measurable scalar functions $\phi : X \to \mathbb{R}$ (as usual, we identify functions that are equal μ–a.e.). The topology of Λ can be defined by a family of continuous seminorms Q such that whenever ϕ is in Λ and ψ is a measurable function satisfying $|\psi(\cdot)| \leq |\phi(\cdot)|$ then $\psi \in \Lambda$ and $q(\psi) \leq q(\phi)$ for every $q \in Q$. If we denote by χ_Y the characteristic function of a measurable set Y, then the set of all continuous linear mappings

$$\phi \in \Lambda \to \chi_Y \cdot \phi \in \Lambda \qquad (Y \in \Sigma)$$

is equicontinuous.

* The second and third named authors acknowledge support from La Consejería de Educación y Ciencia de la Junta de Andalucía.

Examples of Λ are the classical Lebesgue spaces $L_p(\mu)$ $(1 \le p \le +\infty)$; Orlicz spaces $L_\Phi(\mu)$ and Köthe function spaces as defined by Dieudonné [**1**]. These and more examples can be found in [**4**], [**6**], [**7**], [**8**], [**12**], [**14**], or [**15**].

The space Λ is said to be order–continuous if

$$\lim_{\mu(Y)\to 0} \chi_Y \cdot \phi = 0, \qquad \text{for all } \phi \in \Lambda.$$

This is the case of $L_p(\mu)$ for $1 \le p < +\infty$ and Orlicz spaces $L_\Phi(\mu)$ when Φ satisfies the Δ_2 condition. When Λ is a perfect Köthe space endowed with its strong topology $\beta(\Lambda, \Lambda^\times)$, then (under certain conditions on the measure space) Λ is order–continuous if and only if its topological dual Λ' equals its Köthe dual $\Lambda^\times$ [**14**], [**15**, Ch. 15].

Definition. Let E be a normed space (not necessarily complete). We define $\Lambda(E)$ as the space of (classes of) strongly measurable functions $f : X \to E$ such that the composition

$$t \in X \to f(t) \in E \to \|f(t)\| \in \mathbb{R}$$

is a function in Λ. To avoid confusion we shall denote the composition above by $N(f)$, thus $N(f)(t) := \|f(t)\|$ for $t \in X$ and $N(f)$ is a function in Λ.

We shall consider $\Lambda(E)$ endowed with the topology defined by the seminorms

$$f \in \Lambda(E) \to q(N(f)) \in \mathbb{R} \qquad (q \in Q).$$

If Λ is order–continuous, then we have the following useful property

$$\lim_{\mu(Y)\to 0} \chi_Y \cdot f = 0, \qquad \text{for all } f \in \Lambda(E).$$

When $\Lambda = L_p(\mu)$ with $1 \le p < +\infty$, we obtain the space $L_p(\mu, E)$ of p–Bochner integrable functions. $L_\infty(\mu, E)$ is the space of strongly measurable, essentially bounded functions from X into E with the ess–sup norm.

Back in our $\Lambda(E)$, note that if $\phi \in \Lambda$ and $h \in U_1(L_\infty(\mu, E))$, the unit ball of $L_\infty(\mu, E)$, then the function f given by $f(t) := \phi(t) \cdot h(t)$ $(t \in X)$ is in $\Lambda(E)$ and the mapping $\phi \to \phi \cdot h$ is continuous. On the other hand, every function $f \in \Lambda(E)$ can be decomposed in the same way, namely $\phi = N(f)$ and

$$h(t) = \begin{cases} (N(f)(t))^{-1} f(t) & \text{if } N(f)(t) \neq 0; \\ 0 & \text{otherwise.} \end{cases}$$

Macdonald [**9**] and Phuong–Các [**11**] studied several aspects of spaces $\Lambda(E)$ when E is a Banach space and Λ is a perfect Köthe space endowed with its strong topology $\beta(\Lambda, \Lambda^\times)$. In this case, $\Lambda(E)$ is complete.

Our purpose here is to give conditions for $\Lambda(E)$ to be quasi–barrelled or barrelled. Our barrelledness result, Theorem 2 below, is based on a sliding–hump technique that we used to prove that the space of Pettis integrable functions and other metrizable spaces are barrelled [**3**]. In the framework of this paper, the spaces Λ and $\Lambda(E)$ are not metrizable in general. That is why we start by giving the following quasi–barrelledness result.

Theorem 1. *Let Λ be a complete, solid, locally convex lattice of measurable functions defined over a finite measure space (X, Σ, μ). Let E be a normed space. If Λ is quasi–barrelled then $\Lambda(E)$ is also quasi–barrelled.*

Proof. Let M be a $\beta(\Lambda(E)', \Lambda(E))$–bounded subset of $\Lambda(E)'$, the topological dual of $\Lambda(E)$. We have to prove that M is equicontinuous.

For every $u \in M$ and every function $h \in U_1(L_\infty(\mu, E))$ we consider the composition

$$T(u,h) : \phi \in \Lambda \to h \cdot \phi \in \Lambda(E) \to \langle u, h \cdot \phi \rangle \in \mathbb{R}.$$

It is clear that $T(u,h)$ is a continuous linear form on Λ. Now, if C is a bounded set in Λ, then it is easy to see that

$$\{\phi \cdot h : \; h \in U_1(L_\infty(\mu, E)), \; \phi \in C\}$$

is a bounded subset of $\Lambda(E)$, and since M is strongly bounded we have that

$$\sup\{|\langle T(u,h), \phi\rangle| = |\langle u, h \cdot \phi\rangle| : \; u \in M, \; h \in U_1(L_\infty(\mu, E)), \; \phi \in C\} < +\infty.$$

This shows that

$$N := \{T(u,h) : \; u \in M, \; h \in U_1(L_\infty(\mu, E))\}$$

is a $\beta(\Lambda', \Lambda)$–bounded subset of Λ'. Since Λ is quasi–barrelled it follows that N is equicontinuous. Therefore, there is a continuous seminorm $q \in Q$ such that

$$\sup\{|\langle T(u,h), \phi\rangle| : \; u \in M, \; h \in U_1(L_\infty(\mu, E))\} \leq q(\phi), \qquad \text{for all } \phi \in \Lambda.$$

As we pointed out, every function $f \in \Lambda(E)$ can be written as $f = N(f) \cdot h$ where $h \in U_1(L_\infty(\mu, E))$. Then we have the following

$$\sup\{|\langle u, f\rangle| : u \in M\} \leq \sup\{|\langle T(u,h), N(f)\rangle| : u \in M,\, h \in U_1(L_\infty(\mu, E))\} \leq q(N(f)),$$

and this proves that M is equicontinuous. Q.E.D.

Remark. Note that we do not use the hypothesis that Λ is complete, so that the theorem is true in that slightly more general case.

For our second result, we shall use a simple, useful fact that we have not found in the literature: Let F be a locally convex space. A sequence (x_n) in F is said to be locally convergent to x if there is a disc B in F such that (x_n) converges to x in F_B, the linear span of B normed with the gauge p_B of B. See e. g. [**10**, 3.2 and 5.1] or [**5**, 8.3 and 10.1]. A sequence is locally null if it is locally convergent to 0. If M is a $\sigma(F', F)$–bounded subset of F', we know that M is $\beta(F', F)$–bounded if and only if M is uniformly bounded on every null sequence in F [**10**, 3.1.3]. The fact announced is the following slightly stronger result:

Fact. *Let F be a locally convex space and M a $\sigma(F', F)$–bounded subset of F'. Then M is $\beta(F', F)$–bounded if and only if*

$$\sup\{\langle u, x_n\rangle : u \in M,\ n \in \mathbb{N}\} < +\infty$$

whenever (x_n) is a locally null sequence in F.

Proof. Note that if (z_n) is a null sequence in F and $\sup\{\langle u, z_n\rangle : u \in M\} > n^2$, then the sequence given by $x_n := n^{-1} z_n$ is locally null and $\sup\{\langle u, x_n\rangle : u \in M\} > n$. Q.E.D.

We will also use the following lemma about locally null sequences in $\Lambda(E)$.

Lemma. *Let (f_n) be a locally null sequence in $\Lambda(E)$. Then*

(i) $(N(f_n))$ is a locally null sequence in Λ.

(ii) If (Y_n) is a sequence of measurable sets, then $(\chi_{Y_n} \cdot f_n)$ is locally null in $\Lambda(E)$.

Proof. Let B be a disc in $\Lambda(E)$ such that $p_B(f_n) \to 0$. For the proof of (i) consider the set C defined as the absolutely convex and closed hull of

$$\{\phi \in \Lambda : |\phi(\cdot)| \leq \|f(\cdot)\| \text{ for some } f \in B\}.$$

The set C is bounded in Λ because B is bounded, hence C is a disc in Λ. Now, for every nonzero $f \in \Lambda(E)_B$ we have $(p_B(f))^{-1}f \in B$ hence

$$N((p_B(f))^{-1}f) = (p_B(f))^{-1}N(f) \in C$$

and it follows that $p_C(N(f)) \leq p_B(f)$. Therefore $p_C(N(f_n)) \to 0$.

(ii) The set D defined as the absolutely convex and closed hull of

$$\{\chi_Y \cdot f : f \in B, Y \in \Sigma\}$$

is a disc in $\Lambda(E)$; note that D is bounded because B is bounded and $\{\chi_Y : Y \in \Sigma\}$ is equicontinuous. Now, if $Y \in \Sigma$ and $f \in rB$ for some $r > 0$ then $\chi_Y \cdot f \in rD$ so that

$$\{r > 0 : f \in rB\} \subseteq \{\rho > 0 : \chi_Y \cdot f \in \rho D\};$$

taking infima yields $p_D(\chi_Y \cdot f) \leq p_B(f)$. Therefore $p_D(\chi_{Y_n} \cdot f_n) \to 0$. Q.E.D.

Theorem 2. *Let Λ be a complete, solid, order–continuous locally convex lattice of measurable functions defined over a finite measure space (X, Σ, μ). Let E be a normed space. If the measure μ is atomless and Λ is barrelled, then $\Lambda(E)$ is barrelled.*

Proof. By Theorem 1, $\Lambda(E)$ is quasi–barrelled so that, according to the Fact above, we have to prove that if $M \subset \Lambda(E)'$ is $\sigma(\Lambda(E)', \Lambda(E))$-bounded and $(f_m)_m$ is a locally null sequence in $\Lambda(E)$, then

$$\sup\{|\langle u, f_m\rangle| : m \in \mathbb{N}, u \in M\} < +\infty.$$

Suppose, on the contrary, that this supremum is $+\infty$. We start by making a sliding–hump type inductive construction:

(Step 1) Call $X_0 := X$; on account of our assumption, we have:

$$\sup\{|\langle u, \chi_{X_0} \cdot f_m\rangle| : m \in \mathbb{N}, u \in M\} = +\infty \tag{1}$$

and, therefore, we can find $u_1 \in M$ and $m(1) \in \mathbb{N}$ such that

$$\alpha := \left|\left\langle u_1, \chi_{X_0} \cdot f_{m(1)} \right\rangle\right| > 1.$$

Apply the fact that Λ is order–continuous: we can find $\delta > 0$ such that whenever $\mu(C) < \delta$ then

$$\left|\left\langle u_1, \chi_C \cdot f_{m(1)} \right\rangle\right| < \alpha - 1.$$

We can write X_0 as a finite union of disjoint measurable sets of measure at most δ because the measure space is atomless. By (1), on one of these sets, say X_1, the corresponding supremum must be infinite, thus having:

$$\sup\{|\langle u, \chi_{X_1} \cdot f_m \rangle| : m > m(1),\ u \in M\} = +\infty. \tag{2}$$

Set $Y_1 := X_0 \backslash X_1$; then, because $\mu(X_1) < \delta$, it follows:

$$\begin{aligned} \left|\left\langle u_1, \chi_{Y_1} \cdot f_{m(1)} \right\rangle\right| &= \left|\left\langle u_1, \chi_{X_0} \cdot f_{m(1)} \right\rangle - \left\langle u_1, \chi_{X_1} \cdot f_{m(1)} \right\rangle\right| \geq \\ &\geq \left|\left\langle u_1, \chi_{X_0} \cdot f_{m(1)} \right\rangle\right| - \left|\left\langle u_1, \chi_{X_1} \cdot f_{m(1)} \right\rangle\right| > \\ &> \alpha - (\alpha - 1) = 1. \end{aligned}$$

(Step 2) We start from (2) by finding $u_2 \in M$ and $m(2) > m(1)$ such that

$$\left|\left\langle u_2, \chi_{X_1} \cdot f_{m(2)} \right\rangle\right| > 2.$$

Proceeding as in step 1, we can find a set $X_2 \subset X_1$, such that

$$\sup\{|\langle u, \chi_{X_2} \cdot f_m \rangle| : m > m(2), u \in M\} = +\infty,$$

and

$$\left|\left\langle u_2, \chi_{Y_2} \cdot f_{m(2)} \right\rangle\right| > 2,$$

where $Y_2 := X_1 \backslash X_2$ is disjoint with Y_1.

In this way, we are able to find a sequence $(u_n)_n \in M$, a disjoint sequence $(Y_n)_n \in \Sigma$ and a subsequence $(f_{m(n)})_n$ of $(f_m)_m$ such that

$$\left|\left\langle u_n, \chi_{Y_n} \cdot f_{m(n)} \right\rangle\right| > n, \qquad n = 1, 2, \ldots \tag{3}$$

By (ii) in the Lemma above, the sequence $g_n := \chi_{Y_n} \cdot f_{m(n)}$, $n = 1, 2, \ldots$ is locally null in $\Lambda(E)$ and, by (i) in the same Lemma, $(N(g_n))$ is also locally null in Λ. Let C be a disk in Λ such that $p_C(N(g_n)) \to 0$. Then we can take a subsequence $(g_{n(k)})_k$ such that $\sum_k p_C(N(g_{n(k)})) < +\infty$. Since C is a Banach disc, because Λ is complete, we have that $\sum_k N(g_{n(k)})$ is convergent in Λ_C, and consequently in Λ, to some function $\psi \in \Lambda$.

Now, consider the function $g : X \to E$ defined pointwise by $g(\cdot) := \sum_k g_{n(k)}(\cdot)$. This function g is well–defined because the functions $g_{n(k)}$ have disjoint supports $Y_{n(k)}$, and it is strongly measurable. Moreover, for $t \in X$ we have

$$N(g)(t) = \sum_k N(g_{n(k)})(t) = \psi(t),$$

thus $N(g) = \psi \in \Lambda$ and it follows that $g \in \Lambda(E)$. By (3) and the continuity of each $u_{n(k)}$ we have

$$\left|\left\langle u_{n(k)}, \chi_{Y_{n(k)}} \cdot g \right\rangle\right| = \left|\left\langle u_{n(k)}, \chi_{Y_{n(k)}} \cdot g_{n(k)} \right\rangle\right| > n(k) \geq k, \qquad k = 1, 2, \ldots$$

Since M is $\sigma\left(\Lambda(E)', \Lambda(E)\right)$–bounded, the scalar measures defined by

$$m_k : Y \in \Sigma \to m_k(Y) := \left\langle u_{n(k)}, \chi_Y \cdot g \right\rangle, \qquad k = 1, 2, \ldots$$

form a set–wise bounded set, i.e.

$$\sup\{|m_k(Y)| : k = 1, 2, \ldots\} < +\infty, \qquad \text{for all } Y \in \Sigma.$$

Therefore, we may apply the Nikodým Boundedness Theorem [**1**, I.3.1] to deduce that

$$\sup\{|m_k(Y)| : k = 1, 2, \ldots, Y \in \Sigma\} < +\infty$$

in contradiction with the fact that

$$\left|m_k(Y_{n(k)})\right| = \left\langle u_{n(k)}, \chi_{Y_{n(k)}} \cdot g \right\rangle \geq k, \qquad k = 1, 2, \ldots$$

Q.E.D.

Remarks. This result was proved in [**13**] for Λ a Köthe function space and under more restrictive conditions on the measure space and the normed space, namely X locally compact and E' having the Radon–Nikodým Property. The proof there made a strong use of the form of $\Lambda(E)'$.

It is a bit surprising that E is not required to be barrelled. An unexpected consequence of Theorem 2 is that if E is not barrelled and the underlying measure space is atomless, then one cannot localize E as a quotient of $\Lambda(E)$.

REFERENCES

1 J. Diestel and J.J. Uhl, Vector Measures. Mathematical Surveys, no. 15, American Mathematical Society. Providence, 1977.

2 J. Dieudonné, Sur les espaces de Köthe, *J. Analyse Math.* **1** (1951), 81–115.

3 L. Drewnowski, M. Florencio and P.J. Paúl, The space of Pettis integrable functions is barrelled. *Preprint* (1990).

4 I. Halperin, Function Spaces, Proc. Int. Symposium on Linear Spaces. Hebrew Univ., Jerusalem (1960), 242–250.

5 H. Jarchow, Locally Convex Spaces. B.G. Teubner. Stuttgart, 1981.

6 J. Lindenstrauss and L. Tzafriri, Classical Banach Spaces II. Springer–Verlag. Berlin, Heidelberg and New York, 1979.

7 J.A. López Molina, The dual and bidual of an echelon Köthe space, *Collect. Math.* **31** (1980), 159–191.

8 G.G. Lorentz and D.G. Wertheim, Representation of linear functionals on Köthe spaces, *Canad. J. Math* **5** (1953), 568–575.

9 A. Macdonald, Vector valued Köthe function spaces III, *Illinois J. Math.* **18** (1974), 136–146.

10 P. Pérez Carreras and J. Bonet, Barrelled Locally Convex Spaces. North–Holland, Elsevier. Amsterdam, New York, Oxford and Tokyo, 1987.

11 N. Phuong–Các, Generalized Köthe function spaces I, *Proc. Cambridge Phil. Soc.* **65** (1969), 601–611.

12 K. Reiher, Weighted inductive and projective limits of normed Köthe function spaces, *Resultate Math.* **13** (1988), 147–161.

13 C. Sáez, Espacios de Funciones Vectoriales. Tesis Doctoral, Universidad de Sevilla. Sevilla, 1990.

14 R. Welland, On Köthe spaces, *Trans. Amer. Math. Soc.* **112** (1964), 267–277.

15 A.C. Zaanen, Integration. North–Holland, Elsevier. Amsterdam, 1967.

Progress in Functional Analysis
K.D. Bierstedt, J. Bonet, J. Horváth & M. Maestre (Eds.)

On distinguished Fréchet spaces

José Bonet[a] and Susanne Dierolf[b]

Dedicated to Prof. M. Valdivia on the occasion of his 60th *birthday*

[a]Universidad Politécnica de Valencia, E-46071 Valencia, Spain.

[b]FB IV-Mathematik, Universität Trier, D-5500 Trier, Germany.

Abstract

We report on recent results about distinguished Fréchet spaces concerning the behaviour of biduals, stability properties and characterizations in the context of Köthe echelon spaces, of Fréchet spaces of Moscatelli type and projective tensor products. Some new results on an extension of a classical example of Amemiya are also included (see section 5).

1 Introduction

The purpose of this article is to give some information on the structural behaviour of distinguished Fréchet spaces, including recent results in this direction.

Let E be a metrizable locally convex space and let E' denote its topological dual. A very natural topology on E' is the so-called strong topology $\beta(E', E)$ (of uniform convergence on all bounded subsets of E). If E is a Banach space, then $E'_b := (E', \beta(E', E))$ is also a Banach space, but in the general case, the space E'_b is a complete DF-space (in the sense of Grothendieck [27]) and it has a different structure. Its dual $E'' := (E'_b)'$ is called the bidual of E, and it is a Fréchet space with respect to the strong topology $\beta(E'', E')$. On the other hand, given a 0-basis $(U_n)_{n\in\mathcal{N}}$ in E, the polars U_n^o generate Banach spaces $[U_n^o]$ which form an inductive sequence, and their inductive limit ind $[U_n^o]$ is nothing else than E' equipped with an LB-space topology which is stronger than the strong topology $\beta(E', E)$. Since this LB-space topology is independent of the special choice of the 0-basis $(U_n)_{n\in\mathcal{N}}$, we may call it the inductive dual of E and use the notation E'_i.

The identity map $id : E_i' \longrightarrow E_b'$ is continuous and E_i' is a complete, hence regular LB-space. According to [27], E_i' coincides with the "bi-strong" dual $(E', \beta(E', E''))$.

Now a locally convex space E is called distinguished if E_b' is barrelled (which means exactly that $\beta(E', E) = \beta(E', E'')$), and we obtain that whenever a distinguished metrizable locally convex space E is the reduced projective limit of a sequence $(E_n)_{n \in \mathcal{N}}$ of normed spaces, then its strong dual E_b' is the inductive limit of the dual Banach spaces E_n'. For a Fréchet space E one has the equivalence: E is distinguished if and only if for every representation $E =$ proj E_n of E as the reduced projective limit of a sequence of Banach spaces, its strong dual E_b' equals ind E_n'.

The introduction and investigation of distinguished Fréchet spaces goes back to J. Dieudonné and L. Schwartz [22], and the first (very) deep results in this direction were obtained by A. Grothendieck in 1953/54 [27].

Being distinguished is -in some sense- the least good property a Fréchet space may have. Putting this in a humorous way one might say that distinguishedness is the "entrance ticket to good society for Fréchet spaces". In order to indicate how some of the many interesting subclasses of Fréchet spaces, which are under investigation (nowadays), are interrelated which each other and are all contained in the class of distinguished Fréchet spaces, we would like to present the following little chart:

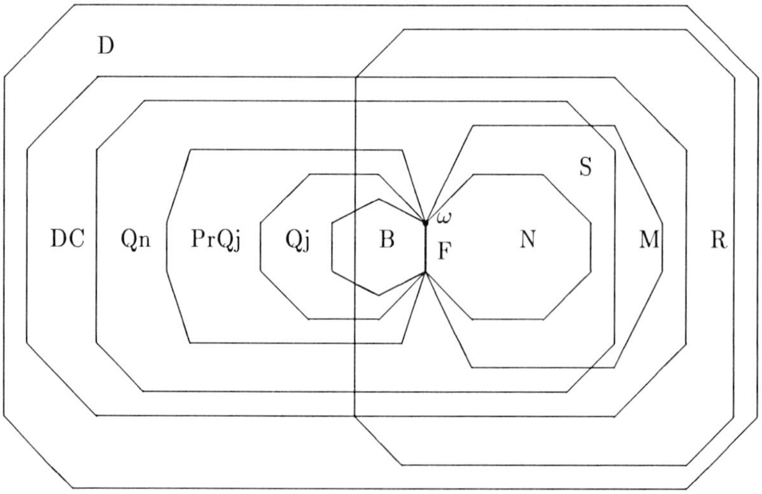

where

F= finite dimensional spaces,

B= Banach spaces,

$\omega = \mathcal{K}^{\mathcal{N}}$,

Qj= quojections, i.e., projective limits of a sequence of Banach spaces with surjective linking maps,

PrQj= prequojections, i.e., Fréchet spaces whose bidual is a quojection,

Qn= quasinormable Fréchet spaces E (i.e. every bounded subset of E'_b gets its topology from a Banach space $[U^o]$ with U a 0-neighbourhood in E),

DC= Fréchet spaces with Stefan Heinrich's density condition (i.e. all bounded subsets of the strong dual are metrizable, see [7]),

R= reflexive Fréchet spaces,

M= Fréchet Montel spaces,

S= Fréchet Schwartz spaces,

N= nuclear Fréchet spaces,

D= distinguished Fréchet spaces.

In connection with this chart, we would like to remark that up to the mid-80's the main interest in Fréchet spaces was concentrated on the nuclear Fréchet spaces and the Fréchet Schwartz spaces, whereas since that time -according to the work of Bierstedt, Diaz, Domanski, Meise, Metafune, Moscatelli, Taskinen, Terzioğlu, Valdivia, Vogt and the authors, among others- also the non-normable left hand side of the chart has gained considerable interest (see the references).

2 Examples of nondistinguished Fréchet spaces

Since according to the above chart, the nondistinguished Fréchet spaces seem to constitute the "rest of the world", we should make sure that the rest of the world is not empty.

The first example of a non-distinguished Fréchet space E was given by Köthe and Grothendieck in 1954 (see [27]). It is an echelon space $\lambda_1(A)$ with a suitable Köthe matrix A and has the property that E'' and $(E'_i)'$ are different.

Another (fairly elaborate) example was presented by Kōmura [29]; it has the somewhat complementary striking property that its bidual coincides with the dual of its inductive dual, whence its strong dual is not a Mackey space.

But, two members to constitute the "rest of the world" is certainly not a lot. Moreover we would like to remark that there are mainly two substantial resources for examples of Fréchet spaces nowadays, namely the Köthe echelon spaces and the so-called Fréchet spaces of Moscatelli type. We would like to start with the second class.

3 Fréchet spaces of Moscatelli type

Given the following two entries

i) a normal Banach sequence space $L = (L, \|\cdot\|)$ (e.g. a space l_p with $p \in [1, \infty]$ or c_0 or a diagonal transform of such spaces) and

ii) two Banach spaces X, Y and a continuous linear map $f : Y \longrightarrow X$ with dense range,

one defines the space

$$E := E(L, Y \xrightarrow{f} X) := \{(y_k)_{k\in\mathcal{N}} \in \prod Y \ : \ (f(y_k))_{k\in\mathcal{N}} \in L(X)\}$$

which is a Fréchet space when provided with the initial topology with respect to the inclusion $i : E \longrightarrow \prod Y$ and the map $E \longrightarrow L(X)$ induced by f. Moreover, E has the following representation as a projective limit of Banach spaces

$$E \ = \ \text{proj} \prod_{k<n} Y \times L((X)_{k\geq n})$$

where $L((X)_{k\geq n}) := \{(x_k)_{k\geq n} \in \prod X \ : \ (\| x_{k+n-1} \|_X)_{k\in\mathcal{N}} \in L\}$ and $L(X) := L((X)_{k\geq 1})$.

It was Moscatelli in 1980 (see [32]) who discovered that "shifting device" to construct twisted Fréchet spaces (i.e. Fréchet spaces without a continuous norm which are not isomorphic to a product of Fréchet spaces each of which has a continuous norm).

So, each Fréchet space of Moscatelli type depends on the two entries i) and ii) and we write it as $E(L, Y \xrightarrow{f} X)$. If every $a = (\alpha_k)_{k\in\mathcal{N}} \in L$ is the limit of its finite sections (which can be equivalently stated in the form $L = \bar{\varphi}$), the dual L' of L, provided with its dual norm, is again a normal Banach sequence space.

The following characterization of distinguished Fréchet spaces of Moscatelli type has been obtained by the authors and C. Fernández (see [12, 14, 25]).

Theorem 1 *Let $E = E(L, Y \xrightarrow{f} X)$ be a Fréchet space of Moscatelli type.*

a) If $L = \bar{\varphi}$ and $L' = \bar{\varphi}$, then E is always distinguished.

b) If $L = \bar{\varphi}$ but $L' \neq \bar{\varphi}$ or if $L \neq \bar{\varphi}$, then E is distinguished if and only if f is surjective.

The typical example for an L in a) is c_0, whereas the typical cases in b) are l_1 and l_∞, respectively. By [12, 2.10] for an $E = E(L, Y \xrightarrow{f} X)$ the surjectivity of f is equivalent to E being a quojection and also to E being quasinormable, or even to E having the density condition.

We would like to indicate the structure of our proof.

Proof.

a) In the case $L = \bar{\varphi}$, the distinguishedness of E is equivalent to the fact that on $E' = \bigoplus Y' + L'(X')$ the quotient topology coming from $\bigoplus Y \times L'(X')$ and the so-called projective hull topology coincide (see [12, 2.3]).

By [12, 2.5] these topologies will coincide if $L' = \bar{\varphi}$; and if $L' \neq \bar{\varphi}$, then they coincide if and only if the transpose $f^t : X' \longrightarrow Y'$ is a topological isomorphism, i.e. if and only if f is surjective (see [12, 2.4] plus some computation).

b) If $L = l_\infty$, one meets the difficulty that there is no nice description of the dual of $l_\infty(X)$ and hence no representation of E'. This difficulty can be overcome with the help of the following

Lemma 2 ([14]) *Given Fréchet spaces G, H such that $G \subset H \subset G''$. If H is distinguished, then the quotient map $q : H \longrightarrow H/G$ "lifts bounded sets with closure" (i.e. $\forall B \subset H/G$ bounded $\exists A \subset H$ bounded : $\overline{q(A)} \supset B$).*

Now one proceeds as follows: Given $f : Y \longrightarrow X$ as in ii) then

$$G := E(c_0, Y \xrightarrow{f} X) \subset H := E(l_\infty, Y \xrightarrow{f} X) \subset G'' = E(l_\infty, Y'' \xrightarrow{f^{tt}} X'').$$

Consequently, if H is distinguished, $q : H \longrightarrow H/G$ will lift bounded sets with closure, from which one derives -via a technical condition- the surjectivity of f (see [14]), i.e. the quasinormability of H.

c) If L is an arbitrary Banach sequence space such that φ is not dense in L, a block construction due to C. Fernández (see [25]) shows that $E(L, Y \xrightarrow{f} X)$ contains $E(l_\infty, Y \xrightarrow{f} X)$ as a complemented subspace. Thus, if $E(L, Y \xrightarrow{f} X)$ is distinguished, one obtains with b) the surjectivity of f and one is done. □

The above characterization of distinguishedness in the class of Moscatelli type Fréchet spaces enabled the authors and C. Fernández to give a negative answer to the following question of A. Grothendieck from 1954 [27]:

<u>Is the bidual of every distinguished</u>
<u>Fréchet space again distinguished?</u>

In fact, let $f : Y \longrightarrow X$ with Y, X Banach be given such that f is linear, continuous and has dense range, but is not open. Then $G := E(c_0, Y \xrightarrow{f} X)$ is distinguished by a) in the theorem, whereas $G'' = E(l_\infty, Y'' \xrightarrow{f^{tt}} X'')$ is not distinguished, as f and hence f^{tt} are not surjective.

For a comparison, we would like to mention how some other properties of Fréchet spaces (see the chart) behave with respect to the formation of the bidual: quasinormability, being a quojection and having the density condition are all stable under the formation of the bidual (as can be shown rather straightforward in all the three cases), whereas the property of having a continuous norm is not (see [20, 37]).

4 Köthe echelon spaces

In order to construct a Köthe echelon space, one again starts with two entries:

i) $p \in 0 \cup [1, \infty]$, an index set I and

ii) a Köthe matrix $A = (a_j^{(n)})_{j \in I,\ n \in \mathcal{N}}$ $\quad 0 < a_j^{(n)} \leq a_j^{(n+1)} \quad (j \in I,\ n \in \mathcal{N})$

and forms

$$\lambda_p(A) := \{(x_j)_{j \in I} \in \mathcal{K}^I \ : \ \forall n \in \mathcal{N} \ \ (a_j^{(n)} x_j)_{j \in I} \in l_p(I)\}$$

(where $l_0(I) = c_0(I)$), which is a Fréchet space with respect to obvious norms. The distinguishedness of echelon spaces has also been completely characterized.

It is well-known that

a) $\lambda_p(A)$ is distinguished for $p = 0$ or $1 < p < \infty$ (for $1 < p < \infty$, $\lambda_p(A)$ is even reflexive); e.g. see [10].

In the cases $p = 1$ or $p = \infty$ we have (cf [3, 4, 7, 9, 40]):

Theorem 1 *Let A a Köthe matrix on I. The following conditions are equivalent:*

(1) A satisfies condition (D) of Bierstedt and Meise (cf [9] or [7]).

(2) $\lambda_1(A)$ is distinguished.

(3) $\lambda_p(A)$ has the density condition ($1 \leq p \leq \infty$ or $p = 0$).

(4) $\lambda_\infty(A)$ is distinguished.

Theorem 2 *Let A be a Köthe matrix on I. The following are equivalent:*

(1) A satisfies condition (ND) of Bierstedt and Meise (cf [9] or [4]).

(2) $\lambda_1(A)$ is not distinguished.

(3) The duals of $\lambda_1(A)'_b$ and $\lambda_1(A)'_i$ do not coincide.

(4) There is a sectional subspace of $\lambda_1(A)$ which is isomorphic to a Köthe echelon space $\lambda_1(\mathcal{N} \times \mathcal{N}, B)$ where the Köthe matrix $B = (b^{(n)})$ satisfies (i) $b_{i,j}^{(n)} = b_{i,j}^{(1)}$ for $n \leq i$, (ii) $\lim_j \ (b_{n,j}^{(n)} / b_{n,j}^{(n+1)}) = 0$ (cf [40]).

The last theorem shows that every non-distinguished $\lambda_1(A)$ space contains a non-distinguished Köthe echelon space with a structure very similar to the Köthe Grothendieck example [27].

Although these two big classes of Fréchet spaces, the echelon spaces and those of Moscatelli type, are fairly different (e.g. the former contains nontrivial Montel spaces and nontrivial quasinormable spaces but no non-normable quojections, whereas the latter contains nontrivial quojections, but only trivial Montel spaces and every quasinormable Fréchet space of Moscatelli type is a quojection), these classes also have a lot of similarity, see, e.g., the above characterizations of distinguishedness. Moreover the intersection of both of them is not empty because the Köthe Grothendieck non-distinguished example belongs to both classes (e.g. [12]).

Non-distinguished Fréchet spaces have been occurring frequently in the work of Taskinen. For instance, his first counterexample to the problem of topologies of Grothendieck was a non-distinguished Fréchet space of Moscatelli type (see [35]). Later, Taskinen and Bonet showed that many Fréchet function spaces of a certain "intersection" type like e.g. $C^k(\Omega) \cap L_1(\Omega)$, $k \in \mathcal{N}_0 \cup \{\infty\}$, Ω an open subset of $\mathcal{R}^n$, are non-distinguished (see [36] and [15]). It is a conjecture of Moscatelli and Metafune that many of these function spaces of intersection type (i.e., roughly speaking, a Banach function space intersected with a nontrivial Fréchet space of functions over the same set) are non-distinguished of Moscatelli type.

5 Stability properties

The stability properties of distinguished spaces are rather poor. One has stability for the trivial settings such as topological products, direct sums and complemented subspaces. On the other hand, quotients of distinguished Fréchet spaces need not be distinguished, since every separable Fréchet space can be represented as a quotient of a suitable Fréchet Montel space and since there exist separable non-distinguished Fréchet spaces. Since every Fréchet space can be represented as a subspace of a product of Banach spaces, distinguishedness is not stable under the formation of closed subspaces.

Furthermore, there is no stability with respect to dense linear subspaces, as can be derived from a classical example of Amemiya [1]. Of course, for this question we must leave the class of Fréchet spaces and pass over the more general class of metrizable locally convex spaces. The idea behind Amemiya's example is the following

Construction. Start with the following entries:

A metrizable locally convex space E, a linear subspace $L \subset E$. Let $q: E \longrightarrow E/L$ denote the quotient map, and let M be another metrizable locally convex space continuously included into E/L. Then $p: E \times M \longrightarrow E/L$, $(x, y) \mapsto q(x) - y$, is linear, continuous and open; hence we get the short exact sequence of metrizable locally convex spaces

$$0 \longrightarrow F := ker\ p \xrightarrow{i} E \times M =: G \xrightarrow{p} E/L \longrightarrow 0$$

Clearly, F is closed (resp. dense) in $E \times M$ if and only if L is closed (resp. dense) in E. Moreover, if there is a bounded set $B \subset M$ such that

$$\overline{q(A) \cap M}^M \not\supset B$$

for all $A \subset E$ bounded, the set $0 \times B \subset E \times M$ is bounded and

$$\overline{C}^{E \times M} \not\supset 0 \times B$$

for all $C \subset F$ bounded.

Now -if in addition L is dense in E- this fact implies that on $G' = F'$ the topology $\beta(G', F)$ is strictly coarser than $\beta(G', G)$, whence F is necessarily non-distinguished. (Note that $\beta(G', G'')$ is an LB-space topology on G' which is stronger than $\beta(G', G)$.)

In order to realize this idea in a concrete example, let X be any non-normable Fréchet space, and let $(B_\gamma)_{\gamma \in \Gamma}$ be a fundamental system of bounded sets in X. Put $E := l_2(X, \Gamma)$. Since X is non-normable, for every $\gamma \in \Gamma$ there is a dense proper hyperplane $H_\gamma \subset X$ such that $B_\gamma \subset H_\gamma$ (in fact if $[\,\overline{B_\gamma}\,]$ is closed in X, it must be of infinite codimension). Fix $x_\gamma \in X \setminus H_\gamma$ for all $\gamma \in \Gamma$. Clearly $L := (\prod H_\gamma) \cap E$ is a dense linear subspace of E and the normed space

$$M := \{(c_\gamma)_{\gamma \in \Gamma} \in l_2(\Gamma) \ : \ \{\gamma \in \Gamma \ : \ c_\gamma \neq 0\} \text{ is finite } \}$$

is included into E/L by mapping the γth unit vector in M onto $q((x_\gamma \delta_\beta)_{\beta \in \Gamma})$, where $q: \ E \longrightarrow E/L$ denotes the quotient map. Assume, there is $A \subset E$ bounded such that $\overline{q(A) \cap M}^M$ contains the unit ball of M. As $l_2(X, \Gamma)$ embeds continuously into $l_\infty(X, \Gamma)$, there is $\beta \in \Gamma$ such that

$$A \subset (\prod_{\gamma \in \Gamma} B_\beta) \cap E \subset ((\prod_{\gamma \neq \beta} X) \times H_\beta) \cap E$$

consequently

$$q(A) \cap M \subset \{(c_\gamma)_{\gamma \in \Gamma} \in M \ : \ c_\beta = 0 \ \}$$

which is a contradiction.

Thus $F := \{(x, y) \in E \times M \ : \ q(x) = y\}$ is a non-distinguished dense linear subspace of $E \times M$ and hence of $E \times \tilde{M} := l_2(X, \Gamma) \times l_2(\Gamma)$ which is the completion $\tilde{F}$ of F. We may choose X to be a non-normable reflexive quojection, by which $\tilde{F}$ will be a reflexive quojection.

Consequence: *A reflexive quojection can contain a non-distinguished dense linear subspace. A fortiori, the bidual of a non-distinguished metrizable locally convex space can be reflexive and hence distinguished.*

This result is connected with the following problem

Can the bidual of a non-distinguished
Fréchet space be distinguished?

which goes back to Grothendieck [27] and which seems to be open.

For a comparison with other properties of Fréchet spaces we would like to mention:

- *A Fréchet space is quasinormable whenever its bidual is quasinormable* (Bonet [11], see also Vogt [39]).
- *There exist proper prequojections, i.e., Fréchet spaces whose biduals are quojections, but that are not quojections themselves* (Behrends, Dierolf, Harmand, Moscatelli [5] and [21]).
- *The question whether there is a Fréchet space without the density condition whose bidual has the density condition also seems to be open.*

The last topic we would like to treat in this section concerns the *three-space-problem*: Given a short exact sequence

$$0 \longrightarrow G \longrightarrow H \longrightarrow H/G \longrightarrow 0$$

of Fréchet spaces such that G and H/G have a certain property (P), is it then true that H also has (P)?

This problem has a positive solution for the properties: quasinormable, reflexive, Montel, Schwartz, nuclear (see [34]) and for being a quojection (see [30]).

In order to investigate this three-space-problem for the properties of being distinguished and having the density condition, a construction of the type presented above turns out to be useful (see Bonet, Dierolf, Fernández [13]).

In fact, one starts with a setting $E \xrightarrow{q} E/L$, $M \xrightarrow{i} E/L$, where E, M are Fréchet spaces, i is a continuous inclusion and L is a closed subspace of E. We would like to choose E, L, M in such a way that E/L is a separable Banach space, $F := \{(x,y) \in E \times M \ : \ q(x) = y\}$ becomes a Montel space, but M is not distinguished.

Then, in the short exact sequence

$$0 \longrightarrow F \longrightarrow E \times M \longrightarrow E/L \longrightarrow 0$$

the spaces F, E/L would both have the density condition and $E \times M$ would not even be distinguished. In [13], C. Fernández and the authors have presented a realization of such a setting.

This negative answer to the three-space-problem for distinguished Fréchet spaces and for Fréchet spaces with the density condition is in some sense "best possible", since according to [13] there is the following partial positive result:

Proposition 1 *Let H be a Fréchet space and G a closed linear subspace of H such that G and H/G are distinguished (resp. have the density condition). If the quotient map $q : H \longrightarrow H/G$ lifts bounded sets with closure, then H is distinguished (resp. has the density condition).*

Furthermore, q will lift bounded sets with closure, if, for instance, G is quasinormable or if H/G is a Montel space.

Concerning the lifting of bounded sets from quotients, which already appeared several times here, the authors have proved very recently the following result that will be published elsewhere.

Theorem 2 *Let H be a Fréchet space and let G be a closed subspace of H. Let $q : H \longrightarrow H/G$ be the quotient map. If q lifts bounded sets of H/G with closure, then q also lifts bounded sets without closure.*

It is well-known that the previous theorem is false for (DF)-spaces H and G.

6 Projective tensor products of distinguished Fréchet spaces

Already Grothendieck [28] considered the question whether the complete projective tensor product $E\hat{\otimes}_\pi F$ of two distinguished Fréchet spaces E and F is also distinguished. He proved that the answer is positive if E or F is nuclear and also if both E and F are quasinormable, and in this case $E\hat{\otimes}_\pi F$ is even quasinormable. The first counterexample was obtained by the second author [19]. She constructed a separable reflexive Fréchet space F of Moscatelli type such that $l_1\hat{\otimes}_\pi F$ is not distinguished. Bierstedt and the first author [7] proved that, for a Fréchet space F, $l_1\hat{\otimes}_\pi F$ is distinguished if and only if F satisfies the density condition. The next result provides a complete characterization of the distinguishedness of $\lambda_1(A)\hat{\otimes}_\pi F$, and it is contained in [7].

Theorem 1 *Let F a Fréchet space and let A be a Köthe matrix on an index set I.*

(i) $\lambda_1(A)\hat{\otimes}_\pi E$ satisfies the density condition if and only if $\lambda_1(A)$ and E do.

(ii) $\lambda_1(A)\hat{\otimes}_\pi E$ is distinguished if and only if

(a) $\lambda_1(A)$ is Montel and E is distinguished or

(b) $\lambda_1(A)$ is distinguished and E has the density condition.

Taskinen [36] gave an example of a Fréchet Montel space F_0 with finite dimensional decomposition such that $F_0\hat{\otimes}_\pi F_0$ is not distinguished. According to Taskinen [35], we say that a pair (E, F) of Fréchet spaces has the *property* (BB) if every bounded subset of $E\hat{\otimes}_\pi F$ is contained in the closed absolutely convex hull of $B_1 \otimes B_2$ with B_1 bounded in E and B_2 bounded in F. In Taskinen's example, (F_0, F_0) does not satisfy the property (BB). The next result proved in [8] shows that the failure of property (BB) is the only obstruction for the stability of the density condition in complete projective tensor products of Fréchet spaces. This result was useful in infinite holomorphy (see [23, 26]).

Proposition 2 *Let E and F be Fréchet spaces with the density condition such that the pair (E, F) has the property (BB). Then $E\hat{\otimes}_\pi F$ has the density condition.*

Concerning proposition 2 (which extends theorem 1 (i) because $(\lambda_1(A), F)$ has property (BB)), we should add the following remark:

There are pairs (E, F) of Fréchet spaces such that $E\hat{\otimes}_\pi F$ has the density condition but which do not satisfy the property (BB). In fact in [16] examples of quojections E are given such that (E, l_2) does not have property (BB), but $E\hat{\otimes}_\pi l_2$ is quasinormable by Grothendieck's result mentioned before. More recently Diaz and Metafune [17] succeded in characterizing, under some mild assumptions, those quojections E of Moscatelli type such that (E, Z) has the property (BB) for every Banach space Z. Moreover, their approach is interesting because it does not depend on the geometry of finite dimensional spaces as in the original examples of Taskinen.

More progress concerning the projective tensor product of distinguished Fréchet spaces was obtained recently by Diaz and Miñarro [18]. They show that for distinguished Fréchet spaces E and F, the space $E\hat{\otimes}_\pi F$ is also distinguished in the following cases

(1) E or F is Fréchet Schwartz with the bounded approximation property

(2) E or F is a Fréchet Montel space with the bounded approximation property and the pair (E, F) has the property (BB).

References

[1] I. Amemiya, Some examples of (F) and (DF) spaces, Proc. Japan Acad. **33** (1957), 169-171.

[2] F. Bastin, Weighted spaces of continuous functions, Bull. Soc. Roy. Sci. Liège **59** (1990), 1-81.

[3] F. Bastin, Distinguishedness of weighted Fréchet spaces of continuous functions, Proc. Edinburgh Math. Soc. (to appear).

[4] F. Bastin, J. Bonet, Locally bounded noncontinuous linear forms on strong duals of nondistinguished Köthe echelon spaces, Proc. Amer. Math. Soc. **108** (1990), 769-774.

[5] E. Behrends, S. Dierolf, P. Harmand, On a problem of Bellenot and Dubinsky, Math. Ann. **275** (1986), 337-339.

[6] S.F. Bellenot, E. Dubinsky, Fréchet spaces with nuclear Köthe quotients, Trans. Amer. Math. Soc. **273** (1982), 579-594.

[7] K.D. Bierstedt, J. Bonet, Stefan Heinrich's density condition for Fréchet spaces and the characterization of the distinguished Köthe echelon spaces, Math. Nach. **135** (1988), 149-180.

[8] K.D. Bierstedt, J. Bonet, Density conditions in Fréchet and (DF) spaces, Rev. Matem. Univ. Complutense Madrid **2**, no. suplementario (1989), 59-76.

[9] K.D. Bierstedt, R. Meise, Distinguished echelon spaces and the projective description of weighted inductive limits of type $\mathcal{V}C(X)$, p. 169-226 in: Aspects of Mathematics and its Applications, North-Holland Math. Library, 1986.

[10] K.D. Bierstedt, R.G. Meise, W.H. Summers, Köthe sets and Köthe sequence spaces, p.27-91 in: Funct. Anal., Holomorphy and Approx. Theory, North-Holland Math. Studies **71**, Amsterdam, 1982.

[11] J. Bonet, A question of Valdivia on quasinormable Fréchet spaces, Canadian Math. Bull. (to appear).

[12] J. Bonet, S. Dierolf, Fréchet spaces of Moscatelli type, Rev. Math. Univ. Complutense Madrid **2** (1989), 77-92.

[13] J. Bonet, S. Dierolf, C. Fernández, On the three-space-problem for distinguished Fréchet spaces, Bull. Soc. Roy. Sci. Liège **59** (1990), 301-306.

[14] J. Bonet, S. Dierolf, C. Fernández, The bidual of a distinguished Fréchet space need not be distinguished, Archiv Math. (to appear).

[15] J. Bonet, J. Taskinen, Non-distinguished Fréchet function spaces, Bull. Soc. Roy. Sci. Liège **58** (1989), 483-490.

[16] J. Bonet, J. Taskinen, Quojections and the problem of topologies of Grothendieck, Note di Mat. (to appear).

[17] J.C. Diaz, G. Metafune, The problem of topologies of Grothendieck for quojections, Results of Math. (to appear)

[18] J.C. Diaz, M.A. Miñarro, Distinguished Fréchet spaces and projective tensor products, DOGA Tr. J. Math. **14** (1990), 191-208.

[19] S. Dierolf, On spaces of continuous linear mappings betwen locally convex spaces, Note di Mat. **5** (1985), 147-255.

[20] S. Dierolf, V.B. Moscatelli, A Fréchet space which has a continuous norm but whose bidual does not, Math. Z. **191** (1986), 17-21.

[21] S. Dierolf, V.B. Moscatelli, A note on quojections, Funct. et Approx. **17** (1987), 131-138.

[22] J. Dieudonné, L. Schwartz, La dualité dans les espaces (F) et (LF), Ann. Inst. Fourier **1** (1949), 61-101.

[23] S. Dineen, Holomorphic functions on Fréchet-Montel spaces, Complex Anal. Appl. (to appear).

[24] P. Domanski, $\mathcal{L}_p$-spaces and injective locally convex spaces, Dissertationes Math. **298**, 1990.

[25] C. Fernández, On distinguished Fréchet spaces of Moscatelli type, DOGA Tr. J. Math. (to appear).

[26] P. Galindo, D. García, M. Maestre, Coincidence of τ_0 and τ_ω for some classes of Fréchet Montel spaces, Proc. Roy. Ir. Acad. (to appear).

[27] A. Grothendieck, Sur les espaces (F) et (DF), Summa Brasil Math. **3** (1954), 57-123.

[28] A. Grothendieck, Produits tensoriels topologiques et espaces nucléaires, Mem. Amer. Math. Soc. **16** (1955).

[29] Y. Kōmura, Some examples of linear topological spaces, Math. Ann. **153** (1962), 150-162.

[30] G. Metafune, V.B. Moscatelli, On the three-space-problem for locally convex spaces, Collect. Math. **37** (1986), 287-296.

[31] G. Metafune, V.B. Moscatelli, Quojections and prequojections, p.235-254 in Advances in the Theory of Fréchet Spaces, Kluwer Acad. Publ., Dordrecht 1989.

[32] V.B. Moscatelli, Fréchet spaces without continuous norm and without basis, Bull. London Math. Soc. **12** (1980), 63-66.

[33] S. Önal, T. Terzioğlu, Unbounded linear operators and nuclear Köthe quotients, Archiv Math. **54** (1990), 576-581.

[34] W. Roelcke, S. Dierolf, On the three-space-problem for topological vector spaces, Collect. Math. **32** (1981), 13-35.

[35] J. Taskinen, Counterexamples to "problème des topologies" of Grothendieck, Ann. Acad. Sci. Fenn., Ser. A I Math. Dissertationes **63** (1986).

[36] J. Taskinen, Examples of nondistinguished Fréchet spaces, Ann. Acad. Sci. Fenn., Série A.I. **14** (1989), 75-88.

[37] T. Terzioğlu, D. Vogt, A Köthe space which has a continuous norm but whose bidual does not, Archiv Math. **54** (1990), 180-183.

[38] M. Valdivia, A characterization of totally reflexive Fréchet spaces, Math. Z. **200** (1989), 327-346.

[39] D. Vogt, On two problems of Mitiagin, Math. Nachr. **141** (1989), 13-25.

[40] D. Vogt, Distinguished Köthe spaces, Math. Z. **202** (1989), 143-146.

Progress in Functional Analysis
K.D. Bierstedt, J. Bonet, J. Horváth & M. Maestre (Eds.)

Prequojections and their duals

Giorgio Metafune [1] and Vincenzo B. Moscatelli [2]

[1] Dipartimento di Matematica, II Università, v. O. Raimondo, 00173 Roma, Italy

[2] Dipartimento di Matematica, Università C.P. 193, 73100 Lecce, Italy

Dedicated to Manuel Valdivia on his sixtieth birthday.

Abstract

We prove that, under very natural assumptions, the dual of a quojection, is also the dual of a Fréchet space with continuous norm (prequojection). Several consequences of this result are explored and applications are also given.

INTRODUCTION

The aim of this paper is to prove that, under very natural assumptions, the dual of a quojection is also the dual of a countably normed prequojection (see §1 for the definitions). This shows that the class of prequojections is very large indeed.

The problem of the existence of non-trivial prequojections (that is, prequojections not isomorphic to quojections) arose in a natural way in [3] and it has been solved, in the positive sense, in various papers (see [2] and also [9] + [1], [18], [19]). In particular, in [19] a countably normed prequojection is constructed whose dual is a countable direct sum of copies of l^1. All the above papers employed the same method, which consists in the construction of so-called "strongly non-norming" subspaces in the duals of suitable Banach spaces, the link with prequojections being provided by Theorem 3 of [9]. What

[1,2] Research supported by the Italian MURST.

one obtains in this way is the dual of a prequojection, rather than the prequojection itself, which can be difficult to describe (cf. our survey article [15] for further information and unexplained terminology).

Here we propose an alternative method for constructing prequojections which has the advantage of being direct, the underlying idea being dual to that of [9].

The paper is organized as follows. After a preliminary §1, in §2 we state and prove our main result: starting with an arbitrary quojection $E = \mathrm{quoj}_n(E_n, R_n)$ such that $\ker R_n$ is non-quasi-reflexive for infinitely many n, we produce by a pertubation argument Banach spaces F_n and operators $S_n\colon F_{n+1} \longrightarrow F_n$ such that $F = \mathrm{proj}_n(F_n, S_n)$ is countably normed and satisfies $F' = E'$. The condition on the kernels of the maps R_n is also proved to be necessary. In §3 we make a few comments on our result and its proof, while applications within the context of quojections of Moscatelli type are treated in §4. The final §5 contains further results involving exact sequences of prequojections.

1. PRELIMINARIES

Our notation for Banach and Fréchet spaces is standard (cf., e.g., [10]), except that, for simplicity, the strong dual and the strong bidual of a Fréchet space are denoted by F' and F''.

ω and φ will stand, respectively, for the product and the direct sum of countably many copies of the scalar field with their natural topologies, while topological isomorphism is simply denoted by " = ".

A *quojection* $E = \mathrm{quoj}_n(E_n, R_n)$ is the projective limit of a sequence of Banach spaces E_n and surjective maps $R_n\colon E_{n+1} \longrightarrow E_n$.

A quojection is *twisted* [17] if it is not isomorphic to a product of Banach spaces. Further notation concerning quojections is needed in §4 and will be introduced there.

A *prequojection* is a Fréchet space F such that F'' is a quojection and it is *non-trivial* if it is not itself a quojection. Equivalently, F is a prequojection if and only if F' is a strict (LB)-space. Alternative characterizations of a prequojection F are, e.g.: (i) if and only if every quotient of F with a continuous norm and the bounded approximation property is Banach ([13] + [6]), or (ii) if and only if F has no nuclear quotient with a continuous norm ([9] + [3] + [23]) (see [20] in place of [3] for the non-separable case). Moreover, a prequojection is quasinormable [23] and, if non-trivial, it does not have the bounded approximation property ([13], [6]). For these and related properties of prequojections and quojections the reader is referred to [15].

A Fréchet space F is *countably normed* if it is the intersection of a sequence of Banach spaces; in particular, such an F has a continuous norm. All the prequojections constructed with the method of [9, Theorem 3] are countably normed and so will be also the ones constructed in this paper.

Finally, we recall that a Banach space X is non-quasi-reflexive if dim $X''/X = \infty$.

2. THE MAIN RESULT

We shall need the following lemmas. The first one is well-known and we report here the proof for the sake of completeness.

Lemma 1. *Let X be an infinite-dimensional Banach space. There exist sequences $(x_n) \subset X$ and $(u_n) \subset X'$ such that:*

(a) $\|u_n\| = 1$ *and (u_n) is a basic sequence;*

(b) $\|x_n\| < 3$ *and $\langle x_n, u_k\rangle = \delta_{nk}$.*

Proof. Let (ε_i) be such that $0 < \varepsilon_i < 1$ and $\prod_i(1+\varepsilon_i) \leq 2$. Take $u_1 \in X'$, $\|u_1\| = 1$ and choose $x_1 \in X$ with $\|x_1\| \leq 2$ and $\langle x_1, u_1\rangle = 1$.

Suppose that, for $k \leq n$, x_k and u_k have been defined so that $\|x_k\| < 3$, $\|u_k\| = 1$, $\langle x_k, u_j\rangle = \delta_{kj}$ and $\|v\| \leq (1+\varepsilon_n)\, \|v + \lambda u_n\|$ for all real λ and all $v \in \text{span}(u_1, \ldots, u_{n-1})$. We choose $u_{n+1} \in X'$ such that $\|u_{n+1}\| = 1$, $u_{n+1} \in \{x_1, \ldots, x_n\}^0$ and $\|v\| \leq (1 + \varepsilon_{n+1})\, \|v + \lambda u_{n+1}\|$ for all real λ and all $v \in \text{span}(u_1, \ldots, u_n)$ (cf. the proof of [11, Lemma 1.a.6]). Now, for $w = \sum_{i=1}^{n+1} a_i u_i$ put $f(w) = a_{n+1}$; then

$$f \in [\text{span}(u_1, \ldots, u_{n+1})]' = \frac{X}{\{u_1, \ldots, u_{n+1}\}^0}$$

and we have

$$|f(w)| = \|a_{n+1}\, u_{n+1}\| = \left\| w - \sum_{i=1}^{n} a_i u_i \right\| \leq \|w\| + (1 + \varepsilon_{n+1})\, \|w\| < 3\, \|w\|\,.$$

Thus, we may find $x_{n+1} \in X$ such that $\|x_{n+1}\| < 3$ and $\langle x_{n+1}, u_j \rangle = f(u_j) = \delta_{j,n+1}$ for $j \leq n+1$. By induction, we obtain sequences (x_n) and (u_n) with the desired properties.

Lemma 2. *Let X be a separable and non-quasi-reflexive Banach space. There are sequences $(z_n) \subset X''$ and $(x'_n) \subset X'$ with the following properties:*

(a) $\|z_n\| = 1$, (z_n) *is basic and* $X \cap [z_n] = \{0\}$;

(b) $\|x'_n\| \leq 4$, (x'_n)*is total over* $X + [z_n]$ *and* $\langle x'_n, z_k \rangle = \dfrac{\delta_{nk}}{n}$.

Here $[z_n]$ denotes the closed linear span of (z_n).

Proof. Since X is non-quasi-reflexive, there is by [7] a closed and w^*-dense subspace M of X' for which $\dim(X'/M) = \infty$. Let (u_n) be a w^*-dense sequence in $B_M = M \cap B_{X'}$. Applying Lemma 1 to X'/M we obtain a biorthogonal system (y_n, z_n), with $(y_n) \subset X'/M$, $(z_n) \subset M^0 \subset X''$, $\|y_n\| < 3$, $\|z_n\| = 1$ and (z_n) is basic. Since $X \cap M^0 = \{0\}$, the sequence (z_n) satisfies (a). Let $\overline{y}_n$ be pre-images of y_n in X' with $\|\overline{y}_n\| \leq 3$ and put $v_n = \overline{y}_n/n$. Next, let $s: \mathbf{N} \longrightarrow \mathbf{N}$ be a surjective map such that $s^{-1}(j)$ is an infinite set for all j and put $x'_n = v_n + u_{s(n)}$; clearly $\|x'_n\| \leq 4$. Since $\langle u_n, z_k \rangle = 0$ for all n, k, we have

$$\langle x'_n, z_k \rangle = \langle v_n, z_k \rangle = \frac{\delta_{nk}}{n}. \tag{1}$$

We show that (x'_n) is total over $X + [z_n]$. Let $w = x + z \in X + [z_n]$ and suppose that, for all n,

$$0 = \langle w, x'_n \rangle = \langle x, u_{s(n)} \rangle + \langle x + z, v_n \rangle. \tag{2}$$

Fix j; then if $s(n) = j$ we have $0 = \langle x, u_j \rangle + \langle x + z, v_n \rangle$. Letting $n \longrightarrow \infty$ through $s^{-1}(j)$ and recalling that $v_n \longrightarrow 0$ we obtain $\langle x, u_j \rangle = 0$. This holds for all j and hence $x = 0$, since M is w^*-dense in X'. Now, writing $w = z = \sum_k a_k z_k$ for suitable scalars a_k, it follows from (1) and (2),

$$0 = \langle z, v_n \rangle = \sum_k a_k \langle z_k, v_n \rangle = \frac{a_n}{n}$$

and, therefore, $a_n = 0$ for all n. This shows that $w = 0$ and completes the proof of the lemma.

Remark 1. The bound on $\|x_n\|$ in Lemma 1(b) is not really necessary for the proof of Lemma 2(b). In fact, if we do not care for such a bound we may simply take

$x_{n+1} \in \{u_1, \ldots, u_n\}^0$, with $\langle x_{n+1}, u_{n+1}\rangle = 1$. Then, in the proof of Lemma 2(b) it suffices to put $v_n = \dfrac{\overline{y}_n}{n\,\|\overline{y}_n\|}$ to obtain a sequence (x'_n) satisfying $\|x'_n\| \le 2$ and $\langle x'_n, z_k\rangle = \dfrac{\delta_{nk}}{n\,\|\overline{y}_n\|}$.

Before giving our main result we have to make the following observation. Let $E = \operatorname{quoj}_n(E_n, R_n)$ be a quojection, with $R_n: E_{n+1} \longrightarrow E_n$, and put $X_n = \ker R_n$. Then one and only one of the following statements is true:

(i) X_n is non-quasi-reflexive for infinitely many n,

(ii) X_n is quasi-reflexive for all $n \ge n_0$, with a suitable n_0.

Using the three-space property for quasi-reflexive Banach spaces, it is not difficult to see that if $\operatorname{quoj}_n(F_n, S_n)$ is another representation of E (and this means, by definition, that each F_n is Banach and each $S_n: F_{n+1} \longrightarrow F_n$ is surjective) and $Y_n = \ker S_n$, then X_n and Y_n jointly satisfy either (i) or (ii). The same happens for the cases (i') and (ii') which are obtained from (i) and (ii) replacing "quasi-reflexive" by "separable".

After these remarks, we are now in the position to state and prove the following

Theorem 1. *Let $E = \operatorname{quoj}_n(E_n, R_n)$ with $X_n = \ker R_n$ separable for all n large enough. There exists a (countably normed) prequojection F with $F' = E'$ if and only if X_n is non-quasi-reflexive for infinitely many n. In such a case there is an exact sequence*

$$(3) \qquad O \longrightarrow N \longrightarrow F \longrightarrow E \longrightarrow O$$

with N nuclear (and, obviously, countably normed).

Proof. Step 1. Suppose that X_n is not quasi-reflexive for infinitely many n. By going on to a subsequence, if necessary, we may assume that X_n is not quasi-reflexive for all n. Without loss of generality, we may also assume that all X_n are separable and that all the maps R_n are quotient maps, so that their duals R'_n are isometries. For each n let $(z_k^n) \subset X''_n \subset E''_{n+1}$ and $(x'^n_k) \subset X'_n$ be sequences as in Lemma 2 and let $y'^n_k \in E'_{n+1}$ be norm-preserving extension of x'^n_k to all of E_{n+1}. For $n \ge 2$ put $W_n = [z_k^{n-1}]$ and $F_n = E_n \bigoplus W_n$, so that F_n is continuously injected into $\overline{E_n + W_n} \subset E''_n$, since $E_n \cap W_n = E_n \cap X''_{n-1} \cap W_n = X_{n-1} \cap W_n = \{0\}$. Now define a continuous linear map $J_n: E_{n+1} \longrightarrow W_n$ by $J_n = \sum_k \varepsilon_k y'^n_k \otimes z_k^{n-1}$, where $\varepsilon_k > 0$ and $\sum_k \varepsilon_k \le 1/8$; J_n is nuclear and $\|J_n\| \le 1/2$. Further, let the map $i_n: E'_n \longrightarrow E'_n \bigoplus W'_n$ be given by

$i_n(x') = x' + x'|W_n$, i.e.

$$(4) \qquad \langle i_n(x'), x + w\rangle = \langle x', x\rangle + \langle x', w\rangle \qquad \text{for } x' \in E_n', \ x \in E_n, \ w \in W_n.$$

We put $\widetilde{J}_n = \sum_k \varepsilon_k i_{n+1}(y_k'^n) \otimes z_k^{n-1}$, so that $\widetilde{J}_n : E_{n+1} \oplus W_{n+1} \longrightarrow W_n$ and $\left\| \widetilde{J}_n \right\| \leq 1/2$. Since $(i_{n+1}(y_k'^n))_k$ is total over $X_n \oplus W_{n+1}$ by construction and the sequence (z_k^{n-1}) is basic, $\widetilde{J}_n$ is one-to-one on $X_n \oplus W_{n+1}$. Moreover, by (4) and Lemma 2(b) we have, since $(z_j^n) \subset X_n''$,

$$\langle i_{n+1}(y_k'^n), z_j^n \rangle = \langle y_k'^n, z_j^n \rangle = \langle x_k'^n, z_j^n \rangle = \frac{\delta_{kj}}{k},$$

showing that $\widetilde{J}_n(W_{n+1})$ is dense in W_n.

Define a continuous linear map $S_n : E_{n+1} \oplus W_{n+1} \longrightarrow E_n \oplus W_n$ by

$$(5) \qquad S_n(x + w) = R_n\, x + \widetilde{J}_n(x + w) \qquad \text{for } x \in E_{n+1}, \ w \in W_{n+1}.$$

If $S_n(x + w) = 0$, then $R_n\, x = \widetilde{J}_n(x + w) = 0$ and hence $x \in X_n$ and $x + w \in X_n \oplus W_{n+1}$. Since $\widetilde{J}_n$ is one-to-one on $X_n \oplus W_{n+1}$, it follows that $x + w = 0$, i.e. S_n is one-to-one. Note that $S_n\,(E_{n+1} \oplus W_{n+1}) \supset S_n(W_{n+1}) = \widetilde{J}_n(W_{n+1})$ which is dense in W_n. Now let $x \in E_n$ and let $y \in E_{n+1}$ be such that $R_n\, y = x$. Given $\varepsilon > 0$ there is $w \in W_{n+1}$ for which $\left\| \widetilde{J}_n\, y - \widetilde{J}_n\, w \right\| < \varepsilon$. Since $S_n(y - w) = R_n\, y + \widetilde{J}_n(y - w)$, we have $\|S_n(y - w) - x\| < \varepsilon$ and $S_n(E_{n+1} \oplus W_{n+1})$ is dense in E_n. It follows that S_n has a dense range. Thus, recalling that $F_n = E_n \oplus W_n$, if we let $F = \operatorname{proj}_n(F_n, S_n)$, then the projective limit is reduced and F is countably normed.

Step 2. We show that $F' = E'$. Let $x' + w' \in E_n' \oplus W_n'$ and $x + w \in E_{n+1} \oplus W_{n+1}$; we have

$$(6) \quad \langle S_n'(x' + w'), x + w\rangle = \left\langle x' + w', R_n x + \widetilde{J}_n(x + w)\right\rangle = \langle R_n' x', x\rangle + \left\langle \widetilde{J}_n' w', x + w\right\rangle.$$

But $\widetilde{J}_n' = i_{n+1} J_n'$ and

$$(7) \qquad \langle i_{n+1} R_n'\, x', x + w\rangle = \langle R_n'\, x', x\rangle + \langle x', R_n''\, w\rangle = \langle R_n'\, x', x\rangle,$$

since $w \in X_n'' = X_n^{oo} = \ker R_n''$. Hence, from (6) and (7),

$$S_n'(x' + w') = i_{n+1}(R_n'\, x' + J_n'\, w').$$

Let then $A_n: E'_n \bigoplus W'_n \longrightarrow E'_{n+1}$ be defined by $A_n(x'+w') = R'_n\, x' + J'_n\, w'$; from above we have

(8) $$S'_n = i_{n+1} A_n.$$

Consider the map $A_n\, i_n: E'_n \longrightarrow E'_{n+1}$ and introduce the restriction map $r_n: E'_n \longrightarrow W'_n$ defined by $r_n(x') = x'|_{W_n}$; clearly

$$A_n\, i_n(x') = A_n\,(x' + r_n(x')) = R'_n\, x' + J'_n\, r_n(x')$$

and hence

(9) $$A_n\, i_n = R'_n + J'_n\, r_n.$$

Note that the map $J'_n\, r_n: E'_n \longrightarrow E'_{n+1}$ is not the dual of any map from E_{n+1} to E_n.

Now from (8) and (9) we have that

(10) $$\text{ind}_n(F'_n, S'_n) = \text{ind}_n(E'_n, R'_n + J'_n\, r_n).$$

Let T_2 be the identity map of E'_2. Since R'_2 in an isometry of E'_2 into E'_3, there exists a nuclear extension $L_3: E'_3 \longrightarrow E'_3$ of $J'_2\, r_2$ via R'_2, i.e. $L_3\, R'_2 = J'_2\, r_2$ and $\|L_3\| \leq 1/2$. If I_n is the identity map of E'_n, put $T_3 = I_3 + L_3$; T_3 in an automorphism of E'_3 and $T_3\, R'_2 = (R'_2 + J'_2\, r_2)T_2$. As before, let $L_4: E'_4 \longrightarrow E'_4$ be a nuclear map such that $\|L_4\| \leq 1/2$ and $L_4\, R'_3 = J'_3\, r_3$. Next, let $N_4: E'_4 \longrightarrow E'_4$ be another nuclear map satisfying $\|N_4\| \leq 1/2$ and $N_4\, R'_3 = R'_3\, L_3$. Defining T_4 by $T_4 = (I_4 + L_4)(I_4 + N_4)$ we see that T_4 is an automorphism of E'_4 such that $T_4\, R'_3 = (R'_3 + J'_3\, r_3)T_3$. Inductively, for every $n \geq 3$ we find an automorphism T_n of E'_n satisfying $(R'_n + J'_n\, r_n)T_n = T_{n+1}\, R'_n$ and hence we may assert by (10) that

(11) $$\text{ind}_n(F'_n, S'_n) = \text{ind}_n(E'_n, R'_n) = E'.$$

This shows that the left-hand side of (11) is a strict inductive limit, hence by [23, Lemma 0.2] it coincides with F' and we conclude that $F' = E'$.

Step 3. Recall that $\tilde{J}_n$ is a one-to-one, nuclear map from $E_{n+1} \bigoplus W_{n+1}$ into W_n such that $\tilde{J}_n(W_{n+1})$ is dense in W_n. Denoting by $\hat{J}_n$ the restriction of $\tilde{J}_n$ to W_{n+1} we put

(12) $$N = \text{proj}_n(W_n, \hat{J}_n).$$

Clearly N is a nuclear subspace of F. Moreover, since the projective limit in (12) is reduced, we have

$$\frac{F}{N} = \operatorname{proj}_n \left(\frac{E_n \bigoplus W_n}{W_n}, \hat{S}_n \right), \tag{13}$$

where $\hat{S}_n$ is the map induced by S_n on the quotient space. Since $\dfrac{E_n \bigoplus W_n}{W_n} = E_n$ and $\hat{S}_n = R_n$, (13) yields $F/N = E$ and hence the exact sequence (3).

Step 4. Recalling the alternative (i) – (ii) just before the theorem, suppose now that X_n is quasi-reflexive for all $n \geq n_0$, for some n_0. Without loss of generality, we may assume that X_n is quasi-reflexive for all n. Let $q: E \longrightarrow E_2$ be the canonical quotient map and $E_0 = q^{-1}(X_1)$ so that $E/E_0 = E_1$. By the three-space-property for quasi-reflexive Banach spaces, E_0 is the strict projective limit of a sequence M_n of quasi-reflexive subspaces of X_n and hence E_0''/E_0 is a closed subspace of ω. Since E_0 is a quojection, hence quasinormable, we have (cf. [8]):

$$E_1'' = \left(\frac{E}{E_0}\right)'' = \frac{E''}{E_0''}$$

and $E + E_0''$ is closed in E''. But then the equality

$$\frac{E''}{E + E_0''} = \frac{\dfrac{E''}{E_0''}}{\dfrac{E + E_0''}{E_0''}}$$

shows that $\dfrac{E''}{E + E_0''}$ is a (possibly finite-dimensional) Banach space, say Y, and we obtain

$$Y = \frac{E''}{E + E_0''} = \frac{\dfrac{E''}{E}}{\dfrac{E + E_0''}{E}}.$$

Since $(E + E_0'')/E = E_0''/E_0$ it follows that E''/E is isomorphic to $\omega_0 \times Y$, with ω_0 a closed subspace of ω (and hence $\dim \omega_0 < \infty$ or ω_0 isomorphic to ω). The fact that E is a quojection implies that $(E''/E)' = E^0 = E'''/E'$, where the polar is taken in E''', and hence E'''/E' is isomorphic to $\varphi_0 \bigoplus Z$ with $Z = Y'$ and φ_0 a closed subspace of φ.

Now suppose that F is a Fréchet space for which there is an isomorphism T of F' onto E'. Then F' is a strict (LB)-space, hence F is a prequojection and, therefore, quasinormable [23, Lemma 0.2]. Since $F''' = F' \bigoplus F^0$, we have

$$E''' = T''(F''') = T''(F') \bigoplus T''(F^0) = E' \bigoplus T''(F^0)$$

and hence

$$\frac{F'''}{F'} = F^0 = T''(F^0) = \frac{E'''}{E'} = \varphi_0 \bigoplus Z.$$

This shows that $(F''/F)' = \varphi_0 \bigoplus Z$ topologically, since F''/F is a quojection and since $(F''/F)' = F^0$ by virtue of the fact that F is quasinormable [8] (or because both natural topologies on F^0 are (LB)–space topologies). Now, if the dual of the quojection F''/F is $= \varphi_0 \bigoplus Z$, then $F''/F = \omega_0 \times \tilde{X}$ for some Banach space $\tilde{X}$ (possibly finite–dimensional). Summing up, we have shown that either $\dim F''/F < \infty$ or $F''/F = \omega$, X or $\omega \times X$ for a suitable Banach space X with $\dim X = \infty$. By using [23, Proposition 3.1] and [16, Proposition 3.5] we conclude in all cases that F is a quojection.

The theorem is now completely proved.

Remark 2. Note that the condition that the kernels X_n be quasi–reflexive for n sufficiently large is equivalent to the quotient E''/E being isomorphic to a closed subspace of $\omega \times Y$, with Y Banach, as shown in step 4 of the proof of Theorem 1. In turn, this latter case is equivalent to the existence of a continuous seminorm p on E for which $(\ker p)''/\ker p$ is isomorphic to a closed subspace of ω.

3. REMARKS

In this section we make a few comments on Theorem 1.

Remark 3. It is easier to find sequences (z_n) and (x'_n) as in Lemma 2 (with (x'_n) linearly independent) if these are not required to be biorthogonal. This may be quickly seen by using Helly's lemma. Then, if in Theorem 1 we use such sequences, the first assertion continues to hold, but not the second.

In Theorem 1 the assumption that all X_n be separable for n sufficiently large (needed only to prove the sufficiency) is used to guarantee the existence of injective maps on $X_n \bigoplus W_n$ and, at first sight, it may seem to be only a technical one. In actual fact, this is not the case as the following proposition shows.

Proposition 1. *There is a quojection $E = \mathrm{quoj}_n(E_n, R_n)$, with $\ker R_n$ non-quasi-reflexive for all n, for which there does not exist a Fréchet space F with continuous norm such that $F' = E'$.*

Proof. Let (d_n) be an increasing sequence of cardinal numbers with the property that for every n there exists m such that $2^{d_n} \leq d_m$. We take a sequence (X_n) of Banach spaces such that dens $(X_n'') = d_n$ and form the quojection $E = \prod_n X_n$. Supposing that there exists a Fréchet space F with continuous norm such that $F' = E'$, we must have $F'' = E''$. Thus, if $d = \mathrm{card}(F)$, then $\mathrm{card}(E'') = \mathrm{card}(F'') \leq 2^{2^d}$. Now regard F as a subspace of $E'' = \prod_n X_n''$. Since F has a continuous norm, there must exist $k > 1$ for which $F \cap \prod_{n>k} X_n'' = \{0\}$. It follows that the quotient map $E'' \longrightarrow \prod_{n=1}^{k} X_n''$ is one-to-one on F, hence $d \leq d_k$ and, therefore, $\mathrm{card}(E'') \leq 2^{2^{d_k}}$. But this is absurd, since $\mathrm{card}(E'') > 2^{2^{d_n}}$ for all n; hence no such an F can exist.

Remark 4. Clearly the proof of Proposition 1 may be applied, more generally, to any Fréchet space E such that, if (p_n) is a fundamental sequence of seminorms defining the topology of E'', then the cardinal numbers $d_n = \mathrm{dens}\,(E''/\ker p_n)$ satisfy $2^{d_n} \leq d_m$ for suitable $m = m(n)$.

Remark 5. The proof of Proposition 1 leads us to think that Theorem 1 may still hold if dens (X_n) is constant for n large enough. We do not know if this is true.

Finally, we note that the nuclear subspace N in the exact sequence (3) may be made more than just countably normed. Precisely, we have

Proposition 2. *Let $E = \mathrm{quoj}_n(E_n, R_n)$ be a quojection with $X_n = \ker R_n$ separable for all n large enough and non-quasi-reflexive for infinitely many n. For every nuclear Köthe space K with continuous norm there are a countably normed prequojection F_K, with $F_K' = E'$, and an exact sequence*

$$0 \longrightarrow K \longrightarrow F_K \longrightarrow E \longrightarrow 0.$$

Proof. Theorem 1 yields the exact sequence (3) with N nuclear and $F' = E'$. If $K = K(a)$ is a nuclear Köthe space with continuous norm, let us show how we can get $N = K$ (then F_K will be different from F, in general).

Let $a = (a_{nk})$ be a nuclear matrix, i.e.

$$a_{nk} > 0 \qquad \text{and} \qquad \sum_k \frac{a_{nk}}{a_{n+1,k}} < \infty \qquad \text{for all } n.$$

For every n let (b_{nk}) be a sequence such that

$$b_{nk} > 0, \qquad \lim_k b_{nk} = +\infty \qquad \text{and} \qquad \sum_k b_{nk} \frac{a_{nk}}{a_{n+1,k}} < \infty.$$

Now in the proof of Theorem 1 for each n we choose $(z_k^n) \subset X_n''$ and $(x_k'^n) \subset X_n'$ as in Lemma 2, but with $\langle x_k^n, z_j^n \rangle = \frac{\delta_{kj}}{b_{nk}}$. Next, we put $\varepsilon_{nk} = c_n b_{nk} \frac{a_{nk}}{a_{n+1,k}}$, with c_n such that $\sum_k \varepsilon_{nk} \leq 1/8$, and define $\tilde{J}_n : E_{n+1} \bigoplus W_{n+1} \longrightarrow W_n$ by $\tilde{J}_n = \sum_k \varepsilon_{nk} i_{n+1}(y_k'^n) \bigoplus z_k^{n-1}$, where, as before, each $y_k'^n$ is a norm-preserving extension of $x_k'^n$ to all of E_{n+1}. With this choice, Theorem 1 gives us an $F = F_K$, with $F_K' = E'$, and an $N = N_K$ for which we have the exact sequence (3). Since $N_K = \text{proj}_n(W_n, \tilde{J}_n)$, with $W_n = [z_k^{n-1}]$ and $\tilde{J}_n(z_k^n) = (\varepsilon_{nk}/b_{nk})\, z_k^{n-1} = c_n(a_{nk}/a_{n+1,k}) z_k^{n-1}$, it is clear that N_K is isomorphic to K.

The above proposition then implies the following noteworthy

Corollary 1. *To have the bounded approximation property is not a three-space property within the class of Fréchet spaces.*

4. PREQUOJECTIONS ARISING FROM QUOJECTIONS OF MOSCATELLI TYPE

Let us recall that an example of a countably normed prequojection whose dual is $(l^1)^{(\mathbb{N})}$ is already known (cf. [19]). The following result then completely clarifies the situation in the separable case.

Proposition 3. *Let (X_n) be a sequence of separable Banach spaces. There exists a (countably normed) prequojection F such that $F' = \bigoplus_n X_n'$ if and only if X_n is non-quasi-reflexive for infinitely many n. Moreover, in such a case for every k there is a prequojection subspace G_k of F such that*

$$F = \left(\bigoplus_{n=1}^{k} X_n \right) \bigoplus G_k \qquad \text{and} \qquad G_k' = \bigoplus_{n>k} X_n'.$$

Proof. The first part follows directly from Theorem 1 applied to $E = \prod_n X_n$. For the second part, note that we can always write, in the case indicated,

$$F = \text{proj}_n \left(X_1 \bigoplus \cdots \bigoplus X_{n-1} \bigoplus W_n, S_n \right);$$

here $W_n \subset X''_{n-1}$ and S_n are chosen as in the proof of Theorem 1 (cf. [7]) with $R_n: \bigoplus_{j=1}^{n} X_j \longrightarrow \bigoplus_{j=1}^{n-1} X_j$ the canonical projection and $\tilde{J}_n$ acting only on $X_n \bigoplus W_{n+1}$, having chosen extensions $y_j'^n$ of $x_j'^n$ which vanish on $\bigoplus_{j=1}^{n-1} X_j$. From this, the second assertion in the proposition easily follows.

Remark 6. Note that Proposition 3 yields the surprising fact that, e.g., $(l^1)^{(\mathbb{N})} \times (l^2)^{(\mathbb{N})}$ *is* the dual of a non-trivial prequojection, while $l^1 \times (l^2)^{(\mathbb{N})}$ *is not*.

Now we shall investigate those prequojections which are related to quojections of Moscatelli type. Recall that, if (X_n) is a sequence of Banach spaces, if Y_n is a closed subspace of X_n for every n and if L is a normal Banach sequence space, then the *quojection of Moscatelli type* $Q((X_n), (X_n/Y_n); L)$ is the quojection obtained by forming the projective limit of the Banach spaces

$$E_k = \left(X_1 \bigoplus \dots \bigoplus X_{k-1} \bigoplus \frac{X_k}{Y_k} \bigoplus \frac{X_{k+1}}{Y_{k+1}} \bigoplus \dots \right)_L$$

with respect to the (surjective) maps which arise from the identities of the spaces X_n and X_n/Y_n and the quotient maps $X_n \longrightarrow X_n/Y_n$.

Dually, we define the (LB)-*space of Moscatelli type* $\mathrm{LB}((X_n), (Y_n); L)$ as the strict (LB)-space obtained as the inductive limit of the Banach spaces

$$M_k = \left(X_1 \bigoplus \dots \bigoplus X_{k-1} \bigoplus Y_k \bigoplus Y_{k+1} \bigoplus \dots \right)_L .$$

There is a natural duality between these two classes of spaces, since they correspond one to the other by duality.

For properties of quojections and (LB)-spaces of the above type we refer to [17], [12], [4], [5], [14] and [16]. Here we just recall from [17] that $Q((X_n), (X_n/Y_n); L)$ is twisted if and only if Y_n is not complemented in X_n for infinitely many n, the same being true of the space $\mathrm{LB}((X_n), (Y_n); L)$, for which *twisted* means not isomorphic to a direct sum of Banach spaces. We also recall that there exist both twisted quojections and twisted (LB)-spaces of Moscatelli type with non-twisted duals (this was already implicitely contained in [17]). On the other hand, we have the following lemma, where the Banach sequence space L is assumed to be reflexive.

Lemma 3. *If E is a quojection (resp., (LB)-space) of Moscatelli type for which E' is twisted, then all iterated duals $E^{(n)}$ of E are twisted.*

Proof. Let $E = Q((X_n),(X_n/Y_n);L)$; then $E' = \mathrm{LB}((X_n'),(Y_n^0);L')$. Supposing E' twisted, by what we said above we must have an increasing sequence (n_k) of positive integers for which $Y_{n_k}^0$ is not complemented in X_{n_k}' (and hence Y_{n_k} is not complemented in X_{n_k}, so that E itself is twisted). Now $E'' = Q((X_n''),(X_n''/Y_n'');L)$, hence, if E'' were not twisted, $Y_n'' = Y_n^{00}$ would be complemented in X_n'' for all $n \geq n_0$ for a suitable n_0. Then Y_n^{000} would be complemented in X_n''' for $n \geq n_0$ and hence so would Y_n^0, since the latter is complemented in Y_n^{000}. But $Y_n^0 \subset X_n'$, hence Y_n^0 would be complemented in X_n' for all $n \geq n_0$ and we would get a contradiction. Thus E'' is twisted. The same argument applies if we start with an (LB)–space of Moscatelli type M, thereby getting that if M' is twisted, then so is M''. Now, going back to our quojection E, apply the latter conclusion to the (LB)–space E' to obtain that also E''' must be twisted, etc.

Now Theorem 1 may be considerably strengthened as follows.

Theorem 2. *Let $E = Q((X_n),(X_n/Y_n);L)$ be a quojection of Moscatelli type with Y_n separable for all n sufficientely large. There exists a (countably normed) prequojection F with $F' = E'$ if and only if Y_n is non-quasi-reflexive for infinitely many n. Moreover, in such a case there are a countably normed prequojection G with $G' = \bigoplus_n Y_n'$ and exact sequences*

$$0 \longrightarrow G \longrightarrow F \longrightarrow L\left(\frac{X_n}{Y_n}\right) \longrightarrow 0, \tag{17}$$

$$0 \longrightarrow L(Y_n) \longrightarrow G \times L(X_n) \longrightarrow F \longrightarrow 0. \tag{18}$$

Proof. The first part is direct consequence of Theorem 1.

For the second part we write, with $W_n \subset Y_{n-1}^{00} \subset X_{n-1}''$ and chosen as in the proof of Theorem 1,

$$F = \mathrm{proj}_n \left[\left(X_1 \bigoplus \cdots \bigoplus X_{n-2} \bigoplus \left(X_{n-1} \bigoplus W_n\right) \bigoplus \frac{X_n}{Y_n} \bigoplus \cdots\right)_L, S_n\right], \tag{19}$$

where S_n is defined as in (5), with $\tilde{J}_n$ acting only on $X_n \bigoplus W_{n+1}$. From (19) it is then easy to see that the prequojection

$$G = \mathrm{proj}_n \left(Y_1 \bigoplus \cdots \bigoplus Y_{n-2} \bigoplus \left(Y_{n-1} \bigoplus W_n\right), S_n\right) \tag{20}$$

is a subspace of F and that $F/G = L(X_n/Y_n)$. This establishes (17), since $G' = \bigoplus_n Y_n'$ (recall the proof of Proposition 2).

Now define maps

$$T_n: \left(Y_1 \bigoplus \cdots \bigoplus Y_{n-2} \bigoplus \left(Y_{n-1} \bigoplus W_n\right)\right) \times L(X_k) \longrightarrow \left(X_1 \bigoplus \cdots \bigoplus X_{n-2} \bigoplus \left(X_{n-1} \bigoplus W_n\right) \bigoplus \frac{X_n}{Y_n} \bigoplus \cdots\right)_L$$

as follows:

$$(21) \quad T_n\left[(y_1, \ldots, y_{n-1}, w), (x_k)\right] = \left(y_1 + x_1, \ldots, y_{n-1} + x_{n-1}, w + \tilde{J}_n x_n + \ldots + \tilde{J}_n \ldots \tilde{J}_{n+i} x_{n+i} + \ldots, q_n(x_n), \ldots\right).$$

Here $y_j \in Y_j$ for $j < n$, $w \in W_n$, $(x_k) \in L(X_k)$, $q_n: X_n \longrightarrow X_n/Y_n$ is the quotient map and i runs through the set of non-negative integers. Note that the series in the right-hand side of (21) is absolutely convergent, since $\left\|\tilde{J}_n\right\| \leq 1/2$ for all n, by construction. One verifies directly that the above maps T_n are surjective and that they commute with the maps S_n appearing in (19) and (20). Therefore, the sequence (T_n) defines a continuous map $T: G \times L(X_k) \longrightarrow F$. Since

$$\ker T_n = \left[(y_1, \ldots, y_{n-1}, \tilde{J}_n y_n + \ldots + \tilde{J}_n \ldots \tilde{J}_{n+i} y_{n+i} + \ldots), (-y_k)\right],$$

with $(y_k) \in L(Y_k)$, denoting by I the identity map of $L(X_k)$ we have $(S_n \times I)(\ker T_{n+1}) = \ker T_n$ and hence T is surjective.

Since $\ker T = \operatorname{proj}_n(\ker T_n, S_n \times I)$ and $S_n \times I: \ker T_{n+1} \longrightarrow \ker T_n$ is an isomorphism, (18) follows upon noticing that $\ker T_n$ is isomorphic to $L(Y_k)$ for all n.

Remark 7. The "moreover" part of Theorem 2 is the exact analogue of what happens for quojections of Moscatelli type. In fact, if $E = Q\left((X_n), (X_n/Y_n); L\right)$, then we have exact sequences

$$0 \longrightarrow \prod_n Y_n \longrightarrow E \longrightarrow L\left(\frac{X_n}{Y_n}\right) \longrightarrow 0,$$

$$0 \longrightarrow L(Y_n) \longrightarrow \prod_n Y_n \times L(X_n) \longrightarrow E \longrightarrow 0,$$

for which we refer to [12, Theorem 2.2] and [16, Corollary 3.1].

Remark 8. We note that, by choosing Y_n so that Y_n^0 is not complemented in X_n' for infinitely many n, the quojection E in Theorem 2 has the property that E' is twisted and hence, by Lemma 3, that $E^{(n)}$ is twisted for all n. Thus, applying Theorem 2 we obtain the following counterpart to Proposition 3:

Proposition 4. *There exist countably normed prequojections F whose iterated duals $F^{(n)}$ are all twisted.*

5. EXACT SEQUENCES OF PREQUOJECTIONS

Motivated by Theorem 2 we shall give here some results on exact sequences of prequojections. This means that now we are not interested in exact sequences like (3), where the kernel is nuclear, but in the "opposite" case, i.e. when the kernel is a prequojection. For this purpose, the following easy proposition will be useful.

Proposition 5. *Let E and F be Fréchet spaces. If both E and F are prequojections (resp., quojections), then so is $E\tilde{\otimes}_\pi F$.*

Proof. The assertion about quojections is an immediate consequence of the properties of the projective tensor product. If E and F are prequojections, they satisfy Vogt's property Ω_1 (cf. [23, Lemma 0.2]). Then it is easy to see that $E\tilde{\otimes}_\pi F$ has the same property, hence is a prequojection by the same result.

Now we have the following theorem which complements our universality result on quojections [16, Theorem 1.1].

Theorem 3. *Let F be a non-normable, separable and countably normed prequojection. Then $G = l^1\tilde{\otimes}_\pi F$ is a separable and countably normed prequojection with the property that every separable quojection is isomorphic to a quotient of G by a prequojection subspace.*

Proof. Let $L \subset F$ be such that $F/L = \omega$. By [23, Proposition 3.2] L is a prequojection. If $q: F \longrightarrow \omega$ is a surjective map with $\ker q = L$, then

$$Q = \mathrm{id} \bigotimes q: l^1 \tilde{\bigotimes}_\pi F \longrightarrow l^1 \tilde{\bigotimes}_\pi \omega = (l^1)^{\mathrm{N}}$$

is an onto map whose kernel $l^1\tilde{\otimes}_\pi L$ is a prequojection. Let E be a separable quojection and let $T: (l^1)^{\mathrm{N}} \longrightarrow E$ be a surjection whose kernel is a quojection (cf. [16, Theorem 1.1]). Then the map $TQ: G \longrightarrow E$ is onto and $\ker TQ/\ker Q = \ker T$. Now it follows from the three-space property for prequojections (cf. [15]) that $\ker TQ$ is a prequojection (obviously, countably normed).

Corollary 2. *Let F be as in Theorem 3. For every separable Fréchet space H, there* *exists an exact sequence*

$$0 \longrightarrow M \longrightarrow L \longrightarrow H \longrightarrow 0,$$

with L a subspace of $l^1 \tilde{\otimes}_\pi F$ and M a prequojection. Moreover, L is a prequojection if and only if so is H.

Proof. Embed H as a subspace of $C([0,1])^{\mathbf{N}}$, which is a quotient of $l^1 \tilde{\otimes}_\pi F$ by a prequojection subspace thanks to Theorem 3. If

$$q: l^1 \tilde{\bigotimes}_\pi F \longrightarrow C([0,1])^{\mathbf{N}}$$

is the quotient map, then $L = q^{-1}(H)$ is the required subspace.

Corollary 3. *Let F be as in Theorem 3 and let N be a Fréchet space which is either nuclear and countably normed or a Köthe space $K(a)$ with continuous norm. Then N is a subspace of $l^1 \tilde{\otimes}_\pi F$.*

Proof. Let $0 \longrightarrow M \longrightarrow L \longrightarrow N \longrightarrow 0$ be an exact sequence as in Corollary 2. Since M is a prequojection, this sequence splits (cf. [22] and also [21]), hence $L = M \oplus N$ and N is a subspace of $L \subset l^1 \tilde{\otimes}_\pi F$.

Corollary 4. *Let N be a nuclear Fréchet space. The following assertions are equivalent:*

(i) *N is countably normed;*

(ii) $\mathrm{Ext}^1(N, M) = 0$ *for every separable (countably normed) prequojection M.*

Proof.

(i)$\Longrightarrow$(ii) : Follows from the results in [21].

(ii)$\Longrightarrow$ (i) : By Corollary 2 we have an exact sequence $0 \longrightarrow M \longrightarrow L \longrightarrow N \longrightarrow 0$, where M is a prequojection and L is countably normed. If (ii) holds, then the sequence splits and (i) follows.

Corollary 5. *There exists a countably normed prequojection U such that every separable prequojection is a quotient of U by a prequojection subspace.*

Proof. Let F be a separable and countably normed prequojection and let U be the l^1–sum of all prequojection subspaces of $l^1 \tilde{\otimes}_\pi F$. By Corollary 2, U is the required prequojection.

REFERENCES

1. Banach, S.: Théorie des operations linéaires (2nd ed.), New York: Chelsea 1955.
2. Behrends, E., Dierolf, S., Harmand P.: On a problem of Bellenot and Dubinsky, Math. Ann. **275**, 337–339 (1986).
3. Bellenot, S. F., Dubinsky, E.: Fréchet spaces with nuclear Köthe quotients, Trans. Amer. Math. Soc. **273**, 579–594 (1982).
4. Bonet, J., Dierolf, S.: On LB–spaces of Moscatelli type, Doğa Mat. **13**, 9–33 (1989).
5. Bonet, J., Dierolf, S.: Fréchet spaces of Moscatelli type, Revista Mat. Univ. Complut. Madrid **2** (suppl.), 77–92 (1989).
6. Bonet, J., Galbis, A.: The identity L(E, F) = LB(E, F), tensor products and inductive limits, Note Mat. (Lecce) **9**, (1989).
7. Davis, W. J., Lindenstrauss, J.: On total nonnorming subspaces, Proc. Amer. Math. Soc. **31**, 109–111 (1972).
8. De Wilde, M.: Sur le relèvement des parties bornées d'un quotient d'espaces vectoriels topologiques, Bull. Soc. Roy. Sci. Liege **43**, 299–301 (1971).
9. Dierolf, S., Moscatelli, V. B.: A note on quojections, Funct Approx. Comment. Math. **17**, 131–138 (1987).
10. Jarchow, H.: Locally Convex Spaces, Stuttgart: Teubner 1981.
11. Lindenstrauss, J., Tzafriri, L.: Classical Banach Spaces I, Berlin: Springer 1977.
12. Metafune, G., Moscatelli, V. B.: On the three–space property for locally convex spaces, Collect. Math. **37**, 287–296 (1986).
13. Metafune, G., Moscatelli, V. B.: Generalized prequojections and bounded maps. Resultate Math. **15**, 172–178 (1989).
14. Metafune, G., Moscatelli, V. B.: On twisted Fréchet and (LB)–spaces, Proc. Amer. Math. Soc. **108**, 145–150 (1990).
15. Metafune, G., Moscatelli, V. B.: Quojections and prequojections. Advances in the Theory of Fréchet Spaces (ed.: T. Terzioğlu), Kluwer Acad. Publ., 235–254 (1989).

16. Metafune, G., Moscatelli, V. B.: On exact sequences of quojections, Revista Mat. Univ. Complut. Madrid (to appear).

17. Moscatelli, V. B.: Fréchet spaces without continuous norms and without bases, Bull. London Math. Soc. **12**, 63–66 (1980).

18. Moscatelli, V. B.: On strongly non-norming subspaces, Note Mat. (Lecce) **7**, 311–314 (1987).

19. Moscatelli, V. B.: Strongly non-norming subspaces and prequojections, Studia Math. **95**, 249–254 (1990).

20. Önal, S., Terzioğlu, T.: Unbounded linear operators and nuclear Köthe quotients. Arch. Math. **54**, 576–581 (1990).

21. Vogt, D.: Some results on continuous linear maps between Fréchet spaces. Functional Analysis: Surveys and Recent Results III (eds.: K. D. Bierstedt and B. Fuchssteiner), North-Holland Math. Stud. **90**, 349–381 (1984).

22. Vogt, D.: On the functors $\mathrm{Ext}^1(E,F)$ for Fréchet spaces, Studia Math. **85**, 163–197 (1987).

23. Vogt, D.: On two problems of Mityagin, Math. Nachr. **141**, 13–25 (1989).

Progress in Functional Analysis
K.D. Bierstedt, J. Bonet, J. Horváth & M. Maestre (Eds.)

PROBLEMS FROM THE PÉREZ CARRERAS/BONET BOOK

Stephen A. Saxon

Mathematics Department, Lehigh University, Bethlehem, PA 18015, USA and
Mathematics Department, University of Florida, Gainesville, FL 32611, USA

While writing the review [**11**] for the Pérez Carreras/Bonet book [**9**], I discovered complete and partial answers to a number of the interesting problems/questions which constitute Chapter 13. Proofs, often inspired by the book itself, are given in separate papers [**12-18**]. The questions range from very easy to very difficult, and the answers can be quite useful.

Question 13.8.16 has a positive answer [**12**]: ℓ^∞-barrelled spaces affirm the 3-space problem; i.e., *If* E/M *and* M *are* ℓ^∞*-barrelled, so is* E. Y. Melendez and others have also observed this. Its delightfully easy Hahn-Banach Theorem proof also yields Roelcke/Dierolf's corresponding result for barrelled spaces (1981).

Answering Question 13.2.2 has led to a more general result [**13**],[**14**]: *Every countable-codimensional subspace of an infinite-dimensional [non- normable] metrizable <proper> (LF)-space has an infinite-dimensional <separable> Fréchet quotient [isomorphic to* $\omega(\equiv K^N)$*]*. The Question is also answered in [**15**] by a sequence space method that yields the titled extension of the Banach-Mazur result (1933). Question 13.4.2, "Does every infinite-dimensional Banach space E have a dense non-barrelled subspace?" seems to be a very difficult, related question: in a joint paper with A. Wilansky ('77) I showed that the answer is "Yes" if and only if E has a (properly) separable quotient, and the separable quotient problem has persisted from Banach's era.

Juxtaposed is Question 13.4.1: Let B be barrelled with $E' \neq E^*$; does E have a barrelled countable enlargement (BCE)? Robertson, Tweddle, and Yeomans (1980) have given partial answers, including: *If* E *has a dense barrelled subspace* F *with* $dim(E/F) \geq c$*, then* E *has a* BCE. In both 13.4.2 and 13.4.1, then, a dense subspace of codimension $\geq c$ is desireable, non-barrelled in the one case, barrelled in the other!

More partial answers are in my joint paper with W. Robertson (1989), including: *Every properly separable barrelled space has a BCE, assuming the Continuum Hypothesis (CH).* The proof shows that such a space is barrelledly fit. [*E is <barrelledly> fit* if there is a dense <and barrelled> subspace F with dim(E/F) = dim(E).] Assuming a condition weaker than the Generalized CH, I can prove that: *Every barrelled space is the direct sum of two subspaces, one fit and the other with its strongest l.c. topology.* This quickly suggests an alternate to Question 13.4.1: Is every barrelled, fit space barrelledly fit?

Also related to Question 13.4.2 is the very first one, Question 13.1.1: Does every infinite-dimensional Banach space contain a dense, non-Baire hyperplane? Assuming Martin's Axiom (MA), Arias de Reyna (1980) answered "Yes" for properly separable Banach spaces, thereby denying the Wilansky-Klee Conjecture, and M. Valdivia (1983) gave the more general theorem: *Every properly separable Baire space has a dense, non-Baire hyperplane, assuming MA.* A yet more general theorem [**17**] affirms Question 13.1.1. In fact, *Every (Hausdorff) locally convex space* E *with* $E' \neq E^*$ *contains a dense, non-Baire hyperplane, assuming* MA.

Joint papers with P. P. Narayanaswami show that many Banach spaces contain dense proper (LF)-subspaces (1989), partially answering 13.8.4; and that: *A Banach space has a properly separable quotient if and only if it has a dense subspace dominated by a proper* (LF)*-space* (1981).

Question 13.4.3 asks: "If E is an infinite-dimensional Fréchet space and F a subspace of E strictly dominated by an (LF)-space, is there a closed infinite-dimensional subspace H of E transverse to F?" Saxon/Levin and M. Valdivia simultaneously (1971) published the fact that every countable-codimensional subspace of a barrelled space is barrelled, and the S/L paper mentions that techniques of Saxon's dissertation prove, without use of the CH, that every subspace of codimension $< c$ in a Fréchet space E is barrelled. M. Valdivia (1984) subsequently proved this for more general E by affirming 13.4.3 for *im*proper (LF)-spaces (i.e., Fréchet spaces). By affirming the general Question 13.4.3 [**16**], I obtain the corollary (cf. [**18**]): *Every subspace of codimension* $< c$ *in a Fréchet space is totally barrelled* (TB). One may improve Proposition 8.6.8(iii) in [**9**] to read: *Every subspace of codimension* $< c$ *in an* (LF)*-space is a limit subspace.* These combine: *Every subspace of codimension* $< c$ *in an* (LF)*-space is the inductive limit of an increasing sequence of* TB *spaces*, which yields, via standard closed graph techniques, a maximal generalization of Köthe's Homomorphism Theorem [**18**], going from finite codimension to that $< c$. Similarly, Proposition 8.6.15 of P.C./B. may be improved; in particular, *A subspace of codimension $< c$ in an (LF)-space is itself an (LF)-space if and only if it is closed, and then in fact it must be countable-codimensional.*

Our lately mourned colleague José Maria Garcia Lafuente published [**8**] a tantalizing but (cf. MR 89e:46004) false proof affirming Question 13.8.1: "Is the completion of an (LB)-space again an (LB)-space?" Nevertheless, it may be shown that if the answer to 13.8.1 is negative, then so is the answer to (Grothendieck's) Question 13.8.6, for any non-(LB)-space completion of an (LB)-space would necessarily be dominated by a non-complete regular (LB)-space. This follows from the standard methods of Grothendieck and others found in the book [**9**]. (Observation 8.5.19 of [**9**]: A [Hausdorff] (LF)-space is regular if and only if it is locally complete.)

THEOREM. *Suppose $(E, \mathcal{T})$ is an (LB)-space whose completion $(\hat{E}, \hat{\mathcal{T}})$ is not an (LB)-space. Then there is a finer topology τ such that $(\hat{E}, \tau)$ is a non-complete regular (LB)-space.*

PROOF: By hypothesis, $(E, \mathcal{T})$ is a Hausdorff barrelled space covered by an increasing sequence $(B_n)_n$ of Banach discs (cf. [**9**], Chapter 3). Taking closures in $(\hat{E}, \hat{\mathcal{T}})$, note that $(\overline{B}_n)_n$ is an increasing sequence of Banach discs (Corollary 3.2.5 of [**9**]) whose union spans a subspace F of $(\hat{E}, \hat{\mathcal{T}})$ which is barrelled, since $E \subset F \subset \hat{E}$ and E is barrelled. Thus, by Proposition 8.2.27 of [**9**], F is closed, so that $F = \hat{E}$, and the sequence $(\hat{E}_{\overline{B}_n})_n$ of Banach spaces defines an (LB)-space topology τ on $\hat{E}$ properly finer than $\hat{\mathcal{T}}$, although both induce $\mathcal{T}$ on E. Let U be a closed neighborhood of 0 in $(\hat{E}, \tau)$ which is not a $\hat{\mathcal{T}}$-neighborhood of 0. Thus U fails to contain the $\hat{\mathcal{T}}$-closure of $U \cap E$, and there exists a Cauchy net in $U \cap E$ which is $\hat{\mathcal{T}}$-convergent but not τ-convergent, so that $(\hat{E}, \tau)$ is not complete. Finally, suppose A is a bounded subset of $(\hat{E}, \tau)$. Let B be the absolutely convex, closed hull of A in $(\hat{E}, \hat{\mathcal{T}})$, so that $\hat{E}_B$ is a Banach space. By uniqueness of limits and Grothendieck's Factorization Theorem ([**9**], 1.2.2(i)), $\hat{E}_B$ is continuously contained in some $\hat{E}_{\overline{B}_p}$, so that A is both contained and bounded in one of the defining Banach spaces, and $(\hat{E}, \tau)$ is regular. ∎

Many thanks to Professor José Bonet, who reports the interesting solutions below:

13.5.1. Bonet proved in [**3**] that if E is a Fréchet space such that its strong dual E'_b has the Mackey convergence condition, then E is quasinormable and hence E'_b has the strict Mackey condition.

13.8.2. Piñeiro proved in [**10**] that, for (DF)-spaces E, $c_o(E)$ is quasibarrelled $\Leftrightarrow c_o(E)$ is $\aleph_0$-quasibarrelled $\Leftrightarrow E'_\beta$ has Pietsch's property (B).

13.8.3. Bierstedt, Bonet and Schmets [**2**] proved that if X is a completely regular Hausdorff space and E is a (DF)-space, then the space $CB(X, E)$ of all bounded continuous functions on X with values in E is also (DF).

13.8.10. Bonet and Dierolf [**4**] solved in the negative a problem of Grothendieck exhibiting examples of strict (LF)-spaces E such that their strong bidual is not an (LF)-space.

13.8.15. Vogt (personal communication) has proved that every Fréchet Schwartz space is the quotient of a Schwartz Köthe echelon space.

13.11.2, 13.11.6, 13.11.7, 13.11.12. are all solved in the negative by the important contributions of Taskinen [**19**],[**20**] on the problem of topologies of Grothendieck and related problems.

13.11.3. There are many related results. See [**1**],[**7**] and more references given in Bonet/Dierolf's contribution [**5**] to these Proceedings.

13.11.8. The complete solution for Banach spaces E is contained in the following result of Defant, Floret and Taskinen [**6**]: *For a Banach space E the following conditions are equivalent: (1) E is an $\mathcal{L}_\infty$-space, (2) $E \otimes_\epsilon F$ is a (DF)-space for every (DF)- space F, (3) $E \otimes_\epsilon F$ is a quasibarrelled (DF)-space for every quasibarrelled (DF)-space F, (4) $E \otimes_\epsilon \text{ind } F_n = \text{ind}(E \otimes_\epsilon F_n)$ for each inductive limit ind F_n.*

Hopefully, in time, solutions will be found sufficient to persuade Professors Pérez Carreras and Bonet to contribute a sequel.

References

1. K.D. Bierstedt, J. Bonet, *Stefan Heinrich's density condition for Fréchet spaces and the characterization of the distinguished Köthe echelon spaces*, Math. Nachr. **135** (1988), 149-180.

2. K.D. Bierstedt, J. Bonet, J. Schmets, *(DF)-spaces of type $CB(X, E)$ and $C\overline{V}(X, E)$*, Note di Mat. Köthe commemorative volume (to appear).

3. J. Bonet, *A question of Valdivia on quasinormable Fréchet spaces*, Bull. Canadian Math. Soc. (to appear).

4. J. Bonet, S. Dierolf, *A note on biduals of strict LF- spaces*, Results of Math. **13** (1988), 23-32.

5. J. Bonet, S. Dierolf, *On distinguished Fréchet spaces*, Progress in Functional Analysis (eds. J. Bonet et al.).

6. A. Defant, K. Floret, J. Taskinen, *On the injective tensor product of* (DF)*-spaces*, Arch. Math. (to appear).

7. J.C. Diaz, M. A. Miñarro, *Distinguished Fréchet spaces and projective tensor products*, Doga Tr. Math. J. **14** (1990), 191-208.

8. J.M. Garcia-Lafuente, *On the completion of* (LF)*-spaces*, Mh. Math. **103** (1987), 115-120.

9. P. Pérez Carreras and J. Bonet, *Barrelled Locally Convex Spaces*, Math. Studies **131** (1987), North-Holland, Amsterdam.

10. C. Piñeiro, $\aleph_0$*-quasibarrelledness of* $c_o(E)$, Arch. Math. **53** (1989), 61-64.

11. S. A. Saxon, *a review of [2]*, Bull. Amer. Math. Soc. **April** (1991).

12. S. A. Saxon, *Affirming the three-space problem for* ℓ^∞*-barrelled spaces*, in preparation.

13. S. A. Saxon, *Fréchet, Banach and other quotients of* $(LF)-, (LN)-$*and other spaces by limit subspaces*, in preparation.

14. S. A. Saxon, *Every countable-codimensional subspace of an infinite-dimensional [non-normable] Fréchet space has an infinite- dimensional Fréchet quotient [isomorphic to* ω*]*, Bull. Pol. Acad. Sciences **2** (1991) (to appear).

15. S. A. Saxon, *All separable Banach spaces are quotients of any countable-codimensional subspace of* ℓ_1, Bull. Pol. Acad. Sciences **2** (1991) (to appear).

16. S. A. Saxon, *The questions of* (LF)*-space dominance:* $(u)d, b$ *or not* $(u)d, b$*?*, in preparation.

17. S. A. Saxon, *Non-Baire hyperplanes in non-separable Baire spaces*, J. Math. Analysis and Applications (to appear).

18. S. A. Saxon, *A maximal extension of Köthe's Homomorphism Theorem*, Note di Matematica Köthe commemorative vol. (to appear).

19. J. Taskinen, *Counterexamples to "problème des topologies" of Grothendieck*, Ann. Acad. Sci Fenn. serie A **63** (1986).

20. J. Taskinen, *The projective tensor product of Fréchet Montel spaces*, Studia Math **91** (1988), 17-30.

Progress in Functional Analysis
K.D. Bierstedt, J. Bonet, J. Horváth & M. Maestre (Eds.)

INTERIOR PROPERTIES AND FIXED POINTS OF CERTAIN DISCONTINUOUS OPERATORS

W. R. Derrick * and L. Nova G. **

* Department of Mathematics, University of Montana
Missoula, Montana, U.S.A. 59812
** Departamento de Matemáticas, Universidad Nacional
Bogotá 1, Colombia

Abstract. The fixed point properties of four classes of operators mapping a Banach space into itself are considered. These classes include contraction and nonexpansion mappings, but also include discontinuous operators for certain parameter values of the classes. A cone ordering scheme is devised for one of these classes, while another class is shown to have open mapping properties similar to the Koebe 1/4 theorem of complex analysis.

Key Words: Discontinuous Operators in a Banach Space, Fixed Points, Open Mappings, Cones.

AMS Classification: Primary 47H10, Secondary 46A30

1. Introduction. Let T be a mapping of a Banach space X, or some closed subset K of X, into itself satisfying one of the following conditions for any two points x and y and a, b $\geq$ 0:

(A) $\| (Tx - Ty) - b[(x - Tx) + (Ty - y)] \| \leq a \| x - y \|$,

(B) $\| (Tx - Ty) - b(x - Tx) \| \leq a \| x - y \| + b \| y - Ty \|$,

(C) $\| (Tx - Ty) - a(x - y) \| \leq b[\| x - Tx \| + \| y - Ty \|]$,

(D) $\| Tx - Ty \| \leq a \| x - y \| + b[\| x - Tx \| + \| y - Ty \|]$,

These conditions, and several other possible variants not listed above, all become contraction mappings when we set b = 0 and let 0 < a < 1. We shall say a mapping is of class A(a,b) when it satisfies condition (A) above, and similarly for classes B(a,b), C(a,b), and D(a,b). Mappings of type C(0,b) have been studied by Kannan ([5] and [6]), where it was proved that T has a unique fixed point if $0 < b < \frac{1}{2}$. Kannan also proved the uniqueness of fixed points with $b = \frac{1}{2}$ in uniformly convex spaces under certain restrictions. The authors [1] , [7] have examined the fixed point properties of class D(a,b); Hardy & Rogers [4] and Goebel, Kirk, & Shimi [3] have studied a generalized version of class D(a,b) on a complete metric space. The survey articles by Rhoades [8], [9] list many other papers written on mappings of these types.

In this paper we will show the similarities and contrast the differences among these four classes. One common similarity is that if T has a fixed point, then this fixed point must be unique whenever $0 \leq a < 1$. Observe that any mapping of classes A(a,b), B(a,b), or C(a,b), is of class D(a,b) by a trivial application of the triangle inequality. We will show, however, that the former classes are "smoother" than class D(a,b), which includes discontinuous operators (our motivation for using the letter D) whenever b > 0. In particular, class D(1,1) contains all operators from X onto itself, since

$$\| Tx - Ty \| \leq \| Tx - x \| + \| x - y \| + \| y - Ty \|$$

by the triangle inequality. Because all mappings of the first three classes are included in the fourth class, it is essential to discuss class D(a,b)'s properties first.

2. Properties of class D(a,b). Observe that class D(a',b') contains class D(a,b) whenever $a' \geq a$ and $b' \geq b$. This is a cone order with the first quadrant as the cone. Since all operators (continuous or not) are in D(1,1), we need only consider the pairs (a,b) in the first quadrant where either a or $b < 1$ or $a = b = 1$. If $b = 0$, then D(a,0) is a contraction for $0 \leq a < 1$, nonexpansive mapping if $a = 1$, and a Lipschitz mapping if $a > 1$.

Suppose T belongs to classes D(a,b) and D(c,d). Then, multiplying the D(a,b) inequality by t and the D(c,d) inequality by 1-t we obtain that T belongs to class D(ta+(1-t)c,tb+(1-t)d), for $0 \leq t \leq 1$. We do not know whether this convexity result is best possible: if D(a,b) and D(c,d) cannot be improved upon, is this also the case for D(ta+(1-t)c,tb+(1-t)d)?

Theorem 1. Let T be in class D(a,b).

(i) If $0 \leq a < 1$ and $\inf \| x - Tx \| = 0$, then there exists a convergent sequence $\{x_n\}$ for which $x_n - Tx_n \to 0$.

(ii) If $0 \leq b < 1$ and $x_n - Tx_n \to 0$ for a convergent sequence $\{x_n\}$, then T has a fixed point.

Proof. (i) By hypothesis select a sequence $\{x_n\}$ for which $x_n - Tx_n \to 0$. Substitute x_m and x_n in the D(a,b) condition and apply the triangle inequality to the first term on the right side of the D(a,b) inequality to get

$$0 \leq (1 - a) \| Tx_m - Tx_n \| \leq (a + b) [\| Tx_m - x_m \| + \| Tx_n - x_n \|],$$

from which it follows that $\{Tx_n\}$ is a Cauchy sequence. Hence $Tx_n \to p$, and

$$\| x_n - p \| \leq \| x_n - Tx_n \| + \| Tx_n - p \| \to 0, \quad \text{as } n \to \infty.$$

(ii) Suppose $x_n \to p$, then by the triangle inequality and applying D(a,b) to the last term on the right side

$$\| p - Tp \| \leq \| p - x_n \| + \| x_n - Tx_n \| + \| Tx_n - Tp \|$$
$$\leq (1 + a) \| p - x_n \| + (1 + b) \| x_n - Tx_n \| + b \| p - Tp \| .$$

Hence, $0 \leq (1 - b) \| p - Tp \| \leq (1 + a) \| p - x_n \| + (1 + b) \| x_n - Tx_n \| \to 0.$

The following example shows that Theorem 1 is sharp: let $X = c_0$ be the Banach space of real sequences converging to zero with sup norm and let $T\{x_1, x_2, x_3, \dots \} = \{1, x_1, x_2, \dots\}$. Then T belongs to class D(1,0), and $\inf \| x - Tx \| = 0$ since $x_n = \{(n-1)/n, (n-2)/n, \dots , 1/n, 0, 0, \dots \}$ satisfies $\| x_n - Tx_n \| = 1/n \to 0$, but no converging sequence satisfies $x_n - Tx_n \to 0$, and T has no fixed points in X.

It follows from (i) and (ii) that if $0 \leq a , b < 1$ and $\inf \| x - Tx \| = 0$, then T has a unique fixed point. We can omit the hypothesis that $\inf \| x - Tx \| = 0$ if we require that $a + 2b < 1$: select any x_0 and define the Picard iterates $x_{n+1} = Tx_n$, for $n = 0, 1, 2, \dots$. Then

$$\| x_{n+1} - Tx_{n+1} \| = \| Tx_n - Tx_{n+1} \|$$
$$\leq a \| x_n - x_{n+1} \| + b [\| x_n - Tx_n \| + \| x_{n+1} - Tx_{n+1} \|]$$

or

$$(1 - b) \| x_{n+1} - Tx_{n+1} \| \leq (a + b) \| x_n - Tx_n \| .$$

But $a + b < 1 - b$, so that $\| x_n - Tx_n \| \to 0$, implying that $\inf \| x - Tx \| = 0$. Thus

<u>Theorem</u> <u>2.</u> If T is in D(a,b) with $a + 2b < 1$, then T has a unique fixed point.

A standard argument shows that if $T: K \to K$ is in D(a,b), with $a + 2b = 1$, where K is a closed bounded convex subset of a uniformly convex Banach space, then T has a convex set of fixed points in K.

In [1] we proved that if $T : K \to K$ is in D(a,b), where K is a compact convex subset of $\mathbb{R}^1$ and $0 \leq b < 1$, then T has a fixed point for any $a \geq 0$. We conjecture that this Schauder-type result is true for any Banach space.

3. Properties of the other classes. Although cone ordering is lost for classes A(a,b), B(a,b), and C(a,b), it is trivial to verify that A(a',b) and B(a',b) contain A(a,b) and B(a,b) respectively, when $a' \geq a$, and C(a,b') contains C(a,b) for $b' \geq b$. The convexity result still holds: for example, if T belongs to B(a,b) and B(c,d), then by the triangle inequality it follows that T belongs to class B(ta+(1-t)c,tb+(1-t)d), but is this best possible when B(a,b) and B(c,d) are best possible?

Theorem 2 now holds whenever $0 \leq a < 1$, for any $b \geq 0$, for classes A(a,b) and B(a,b). Note that (A) can be rewritten as

(A)' $$\| (1 + b)(Tx - Ty) - b (x - y) \| \leq a \| x - y \|,$$

so that

$$\| Tx - Ty \| \leq \alpha \| x - y \|, \qquad \alpha = (a + b)/(1 + b) < 1,$$

implying that T is a contraction mapping, and thus has a unique fixed point. For the second class, apply B(a,b) to the Picard iterates getting

$$\begin{aligned}(1 + b) \| x_{n+1} - x_n \| &= \| (Tx_n - Tx_{n-1}) - b(x_n - Tx_n) \| \\ &\leq a \| x_n - x_{n-1} \| + b \| x_{n-1} - Tx_{n-1} \| = (a + b) \| x_n - x_{n-1} \|,\end{aligned}$$

from which it follows that $x_n - Tx_n \to 0$. Using the geometric series

$$\begin{aligned}\| Tx_m - Tx_n \| &\leq \Sigma^n{}_{k=m+1} \| Tx_k - Tx_{k-1} \| \\ &\leq \alpha^{m+1}((1 + b)/(1 - a)) \| x_1 - x_0 \| \to 0,\end{aligned}$$

we have that $\{Tx_n\}$ is a Cauchy sequence, $x_{n+1} = Tx_n \to p$, and

$$\begin{aligned}\| (1 + b)(Tp - p) + (p - Tx_n) \| &= \| (Tp - Tx_n) - b(p - Tp) \| \\ &\leq a \| x_n - p \| + b \| x_n - Tx_n \| \to 0,\end{aligned}$$

implying that p is the unique fixed point.

Condition (A)' implies that class A(a,b) operators are at least Lipschitz smooth.

One can assume, without loss of generality, that $T0 = 0$, since the operator $T'x = Tx - T0$ also satisfies A(a,b). The next argument is adapted from Graves [2].

Theorem 3. If T is in A(a,b) with $a < b$, then T is onto.

Proof. Select any y in X, let $x_0 = 0$ and $y_0 = y$ and define recursively

$$x_n = x_{n-1} + ((1+b)/b)\, y_{n-1}, \qquad y_n = y_{n-1} - (Tx_n - Tx_{n-1}).$$

We show that $Tx_n \to y$ and $y_n \to 0$. Note that

$$\| x_n - x_{n-1} \| = ((1+b)/b)\ \| y_{n-1} \| = b^{-1} \| (1+b)(Tx_{n-1} - Tx_{n-2}) - b(x_{n-1} - x_{n-2}) \|$$
$$\leq (a/b)\ \| x_{n-1} - x_{n-2} \| \leq \ldots \leq (a/b)^n\ \| x_1 - x_0 \| \to 0,$$

hence $y_n \to 0$ and $\{x_n\}$ is a Cauchy sequence, so $x_n \to p$. Finally, by continuity and $T0 = 0$,

$$y = -\sum_{k=0}^{\infty} (y_{k+1} - y_k) = \sum_{k=0}^{\infty} (Tx_{k+1} - Tx_k) = \lim_n Tx_n = Tp.$$

Theorem 4. If $T \in A(a,b)$ with $a < b$, then the open ball centered at the origin of radius $(b-a)/(1+b)$ is contained in the image under T of the open unit ball.

Proof. Select any y in the open ball centered at the origin of radius $(b-a)/(1+b)$. By Theorem 3 there is a point x such that $Tx = y$. Apply (A)' to x and 0 and use the triangle inequality:

$$(b-a)/(1+b)\ \| x \| \leq \| Tx \| = \| y \| < (b-a)/(1+b).$$

If $b = 1 > a = \frac{1}{2}$, then the open ball centered at the origin of radius 1/4 is contained in the image under T of the open unit ball, so we have a Köebe 1/4 theorem in this situation. The proof of theorem 4 can be localized and shows why C^1 operators are locally open at nonsingular points, since we then have an (A)' condition:

$$\| (Tx - Ty) - dT\,(x - y) \| \leq \epsilon\ \| x - y \|.$$

References

[1] W. R. Derrick & L. Nova, Fixed Point Theorems for Discontinuous Operators, Glasnik Matemat. 24 (1989), 339-348.

[2] L. M. Graves, Some Mapping Theorems, Duke Math. J., 17 (1950),111-114.

[3] K. Goebel, W. A. Kirk, & T. N. Shimi, A fixed point theorem in uniformly convex spaces, Boll. Un. Mat. Ital.,(4) 7 (1973),63-75.

[4] G. Hardy & T. Rogers, A generalization of a fixed point theorem of Reich, Canad. Math. Bull., 16 (1973), 201-206.

[5] R. Kannan, Some results on fied points II, Amer. Math. Monthly, 76 (1969),405-408.

[6] R. Kannan, Some results on fixed points III, Fund. Math., 70 (1971),169-177.

[7] L. Nova, Fixed Point Theorems for some Discontinuous Operators, Pacific Journal of Mathematics, 123 (1986),189-196.

[8] B. E. Rhoades, A Comparison of Various Definitions of Contractive Mappings, Trans. Amer. Math. Soc., 226 (1977),257-290.

[9] B. E. Rhoades, Contractive Definitions Revisited, Contemporary Mathematics, 21 (1983),189-205.

Progress in Functional Analysis
K.D. Bierstedt, J. Bonet, J. Horváth & M. Maestre (Eds.)

Functional analytic aspects of geometry. Linear extending of metrics and related problems

Sixty is ten times a perfect number (W. Sierpiński [10])

C. Bessaga
Institute of Mathematics, Warsaw University, ul. Banacha 2, 00-913 Warsaw, Poland.

Abstract.
For some pairs (X, A), where X and A are compacta and $A \subseteq X$, linear (meaning: additive and positive-homogeneous) operators extending continuous pseudometrics and metrics are constructed. They are defined by explicit analytic formulas, with aid of, so called, 'squeezed cones'. Main results are Theorems 2 and 4 saying that, if A is an ANR, then such an extending operator exists for every X, and that the assumption of A being ANR can be weakened. The paper includes a list of open problems.

Introduction

The last two decades is a period of intensive studies of geometrical aspects of functional analysis, including the wide area of ideas and results appearing under the name 'geometry of Banach spaces'. Many spectacular results connecting this area with harmonic, real and complex analysis and probability theory, have been obtained.

There are instances of research in the opposite direction: looking for functional analytic aspects of geometry, especially geometry of convex sets and distance geometry. As an example of the latter may serve the study of different kinds of embeddings of metric spaces into functional spaces. Despite the first part of the title of the paper, it is neither my intention to survey the results in this direction, nor to describe uncultivated areas and set up any program of research.

My interest concerns a particular problem related to the second part of the title. This is an attempt of a marriage between the classical result of Hausdorff [7] on extending metrics with Borsuk–Dugundji theorem ([2],[6]) on existence of a linear operator extending continuous functions. The main result, I was able to get in this direction (see [1]), is slightly stronger than the following:

If X is an absolute neighbourhood retract for metric spaces and A is a closed subset of X, then there exists an additive and positive-homogeneous operator T which extends admissible metrics for A to those for X (an admissible metric is one that induces the *topology of the space), and T is continuous with respect to the uniform convergence and with respect to pointwise convergence of metrics.*

In the present paper I discuss the situation for compact pairs (X, A) only. This allows to emphasize the main idea of the construction and avoid technical difficulties.

Very stimulating and helpful for my discussion was the paper [9] of Nguyen Van Kue and Nguyen To Nhu. These authors were the first who constructed continuous, but merely sublinear, operators extending metrics.

More information on the history of the subject together with references can be found in Luukkainen's paper [8].

Aknowledgement. I am very indebted to Mr. Lauri Leppik, the discussion with whom in 1988 (when he was a student of Tallin State Pedagogical Institute visiting Warsaw University) engrowed my interest in the subject of extending metrics and helped to cristalize the concept of 'squeezed cone'.

1. Magic formula and 'squeezed cone' extending operator

The letter $\mathbf{R}$ will stand for the real line, $\mathbf{N}$ for the set of positive integers, and $\mathbf{I} = [0, 1]$ the closed unit interval.

We define the *squeezed cone* construction. Let A be an arbitrary nonempty set, $o \in A$ a point, and $f : A \times A \to \mathbf{R}$. Consider the quotient set

$$\mathrm{Sc}\,(A, o) = A \times \mathbf{I}/(A \times 0 \cup o \times \mathbf{I});$$

the subset $A \times 0 \cup o \times \mathbf{I}$ of $A \times \mathbf{I}$ is regarded as single point $\mathbf{o}$ of $\mathrm{Sc}\,A$. Each of the expressions: '$(a, u) = \mathbf{o}$' and '$(a, u) \in \mathbf{o}$' means that (a, u) is a representative of $\mathbf{o} \in \mathrm{Sc}\,(A, o)$, i.e., either $a = o$ or $u = 0$. Each point $a \in A \backslash \{o\}$ will be identified with $(a, 1) \in \mathrm{Sc}\,(A, o)$ and $o \in A$ with $\mathbf{o} \in \mathrm{Sc}\,(A, o)$. Hence

$$A = (A \backslash \{o\}) \times 1 \cup \{\mathbf{o}\} \subseteq \mathrm{Sc}\,(A, o).$$

Assume that $x = (a, u)$ and $y = (b, v)$ are points of $A \times \mathbf{I}$. The following formula

$$\mathrm{sc}_o\, f(x, y) = \frac{1}{2}[f(a, b)(u + v - |u - v|) + f(a, o)(u - v + |u - v|) + f(b, o)(v - u + |u - v|)]$$

will be referred to as *magic formula*.

In the situation where only one base point o is considered we shall write $\mathrm{Sc}\,A$ and $\mathrm{sc}\,f$ instead of $\mathrm{Sc}\,(A, o)$ and $\mathrm{sc}_o\, f$.

Denote by $P(A)$ the cone of all pseudometrics for A, and by $M(A)$ its subset consisting of metrics.

Assume that $p \in P(A)$; $x, y \in A \times \mathbf{I}$; $z, w \in A \times 0 \cup o \times \mathbf{I}$; $a, b \in A$. According to the magic formula,

$$\operatorname{sc} p(x,y) = \operatorname{sc} p(y,x), \quad \operatorname{sc} p(x,z) = \operatorname{sc} p(x,w), \quad \operatorname{sc} p(z,w) = 0,$$

and

$$\operatorname{sc} p((a,1),(b,1)) = p(a,b).$$

Hence it is legal to write

$$\operatorname{sc} p : \operatorname{Sc} A \times \operatorname{Sc} A \to \mathbf{R}\,.$$

Proposition 1. *If $p \in P(A)$, then $\operatorname{sc} p \in P(\operatorname{Sc} A)$.*

Proof. Since the triangle inequality involves three points of the space, it in enough to consider the case, where $A = \{1, 2, 3, o\}$, $\operatorname{card} A = 4$. In this case each $p \in P(A)$ is a non-negative linear combination of seven 'Hamming pseudometrics' [5]:

$$p_B(a,b) = \chi_B(a)\chi_{A\setminus B}(b) + \chi_B(b)\chi_{A\setminus B}(a),$$

for $B \subseteq \{1, 2, 3\}$ nonempty , where χ_C is the characteristic function of the set C. The proof is completed by routine verification that each $\operatorname{sc} p_B$ is a pseudometric.

If C and D are cones in [normed] linear spaces, a map $T : C \to D$ is said to be a [*bounded*] *linear operator* if T is additive, positive-homogeneous [and satisfies the estimate $\|Tx - Ty\| \le K\|x - y\|$ with a positive constant K].

The routine argument yields the following results:

Theorem 1. *The map $\operatorname{sc} : P(A) \to P(\operatorname{Sc} A)$ is a linear operator; $\operatorname{sc}(M(A)) \subseteq M(\operatorname{Sc} A)$; $\operatorname{sc} p|(A \times A) = p$; if $x = (a,u)$, $y = (b,v)$ are in $A \times \mathbf{I}$, $p, q \in P(A)$, and $f = p - q$, then*

$$|\operatorname{sc} f(x,y)| \le |f(a,b)| + |f(a,o)| + |f(b,o)|.$$

Convergence Criterion. *Assume that $p \in P(A)$ is a bounded pseudometric, $x_n = (a_n, u_n) \in \operatorname{Sc} A$ for $n = 0, 1, 2, \ldots$, and $x_o = (a, u) \neq \mathbf{o}$. Then*

(i) $\operatorname{sc} p(x_n, x_o) \to 0$ *iff* $\max(p(a_n, a_o), |u_n - u_o|) \to 0$.

(ii) $\operatorname{sc} p(x_n, \mathbf{o}) \to 0$ *iff* $\min(p(a_n, o), |u_n|) \to 0$.

Let us notice that (ii) follows from the identity: $\operatorname{sc} p((a,u), \mathbf{o}) = p(a,o)|u|$. The same identity shows that, for (ii), the assumption of boundedness of p is indispensable.

If A is a metrizable topological space, then, for each bounded admissible metric $p \in P(A)$, each metric $\operatorname{sc} p$ yields the same topology for $\operatorname{Sc} A$; this topology will be referred to as *the topology* of $\mathrm{Sc}\,A$. If A is a *compactum* (i.e. a metrizable compact space), then the topology of $\mathrm{Sc}\,A$ coincides with the quotient topology of $(A\times\mathbf{I})/\mathbf{o}$. We recall that an admissible metric is one that determines the topology of the space.

Example 1.1. *Let* $A=\{o,c\}$, $d\in M(A)$, $d(o,c)=\delta>0$. *Then* $\mathrm{Sc}\,A$ *is isometric to the segment* $[0,\delta]$ *by the isometry:* $(c,u)\to\delta u$. *The set* $\mathrm{Sc}\,\{o,c\}$ *will be referred to as the string fixed at* o *and* c.

Example 1.2. *Let* $S=\{x\in\mathbf{R}^n:|x-e|=\delta\}$ *be a Euclidean sphere. Then the map* $h:\mathrm{Sc}\,S\to B=\{x\in\mathbf{R}^n:|x-e|\le\delta\}$ *given by the formula* $h(x,u)=o+\delta u(x-o)$ *is a homeomorphism of* $\mathrm{Sc}\,S$ *onto* B *and is the identity on* S .

* * *

The magic formula is a special case of the following *squeezed product* formula:

$$f*g((a,u),(b,v))=\tfrac{1}{2}[f(a,b)(g(u,O)+g(v,O)-g(u,v))$$
$$+f(a,o)(g(u,O)-g(v,O)+g(u,v))+f(b,o)(g(v,O)-g(u,O)+g(a,b))],$$

where (A,o), (T,O) are pointed spaces, $f:A\to\mathbf{R}$, $g:T\to\mathbf{R}$. For the properties of *squeezed products of pseudometrics and spaces* see [1].

2. ANR property implies AREP

For a metrizable topological space X denote by $Pc(X)$ and $Ma(X)$ the cone of all continuous pseudometrics defined on $X\times X$ and its subset consisting of admissible metrics for X. If X is a compactum then $Pc(X)$ is a cone in a Banach space $C(X\times X)$, every $p\in Pc(X)$ is bounded, and $Ma(X)=Pc(X)\cap M(X)$.

DEFINITION 2. Assume that A and X are compacta and $A\subseteq X$. A bounded linear operator $T:Pc(A)\to Pc(X)$ is called a *regular extendor* for the pair (X,A), if $Tp(a,b)=p(a,b)$ for every $p\in Pc(A)$ and for every $a,b\in A$; $T(Ma(A))\subseteq Ma(X)$ and, additionally, T is continuous with respect to the pointwise convergence of pseudometrics. A compactum A has the *absolute regular extendor property*, briefly: $A\in$ AREP if, for every compactum Y containing A, the pair (Y,A) admits a regular extendor.

An easy consequence of Theorem 1 and Convergence Criterion is

Corollary 2. *If* A *is a compactum,* $o\in A$, *then the operator* $S=\mathrm{sc}_o\,|(Pc(A)\times Pc(A))$ *is a regular extendor for the pair* $\mathrm{Sc}\,(A,A)$.

Recall that an ANR space is a compactum X such that, for every compactum Y containing X, there are an open neighbourhood U of X and a continuous retraction α of U onto X.

The last corollary and the three lemmas below yield the result of the section heading.

Lemma 2.1. *If (X, A) is a compact pair such that there exists a continuous retraction α of an open neighbourhood U of A onto A, an if $o \in A$, then* $\mathrm{Sc}\, A$ *is a retract of* $\mathrm{Sc}\, X$.

The proof copies the classical argument concerning the topological cones. We pick an Urysohn function $g : X \to \mathbf{I}$ such that $g^{-1}(1) = A$, $g^{-1}(0) = X \backslash U$, and a (possibly discontinuous) map $\gamma : X \to \mathbf{I}$ such that γ extends α. Then the formula

$$\beta(x, u) = (\gamma(x), g(x)u)$$

defines the required continuous retraction $\beta : \mathrm{Sc}\, X \to \mathrm{Sc}\, A$.

Lemma 2.2. *If (Y, Z) is a compact pair such that Z is a retract of Y, then (Y, Z) admits a regular extendor.*

Proof. If Z consists of single point, the assertion is obvious. Assume that $o, c \in Z$ are distinct. Let d be a continuous metric for the quotient space Y/Z, regarded as pseudometric $d \in Pc(Y)$ which separates the points in $Y \backslash Z$, but glues together all those of Z. Then the required regular extendor T from $Pc(Z)$ into $Pc(Y)$ can be defined by the formula:

$$Tp(x, y) = p(\alpha x, \alpha y) + p(o, c)d(x, y),$$

where α is a continuous retraction of Y onto Z.

Lemma 2.3. *If A, X, Y, Z are compacta such that $A \subseteq X \subseteq Y$, $A \subseteq Z \subseteq Y$, and such that the pairs (Z, A), (Y, Z) admit regular extendors. Then the pair (X, A) does admit too.*

Proof. Let $T : Pc(A) \to Pc(Y)$ be the composition of regular extendors: $Pc(A) \to Pc(Z) \to Pc(Y)$. The required regular extendor $S : Pc(A) \to Pc(X)$ can be defined by

$$Sp(x, y) = Tp(x, y) \; for \quad x, y \in X, \quad p \in Pc(A).$$

Theorem 2. *If A is an ANR space, then (A, X) admits a regular extendor for every compactum X containing A.*

Proof. Pick a base point $o \in A \subseteq X$. Let $Z = \mathrm{Sc}\, A$, $Y = \mathrm{Sc}\, X$. Clearly, $A \subseteq X \subseteq Y$ and $A \subseteq Z \subseteq Y$. By Corollary 2, the pair (Z, A) admits a regular extendor, and by

Lemmas 2.1 and 2.2, the pair (Y, Z) does. The proof is completed by applying Lemma 2.3.

We note that Theorem 1 generalizes to the case of absolute neighbourhood retracts for metric spaces, see [1]. However the proof is technically much more complicated.

3. Hats, multihats and harps

Given a triple (X, A, o), where $o \in A \subseteq X$. Define

$$\mathrm{Hat}\,(X, A, o) = (X, A, o) \cup \mathrm{Sc}\,(A, o),$$

which can be regarded as subset of $\mathrm{Sc}\,(X, o)$. For $p \in P(X)$ we let

$$\mathrm{hat}\, p = \mathrm{sc}_o\, p|(\mathrm{Hat}\,(X, A, o) \times \mathrm{Hat}\,(X, A, o)).$$

Obviously hat $: P(X) \to P(\mathrm{Hat}\,(X, A, o))$ is a linear operator which extends pseudometrics.

The $\mathrm{Hat}\,(X, A, o)$ can also be expressed as disjoint union of X and the *crown* of the hat:

$$\mathrm{Crown}\,(X, A, o) = \mathrm{Sc}\,(A, o)\backslash A.$$

A hat with many crowns is called *multihat*. More precisely, given a pair $(X, \mathcal{A})$ where X is a hat, and $A = [(A_\lambda, o_\lambda)]_{\lambda \in \Lambda}$ a family of pointed subsets of X, define

$$\mathrm{Mht}\,(X, \mathcal{A}) = X \cup C,$$

where C is **disjoint** union of all the crowns $\mathrm{Crown}\,(X, A_\lambda, o_\lambda)$, $\lambda \in \Lambda$.

For a $p \in P(X)$ we define $\mathrm{mht}\, p : (X \times X) \to \mathbf{R}$ as follows:

If $\mathbf{x} = (a, u) \in \mathrm{Hat}\,(X, A_\lambda, o_\lambda)$ and $\mathbf{y} = (b, v) \in \mathrm{Hat}\,(X, A_\mu, o_\mu)$, we let $u' = 1 - u$, $v' = 1 - v$, and define

(mht 1) $$\mathrm{mht}\, p(\mathbf{x}, \mathbf{y}) = \mathrm{sc}_{o_\lambda}\, p((a, u), (b, v)) \quad if\ \lambda = \mu,$$

(mht 2) $$\mathrm{mht}\, p(\mathbf{x}, \mathbf{y}) = p(a, b)uv + p(a, o_\mu)uv' + p(b, o_\lambda)u'v + p(o_\lambda, o_\mu)u'v' \quad if\ \lambda \neq \mu.$$

Proposition 3. *If* $p \in P(X)$, *then* $\mathrm{mht}\, p \in P(\mathrm{Mht}\,(X, \mathcal{A}))$.

Proof. It is enough to consider the case, where $\Lambda = \{1, 2, 3\}$. Then, by (mht1) and (mht2), the function $\mathrm{mht}\, p$ is a restriction of the pseudometric $\mathrm{sc}_{o_3} \circ \mathrm{sc}_{o_2} \circ \mathrm{sc}_{o_1} p$, and therefore satisfies the triangle inequality.

DEFINITION 3. Let (X,d) be a metric space. A family $\mathcal{A} = [(A_\lambda, o_\lambda)]_{\lambda\in\Lambda}$ of pointed subsets of X is called a *shrinking family in* X, if all the sets A_λ are compact and, for every $\epsilon > 0$, there are only finitely many indices λ such that the diameter of A_λ is greater than ϵ. If X is compact the property above does not depend on the choise of $d \in Ma(X)$.

Theorem 3. *If X is compact and $\mathcal{A}$ is shrinking then, for every $p \in \mathrm{Mht}\,(X, \mathcal{A})$, the multihat $Y = \mathrm{Mht}\,(X, \mathcal{A})$ is compact in the metric $d = \mathrm{mht}\,p$ and the topology of Y induced by $\mathrm{mht}\,p$ does not depend on the choise of $p \in Ma(X)$.*

Proof. Assume that (y_n) is a sequence of elements of Y. From Theorem 1 and Convergence Criterion it follows that each $Y_\lambda = \mathrm{Sc}\,(A_\lambda, o_\lambda)$ and, therefore, each hat $X \cup A_\lambda$ is a compact subset of Y. Hence, if one of the hats contains infinitely many elements y_n, then there is a convergent subsequence of the sequence (y_n). Otherwise, we let $x_n = y_n$ if $y_n \in X$ and $x_n = o_\lambda$ if $y_n \in \mathrm{Crown}\,(X, A_\lambda, o_\lambda)$. Then $d(x_n, y_n) \to 0$. Since (x_n), which is a sequence of elements of X, admits a convergent subsequence, the corresponding subsequence of (y_n) is also convergent. We have proved that Y is compact.

Assume that $p_1, p_2 \in P(X)$ and $d_1 = \mathrm{mht}\,p_1$, $d_2 = \mathrm{mht}\,p_2$. Clearly, $d_1 + d_2 = \mathrm{mht}\,(p_1 + p_2)$, whence, as we have just shown, the metric space $(Y, d_1 + d_2)$ is compact. For $j = 1, 2$, the identity map h_j from $(Y, d_1 + d_2)$ to (Y, d_j) is continuous, and since the two spaces are compact, is a homeomorphism. Hence the identity map from (Y, d_1) to (Y, d_2), equal to $h_2^{-1} \circ h_1$, is a homeomorphism. This completes the proof of the theorem.

For the pair $(X, \mathcal{A})$ satisfying the assumptions of Theorem 3, the set $\mathrm{Mht}\,(X, \mathcal{A})$ will be regarded as (compact) topological space equipped with the topology desribed in the Theorem.

Corollary 3. *If X is a compactum and $\mathcal{A}$ is a shrinking family in X, then the operator* mht , *restricted to $Pc(X)$, is a regular extendor for the pair* $(\mathrm{Mht}\,(X, \mathcal{A}), X)$.

The following lemma will be applied in the next section:

Lemma 3. *If X is a compactum with a shrinking family $\mathcal{A}$ such that* $\mathrm{Mht}\,(X, \mathcal{A}) \in$ AREP , *then* $X \in$ AREP .

Proof. Assume that W is a compactum containing X. Taking into account the assumption of the Lemma and Corollary 3, and applying Lemma 2.3 to the chains of inclusions

$$X \subseteq W \subseteq \mathrm{Mht}\,(W, \mathcal{A}); \qquad X \subseteq \mathrm{Mht}\,(X, \mathcal{A}) \subseteq \mathrm{Mht}\,(W, \mathcal{A}),$$

we conclude that the pair (X, W) admits a regular extendor.

* * *

If $\mathcal{A} = [(A_\lambda, o_\lambda)]_{\lambda \in \Lambda}$, where each $A_\lambda = \{a_\lambda, o_\lambda\}$ is a two-point subset of X, then the multihat $\mathrm{Mht}\,(X, \mathcal{A})$ is called a *harp*, the set X is the *frame* of the harp, the sets $\mathrm{Sc}\,(A_\lambda, o_\lambda)$ are the strings, cf. Example 1.1.

If the harp is regarded as metric space equipped with the metric $d = \mathrm{mht}\, p$, then each string is well-spanned, i.e., homeomorphic to the line segment of length equal to the distance between the ends of the string. Even in the case where all strings are fixed at the same ends, they do not touch each other in the middle points (see the formula (mht2)). Such an harp cannot be isometrically embedded into any Euclidean space, and therefore cannot be played by human beings. It will be referred to as an *angelic harp*.

4. AREP does not imply ANR property. The Swiss Cheese counterexample

Before giving the precise definition of the Swiss Cheese observe that there is one sort of the cheese, very much appreciated by topologists: the Cantor set.

Consider the n-cube $\mathbf{I}^n$ and assume that $[(B_k)]_{k \in \mathbf{N}}$ is a countable family of pair-wise disjoint closed Euclidean balls contained in $\mathbf{R}^n$. The compactum $\mathbf{C} = \mathbf{I}^n \setminus \bigcup_{k \in \mathbf{N}} \mathrm{int}\, B_k$ is referred to as *Swiss Cheese*.

Let $A_k = \mathrm{bd}\, B_k$, $o_k \in B_k$ for $k \in \mathbf{N}$. Consider the shrinking family $\mathcal{A} = [(a_k, o_k)]_{k \in \mathbf{N}}$.

Proposition 4. *There is a homeomorphism from the multihat* $\mathrm{Mht}\,(\mathbf{C}, \mathcal{A})$ *onto the n-cube* $\mathbf{I}^n$ *which is the identity on* $\mathbf{C}$.

Proof. The required homeomorphism is obtained by pasting together the homeomorphisms of Example 1.2 with aid of the following elementary lemma:

Lemma 4. *Let X and Y be compacta, $\mathcal{A} = [(A_\lambda, o_\lambda)]_{\lambda \in \Lambda}$ a shrinking family in X. If $g : X \to Y$ is a bijective map such that the restriction of g to each A_λ and the restriction to $X \setminus \bigcup_{\lambda \in \Lambda} \mathrm{int}\, A_\lambda$ are continuous, then g is a homeomorphism of X onto .*

Theorem 4. *The Swiss Cheese* $\mathbf{C}$ *is an absolute regular extendor without being an* ANR *space.*

Proof. The assertion: $\mathbf{C} \in \mathrm{AREP}$ follows immediately from Lemma 3, Proposition 4 and the fact that $\mathbf{I}^n$ is an absolute retract (Tietze's theorem). Since $\mathbf{C}$ is compact, there is a cluster point, say y, for the family $[A_n]$ of spheres, and therefore **C** is not locally contractible to the point y. Hence **C** is not an ANR space (cf. Borsuk [2]).

To establish the particular case of the Theorem, where **C** is the Cantor set it is enough to play with harps.

5. Open problems

To keep in the frames of Banach spaces, we ask the questions about pseudometrics for compacta, although analogous questions can be asked in the case of general metrizable spaces.

Problem 1. Does every compact pair (X, A) admit a regular extendor? Do there exist bounded linear extending operators from $Pc(A)$ to $Pc(X)$?

Problem 2. Does $X \in$ AREP imply Sc $X \in$ AREP ?

Problem 3. Describe all the edge vectors of the polyhedral cone $P(\{1, \ldots, n\})$.

As far as I know the question has been answered only for small values of n.

Problem 4. Let X be a compactum. Investigate isomorphic properties of the Banach space $E(X)$ generated in the space $C(X \times X)$ by $Pc(X)$. Does $E(X)$ always have the approximation property or a basis?

By a *Dugundji operator* for a compact pair (Z, Y) we mean a bounded linear operator $T : C(Y) \to C(Z)$ such that $Tf|Y = f$ for $f \in C(Y)$.

Problem 5. Given a compactum A. Let $X = \operatorname{Sc} A$. Does the operator sc : $Pc(A) \to Pc(X)$ admit an extension $T : C(A \times A) \to C(X \times X)$ which is a Dugundji operator?

Problem 6. Let A and B be compacta with the same Borsuk shape (cf. [4]). If $A \in$ AREP , does it B ?

Similar problems to those of linear extending of metrics can be stated in the equivariant context for: (1) spaces with group actions, (2) pointed spaces with group actions fixing the base point.

Let X be a Banach space. Denote by $Na(X)$ and $Mainv(X)$ the sets of all admissible norms and admissible invariant metrics for X. The $Na(X)$ is regarded as metric space with the Banach–Mazur metric:

$$d(p, q) = \sup\{|\log p(x) - \log q(x)| : x \in X, x \neq 0\}$$

for $p, q \in X$.

Problem 7. Let Y be a closed linear subspace of X. Does there exist a [continuous] linear operator $F : Na(Y) \to Na(X)$ which extends norms?

Problem 8. Does there exist a retraction $r : Mainv(X) \to Na(X)$ which is continuous in any sense?

6. Is the 'magic formula' magic ?

Here is the background of the particular procedure of 'taking the rabbit out of a hat' of sect. 1.

When I started studying linear extensions of metrics, one of the first questions was:

Does there exist a linear operator extending continuous metrics for the circle to those for the disc?

The disc is the cone over the circle. So, it is natural to begin with the simplest cone, the one over two points: x, y. Consider x and y as the points of the plane $\mathbf{R}^2$ such that x, y, 0 are not colinear. The cone is then the union of two segments A and B joining, respectively, x and y with 0. The points of the cone $A \cup B$ are of the form: ux or vy, where $u, v \in \mathbf{I}$. Every pseudometric p defined on the set $\{x, y, 0\} \times \{x, y, 0\}$ is a non-negative linear combination of the pseudometrics p_0, p_x, p_y, where $p_0(x, y) = 0$, $p_0(0, x) = p_0(0, y) = 1$ and the other two are obtained by permuting the symbols $0, x, y$. Of course, it is natural to extend these pseudometrics to the pseudometrics d_0, d_x, d_y for the whole cone $A \cup B$ in the affine way, by the formulas:

$$d_0(ux, vy) = d_0(ux, vx) = d_0(uy, vy) = |u - v|;$$

$$d_x(ux, vy) = u, \; d_x(ux, vx) = |u - v|, \; d_x(uy, vy) = 0;$$

$$d_y(ux, vy) = v, \; d_y(uy, vy) = |u - v|, \; d_y(ux, vx) = 0.$$

Since the original three pseudometrics are linearly independent, the correspondence $p \to d$ uniquely extends to the linear extension operator from $P(\{x, y, 0\})$ to $P(A \cup B)$, and this is exactly the 'magic formula'.

I must admit that finding this extremely simple path leading to the 'magic formula' took me several months. This is psychologically interesting: it illustrates how difficult it is to pass to $\mathcal{NEW\ THINKING}$.

References

[1] C. Bessaga, On linear operators and functors extending pseudometrics, submitted to Fund. Math.

[2] K. Borsuk, Über der Isomorphie der Funktionalräume, Bull. Inter. Polon. Sci., Ser A. (1933), 1–10.

[3] — , Theory of retracts, PWN, Warszawa 1967.

[4] — , Theory of shape, PWN, Warszawa 1973.

[5] M. Deza (Tylkin), On Hamming geometry of unitary cubes, Dokl. Akad. Nauk USSR 134 (1960), 1037–312 (in Russian).

[6] J. Dugundji, An extension of Tietze's theorem, Pacific J. Math. 1 (1961), 353–367.

[7] F. Hausdorff, Erweiterung einer Homöomorphie, Fund. Math. 16 (1930), 353–360.

[8] J. Luukkainen, Extension of spaces, maps, and metrics in Lipschitz topology, Ann. Acad. Sci. Fenn., Ser. AI Math. Dissertationes 17 (1978), 1–62.

[9] Nguyen Van Kue and Nguyen To Nhu, Two extensors of metrics, Bull. Acad. Polon. Sci. 29 (1981), 285–291.

[10] W. Sierpiński, Elementary theory of numbers, PWN & North Holland (1987), 182–183.

Progress in Functional Analysis
K.D. Bierstedt, J. Bonet, J. Horváth & M. Maestre (Eds.)

Lototsky-Schnabl operators on the unit interval and degenerate diffusion equations

F. Altomare

Dipartimento di Matematica, Università di Bari, Traversa 200 Via Re David, 4, 70125 Bari, Italy

Dedicated to Professor Manuel Valdivia on the occasion of his 60th birthday

Abstract

The article concerns a sequence of positive linear operators on $\mathscr{C}([0,1])$, called Lototsky-Schnabl operators, which are associated with a continuous function $\lambda:[0,1]\to[0,1]$. We show that, if λ is a strictly positive polynomial, then there exists a positive contraction semigroup that can be expressed in terms of iterates of these operators. We give some applications of this result dealing with initial value problems associated with particular degenerate diffusion equations.

1. INTRODUCTION

This paper is devoted to the study of the iterates of a sequence $(L_{n,\lambda})_{n\in\mathbb{N}}$ of positive linear operators acting on $\mathscr{C}([0,1])$, which are called Lototsky-Schnabl operators. These operators are defined by putting for every $n\geq 1$, $f\in\mathscr{C}([0,1])$ and $x\in[0,1]$

$$L_{n,\lambda}(f)(x) = \sum_{r=0}^{n} \sum_{h=0}^{r} \binom{n}{r}\binom{r}{h} \lambda(x)^r(1-\lambda(x))^{n-r} f(\tfrac{h}{n}+(1-\tfrac{r}{n})x) x^h(1-x)^{r-h},$$

where $\lambda:[0,1]\to[0,1]$ is a fixed continuous function.

In fact, these operators can be defined in a more general setting (see Section 1) and, in this context, we know the behavior of the iterates only when λ is a constant function (see [2]).

Extending these results, we show that, in the case of the unit interval and when λ is a strictly positive polynomial, then there exists a positive contraction semigroup that can be expressed in terms of iterates of the operators $L_{n,\lambda}$.

In order to achieve this result we prove a Voronovskaja-type result which also holds for other convex compact subsets of $\mathbb{R}^p$ $(p \geq 1)$ (cf. Theorem 3.3).

Furthermore, we study some properties of this semigroup; in particular, we show that it is mean ergodic and it strongly converges, as $t \to +\infty$, to the projection $T: \mathscr{C}([0,1]) \to \mathscr{C}([0,1])$ defined by

$$T(f)(x) = (1-x)f(0) + xf(1).$$

Moreover, the infinitesimal generator of the semigroup is shown to be the degenerate differential operator $A: D(A) \to \mathscr{C}([0,1])$ defined by

$$A(u)(x) = \frac{x(1-x)}{2} \lambda(x)\, u''(x) \qquad \text{if } 0 < x < 1,$$

$$A(u)(x) = 0 \qquad \text{if } x=0 \text{ or } x=1.$$

for every $u \in D(A) = \{u \in \mathscr{C}([0,1]) \mid u \in \mathscr{C}^2(]0,1[),\ \lim\limits_{x \to 0^+} \frac{x(1-x)}{2} \lambda(x)\, u''(x) = \lim\limits_{x \to 1^-} \frac{x(1-x)}{2} \lambda(x)\, u''(x) = 0\}$.

As an application, we show that it is possible to represent in terms of iterates of suitable Lototsky-Schnabl operators, the solution of the initial value problem

$$\begin{cases} \dfrac{\partial u}{\partial t}(x,t) = \alpha(x) \dfrac{\partial^2 u}{\partial x^2}(x,t) & 0 < x < 1,\ t \geq 0, \\ \lim\limits_{x \to 0^+} \alpha(x) \dfrac{\partial^2 u}{\partial x^2}(x,t) = \lim\limits_{x \to 1^-} \alpha(x) \dfrac{\partial^2 u}{\partial x^2}(x,t) = 0 & t \geq 0, \\ u(x,0) = u_0 & u_0 \in \mathscr{C}([0,1]) \cap \mathscr{C}^2(]0,1[), \\ & \lim\limits_{x \to 0^+} \alpha(x) u_0''(x) = \lim\limits_{x \to 1^-} \alpha(x) u_0''(x) = 0. \end{cases}$$

where $\alpha: [0,1] \to \mathbb{R}$ is a positive polynomial which vanishes only at 0 and 1 and for which 0 and 1 are simple roots.

2. LOTOTSKY-SCHNABL OPERATORS ON COMPACT CONVEX SETS

Let X be a metrizable convex compact subset of some locally convex Hausdorff space and let us denote by $\mathscr{C}(X,\mathbb{R})$ the Banach lattice of all real continuous functions on X, endowed with the sup-norm topology and the natural order.

Let us consider a linear positive projection $T:\mathscr{C}(X,\mathbb{R})\to\mathscr{C}(X,\mathbb{R})$ (i.e. T is a linear positive operator such that $T^2 = T$).

We set $H = T(\mathscr{C}(X,\mathbb{R}))$ and we suppose that the space A(X) of all real continuous affine functions on X is contained in H.

Moreover we also assume that for every $\overline{x}\in X$, $\lambda\in[0,1]$ and $h\in H$ the function

$$x \in X \mapsto h((1-\lambda)\overline{x} + \lambda x)$$

belongs to H.

For every $x\in X$ we shall denote by μ_x the probability Radon measure on X defined by putting for every $f\in\mathscr{C}(X,\mathbb{R})$,

$$\mu_x(f) = T(f)(x). \tag{2.1}$$

For every $n\in\mathbb{N}$, $n\geq 1$, the n-th *Bernstein-Schnabl operator associated with the projection* T is defined to be the positive linear operator $B_n:\mathscr{C}(X,\mathbb{R})\to\mathscr{C}(X,\mathbb{R})$ defined by putting for every $f\in\mathscr{C}(X,\mathbb{R})$ and $x\in X$

$$B_n(f)(x) = \int_{X^n} f\left(\frac{x_1+\ldots+x_n}{n}\right) d\left(\bigotimes_{i=1}^{n}\mu_{x,i}\right), \tag{2.2}$$

where $\mu_{x,i} = \mu_x$ for every $i=1,\ldots,n$.

Given a continuous function $\lambda:X\to[0,1]$, for every $x\in X$ we also consider the probability Radon measure υ_x defined as

$$\upsilon_x(f) = \lambda(x)\,\mu_x(f) + (1-\lambda(x))\,f(x) \tag{2.3}$$

for every $f\in\mathscr{C}(X,\mathbb{R})$. Then, the n-th *Lototsky-Schnabl operator associated with the projection* T *and the function* λ is, by definition, the positive linear operator $L_{n,\lambda}:\mathscr{C}(X,\mathbb{R})\to\mathscr{C}(X,\mathbb{R})$ defined by putting for every $f\in\mathscr{C}(X,\mathbb{R})$ and $x\in X$,

$$L_{n,\lambda}(f)(x) = \int_{X^n} f\left(\frac{x_1+\ldots+x_n}{n}\right) d\left(\bigotimes_{i=1}^{n}\upsilon_{x,i}\right), \tag{2.4}$$

where $\upsilon_{x,i} = \upsilon_x$ for every $i=1,\ldots,n$.

Lototsky-Schnabl operators were first introduced by Schempp [15] and subsequently studied, with some variants and in different context, by Grossman [5], Nishishiraho [11], [12] and Altomare and Romanelli [2]. For more details about Bernstein-Schnabl operators we also refer to [1].

Lototsky-Schnabl operators are strongly related to Bernstein-Schnabl operators because, for every $f\in\mathscr{C}(X,\mathbb{R})$ and $x\in X$, we have

$$L_{n,\lambda}(f)(x) = \sum_{r=0}^{n} \binom{n}{r} \lambda(x)^r(1-\lambda(x))^{n-r}B_r(f_{n,r,x})(x), \tag{2.5}$$

where $f_{n,r,x}\in\mathscr{C}(X,\mathbb{R})$ is defined as

$$f_{n,r,x}(y) = f(\tfrac{r}{n}\, y + (1-\tfrac{r}{n})x), \quad (y\in X). \tag{2.6}$$

By using (2.5) and examples 2.1 and Section 3 of [1] we can more explicitly describe the operators $L_{n,\lambda}$ in some particular case.

2.1 Examples

1. Set $X = [0,1]$ and consider the linear positive projection $T:\mathscr{C}([0,1],\mathbb{R})\to\mathscr{C}([0,1],\mathbb{R})$ defined by putting

$$T(f)(x) = (1-x)\, f(0) + xf(1), \quad (f\in\mathscr{C}([0,1],\mathbb{R}),\ x\in[0,1]). \tag{2.7}$$

In this case, for every $f\in\mathscr{C}([0,1],\mathbb{R})$ and $x\in[0,1]$ we have

$$L_{n,\lambda}(f)(x) = \sum_{r=0}^{n}\sum_{h=0}^{r} \binom{n}{r}\binom{r}{h} \lambda(x)^r(1-\lambda(x))^{n-r}\, f(\tfrac{h}{n} + (1-\tfrac{r}{n})x)\, x^h(1-x)^{r-h}. \tag{2.8}$$

2. Given $p\in\mathbb{N}$, $p\geq 1$, set $X = X_p = \{(x_1,\dots,x_p)\in\mathbb{R}^p \mid 0\leq x_i,\ i=1,\dots,p$ and $x_1+\dots+x_p \leq 1\}$. Consider $T:\mathscr{C}(X_p,\mathbb{R})\to\mathscr{C}(X_p,\mathbb{R})$ defined by putting for every $f\in\mathscr{C}(X_p,\mathbb{R})$ and $(x_1,\dots,x_p)\in X_p$

$$T(f)(x_1,\dots,x_p) = \sum_{\substack{0\leq h_1+..+h_p\leq 1\\ h_i\in\mathbb{N}}} \alpha_f(h_1,\dots,h_p)x_1^{h_1}\dots x_p^{h_p}(1-\sum_{i=1}^{p} x_i)^{1-\sum_{i=1}^{p} h_i}, \tag{2.9}$$

where

$$\alpha_f(h_1,..,h_p) = f(\delta_{h_1 1},\dots,\delta_{h_p 1}),$$

and $\delta_{h_i 1}$ is the Kronecker symbol.

In this case we have

$$L_{n,\lambda}(f)(x_1,\dots,x_p) =$$

$$= \sum_{r=0}^{n}\binom{n}{r} \lambda(x)^r(1-\lambda(x))^{n-r} \sum_{\substack{0\leq h_1+..+h_p\leq r\\ h_i\in\mathbb{N}}} \frac{r!}{h_1!..h_p!(r-h_1-..-h_p)!}$$

$$f(\tfrac{h_1}{n} + (1-\tfrac{r}{n})x_1,\dots,\tfrac{h_p}{n} + (1-\tfrac{r}{n})x_p)\ x_1^{h_1}..x_p^{h_p}(1-\sum_{i=1}^{p} x_i)^{r-\sum_{i=1}^{p} h_i}. \tag{2.10}$$

3. Set $X = [0,1]^p$, $p \geq 1$ and define $T: \mathscr{C}(X,\mathbb{R}) \to \mathscr{C}(X,\mathbb{R})$ as

$$T(f)(x_1,\ldots,x_p) = \sum_{h_1,\ldots,h_p=0}^{1} \alpha_f(h_1,\ldots,h_p) x_1^{h_1}(1-x_1)^{1-h_1}\ldots x_p^{h_p}(1-x_p)^{1-h_p}. \tag{2.11}$$

We then obtain

$$L_{n,\lambda}(f)(x_1,\ldots,x_p) =$$

$$= \sum_{r=0}^{n} \binom{n}{r} \lambda(x)^r(1-\lambda(x))^{n-r} \sum_{h_1,\ldots,h_p=0}^{r} \binom{r}{h_1}\cdots\binom{r}{h_p}$$

$$f\left(\frac{h_1}{n} + \left(1-\frac{r}{n}\right)x_1,\ldots,\frac{h_p}{n} + \left(1-\frac{r}{n}\right)x_p\right) x_1^{h_1}(1-x_1)^{r-h_1}\ldots x_p^{h_p}(1-x_p)^{r-h_p}. \tag{2.12}$$

4. Let $X=B(0,1)$ be the closed unit ball of $\mathbb{R}^p$ ($p \geq 1$) with center 0 and consider the linear positive projection $T: \mathscr{C}(X,\mathbb{R}) \to \mathscr{C}(X,\mathbb{R})$ defined by associating to every $f \in \mathscr{C}(X,\mathbb{R})$ the unique solution $T(f)$ of the Dirichlet problem

$$\begin{cases} \Delta u = 0 & \text{on } \mathring{X}, \\ u_{|\partial X} = f_{|\partial X}, \end{cases} \qquad u \in \mathscr{C}(X,\mathbb{R}) \cap \mathscr{C}^2(\mathring{X},\mathbb{R}),$$

where, as usual, $\Delta = \sum_{i=1}^{p} \partial^2/\partial x_i^2$ is the Laplacian on $\mathring{X}$.

Thus, from the Poisson formula for the solutions of the Dirichlet problem for a ball, we have

$$T(f)(x) = \begin{cases} f(x) & \text{if } \|x\|=1, \\ \dfrac{1-\|x\|^2}{\sigma_p} \displaystyle\int_{\partial X} \frac{f(y)}{\|y-x\|^p}\, d\sigma(y) & \text{if } \|x\|<1, \end{cases} \tag{2.13}$$

where σ_p denotes the surface area of the sphere ∂X and σ is the surface measure on ∂X. Hence

$$L_{n,\lambda}(f)(x) = \begin{cases} f(x) & \text{if } \|x\|=1, \\ \displaystyle\sum_{r=0}^{n} \binom{n}{r} \lambda(x)^r(1-\lambda(x))^{n-r}\left(\frac{1-\|x\|^2}{\sigma_p}\right)^r \int_{\partial X}\cdots \\ \cdots\displaystyle\int_{\partial X} \frac{f\left(\frac{x_1+\ldots+x_r}{n} + \left(1-\frac{r}{n}\right)x\right)}{\|x_1-x\|^p\ldots\|x_r-x\|^p}\, d\sigma(x_1)\ldots d\sigma(x_r) & \text{if } \|x\|<1. \end{cases} \tag{2.14}$$

For the sake of more transparency, we collect here some of the main properties of Lototsky-Schnabl operators that will be useful in the sequel.

We begin with two formulas whose proof can be derived by a direct computation or by referring to [2, proof of Theorem 2.3] and [1, proof of Theorem 2.6].

For every $h_1,h_2 \in A(X)$

$$L_{n,\lambda}(h_1,h_2) = h_1 \cdot h_2 + \frac{\lambda}{n} (T(h_1 \cdot h_2) - h_1 \cdot h_2). \tag{2.15}$$

More generally, for every $h_1,\ldots,h_m \in A(X)$ $(m \geq 3)$, we have

$$L_{n,\lambda}\Big(\prod_{i=1}^{m} h_i\Big) = \prod_{i=1}^{m} h_i + \sum_{s=1}^{m_0} \sum_{k=s}^{m-s} n^{k-m}\Big(1-\frac{1}{n}\Big)\ldots\Big(1-\frac{k-1}{n}\Big) \sum_{\substack{r_1+\ldots+r_s=m-k+s \\ r_1,\ldots,r_s \geq 2}} g^*(r_1,\ldots,r_s); \tag{2.16}$$

here m_0 denotes the integer part of $m/2$ and

$$g^*(r_1,\ldots,r_s) = \sum\Big[S\Big(\prod_{1\leq u\leq r_1} h_{j_u^1}\Big)\ldots S\Big(\prod_{1\leq u\leq r_s} h_{j_u^s}\Big) - \prod_{1\leq u\leq r_1} h_{j_u^1}\ldots\prod_{1\leq u\leq r_s} h_{j_u^s}\Big] \prod_{j \neq j_u^{\vee}} h_j, \tag{2.17}$$

where

$$S = \lambda\, T + (1 - \lambda)\, I, \tag{2.18}$$

I is the identity operator on $\mathscr{C}(X,\mathbb{R})$, and the sum in (2.17) is extended to all subsets of integers $j_1^1,\ldots,j_{r_1}^1,\ldots,j_1^s,\ldots,j_{r_s}^s$ between 1 and m such that

$$\{j_1^1,\ldots,j_{r_1}^1\} \cap \ldots \cap \{j_1^s,\ldots,j_{r_s}^s\} = \emptyset.$$

Other properties of the operators $L_{n,\lambda}$ are listed below.

$$L_{n,\lambda}(h) = h \qquad \text{for every } h \in H \text{ and } n \geq 1; \tag{2.19}$$

$$L_{n,\lambda}(f)(x) = f(x) \qquad \text{for every } f \in \mathscr{C}(X,\mathbb{R}) \text{ and } x \in \partial_H X, \tag{2.20}$$

where $\partial_H X$ denotes the Choquet boundary of X with respect to H;

$$f \leq L_{n+1,\lambda}(f) \leq L_{n,\lambda}(f) \leq \lambda\, T(f) + (1-\lambda)\, f \tag{2.21}$$

for every $f \in \mathscr{C}(X,\mathbb{R})$, f convex. (The first two inequalities are due to Rasa [13], [14]; the last one is straightforward.)

The proof of the next Theorem can be found in [2] and [5].

2.2 Theorem *Under the above assumption on* X, T *and* λ, *one has*

1. $\lim_{n\to\infty} L_{n,\lambda}(f) = f$ *uniformly on* X *for every* $f \in \mathscr{C}(X,\mathbb{R})$.
2. *Suppose that* λ *is a constant function and assume that*

   (i) $T(h_1 h_2) \in A(X)$ *for every* $h_1,h_2 \in A(X)$,

or, alternatively, if X *is a subset of some euclidean space,*

(ii) $T(A_m) \subset A_m$ *for every* $m \geq 1$,

where A_m *denotes the linear subspace generated by* $\{\prod_{i=1}^{m} h_i \mid h_i \in A(X),\ i=1,..,m\}$.

Then, there exists a strongly continuous positive contraction semigroup $(T(t))_{t\geq 0}$ *on* $\mathscr{C}(X,\mathbb{R})$ *such that for every* $t \geq 0$ *and for every sequence* $(k(n))_{n\in\mathbb{N}}$ *of positive integers satisfying* $\lim_{n\to\infty} \frac{k(n)}{n} = t$, *one has*

$$\lim_{n\to\infty} L_{n,\lambda}^{k(n)} = T(t) \quad \textit{strongly on } \mathscr{C}(X,\mathbb{R}), \tag{2.22}$$

where, as usual, $L_{n,\lambda}^{k(n)}$ *denotes the iterate of* $L_{n,\lambda}$ *of order* $k(n)$.

Moreover,

$$\lim_{t\to\infty} T(t) = T \qquad \textit{strongly on } \mathscr{C}(X,\mathbb{R}), \tag{2.23}$$

and the generator of the semigroup $(T(t))_{t\geq 0}$ *is the closure of the linear operator* $Z:D(Z)\to\mathscr{C}(X,\mathbb{R})$ *defined by*

$$Z(f) = \lim_{n\to\infty} n(L_{n,\lambda}(f) - f) \tag{2.24}$$

where $D(Z) = \{g\in\mathscr{C}(X,\mathbb{R}) \mid \lim_{n\to\infty} n(L_{n,\lambda}(g)-g)$ *exists in* $\mathscr{C}(X,\mathbb{R})\}$.

Finally, for every $m \geq 1$, $A_m \subset D(Z)$ *and for every* $h_1,..,h_m \in A(X)$

$$Z\Big(\prod_{i=1}^{m} h_i\Big) = \begin{cases} 0, & \text{if } m=1 \\ \lambda\,(T(h_1 h_2) - h_1 h_2), & \text{if } m=2 \\ \lambda \sum_{1\leq i<j\leq m} (T(h_i h_j) - h_i h_j) \prod_{\substack{r=1 \\ r\neq i,j}}^{m} h_r, & \text{if } m\geq 3. \end{cases}$$

Note that, in many concrete situations such as the ones indicated in Examples 2.1, the generator A of the semigroup indicated in Theorem 2.2 is a degenerate elliptic differential operator (cf. [1]). In particular, if X = [0,1], then

$$A(u)(x) = \frac{x(1-x)}{2}\lambda(x)\, u''(x) \qquad \text{if } 0<x<1, \tag{2.25}$$

$$A(u)(x) = 0 \qquad \text{if } x=0 \text{ or } x=1.$$

for every $u\in D(A) = \{u\in\mathscr{C}([0,1]) \mid u\in\mathscr{C}^2(]0,1[),\ \lim_{x\to 0^+} \frac{x(1-x)}{2}\lambda(x)\, u''(x) = \lim_{x\to 1^-} \frac{x(1-x)}{2}\lambda(x)\, u''(x) = 0\}$.

3. LOTOTSKY-SCHNABL OPERATORS ON THE UNIT INTERVAL AND DIFFUSION EQUATIONS

The aim of this Section is to show that, when $X=[0,1]$, then part 2 of Theorem 2.2 also holds for λ non constant. Our method is quite different from the one used in [2]. It is an open problem if a similar result is true in the general context of Section 1 or in some particular cases as indicated in Examples 2.1.

Furthermore, in the last part we briefly indicate some applications to initial value problems associated with particular degenerate diffusion equations.

We begin with two lemmas.

3.1 Lemma *Let* X *be a metrizable convex compact subset of some locally convex Hausdorff space and consider the sequence* $(L_{n,\lambda})_{n\in\mathbb{N}}$ *of Lototsky-Schnabl operators associated with a projection* T *and a continuous function* λ. *Given* $x\in X$ *and four functions* $h_1,\ldots,h_4 \in A(X)$ *vanishing at* x, *one has*

$$|\, n\, L_{n,\lambda}(\prod_{i=1}^{4} h_i)(x)\,| \le \frac{6n-5}{n^2} \prod_{i=1}^{4} \|h_i\|.$$

Proof. From (2.16), for $m=4$, we get

$$L_{n,\lambda}(\prod_{i=1}^{4} h_i)(x) = \sum_{s=1}^{2} \sum_{k=s}^{4-s} n^{k-4}(1-\frac{1}{n})\ldots(1-\frac{k-1}{n}) \sum_{\substack{r_1+\ldots+r_s=4-k+s \\ r_1,\ldots,r_s\ge 2}} g^*(r_1,\ldots,r_s)(x).$$

We note that for every $k>s$ and for every $r_1,\ldots,r_s\ge 2$ satisfying $r_1+\ldots+r_s = 4-k+s$, we have $g^*(r_1,\ldots,r_s)(x) = 0$ by virtue of (2.17) and by our assumptions on the functions h_i.

Therefore,

$$L_{n,\lambda}(\prod_{i=1}^{4} h_i)(x) = \sum_{s=1}^{2} n^{s-4}(1-\frac{1}{n})\ldots(1-\frac{s-1}{n}) \sum_{\substack{r_1+\ldots+r_s=4 \\ r_1,\ldots,r_s\ge 2}} g^*(r_1,\ldots,r_s)(x) =$$

$$= \frac{1}{n^3}\, g^*(4)(x) + \frac{1}{n^2}\,(1-\frac{1}{n})\, g^*(2,2)(x) =$$

$$= \frac{1}{n^3}\Big[S(\prod_{i=1}^4 h_i)(x)-\prod_{i=1}^4 h_i(x)+(n-1)\sum_{1\le i<j\le 4}\Big[S(h_ih_j)(x)\cdot S(\prod_{\substack{\lambda=1\\ \lambda\neq i,j}}^4 h_\lambda)(x)-\prod_{i=1}^4 h_i(x)\Big]\Big] =$$

$$= \frac{1}{n^3}\Big[S(\prod_{i=1}^4 h_i)(x) + (n-1)\sum_{1\le i<j\le 4} S(h_ih_j)(x)\ S(\prod_{\substack{\lambda=1\\ \lambda\neq i,j}}^4 h_\lambda)(x)\Big];$$

consequently, since $\|S\| \le 1$, we get $|n\ L_{n,\lambda}(\prod_{i=1}^4 h_i)(x)| \le \frac{1+6(n-1)}{n^2}\prod_{i=1}^4 \|h_i\|$,
and so the result follows. ■

3.2 Lemma *Let* X *be one of the four convex compact subsets of* $\mathbb{R}^p$, $p\ge 1$, *indicated in the* Examples 2.1 *and consider the respective sequence* $(L_{n,\lambda})_{n\in\mathbb{N}}$ *of Lototsky-Schnabl operators. For every* $x\in X$ *let* $\sigma_x: X\to\mathbb{R}$ *be the continuous function defined as* $\sigma_x(y) = \|y-x\|$ *for every* $y\in X$. *Then*

$$\lim_{n\to\infty} n\ L_{n,\lambda}(\sigma_x^4)(x) = 0$$

uniformly with respect to $x\in X$.

Proof. For every $i=1,\dots,p$ denote by $pr_i: X\to\mathbb{R}$ the restriction to X of the i-th projection on $\mathbb{R}^p$. Then, for a given $x = (x_i)_{1\le i\le p}$ we have

$$\sigma_x^4 = \sum_{i,j=1}^p (pr_i - x_i)^2\ (pr_j - x_j)^2.$$

So, on account of Lemma 3.1,

$$|n\ L_{n,\lambda}(\sigma_x^4)(x)| \le \sum_{i,j=1}^p n\ L_{n,\lambda}((pr_i-x_i)^2(pr_j-x_j)^2)(x) \le \frac{6n-5}{n^2}\ p^2\ \mathrm{diam}(X)^4,$$

where $\mathrm{diam}(X) = \sup_{x,y\in X}\|x-y\|$ and so the proof is complete. ■

In the sequel, if X is a closed subset of $\mathbb{R}^p$, $p\ge 1$, with non empty interior $\overset{\circ}{X}$, we shall denote by $\mathscr{C}^2(X,\mathbb{R})$ the space of all real-valued continuous functions on X which are two times continuously differentiable in $\overset{\circ}{X}$ such that the partial derivatives of order ≤ 2 can be continuously extended to X. If $u\in\mathscr{C}^2(X,\mathbb{R})$, for every $i,j=1,\dots,p$ we shall continue to denote by $\frac{\partial u}{\partial x_i}$ and $\frac{\partial^2 u}{\partial x_i\partial x_j}$ the continuous extensions to X of the partial derivatives $\frac{\partial u}{\partial x_i}$ and $\frac{\partial^2 u}{\partial x_i\partial x_j}$ defined on $\overset{\circ}{X}$.

We are now able to prove a Voronovskaja-type result for the operators $L_{n,\lambda}$.

3.3 Theorem *Let* X *be one of the four convex compact subsets of* $\mathbb{R}^p$, $p \geq 1$, *indicated in the* Examples 2.1 *and consider the respective sequence* $(L_{n,\lambda})_{n\in\mathbb{N}}$ *of Lototsky-Schnabl operators. Then, for every* $u \in \mathscr{C}^2(X,\mathbb{R})$, *we have*

$$\lim_{n\to\infty} n\,(L_{n,\lambda}(u)-u) = A(u) \quad \textit{uniformly on } X, \tag{3.1}$$

where for every $x = (x_i)_{1\leq i\leq p} \in X$,

$$A(u)(x) = \lambda(x)\,\frac{x(1-x)}{2}\,u''(x)\,, \qquad \text{if } p=1 \text{ and } X=[0,1], \tag{3.2}$$

$$A(u)(x)=\lambda(x)\left[\sum_{i=1}^{p}\frac{x_i(1-x_i)}{2}\,\frac{\partial^2 u(x)}{\partial x_i^2} - \sum_{1\leq i<j\leq p} x_i x_j\,\frac{\partial^2 u}{\partial x_i \partial x_j}\right], \qquad \text{if } X=X_p, \tag{3.3}$$

$$A(u)(x)=\lambda(x)\sum_{i=1}^{p}\frac{x_i(1-x_i)}{2}\,\frac{\partial^2 u(x)}{\partial x_i^2}\,, \qquad \text{if } X=[0,1]^p, \tag{3.4}$$

$$A(u)(x)=\lambda(x)\,\frac{1-\|x\|^2}{2}\,\Delta u(x)\,, \qquad \text{if } X=B(0,1). \tag{3.5}$$

Proof. Using the Taylor expansion of u, we know that there exists a function $\omega: X\times X \to \mathbb{R}$ such that for every $x,y\in X$

$$u(y) = u(x)+\sum_{i=1}^{p}\frac{\partial u(x)}{\partial x_i}(y_i-x_i)+\frac{1}{2}\sum_{i,j=1}^{p}\frac{\partial^2 u(x)}{\partial x_i \partial x_j}(y_i-x_i)(y_j-x_j)+\omega(x,y)\|y-x\|^2. \tag{3.6}$$

Moreover, there exists a constant $M\geq 0$ such that

$$|\,\omega(x,y)\,| \leq M \quad \text{for every } x,y\in X, \tag{3.7}$$

and, finally, $\lim_{y\to x} \omega(x,y) = 0$ uniformly with respect to $x\in X$, i.e.

For every $\varepsilon>0$ there exists $\delta>0$ such that $|\,\omega(x,y)\,| \leq \varepsilon$ for every $x,y\in X$ satisfying $\|x-y\| \leq \delta$. (3.8)

For every fixed $x = (x_i)_{1\leq i\leq p} \in X$ consider the following continuous functions on X

$$f_{i,x} = pr_i - x_i, \qquad (i=1,\ldots,p) \tag{3.9}$$

$$f_x = \sum_{i=1}^{p}\frac{\partial u(x)}{\partial x_i}\,f_{i,x}, \tag{3.10}$$

$$g_x = \frac{1}{2}\sum_{i,j=1}^{p}\frac{\partial^2 u(x)}{\partial x_i \partial x_j}\,f_{i,x}f_{j,x}, \tag{3.11}$$

$$\sigma_x(y) = \|y-x\| \qquad (y\in X). \tag{3.12}$$

Clearly,

$$u = u(x)\cdot\mathbb{1} + f_x + g_x + \omega(x,\cdot)\,\sigma_x^2. \tag{3.13}$$

Note that $f_x \in A(X)$ and so $L_{n,\lambda}(f_x) = f_x$ for every $n \geq 1$ (cf. (2.19)). Consequently, from (3.13) for every $n \geq 1$ we get

$$L_{n,\lambda}(u) = u(x)\cdot\mathbb{1} + f_x + L_{n,\lambda}(g_x) + L_{n,\lambda}(\omega(x,\cdot)\sigma_x^2), \tag{3.14}$$

and so,

$$n\,(L_{n,\lambda}(u)(x) - u(x)) = n\,L_{n,\lambda}(g_x)(x) + n\,L_{n,\lambda}(\omega(x,\cdot)\sigma_x^2)(x). \tag{3.15}$$

We prove now that

$$\lim_{n\to\infty} n\,L_{n,\lambda}(\omega(x,\cdot)\sigma_x^2)(x) = 0 \text{ uniformly with respect to } x\in X. \tag{3.16}$$

In fact, note preliminarily that for every $x\in X$, since $\sigma_x^2 = \sum_{i=1}^{p} f_{i,x}^2$ and $f_{i,x}\in A(X)$ for every $i=1,\dots,p$, then from (2.15) we get

$$n\,L_{n,\lambda}(\sigma_x^2)(x) = \sum_{i=1}^{p} n\,L_{n,\lambda}(f_{i,x}^2)(x) = \sum_{i=1}^{p} \lambda(x)\;T(f_{i,x}^2)(x) \leq p\,\mathrm{diam}(X)^2 \tag{3.17}$$

(recall that $\|T\| = \|T(\mathbb{1})\| = 1$)

Now, given $\varepsilon > 0$, by (3.8) there exists $\delta > 0$ such that

$$|\,\omega(x,y)\,| \leq \frac{\varepsilon}{2p\,\mathrm{diam}(X)^2} := \varepsilon_1 \text{ for every } x,y\in X \text{ satisfying } \|x-y\|\leq\delta. \tag{3.18}$$

Moreover, by virtue of Lemma 3.2, there exists $\upsilon\in\mathbb{N}$ such that

$$|\,n\,L_{n,\lambda}(\sigma_x^4)(x)\,| \leq \frac{\varepsilon\delta^4}{2M\,\mathrm{diam}(X)^2} := \varepsilon_2, \tag{3.19}$$

for every $n\geq\upsilon$ and for every $x\in X$.

Now, given $x,y\in X$, if $\|x-y\| \leq \delta$, then

$$|\,\omega(x,y)\;\sigma_x^2(y)\,| \leq \varepsilon_1\,\sigma_x^2(y) \leq \varepsilon_1\,\sigma_x^2(y) + \frac{M\,\mathrm{diam}(X)^2}{\delta^4}\,\sigma_x^4(y),$$

while, if $\|x-y\| > \delta$, on account of (3.7),

$$|\,\omega(x,y)\;\sigma_x^2(y)\,| \leq M\,\mathrm{diam}(X)^2 \leq \frac{M\,\mathrm{diam}(X)^2}{\delta^4}\sigma_x^4(y) \leq \varepsilon_1\sigma_x^2(y) + \frac{M\,\mathrm{diam}(X)^2}{\delta^4}\sigma_x^4(y).$$

Then, for every $x\in X$,

$$|\,\omega(x,\cdot)\;\sigma_x^2\,| \leq \varepsilon_1\,\sigma_x^2 + \frac{M\,\mathrm{diam}(X)^2}{\delta^4}\,\sigma_x^4 \qquad \text{on } X.$$

Consequently, for every $n\geq\upsilon$ and for every $x\in X$,

$$|\,n\,L_{n,\lambda}(\omega(x,\cdot)\sigma_x^2)(x)\,| \leq n\,L_{n,\lambda}(\,|\,\omega(x,\cdot)\,|\,\sigma_x^2)(x) \leq$$

$$\leq \varepsilon_1 \, n \, L_{n,\lambda}(\sigma_x^2)(x) + \frac{M \, \mathrm{diam}(X)^2}{\delta^4} \, n \, L_{n,\lambda}(\sigma_x^4)(x) \leq$$

$$\leq \varepsilon_1 \, p \, \mathrm{diam}(X)^2 + \frac{M \, \mathrm{diam}(X)^2}{\delta^4} \, \varepsilon_2 = \varepsilon,$$

and so the proof of (3.16) is complete.

To accomplish the proof of the Theorem, by virtue of (3.15) and (3.16), it suffices to show that

$$\lim_{n\to\infty} n \, L_{n,\lambda}(g_x)(x) = A(u)(x) \tag{3.20}$$

uniformly with respect to $x \in X$.

In fact, from (2.15) and (3.11) it follows that for every $x \in X$

$$nL_{n,\lambda}(g_x)(x) = \frac{1}{2}\sum_{i,j=1}^{p} \frac{\partial^2 u(x)}{\partial x_i \partial x_j} \, nL_{n,\lambda}(f_{i,x} f_{j,x})(x) = \frac{\lambda(x)}{2}\sum_{i,j=1}^{p} \frac{\partial^2 u(x)}{\partial x_i \partial x_j} \, T(f_{i,x} f_{j,x})(x)$$

and this last term is exactly $A(u)(x)$, on account of (2.7), (2.9), (2.11) and (2.13), respectively (see also formulas (3.2), (3.5) and (3.8) of [1]). ∎

We can now prove the main results of the paper. In the sequel we will write $\mathscr{C}([0,1])$ and $\mathscr{C}^2([0,1])$ instead of $\mathscr{C}([0,1],\mathbb{R})$ and $\mathscr{C}^2([0,1],\mathbb{R})$.

3.4 Theorem *Set* $X = [0,1]$ *and consider the sequence* $(L_{n,\lambda})_{n\in\mathbb{N}}$ *of Lototsky-Schnabl operators* (2.8) *associated with a polynomial* $\lambda:[0,1]\to[0,1]$ *satisfying*

$$\lambda(x) > 0 \quad \textit{for every } x \in [0,1]. \tag{3.21}$$

Then, there exists a strongly continuous positive contraction semigroup $(T(t))_{t\geq 0}$ *on* $\mathscr{C}([0,1])$, *whose generator is the operator* $A:D(A)\to\mathscr{C}([0,1])$ *defined by*

$$A(u)(x) = \frac{x(1-x)}{2} \lambda(x) \, u''(x) \qquad \textit{if } 0<x<1,$$

$$A(u)(x) = 0 \qquad \textit{if } x=0 \textit{ or } x=1.$$

for every $u \in D(A) = \{u \in \mathscr{C}([0,1]) \mid u \in \mathscr{C}^2(]0,1[), \lim_{x\to 0^+} \frac{x(1-x)}{2} \lambda(x) \, u''(x) = \lim_{x\to 1^-} \frac{x(1-x)}{2} \lambda(x) \, u''(x) = 0\}$, *such that for every* $t \geq 0$ *and for every sequence* $(k(n))_{n\in\mathbb{N}}$ *of positive integers satisfying* $\lim_{n\to\infty} \frac{k(n)}{n} = t$ *one has*

$$T(t) = \lim_{n\to\infty} L_{n,\lambda}^{k(n)} \qquad \textit{strongly on } \mathscr{C}([0,1]).$$

In particular,

$$T(t) = \lim_{n\to\infty} L_{n,\lambda}^{[nt]} \quad \textit{strongly on } \mathscr{C}([0,1]), \tag{3.22}$$

[nt] *being the integer part of* nt.

Proof. By virtue of a result of Clément and Timmermans [3, Proposition 1 and Theorem 2] we know that D(A) is dense in $\mathscr{C}([0,1])$ and the operator A generates a strongly continuous contraction semigroup $(T(t))_{t\geq 0}$ on $\mathscr{C}([0,1])$.

Let us consider the linear operator $Z:D(Z)\to\mathscr{C}([0,1])$ defined by putting for every $f\in D(Z)$

$$Z(f) = \lim_{n\to\infty} n(L_{n,\lambda}(f) - f), \tag{3.23}$$

where $D(Z) = \{g\in\mathscr{C}([0,1]) \mid \lim_{n\to\infty} n(L_{n,\lambda}(g)-g) \text{ exists in } \mathscr{C}([0,1])\}$.

By virtue of Theorem 3.3 we have $\mathscr{C}^2([0,1]) \subset D(Z) \cap D(A)$ and

$$A = Z \quad \text{on } \mathscr{C}^2([0,1]). \tag{3.24}$$

In particular, D(Z) is dense in $\mathscr{C}([0,1])$.

We now prove that the range R(I-Z) is dense in $\mathscr{C}([0,1])$, where I denotes the identity operator on $\mathscr{C}([0,1])$, and, to this end, we shall prove that $\mathscr{C}^2([0,1]) \subset R(I-Z)$.

In fact, given $f\in\mathscr{C}^2([0,1])$, there exists $u\in D(A)$ such that

$$u - A(u) = f \tag{3.25}$$

[3, Theorem 2].

Since $A(u)(0) = A(u)(1) = 0$, then $u(0) = f(0)$ and $u(1) = f(1)$.

So,

$$\lim_{x\to 0^+} u''(x) = \lim_{x\to 0^+} 2\,\frac{u(x)-f(x)}{x(1-x)\lambda(x)} = \frac{2}{\lambda(0)}\,(u'(0) - f'(0)),$$

and, analogously,

$$\lim_{x\to 1^-} u''(x) = -\frac{2}{\lambda(1)}\,(u'(1) - f'(1)).$$

Hence, $u\in\mathscr{C}^2([0,1]) \subset D(Z)$ and $u - Z(u) = f$.

Thus, we have proved that R(I-Z) is dense in $\mathscr{C}([0,1])$ and so, taking into account a result of Trotter [16, Theorem 5.3], we conclude that the closure Z of Z is the infinitesimal generator of a contraction semigroup $(S(t))_{t\geq 0}$ and for every $t\geq 0$

$$S(t) = \lim_{n\to\infty} L_{n,\lambda}^{[nt]} \quad \text{strongly on } \mathscr{C}([0,1]). \tag{3.26}$$

In particular, every S(t) is positive. By arguing as in [1, p. 272], it is easy to see that for every sequence $(k(n))_{n\in\mathbb{N}}$ of positive integers satisfying $\lim\limits_{n\to\infty}\frac{k(n)}{n} = t$, one also gets

$$S(t) = \lim_{n\to\infty} L_{n,\lambda}^{k(n)} \qquad \text{strongly on } \mathscr{C}([0,1]).$$

Now the proof will be accomplished if we prove that $A = Z$.

To this purpose, let us denote by $P([0,1])$ the subalgebra of all polynomials on $[0,1]$. $P([0,1])$ is dense in $\mathscr{C}([0,1])$; moreover, since λ is a polynomial, from (2.7), (2.16), (2.17) and (2.18) it follows that $P([0,1])$ is invariant under every $L_{n,\lambda}$ and so, by (3.26), under every S(t).

Consequently, $P([0,1])$ is a core for Z [10, p.5 and Corollary 1.34]. But, from (3.24) it also follows that $Z = Z = A$ on $P([0,1])$, and hence $A = Z$ by virtue of [10, Theorem 1.33]. ■

Before stating some consequences of Theorem 3.4, we present some further properties of the semigroup $(T(t))_{t\geq 0}$.

3.5 Proposition *Under the same hypotheses as in* Theorem 3.4 *we have*

1. *For every* $f\in\mathscr{C}([0,1])$ *the following statements are equivalent:*

(i) f *is convex;*

(ii) $L_{n+1,\lambda}(f) \leq L_{n,\lambda}(f)$ *for every* $n \geq 1$;

(iii) $f \leq L_{n,\lambda}(f)$ *for every* $n \geq 1$;

(iv) $f \leq T(t)\,f$ *for every* $t \geq 0$.

Moreover, if $f\in D(A)$, *then statements* (i)-(iv) *are equivalent to*

(v) $T(t)\,f$ *is convex* *for every* $t \geq 0$.

2. *For every* $f\in\mathscr{C}([0,1])$, *f convex, and for every* $t \geq 0$, *we have*

$$f \leq T(t)\,f \leq T(f).$$

3. $\lim\limits_{t\to+\infty} T(t)\,f = T(f)$ *for every* $f\in\mathscr{C}([0,1])$.

Proof. **1.** (i)⇒(ii): It follows from (2.21).

(ii)⇒(iii): For every $n,m\in\mathbb{N}$ we have $L_{n+m,\lambda}(f) \leq L_{n,\lambda}(f)$, and so, when $n\to\infty$, we get (iii) by virtue of statement 1. of Theorem 2.2.

(iii)⇒(iv): It is a consequence of (3.22).

(iv)⇒(i): Given $t>0$, since $A(\int_0^t T(s)\, f\, ds) = T(t)f - f \geq 0$, the function $\int_0^t T(s)\, f\, ds$ is convex. Consequently, since $f = \lim_{t\to 0^+} \frac{1}{t} \int_0^t T(s)\, f\, ds$, f is convex too.

Suppose now that $f \in D(A)$. If f is convex, then $A(f) \geq 0$ and hence $A(T(t)f) = T(t)(A(f)) \geq 0$. Thus $T(t)f$ is convex.

Conversely, if for every $t \geq 0$ $T(t)f$ is convex, then $\int_0^t T(s)\, f\, ds$ is convex for every $t \geq 0$ too. So $T(t)f - f = A(\int_0^t T(s)\, f\, ds) \geq 0$ and the result follows.

2. We have only to show that $T(t)f \leq T(f)$. In fact, since f is convex, $f \leq T(f)$ and so $\lambda T(f) + (1-\lambda)f \leq T(f)$. From (2.21) it follows that $L_{n,\lambda}(f) \leq T(f)$ for every $n \geq 1$. By iteration, on account of (2.19), we finally have $T(t)f \leq T(f)$.

In order to prove statement **3.**, first fix $f \in \mathscr{C}^2([0,1])$, f convex; set $m = \min_{0 \leq x \leq 1} \lambda(x)$ and denote by $B: D(B) \to \mathscr{C}([0,1])$ the differential operator

$$B(u)(x) = m\, \frac{x(1-x)}{2}\, u''(x) \qquad \text{if } 0<x<1,$$

$$B(u)(x) = 0 \qquad \text{if } x=0 \text{ or } x=1.$$

where $D(B) = \{u \in \mathscr{C}([0,1]) \mid u \in \mathscr{C}^2(]0,1[),\ \lim_{x\to 0^+} \frac{x(1-x)}{2} u''(x) = \lim_{x\to 1^-} \frac{x(1-x)}{2} u''(x) = 0\}$.

Then B generates a strongly continuous positive semigroup $(S(t))_{t\geq 0}$ that can be represented in terms of iterates of suitable Lototsky-Schnabl operators and

$$\lim_{t\to +\infty} S(t) = T \quad \text{strongly on } \mathscr{C}([0,1])$$

(cf. Theorem 2.2 and formula (2.5)).

Clearly $D(B) \subset D(A)$; moreover for every $u \in D(B)$, u convex, we have

$$B(u) \leq A(u),$$

and, for every $t \geq 0$, $S(t)u$ is convex; hence, for sufficiently large $\mu \in \mathbb{R}$ the function

$$R(\mu,B)u = \int_0^{+\infty} e^{-\mu t} S(t)u\, dt$$

is convex, where $R(\mu,B)$ denotes the inverse of $\mu I - B$.

In view of these remarks, we have

$$R(\mu,A)f - R(\mu,B)f = R(\mu,A)(A-B)R(\mu,B)f \geq 0.$$

By iteration, for every $n \geq 1$, we have

$$R(\mu,B)^n f \leq R(\mu,A)^n f,$$

and hence, for every $t \geq 0$,

$$S(t)f = \lim_{t \to +\infty} \left(\frac{n}{t} R\left(\frac{n}{t}, B\right)\right)^n f \leq \lim_{t \to +\infty} \left(\frac{n}{t} R\left(\frac{n}{t}, A\right)\right)^n f = T(t)f.$$

Thus, $S(t)f \leq T(t)f \leq T(f)$ and so $\lim_{t \to +\infty} T(t)f = T(f)$.

From this, it follows in particular that

$$\lim_{t \to +\infty} T(t)(e_2) = T(e_2),$$

where $e_2(x) = x^2$ ($x \in [0,1]$). On the other hand, from (3.22) we get

$$\lim_{t \to +\infty} T(t)(\mathbb{1}) = \mathbb{1} = T(\mathbb{1}),$$

and

$$\lim_{t \to +\infty} T(t)(e_1) = e_1 = T(e_1),$$

where $e_1(x) = x$ for every $x \in [0,1]$. Then by a theorem of Korovkin-type established in [1, Proposition 1.2], we infer that for every $f \in \mathscr{C}([0,1])$,

$$\lim_{t \to +\infty} T(t)(f) = T(f). \qquad \blacksquare$$

We conclude this paper by briefly sketching some applications to initial value problems associated with degenerate diffusion equations.

In fact, from Theorem 3.4 and from the general theory of one-parameter semigroups (see [10], for example), it follows that the Cauchy problem

$$\begin{cases} \dfrac{\partial u}{\partial t}(x,t) = \dfrac{x(1-x)}{2} \lambda(x) \dfrac{\partial^2 u}{\partial x^2}(x,t) & 0 < x < 1, \ 0 \leq t, \\[2ex] \lim\limits_{x \to 0^+} \dfrac{x(1-x)}{2} \lambda(x) \dfrac{\partial^2 u}{\partial x^2}(x,t) = \lim\limits_{x \to 0^+} \dfrac{x(1-x)}{2} \lambda(x) \dfrac{\partial^2 u}{\partial x^2}(x,t) = 0 & t \geq 0, \\[2ex] u(x,0) = u_0 & u_0 \in D(A). \end{cases} \tag{3.27}$$

has a unique (classical) solution given by

$$u(x,t) = T(t)(u_0)(x) = \lim_{n \to +\infty} L_{n,\lambda}^{[nt]}(u_0)(x). \tag{3.28}$$

Moreover (3.28) holds uniformly with respect to $x \in [0,1]$.

This statement can be reformulated in apparently more general terms.

3.6 Corollary *Let* $\alpha:[0,1]\to\mathbb{R}$ *be a polynomial for which* 0 *and* 1 *are simple roots. Moreover, suppose that* $\alpha(x) > 0$ *for every* $x\in]0,1[$.

Then the following problem

$$\begin{cases} \dfrac{\partial u}{\partial t}(x,t) = \alpha(x)\dfrac{\partial^2 u}{\partial x^2}(x,t) & 0<x<1,\ t\geq 0, \\ \lim\limits_{x\to 0^+} \alpha(x)\dfrac{\partial^2 u}{\partial x^2}(x,t) = \lim\limits_{x\to 1^-} \alpha(x)\dfrac{\partial^2 u}{\partial x^2}(x,t) = 0 & t\geq 0, \\ u(x,0) = u_0 & u_0\in\mathscr{C}([0,1])\cap\mathscr{C}^2(]0,1[), \\ & \lim\limits_{x\to 0^+}\alpha(x)u_0''(x)=\lim\limits_{x\to 1^-}\alpha(x)u_0''(x)=0. \end{cases} \tag{3.29}$$

has a unique classical solution given by

$$u(x,t) = \lim_{n\to+\infty} L_{n,\lambda}^{[cnt]}(u_0)(x), \tag{3.30}$$

where

$$c = \max_{0\leq x\leq 1} \frac{2\alpha(x)}{x(1-x)}, \tag{3.31}$$

$\lambda:[0,1]\to[0,1]$ *is the polynomial satisfying*

$$\lambda(x) = \frac{2}{c}\,\frac{\alpha(x)}{x(1-x)} \qquad \textit{for every } x\in]0,1[, \tag{3.32}$$

and the operators $L_{n,\lambda}$ *are defined as in* (2.8).

Moreover, (3.30) *holds uniformly with respect to* $x\in[0,1]$.

Proof. Clearly $\alpha(x) = c\,\dfrac{x(1-x)}{2}\,\lambda(x)$ for every $x\in[0,1]$.

So, if $(T(t))_{t\geq 0}$ is the semigroup associated with the problem (3.27) and the polynomial (3.32), then the semigroup which gives the solution of (3.29) is $(T(ct))_{t\geq 0}$ and this finishes the proof. ∎

The differential equation indicated in (3.29) governs particular one dimensional Markov processes of diffusion type and the associated boundary conditions are also called Wentcel's boundary conditions [17]. According to the terminology introduced by Feller [4, Sect. 11] they mean that 0 and 1 are not entrance boundary points; more precisely, they are exit boundary points. From a probabilistic point of view, this means that the probability that a particle, located in the interior of [0,1], will reach 0 or 1 after a finite lapse of time, is positive.

These kinds of problems were first studied by Feller [4] and, subsequently, by Martini [7], [8] and Clément and Timmermans [3].

The representation formula (3.30) can be useful to obtain qualitative information on the solution $u(x,t)$. Here we indicate some easy consequences that follow directly from Proposition 3.5.

However, we also point-out that, when λ is constant, then the solution $u(x,t)$ (i.e. the semigroup $(T(t))_{t\geq 0}$) can be explicitly expressed in terms of Jacobi polynomials [9, p. 404] (see also [6, (1.10)].

3.7 Proposition *Let* u_0 *and* $u(x,t)$ *be as indicated in* Corollary 3.6. *Then,*

1. *If* $u_0 \geq 0$, *then* $u(x,t) \geq 0$ *for every* $x \in [0,1]$ *and* $t \in [0,+\infty[$;
2. $u(0,t) = u_0(0)$ *and* $u(1,t) = u_0(1)$ *for every* $t \geq 0$;
3. *If* u_0 *is convex, then for every* $x \in [0,1]$ *and* $t \in [0,+\infty[$, $u(x,\cdot)$ *is increasing and* $u(\cdot,t)$ *is convex. Moreover,*

$$u_0(x) \leq u(x,t) \leq T(u_0)(x) = (1-x)\, u_0(0) + x\, u_0(1).$$

for every $x \in [0,1]$ *and* $t \geq 0$.

4. $\lim_{t\to+\infty} u(x,t) = T(u_0)(x)$ *uniformly with respect to* $x \in [0,1]$.

So, $\lim_{t\to+\infty} u(x,t) = 0$ *uniformly with respect to* $x \in [0,1]$ *if and only if* $u_0(0) = u_0(1) = 0$.

Proof. Statements 1. and 2. follow from (3.30) and (2.20), respectively, since in this case H is the subspace generated by the functions $\mathbb{1}$ and x and so $\partial_H[0,1] = \{0,1\}$.

With respect to statement 3., we first note that $A(u_0) \geq 0$, where $A:D(A) \to \mathscr{C}([0,1])$ denotes the differential operator defined as in Theorem 3.4 with λ defined by (3.32).

Thus, if we denote by $(T(t))_{t\geq 0}$ the positive semigroup generated by A, we also get that $T(t)u_0$ is convex, i.e. $u(\cdot,t)$ is convex. On the other hand, $\frac{\partial u}{\partial t}(x,t) = A(T(t))u_0(x) \geq 0$, and hence $u(x,\cdot)$ is increasing.

Statement 4. directly follows from Proposition 3.5, 3. ∎

4. REFERENCES

1 **F. Altomare**, *Limit semigroups of Bernstein-Schnabl operators associated with positive projections,* Ann. Sc. Norm. Sup. Pisa, Serie IV, (16) 2 (1989), 259-279.

2 **F. Altomare** and **S. Romanelli**, *On some classes of Lototsky-Schnabl operators,* to appear in Note di Matematica, 1991.

3 **P. H. Clément** and **C. A. Timmermanns**, *On C_0-semigroups generated by differential operators satisfying Wentcel's boundary conditions,* Indag. Math. 89 (1986), 379-387.

4 **W. Feller**, *The parabolic differential equations and the associated semi-groups of transformations,* Ann. of Math. (2) 55 (1952), 468-519.

5 **M. W. Grossman**, *Lototsky-Schnabl functions on compact convex sets,* J. Math. Anal. Appl. 55 (1976), 525-530.

6 **S. Karlin** and **Z. Ziegler**, *Iteration of positive approximation operators,* J. Approx. Theory 3 (1970), 310-339.

7 **R. Martini**, *A relation between semi-groups and sequences of approximation operators,* Indag. Math. 35 (1973), 456-465.

8 **R. Martini**, *Differential operators degenerating at the boundary as inifinitesimal generators of semi-groups,* Ph.D. Thesis, T. H. Delft, 1975.

9 **R. Martini** and **W. L. Boer**, *On the construction of semi-groups of operators,* Indag. Math. 36 (1974), 392-405.

10 **R. Nagel** ed., *One-parameter semigroups of positive operators,* Lecture Notes in Mathematics, n.1184, Springer-Verlag, Berlin - Heidelberg - New York-Tokyo, 1986.

11 **T. Nishishiraho**, *Quantitative theorems on approximation processes of positive linear operators,* Multivariate Approximation Theory II (Proc. Conf. Math. Res. Inst. Oberwolfach 1982; ed. by W. Schempp and K. Zeller) ISNM Vol. 61, 297-311, Birkhäuser Verlag, Basel - Boston - Stuttgart, 1982.

12 **T. Nishishiraho**, *The order of approximation by positive linear operators,* Tôhoku Math. J. 40 (1988) n. 4, 617-632.

13 **I. Rasa**, *Generalized Bernstein operators and convex functions,* to appear.

14 **I. Rasa**, *On the monotonicity of sequences of Bernstein-Schnabl operators,* Anal. Numér. Théor. Approx. 17 (1988), no. 2, 185-187.

15 **W. Schempp**, *Zur Lototsky-Transformation über kompakten Räumen von Wahrscheinlichkeitsmaßen,* Manuscripta Math. 5 (1971), 199-211.

16 **H. F. Trotter**, *Approximation of semi-groups of operators,* Pacific J. Math. 8 (1958), 887-919.

17 **A. D. Wentcel**, *On boundary conditions for multidimensional diffusion processes,* Theory Prob. Appl. 4 (1959), 164-177.

Progress in Functional Analysis
K.D. Bierstedt, J. Bonet, J. Horváth & M. Maestre (Eds.)

ON WEAKLY LINDELÖF BANACH SPACES

José Orihuela

Departamento de Matemáticas. Universidad de Murcia. Campus de Espinardo.
30100 Espinardo, Murcia (Spain).

A Manuel Valdivia, mi maestro, con admiración y afecto.

Abstract

The purpose of this paper is to show how the method, developed by M.Valdivia, of constructing projections in a weakly compactly generated Banach space X can be used to give a direct proof of the Lindelöf property for the weak topology in X. Since this method extends to large classes of spaces we prove a Lindelöf property for Banach spaces with a Valdivia compact weak* dual unit ball. Our result includes previous ones of K.Alster, R.Pol and S.P.Gul'ko for the space $C(K)$ with the topology of pointwise convergence, where K is a Corson compact space.

If X is a Banach space then the dual X^* has the Radon-Nikodym property if and only if it is Lindelöf in the topology of uniform convergence on the separable bounded subsets of X. Finally, if a dual space X^* is weakly Lindelöf then the bidual unit ball must be a Corson compact space in the weak* topology, so the product $X^* \times X^*$ is weakly Lindelöf, too. This result gives a positive answer to a problem of Corson in the case of dual spaces.

1. INTRODUCTION

Let S be a topological space. We denote by $C_p(S)$ the space of all continuous real valued functions on S endowed with the topology T_p of pointwise convergence on S. A problem of continual interest among topologists is the study of the Lindelöf property of $C_p(S)$, [Ar], [P2], [L]. If we are dealing with a Banach space X and a total subset F of the dual X^*, this problem means to investigate when $X[\sigma(X,F)]$ is a Lindelöf space. It was a conjecture of Corson (which has interested many people) that a Banach space X would be weakly Lindelöf if and only if X is weakly compactly generated (WCG), [C], [L]. M.Talagrand showed that a WCG Banach space is a $K_{\sigma\delta}$ subset of its bidual with the weak* topology, thus X is weakly K-analytic and $X[\sigma(X,X^*)]$ is Lindelöf, [T]. On the other hand, he also gave an example of a compact set K such that the Banach space $C(K)$ is weakly K-analytic and not WCG, which completely solved Corson's conjecture. R.Pol gave another example of a compact space K such that $C(K)$ is weakly Lindelöf and not WCG, [P1].

For any non void set Γ, we denote by $\Sigma(\Gamma)$ the Σ-product of real lines of power Γ, i.e.

$$\Sigma(\Gamma) = \{x \in \mathbb{R}^\Gamma : \{\gamma \in \Gamma : x(\gamma) \neq 0\} \text{ is countable } \}.$$

It is assumed, as usual, that the topology of $\Sigma(\Gamma)$ is the one induced by the product topology of $\mathbb{R}^\Gamma$. A compact space K is called Corson compact if K is homeomorphic to a compact subset of some $\Sigma(\Gamma)$. S.P.Gul'ko proved that the weak* dual unit ball of every weakly K-analytic Banach space is a Corson compact, [Gu2]. K.Alster, R.Pol, and S.P.Gul'ko proved that the space $C_p(K)$ is Lindelöf for every Corson compact space K, [AP], [Gu1]. The class of Corson compact spaces admits quite pathological examples, for instance the Haydon, Kunen and Talagrand example K for which the Banach space $C(K)$ is not weakly Lindelöf under the continuum hypothesis, [N,Th.5.9]. Other examples are included in [AMN], where for a Corson compact space K it is proved that $C(K)$ is weakly Lindelöf if and only if every positive regular Borel measure on K has separable support. Nevertheless, the statement that for every Corson compact space K the Banach space $C(K)$ is weakly Lindelöf is independent of the usual axioms of set theory, [AMN]. I think that no "naive" example of a compact space K with $C_p(K)$ Lindelöf but $C(K)$ not weakly Lindelöf is known.

Turning to stability properties, let us recall the old question of H.H.Corson [C] if the fact that a Banach space X is weakly Lindelöf must already imply that $X \times X$ is weakly Lindelöf, too, [Ar]. Of course, the answer is yes when the weak* dual unit ball $B(X^*)$ is Corson compact. This immediately follows from the stability of the class of Corson compact spaces by finite products and the mentioned results of Alster, Pol and Gul'ko.

A compact space K is called a Valdivia compact, [DG], if K is homeomorphic to a compact subset of the product $\mathbb{R}^\Gamma$ for some set Γ with the property that $K \cap \Sigma(\Gamma)$ is dense in K. In particular, every Corson compact space is Valdivia compact. The cubes $[0,1]^\Gamma$ with uncountable Γ are examples of Valdivia compact spaces which are not Corson compact. Every dyadic compact space and every compact space defined by a family of adequate sets [T] are also natural examples of spaces in the class of Valdivia compacts. Banach spaces $C(K)$ for a Valdivia compact K have recently been studied in [AMN],[DG],[V4-6] and [OV2].

It is our purpose in this paper to show how the method of constructing projections which was developed by M.Valdivia in a sequence of papers [V1-6],[OV1-2] can be adapted to give a direct proof of the Lindelöf property of $C_p(K)$ for a Corson compact space K. Even more, given a Valdivia compact space K, we shall prove that the space $C(K)$ is Lindelöf in the topology of uniform convergence on the separable subsets of $K \cap \Sigma(\Gamma)$. Applying this construction to dual spaces, we give a topological characterization of Asplund spaces. Finally, for dual spaces we obtain an answer for the problem of Corson stated above: Indeed, we shall prove that a dual Banach space is weakly Lindelöf if and only if its weak* dual unit ball is a Corson compact space. So if X is a dual Banach space which is weakly Lindelöf, then $X \times X$ is weakly Lindelöf, too.

Let us briefly describe the organization of this paper and give the statement of the main theorems.

In section 2, we consider Banach spaces in the class $\mathcal{V}$ (see definition below) and point out that they always enjoy "some" Lindelöf property. Roughly speaking, a Banach space is in the class $\mathcal{V}$ if its weak* dual unit ball is a Valdivia compact. If K is a Valdivia compact then $C(K)$ is in the class $\mathcal{V}$. For a Banach space X in the class $\mathcal{V}$ we fix a norming subspace F in X^* (see definition below) and we obtain the following:

THEOREM A. *Let X be a Banach space in the class $\mathcal{V}$. If $\gamma(X,F)$ denotes the topology of uniform convergence on the bounded and weak* separable subsets of F, then $X[\gamma(X,F)]$ is a Lindelöf space. Moreover, every product space $X[\gamma(X,F)]^n$ is a Lindelöf space for $n = 2, 3, ...$*

The proof we present here is based on the separable complementation property of X. More precisely, we have a constructive method to prove the separable complementation property of X, [V6],[OV1-2]; hence in the construction we are able to involve the extra information we have on a $\gamma(X,F)$-open cover of X. Once we get a complemented and separable subspace, we have a countable subcover; then we can work as in the proof of the classical Lindelöf theorem for separable metrizable spaces.

In section 3 we apply the same arguments for dual spaces with the Radon-Nikodym property. We show the following:

THEOREM B.- *Let X be a Banach space and $\gamma(X^*,X)$ the topology of uniform convergence on the bounded separable subsets of X. Then X is an Asplund space if and only if $X^*[\gamma(X^*,X)]$ is a Lindelöf space.*

Moreover, if X is an Asplund space, $X^*[\sigma(X^*,X^{**})]$ is Lindelöf if, and only if, $B(X^{**})$ is a weak* Corson compact space. Since every dual and weakly Lindelöf Banach space has the RNP, [E], we obtain in Section 4 the following positive answer for Corson's question in the case of dual spaces:

THEOREM C.- *Let X be a dual Banach space. Then $X[\sigma(X,X^*)]$ is Lindelöf if, and only if, $(X[\sigma(X,X^*)])^n$ is Lindelöf, $n = 1, 2, ...$*

I have used the classical notation which can be found, for instance, in [D]. The closed unit ball of a Banach space X is denoted by $B(X)$. Only real Banach spaces are considered here. If A (resp. B) is a subspace of X (resp. of X^*), the space orthogonal to A (resp. B) in X^* (resp. X) is denoted by $A^\perp$ (resp. $B^\perp$).

2. BANACH SPACES IN THE CLASS $\mathcal{V}$

A Banach space X is said to be in the class $\mathcal{V}$ if there exists a set I, a norming subspace F in X^* and a continuous injective map J from $B(X^*)$, with the weak* topology, into the product space $\mathbb{R}^I$, with the product topology T_p, such that

$$F \cap B(X^*) \subset \{f \in B(X^*) : J(f) \in \Sigma(I)\}.$$

In the sequel, given a Banach space X in $\mathcal{V}$, F always represents the norming subspace fixed above. The class $\mathcal{V}$ of Banach spaces has been studied in [V6] in connection with the spaces $C(K)$ for a Valdivia compact K. M.Valdivia shows there how to construct simultaneous resolutions of the identity operator in subspaces of $C(K)$, in particular in Banach spaces in the class $\mathcal{V}$. In our joint paper [OV2] we describe a projective generator in the sense of [OV1] which gives the projectional resolution of the identity in a Banach space in the class $\mathcal{V}$, thus yielding an analytic approach to Valdivia's result. In particular, these methods provide a direct construction of the projectional resolution of the identity in Banach spaces with a Corson compact weak* dual unit ball, without using of the results of S.P.Gul'ko, [V2], [OSV]. It is our purpose in this section to give a direct proof of the theorem A stated in the introduction which is only based on the separable complementation property of X. A different proof, based on the existence of an M-basis in X, is given in [OV2].We shall need three lemmas, taken from [OV2], to see how the separable complementation property of X implies the Lindelöf character of $X[\gamma(X,F)]$.

Lemma 1.- *Let X be a Banach space, $A \subset X$ and $B \subset X^*$ $\mathbb{Q}$-linear subspaces with the following properties:*

(i) $\| x \| = \sup\{|< x, y >| : y \in B(X^*) \cap B\}$ *for every* $x \in A$,

(ii) $\overline{B \cap B(X^*)}^{\sigma(X^*,X)} \cap A^\perp = \{0\}$.

Then there is a norm one projection P from X onto the norm closure of A with kernel(P)= $B^\perp$.

Proof.- Let $W := \cup_{n=1}^{\infty} n\,\overline{B \cap B(X^*)}^{\sigma(X^*,X)}$, E the norm closure of A, and R_E the restriction mapping from X^* onto E^*. The condition (i) is also verified for every x in E and it implies that $R_E(B \cap B(X^*))$ is $\sigma(E^*,E)$-dense in $B(E^*)$. The condition (ii) implies that R_E is injective when restricted to W. Hence R_E is a homeomorphism between the compact sets $\overline{B \cap B(X^*)}^{\sigma(X^*,X)}$ and $B(E^*)$ with the weak*-topologies. Consequently we have $W \cap B(X^*) = \overline{B \cap B(X^*)}^{\sigma(X^*,X)}$, the Krein-Smulian theorem says that W is $\sigma(X^*,X)$-closed, and we have $W = \overline{B}^{\sigma(X^*,X)}$. Hence

$$\overline{B}^{\sigma(X^*,X)} \cap E^\perp = \{0\}$$

and by polarity $X = \overline{E + B^{\perp}}$.

If we use again the condition (i) we see that $\| x \| \leq \| x + y \|$ for every x in E and y in $B^{\perp}$, it follows that $E \cap B^{\perp} = \{0\}$ and the projection from $E \oplus B^{\perp}$ onto E along $B^{\perp}$ has norm one. Since X is complete and $X = \overline{E \oplus B^{\perp}}$, we have $X = E \oplus B^{\perp}$.

Lemma 2.- *For a Banach space X in the class $\mathcal{V}$ there is a multivalued map φ from $F \cap B(X^*)$ into the subsets of X with the following two properties:*

(i) $\varphi(g)$ is a countable subset of X for every g in $F \cap B(X^*)$,

(ii) $\overline{B \cap B(X^*)}^{\sigma(X^*,X)} \cap (\varphi(B \cap B(X^*))^{\perp} = \{0\}$ for any $B \subset F$.

Proof.- For every i in I, we denote by h_i the i-th coordinate functional in $\mathbb{R}^I$ and by ξ_i the composition map $h_i \circ J$. Without loss of generality we assume $J(\vec{0}) = \vec{0}$ because of $\vec{0} \in F$. The Stone-Weierstrass theorem gives us a countable subset X_i of X such that ξ_i belongs to the closure of the algebra generated by X_i in $C(B(X^*)[\sigma(X^*, X)])$. Given f in $F \cap B(X^*)$, we define

$$\varphi(f) := \cup\{X_i : \xi_i(f) \neq 0\}.$$

The conditions (i) and (ii) can now be easily verified.

The map φ provides us with a projective generator in X in the sense of [OV1]. Using this map we can then prove the next lemma, where a new multivalued map η is involved which will be used in the proof of theorem A.

Lemma 3.- *Let X be a Banach space in the class $\mathcal{V}$, φ the multivalued map from $F \cap B(X^*)$ into the subsets of X given in Lemma 2, and η be a countably-valued map from X into the subsets of F. Let A_0 and B_0 be countable subsets of X and F, respectively. Then there exist countable $\mathbb{Q}$-linear subspaces A and B in X and F, respectively, with $A_0 \subset A \subset X$, $B_0 \subset B \subset F$, and such that:*

(i) $\| x \| = \sup\{|< x, y >| : y \in B(X^*) \cap B\}$ for every x in A,

(ii) $\varphi(B \cap B(X^*)) \subset A$, and

(iii) $\eta(A) \subset B$.

Proof.- If $x \in X$ we denote by V_x a countable subset of $F \cap B(X^*)$ such that

$$\| x \| = \sup\{|< x, y >| : y \in V_x\}.$$

We shall proceed by recurrence. Assume that for a non negative integer p and every $m < p$ we have constructed the countable subsets $A_m \subset X, B_m \subset$ F.
We write C_m and D_m for the $\mathbb{Q}$-linear hull of A_m and B_m, respectively.
We now define:

$$B_{p+1} := D_p \cup \{V_x : x \in C_p\} \cup \eta(C_p)$$

and

$$A_{p+1} := C_p \cup \{\varphi(f) : f \in D_p \cap B(X^*)\}.$$

If we take

$$A = \cup\{A_n : n = 0, 1, 2, \ldots\} \text{ and } B = \cup\{B_n : n = 0, 1, 2, \ldots\},$$

then the $\mathbb{Q}$-linear spaces A and B verify the required conditions.

If we apply our lemma 1 to this pair of subspaces we obtain

Corollary 4.- *Under the conditions of lemma 3 we have a norm one projection P on X with $P(X) = \overline{A}^{\|\ \|}$ and kernel*(P)= $B^{\perp}$.

We are now ready to prove our main result in this section:

THEOREM A.- *Let X be a Banach space in the class $\mathcal{V}$. If $\gamma(X, F)$ is the topology of uniform convergence on the bounded weak* separable subsets of F, then $X[\gamma(X, F)]$ is a Lindelöf space. Moreover, every product space $X[\gamma(X, F)]^n$ is also Lindelöf for* $n = 2, 3, \ldots$

Proof.- Let $\{V_\gamma : \gamma \in \Gamma\}$ be an open cover of $X[\gamma(X, F)]$. Given x in X, a basis of the filter of neighbourhoods of x in the topology $\gamma(X, F)$ is given by the family of subsets

$$W(x, S, \epsilon) = \{y \in X : \mid f(x) - f(y) \mid \leq \epsilon \text{ for all } f \in S\},$$

where S is a countable subset of $F \cap B(X^*)$ and $\epsilon > 0$. We shall denote by $B(x, \epsilon)$ the open ϵ-ball around x. For every x in X and γ in Γ with $x \in V_\gamma$ we can find a countable subset $S(x, \gamma) \subset F \cap B(X^*)$ and $\epsilon(x, \gamma)$ such that $W(x, S(x, \gamma), \epsilon(x, \gamma)) \subset V_\gamma$. In particular

$$B(x, \epsilon(x, \gamma)) \subset W(x, S(x, \gamma), \epsilon(x, \gamma)) \subset V_\gamma.$$

Given x in X and $r > 0$, we introduce the set $\Gamma(x, r)$ of all those γ in Γ such that there is a countable subset $S(x, \gamma, r)$ of $F \cap B(X^*)$ with $B(x, r) \subset W(x, S(x, \gamma, r), r) \subset V_\gamma$. Observe that $\Gamma(x, r)$ is non void for every x in X and r small enough. If $\Gamma(x, r) \neq \emptyset$ we choose one element $\gamma(x, r) \in \Gamma(x, r)$. We define the countably valued map η from X into the subsets of F given by

$$\eta(x) = \cup\{S(x, \gamma(x, r), r) : \Gamma(x, r) \neq \emptyset \text{ and } r \in \mathbb{Q}^+\}.$$

Let α be an element of Γ such that the origin belongs to V_α, S_0 a countable subset of $F \cap B(X^*)$ and $\epsilon > 0$ with $W(\vec{0}, S_0, \epsilon)$ contained in V_α. We set $A_0 := \varphi(S_0)$ and $B_0 := S_0$. If we apply our lemma 3 we obtain countable $\mathbb{Q}$-linear subspaces A and B in X and F respectively, with

$$A_0 \subset A \subset X, B_0 \subset B \subset F \text{ and such that :}$$

(i) $\| x \| = \sup\{|< x, y >| : y \in B(X^*) \cap B\}$ for every x in A,

(ii) $\varphi(B \cap B(X^*)) \subset A$,

(iii) $\eta(A) \subset$ B.

Corollary 4 gives a norm one projection P from X onto the norm closure of A with kernel(P) equal to $B^\perp$. We shall now prove that

$$X = V_\alpha \cup \{V_{\gamma(z,r)} : z \in A, r \in \mathbb{Q}^+, \Gamma(z,r) \neq \emptyset\}$$

which is a countable subset of the original open cover, and thus gives us the Lindelöf property we want to prove. Indeed, for a given x in X, if $P(x) = 0$ then x belongs to $B^\perp$ and $f(x) = 0$ for every f in S_0, hence $x \in W(\vec{0}, S_0, \epsilon) \subset V_\alpha$.
If $P(x) \neq 0$, we find $\gamma \in \Gamma$ with $P(x) \in V_\gamma$, and

$$W(P(x), S(P(x), \gamma), \epsilon(P(x), \gamma)) \subset V_\gamma$$

for some countable subset $S(P(x), \gamma)$ in $F \cap B(X^*)$ and $\epsilon(P(x), \gamma) > 0$. Let r be a positive rational number with $r \leq \epsilon(P(x), \gamma)$. Since $P(X)$ is the norm closure of A in X we find $z \in A$ with $\| z - P(x) \| < r/2$; then

$$B(z, r/2) \subset W(z, S(P(x), \gamma), r/2) \subset W(P(x), S(P(x), \gamma), \epsilon(P(x), \gamma)) \subset V_\gamma$$

from where it now follows that $\gamma \in \Gamma(z, r/2) \neq \emptyset$. We take the chosen element $\gamma(z, r/2) \in \Gamma(z, r/2)$ (note that it could be different from γ), and we have

$$B(z, r/2) \subset W(z, S(z, \gamma(z, r/2), r/2), r/2) \subset V_{\gamma(z,r/2)}.$$

Since $S(z, \gamma(z, r/2), r/2) \subset \eta(z) \subset B$ we have $f(x) = f(P(x))$ for every f in $S(z, \gamma(z, r/2), r/2)$. Therefore $P(x) \in B(z, r/2)$ implies that

$$x \in W(z, S(z, \gamma(z, r/2), r/2), r/2) \subset V_{\gamma(z,r/2)}$$

which finishes the proof of the Lindelöf property of $X[\gamma(X, F)]$. Moreover, it is clear that every finite product of Banach spaces in the class $\mathcal{V}$ is also in the class $\mathcal{V}$ from where the conclusion for finite products follows. ■

Corollary 5.- *If K is a Valdivia compact, i.e., K is a subspace of a cube I^Γ with $K \cap \Sigma(\Gamma)$ dense in K, then the Banach space $C(K)$ is Lindelöf for the topology of uniform convergence on the separable subsets of $K \cap \Sigma(\Gamma)$. In particular, if K is Corson compact, then $C(K)$ is Lindelöf for the topology of uniform convergence on the separable subsets of K.*

Proof.- For every $\gamma \in \Gamma$, h_γ denotes the γ-th coordinate functional in $\mathbb{R}^\Gamma$ restricted to K. We set

$$L = \{h_\gamma : \gamma \in \Gamma\} \subset C(K).$$

For every positive integer n, put

$$L^n = \{g_1 \cdot g_2 \cdot \ldots \cdot g_n : g_i \in L, i = 1, 2, \ldots, n\}$$

and let I be equal to the subset of $C(K)$ given by $I = \{1\} \cup \{L^n : i = 1, 2, \ldots\}$. The Stone-Weierstrass theorem says that I is a total subset of $C(K)$. If we now define $J : B(C(K)^*) \longrightarrow \mathbb{R}^I$ by $J(\mu) = (\mu(f))_{f \in I}$ and consider the linear hull F of the Dirac measures $\{\delta_x : x \in K \cap \Sigma(\Gamma)\}$, it is clear that F is a norming subspace of $C(K)^*$ and $J(F \cap B(C(K)^*)) \subset \Sigma(\Gamma)$ from where the conclusion follows.

For instance, if $[0, \omega_1]$ is the interval of ordinals numbers between 0 and the first non countable ordinal ω_1, $[0, \omega_1]$ is a Valdivia compact and the interval $[0, \omega_1)$ is represented in the Σ-product, [V4], [DG]. So we have that $C[0, \omega_1]$ is Lindelöf for the topology of uniform convergence on the separable subsets of $[0, \omega_1)$; nevertheless $C[0, \omega_1]$ is not a Lindelöf space with the topology of pointwise convergence [Ar]. The corollary 5 for Corson compact spaces is due to S.P.Gul'ko [Gu1].

Let us recall that a Markushevich basis (M-basis) of a Banach space X is a subset $\{(x_i, f_i) : i \in I\}$ of $X \times X^*$ such that *span* $\{x_i : i \in I\}$ is dense in X, $\{f_i : i \in I\}$ is a total subset of X^* and $< x_i, f_j >= \delta_j^i$. If we bear in mind the M-basis constructed in [V6], [OV2] we can prove:

Corollary 6.- *Let X be a Banach space in $\mathcal{V}$. The following are equivalent:*

(i) $X[\sigma(X, X^*)]$ *is Lindelöf.*

(ii) $B(X^*)$ *is weak* Corson compact.*

(iii) $B(X^*)$ *is angelic.*

(iv) For every f in $B(X^*)$, there is a sequence (f_n) in $F \cap B(X^*)$ which is weak-convergent to f.*

(v) For every f in $B(X^*)$ there is a countable subset D in $F \cap B(X^*)$ with f in $\overline{D}^{\sigma(X^*, X)}$.

Proof.- $ii) \Rightarrow iii) \Rightarrow iv) \Rightarrow v)$ is clear.
$v) \Rightarrow i)$ If we assume $v)$, then $\gamma(X, F) \geq \sigma(X, X^*)$ and the Lindelöf property follows from the theorem.
$i) \Rightarrow ii)$ Let $(x_i, f_i)_{i \in I}$ be the M-basis we have in X, [V6] and [OV2]. $\{x_i : i \in I\} \cup \{0\}$ is weakly closed, so it is weakly Lindelöf, but $\{x_i : i \in I\}$ is weakly discrete, thus for any f in X^* and $\epsilon > 0$ the set

$$\{i \in I : |< f, x_i >| \geq \epsilon\}$$

is countable, from where it follows that

$$\{i \in I : |< f, x_i >| \neq 0\}$$

is a countable subset of I and the mapping T from X^* into $\Sigma(I)$ defined by $T(f) = (< f, x_i >)_{i \in I}$ is one-to-one and weak* to pointwise continuous, therefore $B(X^*)$ is a Corson compact space in the weak* topology.

A result of the same nature for the spaces $C(K)$, where K is a Corson compact, is obtained in [AMN, th.3.5].

3. THE LINDELÖF PROPERTY OF A DUAL BANACH SPACE WITH THE RADON-NIKODYM PROPERTY

Let X be a Banach space. We proved in [OV1] that X is an Asplund space if and only if X^* admits a projective generator defined on the norming subspace X of X^{**}. It is an open problem to know if $B(X^{**})$ with the weak* topology is a Valdivia compact, too. A positive answer for Asplund spaces with density character $\aleph_1$ was given in [DG] and [V5]. In any case, since we have the projective generator we can use the method above to prove the following:

THEOREM B.- *Let X be a Banach space and $\gamma(X^*, X)$ the topology of uniform convergence on the bounded separable subsets of X. Then X is an Asplund space if and only if $X^*[\gamma(X^*, X)]$ is Lindelöf.*

Proof.- Let us suppose that $X^*[\gamma(X^*, X)]$ is Lindelöf. Then for any separable subspace F of X, the strong dual F^* is a continuous image of $X^*[\gamma(X^*, X)]$, so F^* is Lindelöf and consequently norm-separable. This proves that X is an Asplund space, [St].

We suppose now that X is an Asplund space. We consider the multivalued map $\phi : X \to 2^{X^*}$ defined by $\phi(x) = \{u \in B(X^*) :< x, u >=\| x \|\}$. ϕ is $\|\|$ to weak* upper semicontinuous and compact valued and X^* has the RNP. The selection theorem of Jayne and Rogers [JR] gives us a first Baire class selector $f : X \to B(X^*)$ for the norm topologies, i.e. $f(x) \in \phi(x)$ for every x in X, and there exists a sequence of $\|\|$ to $\|\|$ continuous functions $f_n : X \to B(X^*)$ such that $\|\| - \lim_{n\to\infty} f_n(x) = f(x)$ for every x in X. If we denote by

$$\varphi : X \to 2^{X^*}$$

the mapping $\varphi(x) = \{f(x), f_1(x), \ldots, f_n(x), \ldots\}$, we obtain a projective generator in X^*, [OV1]. Hence for every pair A, B of $\mathbb{Q}$-linear subspaces $A \subset X$ and $B \subset X^*$ such that

(i) $\| f \|= \sup\{|< f, x >|: x \in A \cap B(X)\}$ for every $f \in B$, and

(ii) $\varphi(A) \subset B$,

there exists a norm one projection P of X^* onto the norm closure of B with kernel$(P) = A^{\perp}$. Note that for countable subspaces A and B, which is the case we are interested in here, this follows from James' theorem as [FG] proved.

If we use this map we can repeat the proof of lemma 3 above to prove:

Lemma 7.- *Let X be an Asplund space. Let η be a countably valued map from X^* onto the subsets of X. Let A_0 and B_0 be countable subsets of X and X^*, respectively. Then there exist $\mathbb{Q}$-linear countable subspaces A and B in X and X^*, respectively, with*

$$A_0 \subset A \subset X,\ B_0 \subset B \subset X^*,$$

and such that

(i) $\| f \| = \sup\{|< f, x >| : x \in A \cap B(X)\}$ *for every* f *in* B,

(ii) $\varphi(A) \subset B$, *and*

(iii) $\eta(B) \subset A$.

This lemma is the main tool in our proof (below) that $X^*[\gamma(X^*, X)]$ is Lindelöf. With the same notation as in theorem A, for the open-cover $\{V_\alpha : \alpha \in \Gamma\}$ in $X^*[\gamma(X^*, X)]$ we shall find a countable subcover by constructing an adequate separable and complemented subspace of X^*. We define η as in theorem A, where now $\eta : X^* \to 2^X$; we select $\gamma_0 \in \Gamma$ with $\vec{0} \in V_{\gamma_0}$, a countable subset $S_0 \subset B(X)$, and $\epsilon > 0$ with $W(\vec{0}, S_0, \epsilon) \subset V_{\gamma_0}$. We write $A_0 := S_0$ and $B_0 := \varphi(S_0)$. If we apply lemma 7, we obtain countable $\mathbb{Q}$-linear subspaces A and B , $A_0 \subset A \subset X$, $B_0 \subset B \subset X^*$ that verify the conditions above. So there is a projection P from X^* onto the norm closure of B along $A^\perp$. Reasoning as in the proof of theorem A, we see that

$$X^* = V_{\gamma_0} \cup \{V_{\gamma(z,r)} : z \in B, r \in \mathbb{Q}^+, \Gamma(z,r) \neq \emptyset\}$$

from where the conclusion follows.

The dual of an Asplund space always has an M-basis that can be constructed with the projectional resolution of the identity [G, prop.III.7], [V7].

Corollary 8.- *Let X be an Asplund space. The following are equivalent:*

(i) $X^*[\sigma(X^*, X^{**})]$ *is Lindelöf.*

(ii) $B(X^{**})$ *is weak* Corson compact.*

(iii) $B(X^{**})$ *is angelic.*

(iv) *For every f in $B(X^{**})$, there is a sequence (f_n) in $B(X)$ which is weak* convergent to f.*

(v) *For every f in $B(X^{**})$, there is a countable subset $D \subset B(X)$ such that* $f \in \overline{D}^{\sigma(X^*, X)}$.

Proof.- It is the same as the one given in corollary 6. Indeed,
$(ii) \Rightarrow (iii) \Rightarrow (iv) \Rightarrow (v)$ is clear. $(v) \Rightarrow (i)$ because under the assumption (v) the topology $\gamma(X^*, X)$ is finer than $\sigma(X^*, X^{**})$. $(i) \Rightarrow (ii)$: Using the M-basis $(x_i, f_i)_{i \in I}$ in X^*, the weak Lindelöf property of X^* implies that the evaluation map

$$T(f) = (< f, x_i >)$$

is weak* to pointwise continuous from X^{**} into $\Sigma(I)$.

The equivalence $(ii) \Leftrightarrow (iii) \Leftrightarrow (iv) \Leftrightarrow (v)$ was obtained by R.Deville and G.Godefroy [DG, th.III.4]. They also proved that these conditions are equivalent to the fact $B(X^{**})$ does not contain any subset which is weak* homeomorphic to

$[0, \omega_1]$. Here we are interested in the equivalence with the Lindelöf property (i). In the next section we shall apply this equivalence to study the problem of Corson.

4. ON THE PROBLEM OF CORSON

As we said in the introduction, the following question was posed by Corson:
Problem.- If a Banach space X is weakly Lindelöf, is it true that $X[\sigma(X, X^*)] \times X[\sigma(X, X^*)]$ must be Lindelöf, too?

The method of proof which we have used in the former sections for the Lindelöf property of a given Banach space is very well adapted to the study of this problem. Indeed, if $X \in \mathcal{V}$ then $X[\gamma(X, F)] \times X[\gamma(X, F)]$ is a Lindelöf space (Theorem A). When could it be possible to extend this result to the weak topology? Our corollary 6 says that this is the case if and only if $B(X^*)$ is a weak* Corson compact. On the other hand, if X is an Asplund space, then $X^*[\gamma(X^*, X)] \times X^*[\gamma(X^*, X)]$ is also Lindelöf (Theorem B). Our corollary 8 says that the same is true for $X^*[\sigma(X^*, X^{**})]$ if and only if $B(X^{**})$ is a weak* Corson compact space. Fortunately, if X is any weakly Lindelöf dual space, G.A.Edgar [E,prop.1.8] showed that X has the Radon Nikodym property. So we can prove the following result which gives a positive answer to Corson problem for the case of dual spaces:

THEOREM C.- *Let X be a dual Banach space. Then $X[\sigma(X, X^*)]$ is Lindelöf if, and only if, $(X[\sigma(X, X^*])^n$ is Lindelöf for $n = 1, 2, ...$*

Proof.- If X is weakly Lindelöf, X must be a dual space with the Radon-Nikodym property. So we can prove the following result which gives a positive answer to Corson's product of a finite number of Corson compact spaces is Corson compact, too. For every positive integer n, X^n can be renormed to have weak* Corson compact dual unit ball. Thus our corollary 6 says that $X^n[\sigma(X^n, (X^n)^*)]$ is a Lindelöf space, and this space is homeomorphic with $(X[\sigma(X, X^*)])^n$ which is also a Lindelöf space.

The author has been partially supported by DGCYT PS88-0083.

References.

[AP] **K.Alster, R.Pol.** On function spaces of compact subspaces of Σ-products of the real line, Fund. Math. **107** (1980) 135-143.

[AMN] **S.Argyros, S.Mercourakis, S.Negrepontis.** Functional-analytic properties of Corson-compact spaces, Studia Math. **89** (1988) 197-229.

[Ar] **A.V.Arhangel'skii.** A survey of C_p-theory, Q. and A. in General Topology **5** (1987) Special Issue 1-109.

[C] **H.H.Corson.** The weak topology of a Banach space, Trans.Amer.Math.Soc. **101** (1961) 1-15.

[DG] **R.Deville, G.Godefroy.** Some applications of projective resolutions of identity, (1990) Preprint.

[Di] **J.Diestel.** Geometry of Banach spaces. Selected topics. Springer Verlag. L.N.M. 485 (1975).

[E] **G.A.Edgar.** Measurability in Banach spaces, Indiana Univ. Math. J. **26.4** (1977) 663-677.
[FG] **M.Fabian, G.Godefroy.** The dual of every Asplund space admits a projective resolution of identity, Studia Math. **91** (1988) 141-151.
[G] **G.Godefroy.** Five lectures in geometry of Banach spaces. Seminar on Functional Analysis 1987. Notas de Matemática **1**. Universidad de Murcia.
[Gu1] **S.P.Gul'ko.** On properties of subsets of Σ-products, Soviet Math.Dokl. **18-6** (1977) 1438-1442.
[Gu2] **S.P.Gul'ko.** The structure of spaces of continuous functions and their hereditary paracompactness, Russian Math.Surv. **34** (1979) 36-44.
[JR] **J.E.Jayne, C.A.Rogers.** Borel selectors for upper semicontinuous set valued maps, Acta Math. **155** (1985) 41-79.
[L] **J.Lindenstrauss.** Weakly compact sets, their topological properties and the Banach spaces they generate, Ann. Math. Studies **69** (1972), Princeton University Press, 235-276.
[N] **S.Negrepontis.** Banach spaces and topology, Handbook of set theoretic topology, Edited by Kunen-Vaughan, Elsevier Sc.Publ. **23** (1984) 1045-1142.
[OSV] **J.Orihuela, W.Schachermayer, M.Valdivia.** Every Radon-Nikodym Corson compact space is Eberlein compact. To appear in Studia Math.
[OV1] **J.Orihuela, M.Valdivia.** Projective generators and resolutions of identity in Banach spaces, Rev.Mat.Univ.Complutense Madrid **2** Supplementary Issue (1989) 179-199.
[OV2] **J.Orihuela, M.Valdivia.** Resolutions of identity and first Baire class selectors in Banach spaces, (1990) Preprint.
[P1] **R.Pol.** A function space $C(X)$ which is weakly Lindelöf but not weakly compactly generated, Studia Math. **64** (1979) 279-285.
[P2] **R.Pol.** On pointwise and weak topology in function spaces. Warszawski University **(4)84** (1984).
[St] **C.Stegall.** The Radon-Nikodym property in conjugate Banach spaces, Trans. Amer.Math.Soc. **206** (1975) 213-223.
[T] **M.Talagrand.** Espaces de Banach faiblement K-analytiques, Ann. of Math. **110** (1979) 407-438.
[V1] **M.Valdivia.** Espacios de Fréchet de generación débilmente compacta, Coll. Mat. **38** (1987) 17-25.
[V2] **M.Valdivia.** Resolutions of identity in certain Banach spaces, Coll. Mat. **38** (1988) 124-140
[V3] **M.Valdivia.** Resolutions of identity in certain metrizable locally convex spaces, Rev. Real Acad. Ciencias Exactas, Físicas y Naturales, Madrid **83-1** (1989) 75-96.
[V4] **M.Valdivia.** Projective resolutions of identity in $C(K)$ spaces. To appear in Archiv der Math.
[V5] **M.Valdivia.** Resoluciones proyectivas del operador identidad y bases de Markusevich en ciertos espacios de Banach. To appear in Rev. Real Acad. Ciencias Exactas, Físicas y Naturales. Madrid.

[V6] **M.Valdivia.** Simultaneous resolutions of the identity operator in normed spaces, (1990) Preprint
[V7] **M.Valdivia.** Complemented subspaces of certain Banach spaces. To appear in Seminar of Functional Analysis, Notas de Matemática. Universidad de Murcia.

Progress in Functional Analysis
K.D. Bierstedt, J. Bonet, J. Horváth & M. Maestre (Eds.)

Distinguished subsets in Vector Sequence Spaces

Fernando Bombal[1]

Departamento de Análisis Matemático, Facultad de Matemáticas.
Universidad Complutense, 28040 Madrid, Spain.

Dedicated to Professor M. Valdivia, on his 60th birthday

Abstract

We characterize several classes of subsets in the Λ-sum $(\sum E_n)_\Lambda$ of a sequence (E_n) of Banach spaces (where Λ is a Banach sequence ideal space), in terms of the corresponding classes in the spaces E_n. Several stability results are obtained.

INTRODUCTION

Many properties of a Banach space E are, or can be, defined by establishing some non trivial relationships between different classes of subsets of E. Hence, it is always an interesting problem to determine different classes of distinguished subsets in a Banach space.

This paper is devoted to characterize several classes of subsets in the Λ-sum $\Lambda(E_n) = (\sum E_n)_\Lambda$ of a sequence of Banach spaces, where Λ is a Banach sequence ideal space, in terms of the corresponding subsets in the coordinate spaces. As a consequence, we get in an easy and unified way, stability results for a large number of Banach space properties, when passing from the sequence (E_n) to the corresponding Λ-sum. In particular we prove, under suitable conditions on Λ, that $\Lambda(E_n)$ contains a complemented copy of c_o or ℓ_1 if and only if so does at least one of the E_n's. If we specialize to the case $\Lambda = \ell_p$, we obtain the results of $[B3]$ and those announced

[1] Partially supported by DGICYT grant PB88-0141

in [$B6$].

We shall follow the standard notations in Banach space theory, as in the books [D] or [LT]. In particular, if E is a Banach space, $B(E)$ will denote its closed unit ball and E^* its topological dual. Also, a series $\sum x_n$ in E is said to be *weakly unconditionally convergent* (w.u.c. in short) if $\sum |x^*(x_n)| < \infty$, for every $x^* \in E^*$.

1. SOME CLASSES OF SUBSETS OF A BANACH SPACE.

As we mentioned above, many properties of a Banach space can be defined according to the following general scheme: let $\mathcal{H}$ and $\mathcal{G}$ be classes of subsets of Banach spaces (so that, $\mathcal{H}(E)$ and $\mathcal{G}(E)$ are classes of subsets of E, for every Banach space E). Then, we can say that E has *property* $(\mathcal{H}, \mathcal{G})$ if $\mathcal{H}(E) \subseteq \mathcal{G}(E)$. If we denote by $\mathcal{B}$, $\mathcal{W}$, $\mathcal{WC}$ and $\mathcal{K}$ the classes of bounded, weakly relatively compact, weakly conditionally compact (i.e., $A \in \mathcal{WC}$(E) if every sequence in A has a weakly Cauchy subsequence) and relatively norm compact subsets, respectively, we get in this way that finite dimensionality is just property $(\mathcal{B}, \mathcal{K})$, and reflexivity is property $(\mathcal{B}, \mathcal{W})$, whereas Rosenthal's ℓ_1-theorem (see [D], Ch. XI) establishes that a Banach space has property $(\mathcal{B}, \mathcal{WC})$ if and only if it contains no copy of ℓ_1.

Other important classes of subsets of a Banach space E are defined in terms of uniform convergence on some families of sequences in the dual space of E : A subset K of E is said to be *limited* (resp., *Dunford-Pettis*) if for every weak* null (resp., weakly null) sequence (x_n^*) in E^*, the following holds:

$$\lim_{n \to \infty} \sup\{| x_n^*(x) | : x \in K\} = 0. \qquad (*)$$

If (*) holds for every w.u.c. series $\sum x_n^*$ in E^*, K is called a (V^*) *set*. We shall denote by $\mathcal{L}(E), \mathcal{DP}(E)$ and $\mathcal{V}^*(E)$ the families of all limited, Dunford-Pettis and (V^*)-sets of E, respectively.

Dunford-Pettis sets were introduced by Andrews in [A], and the definition of (V^*) sets is due to Pelczynski ([P]). Limited sets were introduced by Köthe in 1938, in the more general setting of locally convex spaces, but their systematic study and applications in Banach space theory starts with the Bourgain and Diestel's paper $[BD]$.

The next propositions collect some useful properties of these classes of subsets:

Proposition 1.1. *Let E be a Banach space and $K \subseteq E$.*

a) ($[SL], [DR]$) *The following statements are equivalent:*

i) K is limited.

ii) $T(K)$ is relatively compact, for every bounded linear operator T from E into c_o.

b) ($[A]$) *The following statements are equivalent:*

i) K is Dunford-Pettis.

ii) $T(K)$ is relatively compact, for every weakly compact linear operator T from E into c_o.

c) ($[E1]$) *The following statements are equivalent:*

i) K is a V^* -set.

ii) $T(K)$ is relatively (weakly) compact, for every continuous linear operator from E into ℓ_1.

Proposition 1.2. ($[B5]$).*Let $\mathcal{H}$ be any of the classes $\mathcal{K}$, $\mathcal{W}$, $\mathcal{WC}$, $\mathcal{L}$, $\mathcal{DP}$ or $\mathcal{V}^*$.*

a) *$\mathcal{H}$ consists of bounded sets, and is preserved by continuous linear images, linear combinations, closed absolutely convex hulls, finite products and passing to subsets.*

b) *If K is a subset of the Banach space E and for every $\epsilon > 0$ there is an $K_\epsilon \in \mathcal{H}(E)$ such that*

$$K \subseteq K_\epsilon + \epsilon B(E), \qquad (**)$$

then $K \in \mathcal{H}(E)$.

Proof. (a) follows easily from the definitions or proposition 1.1. As for (b), the case $\mathcal{H} = \mathcal{W}$ is due to Grothendieck (see, *f.i.*, [D], Ch. XIII, lemma 2). When $\mathcal{H} = \mathcal{K}$, condition (**) implies clearly that K is totally bounded. Finally, the other cases follow from Proposition 1.1 and the case $\mathcal{H} = \mathcal{K}$.

The relationships

$$\mathcal{K} \subseteq \mathcal{W} \subseteq \mathcal{WC}$$

$$\mathcal{K} \subseteq \mathcal{L} \subseteq \mathcal{DP} \subseteq \mathcal{WC} \subseteq \mathcal{V}^* \tag{†}$$

follow from the definitions or proposition 1.1, except the non trivial inclusion $\mathcal{DP} \subseteq \mathcal{WC}$. It can be proved by the arguments used in [*BD*] to show that $\mathcal{L} \subseteq \mathcal{WC}$, taking into account Proposition 1.1(b) and the fact that the natural inclusion of ℓ_1 into c_o is weakly compact (See [*B*5]). In general, all the above inclusions are strict and there are no other relationships between the different classes of subsets considered. If we apply the general scheme stated at the beginning of the section, we obtain, among others, the following known Banach space properties (see [*B*5]):

-Property $(\mathcal{W}, \mathcal{K})$ is the *Schur property.*

-Property $(\mathcal{WC}, \mathcal{W})$ is the *weak sequential completeness.*

-Property $(\mathcal{W}, \mathcal{DP})$ coincides with property $(\mathcal{WC}, \mathcal{DP})$ and is called the *Dunford Pettis Property.*

-Property $(\mathcal{L}, \mathcal{K})$ is called the *Gelfand-Phillips Property.*

-Property $(\mathcal{V}^*, \mathcal{W})$ is called *Pelczynski's* (V^*) *Property.*

-Property $(\mathcal{DP}, \mathcal{W})$ is called the *RDP* Property.*

All these properties are preserved by topological isomorphisms, finite products and passing to complemented subspaces.

2. CHARACTERIZATION OF DISTINGUISHED SUBSETS IN CERTAIN VECTOR SEQUENCE SPACES.

We shall begin by recalling some elementary facts about sequence spaces and establishing some notation (see [*V*] and [*R*].):

A *Banach sequence space* (or *Köthe sequence space*) Λ is a vector subspace of the space of all the scalar sequences $\omega = \mathbf{K}^{\mathbf{N}}$, containing the space φ of all the finitely non zero sequences, and endowed with a complete norm such that $\alpha =$

$(\alpha_n) \in \omega, \beta = (\beta_n) \in \Lambda$ and $|\alpha_n| \leq |\beta_n|$ for all n, implies $\alpha \in \Lambda$ and $\| \alpha \| \leq \| \beta \|$. Clearly, Λ is an order complete Banach lattice and its topology is finer than the pointwise topology, *i.e.*, the inclusion from Λ into ω is continuous.

The *α-dual* $\Lambda^\times$ of Λ is the space of all the sequences $\alpha = (\alpha_n)$ such that $\sum_{n=1}^{\infty} |\alpha_n \beta_n| < \infty$, for all $\beta = (\beta_n) \in \Lambda$. Obviously $\Lambda^\times \subseteq \Lambda^*$ and $\Lambda^\times$ is an ideal of the dual Banach lattice Λ^*. In the norm induced on $\Lambda^\times$ by Λ^*, this space is also a Banach sequence space (See [*LT*], Vol. II, 1.b.17). Often $\Lambda^\times$ is equipped with another natural Banach sequence space norm, of course equivalent to the dual norm by the closed graph theorem. This happens, for instance, in the case of Orlicz sequence spaces ℓ_ϕ endowed with the so called *Luxemburg norm*. Then, $(\ell_\phi)^\times = \ell_\psi$, where ψ is the Young's conjugate function of ϕ. The dual norm on ℓ_ψ is the *Orlicz norm*, equivalent but in general different from the Luxemburg norm.

We also have $\Lambda \subseteq \Lambda^{\times\times}$. If $\Lambda = \Lambda^{\times\times}$, then Λ is called *perfect*. The ℓ_p $(1 \leq p \leq \infty)$ and, in general, the Orlicz sequence spaces ℓ_ϕ, are perfect. If e_n is the nth-unit vector, (e_n) is a monotone basic sequence in Λ, hence a basis for $\Lambda_o = \varphi$.

Let Λ be a perfect Banach sequence space and (E_n) a sequence of Banach spaces. We shall denote by $\Lambda(E_n)$ or $(\sum E_n)_\Lambda$ the space of all the sequences $x = (x_n) \in \prod(E_n)$ such that $p(x) = (\| x_n \|) \in \Lambda$, endowed with the Banach norm $\| x \| = \| p(x) \|_\Lambda$. The notations $\Lambda_o(E_n)$ and $\Lambda^\times(E_n)$ are clear. For every $m \in \mathbf{N}, \pi_m$ will denote the canonical projection $\Lambda(E_n) \ni x = (x_n) \mapsto x_m \in E_m, I_m$ the canonical injection $E_m \ni x \mapsto (0, \ldots, x, 0, \ldots) \in \Lambda(E_n)$ and $S_m = \sum_{i=1}^{m} I_i \circ \pi_i : \Lambda(E_n) \to \Lambda(E_n)$ the mth section. All of them are continuous linear maps. Reasoning as in [*R*], one can see that the dual of $\Lambda_o(E_n)$ can be identified with the space $\Lambda^\times(E_n^*)$, through the isomorphism

$$\Lambda^\times(E_n^*) \ni z = (z_n) \mapsto T_z \in \Lambda_o(E_n)^*,$$

where $T_z(x) = \sum_{n=1}^{\infty} < x_n, z_n >$. (If $\Lambda^\times$ has the dual norm, this map is in fact an isometry: See [*R*], Sect. 5, Th. 4(b).)

From now on, we shall assume that Λ is a perfect Banach sequence space such that $\Lambda_o = \Lambda$ and (E_n) a sequence of Banach spaces.

Theorem 2.1. *Let $\mathcal{H}$ be any of the classes $\mathcal{W}$, $\mathcal{WC}$ or $\mathcal{V}^*$. For a bounded subset K of $F_\Lambda = \Lambda(E_n)$, the following properties are equivalent:*

a) $K \in \mathcal{H}(F_\Lambda)$.

b) *We have*

i) For every $n \in \mathbf{N}$, $\pi_n(K) \in \mathcal{H}(E_n)$,

and

ii) For every $z \in \Lambda^\times(E_n^*)$,

$$\lim_{m \to \infty} \operatorname{Sup} \left\{ \sum_{k=m}^{\infty} |< \pi_k(x), \pi_k(z) >| : x \in K \right\} = 0.$$

Proof. (a) $\Rightarrow$(b): (i) is clear, since π_n is linear continuous. On the other hand, every $z \in \Lambda^\times(E_n^*)$ defines a continuous linear map S_z from $\Lambda(E_n)$ into ℓ_1 by the formula

$$\Lambda(E_n) \ni x \mapsto S_z(x) = (< \pi_k(x), \pi_k(z) >)_{k=1}^{\infty} \in \ell_1.$$

As ℓ_1 has Property V^* ([P]), $S_z(K) \in \mathcal{H}(\ell_1) = \mathcal{W}(\ell_1) = \mathcal{K}(\ell_1)$. Then, (ii) is just the well known characterization of relatively compact subsets of ℓ_1.

(b) $\Rightarrow$ (a): Let us first consider the case $\mathcal{H} = \mathcal{W}$. Let (x_k) be a sequence in K. Passing to a subsequence if necessary, we can suppose from (i) that for every $n \in \mathbf{N}$, $(\pi_n(x_k))$ converges weakly to some a_n in E_n. We assert that $a = (a_n)$ belongs to $\Lambda(E_n)$. For that it will suffice to prove that $\sum_{n=1}^{\infty} | \beta_n ||| a_n \| < \infty$ for all $\beta = (\beta_n)$ in $\Lambda^\times$. Let $\epsilon_n > 0$ be such that $\epsilon = (\epsilon_n) \in \Lambda$ and take norm unit vectors b_n in E_n^* so that $< a_n, b_n > > \| a_n \| - \epsilon_n$. Then,

$$\sum_{n=1}^{\infty} | \beta_n ||| a_n \| \leq \sum_{n=1}^{\infty} |< a_n, \beta_n b_n >| + \sum_{n=1}^{\infty} | \beta_n | \epsilon_n.$$

The second term of the right hand side is finite, because $\beta \in \Lambda^\times$ and $\epsilon \in \Lambda$. As for the first term, if we put $\beta b = (\beta_n b_n)$ we have $\beta b \in \Lambda^\times(E_n^*)$ and, with the notations of (a)$\Rightarrow$(b), the sequence

$$\{S_{\beta b}(x_k) = (< \pi_n(x_k), \beta_n b_n >)_{n=1}^{\infty} : k \in \mathbf{N}\} \subseteq \ell_1$$

is relatively compact (by (ii)) and converges pointwise to $\alpha = (< a_n, \beta_n b_n >)$. Hence, $\alpha \in \ell_1$ (and $\lim_{k\to\infty} \| S_{\beta b}(x_k) - \alpha \|_1 = 0$). This means that $a \in \Lambda(E_n)$ and $\alpha = S_{\beta b}(a)$. If z is now an arbitrary element of $\Lambda^\times(E_n^*)$, we can repeat the previous argument to get

$$\lim_{k\to\infty} \| S_z(x_k - a) \|_1 = 0,$$

i.e., (x_k) converges weakly to a.

The case $\mathcal{H} = \mathcal{WC}$ is proved in the same way.

Finally, let us consider the case $\mathcal{H} = \mathcal{V}^*$: If $K \notin \mathcal{V}^*(F_\Lambda)$ there is a sequence (x_k) in K, a $\delta > 0$ and a w.u.c. series $\sum z_k$ in $(F_\Lambda)^*$ such that

$$|< x_k, z_k >| = |\sum_{n=1}^{\infty} < \pi_n(x_k), \pi_n(z_k) >| > \delta \quad (k = 1, 2, \ldots).$$

As $\sum \pi_n(z_k)$ is a w.u.c. series in E_n^*, (i) implies

$$\inf_k \{|< \pi_n(x_k), \pi_n(z_k) >|\} = 0, \ (n = 1, 2, \ldots).$$

Hence, we can get by induction two subsequences of positive integers $0 = n_1 < n_2 < \ldots$, $1 = k_o < k_1 < k_2 < \ldots$, such that

$$|\sum_{j=1}^{n_{i-1}} < \pi_j(x_{k_i}), \pi_j(z_{k_i}) >| < \delta/4$$

and

$$|\sum_{j>n_i} < \pi_j(x_{k_i}), \pi_j(z_{k_i}) >| < \delta/4.$$

If we put $u_i = z_{k_i}, y_i = (y_{in})$, where $y_{in} = \pi_n(x_{k_i})$ for $n_{i-1} < n \leq n_i$ and 0 otherwise, it turns out that $\sum u_i$ is still a w.u.c. series in $(F_\Lambda)^*$, $|< y_i, u_i >| \geq \delta/2$ and the $y_i's$ have pairwise disjoint supports. Hence, $\{y_i : i \in \mathbf{N}\}$ is a bounded, not V^*-set. Clearly, $\lim_{i\to\infty} \pi_n(y_i) = 0$ for every n and $\{y_i : i \in \mathbf{N}\}$ verifies condition b(ii). Therefore, by the result already proved, (y_i) is a weakly null sequence, hence a V^*-set. This contradiction ends the proof.

For the other classes considered in (†), we get the following characterization:

Theorem 2.2. *Let $\mathcal{H}$ be any of the classes $\mathcal{K}$, $\mathcal{L}$ or $\mathcal{DP}$. For a bounded subset $K \subseteq F_\Lambda = \Lambda(E_n)$, the following properties are equivalent:*

a) $K \in \mathcal{H}(F_\Lambda)$.

b) *We have*

i) For each $n \in \mathbf{N}$, $\pi_n(K) \in \mathcal{H}(E_n)$.

ii)

$$\lim_{n\to\infty} \text{Sup}\,\{\| x - S_n(x) \| : x \in K\} = 0.$$

Proof. (a)⇒ (b): (i) is clear. Suppose (ii) does not hold. Then, there is an $\epsilon > 0$, a strictly increasing sequence $p_1 < q_1 < \ldots < p_j < q_j < \ldots$ of positive integers, and a sequence (x_j) in K, such that

$$\| (S_{q_j} - S_{p_j})(x) \| > \epsilon.$$

For each $j \in \mathbf{N}$, choose a $z_j \in \Lambda^\times(E_n^*)$ such that $\pi_n(z_j) = 0$ if $n < p_j$ or $n > q_j$,

$$< x_j, z_j > = | \sum_{n=p_j}^{q_j} < \pi_n(x_j), \pi_n(z_j) > | > \epsilon,$$

and the corresponding functional $T_{z_j} \in \Lambda(E_n)^*$ has norm 1. Then, because of the isomorphism $\Lambda(E_n)^* \approx \Lambda^\times(E_n^*)$, there is an $M > 0$ such that $\| z_j \|_{\Lambda^\times} \leq M$, for all j. We claim that (z_j) is weakly null. In fact, the sequence is clearly contained in $(\Lambda^\times)_o(E_n^*)$, whose dual is (isomorphic to) $\Lambda^{\times\times}(E_n^{**}) = \Lambda(E_n^{**})$. If $u \in \Lambda(E_n^{**})$, with the notations of the beginning of the section, we have

$$| T_u(z_j) | = \| T_{u_j} \| \| z_j \|_{\Lambda^\times} \leq M \| T_{u_j} \|,$$

where $u_j = (S_{q_j} - S_{p_j})(u)$ converges to 0 (recall that $\Lambda = \Lambda_o$). In consequence, $\lim_{j\to\infty} \| T_{u_j} \| = 0$, and so (z_j) is weakly null. The condition $< x_j, z_j > > \epsilon$ implies then that K is not a Dunford-Pettis set (and, *a fortiori*, neither a limited nor a relatively compact set.).

(b) ⇒ (a): If K satisties b(ii), there is a sequence (ϵ_n) of positive reals, converging to 0 and such that

$$K \subseteq \sum_{i=1}^{n} I_i \circ \pi_i(K) + \epsilon_n B(F_\Lambda).$$

Hypothesis b(i) and Proposition 1.1(a) show that $\sum_{i=1}^{n} I_i \circ \pi_i(K) \in \mathcal{H}(F_\Lambda)$ for every $n > 0$. An appeal to Proposition 1.1(b) concludes the proof.

Remarks 2.3.

a) If $(\Lambda^\times)_o = \Lambda^\times$ (for instance, if $\Lambda = \ell_p, 1 < p < \infty$, or Λ is a reflexive Orlicz sequence space), condition b(ii) of theorem 2.1 is automatically satisfied. In fact, suppose $x \in \Lambda(E_n)$ and $z \in \Lambda^\times(E_n^*)$, and let $\alpha_n = \text{sign}(< \pi_n(x), \pi_n(z) >)$, $u = (\alpha_n \pi_n(z)) \in \Lambda^\times(E_n^*)$. Then

$$\sum_{n=m}^{\infty} |< \pi_n(x), \pi_n(z) >| = \sum_{n=m}^{\infty} < \pi_n(x), \pi_n(u) > \leq \| T_{u - S_m(u)} \| \; \| x \|,$$

and by assumption, $\lim_{n \to \infty} S_m(u) = u$, for every u in $\Lambda^\times(E_n^*)$.

b) Condition b(ii) of theorem 2.2 always implies condition b(ii) of theorem 2.1. In fact, suppose $x \in \Lambda(E_n)$ and $z \in \Lambda^\times(E_n^*)$. With the same notations of part (a), we have

$$\sum_{n=m}^{\infty} |< \pi_n(x), \pi_n(z) >| = \sum_{n=m}^{\infty} < \pi_n(x), \pi_n(u) > \leq \| T_u \| \; \| x - S_m(x) \| .$$

This yields the proof of part (b) of the next corollary.

c) In the particular case $\Lambda = \ell_1$, conditions b(ii) of theorems 2.1 and 2.2 are equivalent. In fact, suppose that $K \subseteq \ell_1(E_n)$ does not satisfy $b(ii)$ of theorem 2.1. Then, there exists $\epsilon > 0, (x_k) \subseteq K$ and a subsequence of positive integers $p_1 < q_1 < p_2 < q_2 < \ldots$ such that

$$\sum_{n=p_k}^{q_k} \| \pi_n(x_k) \| > \epsilon.$$

For every $p_k \leq n \leq q_k$, let us choose an element $z_n \in E_n^*$ such that $\| z_n \| = 1$ and $< \pi_n(x_k), z_n > = \| \pi_n(x) \|$. Then $z = (z_n) \in \ell_\infty(E_n^*) \approx \ell_1(E_n)^*$ and

$$\sum_{n=p_k}^{q_k} |< \pi_n(x_k), \pi_n(z) >| = \sum_{n=p_k}^{q_k} \| \pi_n(x_k) \| > \epsilon.$$

It follows that condition b(ii) of theorem 2.1 does not hold.

This proves part (c) ot the following corollary.

Corollary 2.4. *Let (E_n) be a sequence of Banach spaces and $F_\Lambda = \Lambda(E_n)$.*

a) *Let $\mathcal{H}$ and $\mathcal{G}$ be two classes belonging to the set $\{\mathcal{K}, \mathcal{L}, \mathcal{DP}\}$ (resp., $\{\mathcal{W}, \mathcal{WC}, \mathcal{V}^*\}$). Then, F_Λ has property $(\mathcal{H}, \mathcal{G})$ if and only if every E_n has property $(\mathcal{H}, \mathcal{G})$. In particular, F_Λ is weakly sequentially complete (resp., has the Gelfand-Phillips property or Pelczynski's property (V^*) if and only if so does every E_n.*

b) *Let $\mathcal{H} = \mathcal{L}$ or $\mathcal{DP}$. Then, F_Λ has property $(\mathcal{H}, \mathcal{W})$ if and only if every E_n has property $(\mathcal{H}, \mathcal{W})$. In particular, F_Λ has the RDP* property if and only if so does every E_n.*

c) *Let $\mathcal{H}$ and $\mathcal{G}$ be any of the classes $\mathcal{V}^*$, $\mathcal{WC}$, $\mathcal{W}$, $\mathcal{DP}$, $\mathcal{L}$ or $\mathcal{K}$. Then, $\ell_1(E_n)$ has property $(\mathcal{H}, \mathcal{G})$ if and only if every E_n has property $(\mathcal{H}, \mathcal{G})$. In particular, $\ell_1(E_n)$ is weakly sequentially complete (resp., has the Schur property, the Dunford-Pettis property, the Gelfand-Phillips property, Pelczynski's Property (V^*) or the RDP*-property) if and only if so does every E_n.*

Remark 2.5. Corollary 2.4(c) is not true in general. In fact, it suffices to take $E_n = \mathbf{K}$, the scalar field, for every $n > 0$. Then every E_n has property $(\mathcal{H}, \mathcal{G})$, whatever choice of $\mathcal{H}$ and $\mathcal{G}$ we made. But for $\Lambda = \ell_p (1 < p < \infty)$, $\Lambda(E_n) = \ell_p$ has neither the Schur property, nor the Dunford-Pettis property. Since $\ell_p(E_n)$ contains always a complemented copy of ℓ_p, it *never* has these two properties.

3. COMPLEMENTED COPIES OF ℓ_1 AND c_o IN $\Lambda(E_n)$.

(V^*)-sets and limited sets are especially useful for detecting complemented copies of ℓ_1 and c_o, due to the following result:

Lemma 3.1.

a) ([B4], [E1]) *A bounded subset of a Banach space E is a (V^*)-set if and only if it does not contain a sequence (x_n) equivalent to the usual basis of ℓ_1 and such* *that the closed span $[x_n]$ is complemented in E.*

b) ([*SL*]) *A Banach space contains a complemented copy of c_o if and only if it contains a non limited sequence (x_n), equivalent to the unit c_o-basis.*

From part (a) of the above lemma and Theorem 2.1, we get immediately the following

Theorem 3.2. *Let (E_n) be a sequence of Banach spaces and Λ a perfect Banach sequence space such that $\Lambda = \Lambda_o$ and also $(\Lambda^\times)_o = \Lambda^\times$. Suppose (x_k) is a basic sequence in $\Lambda(E_n)$ equivalent to the unit basis of ℓ_1 and spanning a complemented subspace. Then there exists $n \in \mathbf{N}$ such that $(\pi_n(x_k))$ contains a subsequence equivalent to the unit vector basis of ℓ_1 that spans a complemented subspace of E_n.*

Proof. By lemma 3.1(a), $K = \{x_k : k \in \mathbf{N}\}$ is not a V^*-set. By remark 2.3(a), K satisfies condition *b*(*ii*) of theorem 2.1. Hence it follows that an $n \in \mathbf{N}$ exists such that $\pi_n(K)$ is not a V^*-set in E_n. A new appeal to lemma 3.1(a) gives the result.

When $\Lambda = \ell_p$ $(1 < p < \infty)$, the above result was proved in [*B*3].

In particular, if μ is a σ-finite, purely atomic measure, and E is a Banach space, the space $L_p(\mu, E)$ $(1 < p < \infty)$ of all E-valued Bochner μ-integrable functions (in this case, it is isometric to $\ell_p(E) = \ell_p(E_n), E_n = E$ for all n) contains a complemented copy of ℓ_1 if and only if so does E. Recently Mendoza has proved in [*M*] the same result for a general σ-finite measure, thus completing the partial results obtained in [*B*1] and [*B*2]). By using some strong results of Talagrand, Mendoza reduces the general situation to one studied and solved in [*B*1].

The analogous question for complemented copies of c_o was treated in [*E*2] for a non purely atomic finite measure μ. The result was quite surprising. In fact, in this case, if E contains a (non necessarily complemented) copy of c_o, the space $L_p(\mu, E)$ $(1 \leq p < \infty)$ contains always a *complemented* copy of c_o. As we shall see, for a purely atomic measure the situation is "what it should be".

Theorem 3.3. *Let (E_n) be a sequence of Banach spaces, Λ a perfect Banach*

sequence space such that $\Lambda = \Lambda_o$*, and* $F_\Lambda = \Lambda(E_n)$*. The following assertions are equivalent:*

a) E_n *contains a complemented copy of* c_o*, for some* $n \in \mathbf{N}$.

b) F_Λ *contains a complemented copy of* c_o.

Proof. Let us prove the only non trivial implication (b) $\Rightarrow$ (a): By lemma 3.1, there exists in F_Λ a non limited sequence (x_k), equivalent to the unit c_o-basis, (e_k). Since the dual of c_o has the Schur property, it is obvious that every bounded subset of c_o is Dunford-Pettis. In particular, if $\varphi : c_o \longrightarrow F_\Lambda$ is an isomorphic embedding such that $\varphi(e_k) = x_k$, it turns out that $K = \{x_k : k \in \mathbf{N}\}$ is a Dunford-Pettis subset of F_Λ. Hence, it satisfies condition b(ii) of Theorem 2.2. Since K is not limited, a new appeal to Theorem 2.2 yields the existence of an $n \in \mathbf{N}$ so that $\pi_n(K) = \{\pi_n(x_k) : k \in \mathbf{N}\}$ is not limited. In consequence, a $\delta > 0$, a weak* null sequence $(y_m^*) \subseteq E_n^*$ and a subsequence $(\pi_n(x_{k_m}))$ exist so that

$$< \pi_n(x_{k_m}), y_m^* > \, > \delta, \text{ for every m.}$$

Taking $u_m = x_{k_m}$, we have:

i) $\sum u_m$ is a weakly unconditionally Cauchy series ([D], Ch. V) in F_Λ, as a subseries of $\sum x_k$.

ii) Every subsequence of $(\pi_n(u_m))$ is not limited.

iii) $\| \pi_n(u_m) \| \geq \epsilon$, for some $\epsilon > 0$ and every $m \in \mathbf{N}$.

From (i) it follows that $\sum \pi_n(u_m)$ is also a weakly unconditionally Cauchy series. Therefore, by (iii) and the Bessaga-Pelczynski selection principle (see [D], Ch. V), we can assume (passing to a subsequence if necessary) that $(\pi_n(u_m))$ is a basic sequence. But then, Corollary 7 of [D], Ch. V assures that $(\pi_n(u_m))$ is equivalent to the unit c_o-basis. Hypothesis (ii) and lemma 3.1(b) conclude the proof.

Remark 3.4. Note that we have proved above the following:

If (x_k) *is a complemented* c_o*-basis in* F_Λ*, there exists a subsequence* (u_m) *of* (x_k) *and an* $n \in \mathbf{N}$ *such that* $(\pi_n(u_m))$ *is a complemented* c_o *basis of* E_n.

Corollary 3.5. *Let μ be a σ-finite, purely atomic measure and E a Banach space. The following assertions are equivalent:*

a) *E contains a complemented copy of c_o.*

b) *For every $p, 1 \leq p < \infty$, $L_p(\mu, E)$ contains a complemented copy of c_o.*

c) *There is a p, $1 \leq p < \infty$, such that $L_p(\mu, E)$ contains a complemented copy of c_o.*

Proof. $L_p(\mu, E)$ is isometric to $\ell_p(E)$, and theorem 3.3 applies.

REFERENCES

[A] K. T. Andrews, *Dunford-Pettis sets in the space of Bochner integrable functions.* Math. Ann., **241** (1979), 35-41

[B1] F. Bombal, *On ℓ_1 subspaces of Orlicz vector valued function spaces.* Math. Proc. Cambr. Phil. Soc., **101** (1987), 107-112.

[B2] F. Bombal, *On embedding ℓ_1 as a complemented subspace of Orlicz vector-valued function spaces.* Rev. Mat. Univ. Complutense de Madrid, **1** (1988), 13-17.

[B3] F. Bombal, *On Pelczynski's property (V^*) in vector sequence spaces.* Coll. Math., **39** (1988), 141-148.

[B4] F. Bombal, *On (V^*) sets and Pelczynski's property (V^*).* Glasgow Math. J., **32** (1990), 109-120.

[B5] F. Bombal, *Sobre algunas propiedades de Espacios de Banach.* To appear in Rev. Acad. Ci. Madrid.

[B6] F. Bombal, *Distinguished subsets and complemented copies of c_o in vector sequence spaces.* Extracta Math., **5**, n. 1 (1990), 4-6.

[BD] J. Bourgain and J. Diestel, *Limited operators and strict cosingularity.* Math. Nachr., **119** (1984), 55-58.

[D] J. Diestel, *Sequences and series in Banach spaces*, Graduate Texts in Math., No. 92, Springer, 1984.

[DR] L. Drewnowski, *On Banach spaces with the Gelfand-Phillips property.* Math. Z., **193** (1986), 405-411.

[E1] G. Emmanuele, *On the Banach spaces with the property (V^*) of Pelczynski.* Annali Mat. Pura e Applicata, (1988), 171-181.

[E2] G. Emmanuele, *On complemented copies of* c_o *in* $L^p_X, 1 \leq p < \infty$. Proc. A. M. S., (1988), 785-786.

[LT] J. Lindenstrauss and L. Tzafriri, *Classical Banach Spaces.* Springer-Verlag, Berlin. Vol. I, 1977; Vol. II, 1979.

[M] J. Mendoza, *Complemented copies of* ℓ^1 *in* $L_p(\mu; E)$. Preprint.

[P] A. Pelczynski, *On Banach spaces on which every unconditionally convergent operator is weakly compact.* Bull. Acad. Pol. Sci., **10** (1962), 641-648.

[R] R. C. Rosier, *Dual spaces of certain vector sequence spaces.* Pacific J. of Math., **46**, No. 2, (1973), 487-501.

[SL] T. Schlumprecht, *Limitierte Mengen in Banachräumen.* Thesis. Ludwig Maximilians-Universität, München, 1988.

[V] M. Valdivia, *Topics in Locally Convex Spaces.* North-Holland Mathematics Studies No. 67. North-Holland Pub. Co., Amsterdam, 1982.

Progress in Functional Analysis
K.D. Bierstedt, J. Bonet, J. Horváth & M. Maestre (Eds.)

Weak Topologies on Bounded Sets of a Banach Space. Associated Function Spaces

José G. Llavona[1]

Departamento de Análisis Matemático
Facultad de Matemáticas
Universidad Complutense de Madrid
E-28040 - Madrid - Spain.

In honor of the sixtieth birthday of Manuel Valdivia

This paper is divided into three sections. Section 1 is an exhaustive study concerning the principal spaces of weakly continuous functions on Banach spaces. Since the structure of these function spaces is closely related with properties of different weak topologies, the bounded-weak, the compact-weak and the bounded-weak* topologies are introduced.

Section 2 shows the weakly continuous and weakly differentiable function spaces in relation with the extension of Weierstrass' theorem for infinite dimensional Banach spaces.

Section 3 deals with interpolation of bounded sequences by weakly continuous and weakly differentiable functions. Many of the representation results obtained in the previous sections are used.

Now we start with some notations and definitions:

Let E and F be real or complex Banach spaces; we consider the following linear spaces:

– $L(^nE;F)$, $n \in \mathbf{N}$, all continuous n-linear mappings from E^n to F.

– $L_s(^nE;F)$, $n \in \mathbf{N}$, all symmetric continuous n-linear mappings from E^n to F.

– $L(^0E;F) = L_s(^0E;F) = F$.

– $\mathcal{P}(^nE;F)$, $n \in \mathbf{N}$, all continuous n-homogeneous polynomials from E to F; $\mathcal{P}(^nE;F) = \{p : E \to F : p = A \circ \Delta_n$, Δ_n is the diagonal operator of E into $E\times \overset{n}{\cdots} \times E$, $A \in L(^nE;F)\}$. If $p = A \circ \Delta_n$, we will write $p = \hat{A}$.

[1]Research partially supported by DGICYT (Project PB87-1031) and by DGICYT grant.

– $\mathcal{P}(^0E; F) = F$.

– $\mathcal{P}_f(^nE; F)$, $n \in \mathbf{N}$, all continuous n-homogeneous polynomials of finite type. $\mathcal{P}_f(^nE; F)$ is the subspace of $\mathcal{P}(^nE; F)$ generated by the collection of functions $(\phi^n \otimes y) = \phi^n \cdot y$, $(\phi \in E', y \in F)$, where $(\phi^n \otimes y)(x) = \phi^n(x) \cdot y$, $x \in E$.

– $\mathcal{P}(E; F) = \sum_{n=0}^{\infty} \mathcal{P}(^nE; F)$, all continuous polynomials from E to F.

– $\mathcal{P}_f(E; F) = \sum_{n=0}^{\infty} \mathcal{P}_f(^nE; F)$, all continuous polynomials of finite type from E to F.

If $p \in \mathcal{P}(^nE; F)$, we set $\|p\| = \sup\{\|p(x)\| : x \in E, \|x\| \leq 1\}$.

– $(\mathcal{P}(^nE; F), \|\cdot\|)$ is a Banach space.

– $\mathcal{P}_c(^nE; F)$ will denote the completion of $\mathcal{P}_f(^nE; F)$ with respect to the norm induced by $\mathcal{P}(^nE; F)$, and in general is strictly contained in $\mathcal{P}(^nE; F)$.

– Polarization formula:

The mapping $A \in L_s(^nE; F) \to \hat{A} \in \mathcal{P}(^nE; F)$ is a vector space isomorphism and we have the "formula":

$$A(x_1, \ldots, x_n) = \frac{1}{n!2^n} \sum_{\varepsilon_1 = \pm 1, \ldots, \varepsilon_n = \pm 1} \varepsilon_1 \ldots \varepsilon_n \hat{A}(\varepsilon_1 x_1 + \ldots + \varepsilon_n x_n).$$

1. Weak Topologies. Weakly Continuous Functions.

Let E and F be Banach spaces and $A \subseteq E$. A function $f : A \to F$ is said to be weakly continuous if for each $x \in A$ and $\varepsilon > 0$, there are $\phi_1, \ldots, \phi_n$ in E' and $\delta > 0$ such that if $y \in A, | \phi_i(x - y) | < \delta$ for all $i = 1, \ldots, n$ then $\|f(x) - f(y)\| < \varepsilon$.

We denote the space of all f from E to F which are weakly continuous by $C_w(E; F)$; and by $C_{wb}(E; F)$(respectively $C_{wk}(E; F)$) the space of all functions from E to F which are weakly continuous when restricted to bounded sets (respectively weakly compact sets). Clearly we have $C_w(E; F) \subset C_{wb}(E; F) \subset C_{wk}(E; F)$. If we denote by E_w the space E endowed with the weak topology $\sigma(E; E')$, it follows that $C_w(E; F) = C(E_w; F)$, the space of all continuous functions from E_w to F.

1.1. Definition. *Let E be a Banach space with normed bidual E''. The bounded weak (bw) topology on E (respectively the bounded weak* (bw^*) topology on E'') is the finest topology on E (respectively on E'') which agrees with the weak topology $\sigma(E; E')$ (respectively with the weak*-topology $\sigma(E''; E')$) on bounded sets.*

The space E (respectively E'') endowed with the bw-topology (respectively with the bw^*-topology) will be represented by E_{bw} (respectively E''_{bw^*}). It is easy to see that $C_{wb}(E; F) = C(E_{bw}; F)$, the space of all continuous functions from E_{bw} to F.

As follows from Collins [6], the *bw*-topology is semilinear. Moreover, it can be shown, Ferrera [9], that *bw* is a vectorial topology if and only if it is a locally convex one. It is known that the bw^*-topology, first discussed by Day [8] and Wheeler [21], is locally convex.

If the *cbw*-topology on E denotes the finest locally convex topology which agrees with the weak topology on bounded sets, in Wheeler [21] the following result may be found, characterizing the *cbw*-topology.

1.2. Theorem. *The cbw-topology is the restriction of the bw^*-topology on E'' to E. More specifically, if E is reflexive we have that the bw-topology on E is a locally convex topology.*

Wheeler [21] proves that the *bw*-topology on c_0 is not locally convex. Gómez [11] proves that reflexivity is also a necessary condition.

1.3 Theorem. *Let E be a Banach space. The bw-topology on E is a locally convex topology if, and only if, E is reflexive.*

Now, in order to get a representation of the space $C_{wk}(E; F)$, we are going to introduce the *kw* topology on E.

1.4. Definition. *Let E be a Banach space. The compact weak (kw) topology on E is the finest topology on E which agrees with the weak topology on weakly compact sets.*

If we denote by E_{kw} the space E endowed with the *kw*-topology, if follows that $C_{wk}(E; F) = C(E_{kw}; F)$, the space of all continuous functions from E_{kw} to F.

Gutiérrez [12] proves the following results about the *kw*-topology.

1.5. Theorem. *Let E be a Banach space.*

1. *E has the Schur property if, and only if, the kw-topology on E agrees with the norm topology.*
2. *E contains no subspace isomorphic to ℓ_1 if, and only if, the topologies kw and bw agree on E.*
3. *E is reflexive or Schur if, and only if, the kw-topology on E is locally convex.*

It is obvious that each $f \in C_{wk}(E; F)$ is weakly sequentially continuous in the sense that it takes weakly convergent sequences in E to convergent sequences in F. On the other hand, given a weakly compact set K of $E, A \subseteq K$ and $x \in \overline{A}^w$ (the weak closure of A), since every Banach space is angelic for the weak topology, it follows that there is a sequence (x_n) in A such that (x_n) converges weakly to x; therefore if f is weakly sequentially continuous, $f(x_n)$ converges to $f(x)$ in F, showing that $f(x) \in \overline{f(A)}$. Then if we denote by $C_{wsc}(E;F)$ the space of all weakly sequentially continuous functions from E to F, we have:

1.6. Theorem [10]. *For every pair of Banach spaces E, F , the spaces $C_{wk}(E;F)$ and $C_{wsc}(E;F)$ coincide.*

$C_w(E;F) = C_{wb}(E;F)$ if and only if E is finite dimensional. Indeed, if there exists a sequence $\{\phi_n\}$ of linearly independent continuous linear forms on E such that $\|\phi_n\| = 1$ for every $n \in \mathbf{N}$, then the polynomial $Q = \sum_{n=1}^{\infty} 2^{-n}\phi_n^2$ is not weakly continuous. It is easy to see that $Q \in C_{wb}(E)$.

If we endow the spaces $C_w(E;F), C_{wb}(E;F)$ and $C_{wsc}(E;F) = C_{wk}(E;F)$, with the topology of uniform convergence on weakly compact subsets of E, the following result holds:

1.7. Proposition [10]. *$C_w(E;F)$ is a dense subspace of $C_{wsc}(E;F)$.*

A consequence of Proposition 1.7 is that $C_w(E;F)$ is a dense subspace of $C_{wb}(E;F)$; therefore $C_w(E;F)$ is complete if and only if E has finite dimension. The answer about the completion of $C_w(E;F)$ is given by the following proposition:

1.8. Theorem [10]. *The space $C_{wsc}(E;F)$ is complete. Then $C_{wsc}(E;F)$ is the completion of $C_w(E;F)$.*

A characterization of Banach spaces E containing no subspace isomorphic to ℓ_1 is given in the next theorem.

1.9. Theorem [10], [2]. *Let E be a Banach space.* TFAE:

1. *E contains no subspace isomorphic to ℓ_1.*
2. *There exists a Banach space F such that $C_{wb}(E;F) = C_{wsc}(E;F)$.*

If the space $C(E_{bw};F)$ is endowed with the compact-open topology it follows that the spaces $C_{wb}(E;F)$ and $C(E_{bw};F)$ are topologically isomorphic. Unfortunately, we do not have a general result, eliminating the hypothesis of weak normality, assuring that E_{bw} is always a completely regular space, which is necessary for studying $C(E_{wb};F)$. Nevertheless, if $\tilde{E}_{bw}$ represents the space E endowed with the coarsest topology that makes each function in $C_{wb}(E;F)$ continuous, it follows that the space $\tilde{E}_{bw}$ is completely regular and the following inclusions are continuous:

$$E_{bw} \to \tilde{E}_{bw} \to (E, \sigma(E;E')) = E_w$$

giving equality to the first inclusion, if and only if E_{bw} is completely regular. A consequence of the above technique and the Nachbin-Shirota theorems are the following results:

1.10. Theorem [9]. *Let E be a weakly normal Banach space. Then $C_{wb}(E;F)$ is bornological if and only if E is weakly realcompact.*

Weakly realcompact Banach spaces are described in Corson [7]. Nevertheless, weakly realcompact spaces do exist which are not weakly normal, as the case of ℓ_∞, Corson [7]. In any case, the class of weakly normal and weakly realcompact spaces is broad: specifically every WCG space is weakly Lindelöf, so it satisfies our hypothesis.

The following theorem gives a partial answer to the problem of not necessarily normal spaces.

1.11. Theorem [9]. *If E is the dual of a separable Banach space, then $C_{wb}(E)$ is bornological. In particular $C_{wb}(\ell_\infty)$ is bornological.*

1.12. Corollary [9]. *If E is WCG or E is the dual of a separable Banach space and F is an arbitrary Banach space, then $C_{wb}(E;F)$ is bornological.*

Finally, since $\tilde{E}_{bw}$ is a NS-space, that is, every relatively pseudocompact and closed subset is compact, we obtain:

1.13. Theorem [9]. *Let E, F be Banach spaces. Then $C_{wb}(E;F)$ is barrelled.*

The topology bw^* may be used to describe the space of functions which are weakly uniformly continuous when restricted to bounded sets. If $A \subseteq E$, a function $f : A \to F$ is said to be weakly uniformly continuous if for each $\varepsilon > 0$, there are $\phi_1, \ldots, \phi_n$ in E' and $\delta > 0$ such that if $x, y \in A$, $|\phi_i(x-y)| < \delta$ for all $i = 1, \ldots, n$ then $\|f(x) - f(y)\| < \varepsilon$. We denote as $C_{wbu}(E;F)$ the space of all f from E to F which are weakly uniformly continuous when restricted to bounded sets. Every $f \in C_{wbu}(E;F)$ is compact, in the sense that it maps bounded sets into relatively compact sets. We endow $C_{wbu}(E;F)$ with the locally convex topology τ_b of uniform convergence on bounded subsets of E. $C_{wbu}(E;F)$ is τ_b- complete. Bounded sets for E'' coincide for the norm, the weak* and the bw^* topologies, and that $(E''_{bw^*})' = E'$. Since E''_{bw^*} is also the topological direct limit, $\lim_{n\to\infty}(B''_n,\text{weak}^*)$ (B''_n is the n-ball in E''), a function f belongs to $C(E''_{bw^*};F)$ if and only if for all n, $f|_{B''_n}: (B''_n,\text{weak}^*) \to F$ is continuous. Then we see that this happens precisely when $f|_E \in C_{wbu}(E;F)$. Then we have the following topologically isomorphic representation of $C_{wbu}(E;F)$:

$$C_{wbu}(E;F) \cong C(E''_{bw^*};F)$$

where $C(E''_{bw^*};F)$ is the space of all continuous functions from E''_{bw^*} to F, endowed with the topology of uniform convergence on norm-bounded sets of E''.

It is easy to see that every relatively pseudocompact and closed subset in E''_{bw^*} is compact. Thus, we have the following theorem:

1.14. Theorem [17]. *Let E, F be Banach spaces. Then the space $C_{wbu}(E;F)$ is barrelled.*

A continuous linear mapping $A : E \to F$ is compact if and only if A is weakly (uniformly) continuous on bounded sets. But, a non-linear mapping $A : E \to F$ may be compact, without being weakly uniformly continuous on bounded sets. For example, the mapping $A : \ell_2 \to \mathbf{R}$ given by $A(x) = \sum_{n=1}^{\infty} x_n^2$ is trivially compact although $A(e_n) = 1$ $(n \in \mathbf{N})$ for the usual basis $\{e_n\}$ of ℓ_2, so that $A \notin C_{wbu}(\ell_2)$.

Valdivia [20] shows that a Banach space E is reflexive, if and only if every weakly continuous function $f : E \to \mathbf{R}$ is bounded on balls in E. Since every function in $C_{wbu}(E)$ is bounded on balls in E, we have that $C_{wbu}(E) = C_{wb}(E)$ if and only if E is reflexive. If E is a non-reflexive separable Banach space, the James theorem [13] states that there exists $\phi \in E'$ which does not attain its norm. The function $g : B_1 \to \mathbf{R}$, $g(x) = [\|\phi\| - \phi(x)]^{-1}$, is weakly continuous on B_1, and it is not bounded. Since E is separable, it is WCG and therefore weakly normal. Thus, according to Tietze's theorem, there exists a weakly continuous f on E, which extends g. The function f cannot belong to $C_{wbu}(E)$.

2. On the Completion of the Space of Continuous Polynomials of Finite Type.

This section focuses around uniform approximation on bounded, and on compact sets, by polynomials of finite type. As a natural approximation problem, weakly continuous and weakly differentiable function spaces are introduced.

(a) Continuous case.

Let E, F be Banach spaces and $\mathcal{P}_c(^nE;F)$ the completion of $\mathcal{P}_f(^nE;F)$ with respect to the norm induced by $\mathcal{P}(^nE;F)$; $\mathcal{P}_c(^nE;F)$ is in general strictly contained in $\mathcal{P}(^nE;F)$.

2.1. Proposition [1]. *Let X be a compact Hausdorff space and $C(X)$ the space of all scalar continuous functions on X. If X is dispersed (every closed subset of X contains an isolated point) and $E = C(X)$ with the sup norm, then for every $n \in \mathbf{N}$, $\mathcal{P}(^nE) = \mathcal{P}_c(^nE)$. In particular $\mathcal{P}_c(^nc_0) = \mathcal{P}(^nc_0)$.*

We will denote by $\mathcal{P}_\Phi(^nE;F)$ the space $\mathcal{P}(^nE;F) \cap C_\Phi(E;F)$ for $\Phi = w, wb, wsc$ and wbu. By definition $\mathcal{P}_{wb}(^nE;F) = \mathcal{P}_{wbu}(^nE;F)$ when E is reflexive. But, a polynomial which is weakly continuous on balls is in fact weakly uniformly continuous on balls [4].

One consequence of the above result is the following theorem.

2.2. Theorem [4]. *If E' has the approximation property then $\mathcal{P}_c({}^nE;F) = \mathcal{P}_{wbu}({}^nE;F)$.*

2.3. Remark.

1. $\mathcal{P}_w({}^nE;F) = \mathcal{P}_f({}^nE;F) \subset \mathcal{P}_{wbu}({}^nE;F)$
2. *If E' is separable, $\mathcal{P}_{wsc}({}^nE;F) = \mathcal{P}_{wbu}({}^nE;F)$.*
3. *If $p(x) = \sum_{n=1}^{\infty} x_n^2$, then $p \in \mathcal{P}({}^2\ell_2) \setminus \mathcal{P}_{wsc}({}^2\ell_2)$, and therefore $p \notin \mathcal{P}_{wbu}({}^2\ell_2)$; the same $p \in \mathcal{P}_{wsc}({}^2\ell_1) \setminus \mathcal{P}_{wbu}({}^2\ell_1)$.*
4. *Every polynomial $p(x) = \sum_{n=1}^{\infty} \lambda_n x_n^2$,* with $\lambda = (\lambda_n) \in \ell_1$, *belongs to $\mathcal{P}_{wbu}({}^2\ell_p)$, $1 \le p < \infty$.*
5. [12] *E contains no subspace isomorphic to ℓ_1 if and only if there exists a Banach space F and an integer $n \geq 2$ such that $\mathcal{P}_{wb}({}^nE;F) = \mathcal{P}_{wsc}({}^nE;F)$.*

There are two natural topologies on $\mathcal{P}_f(E;F)$. The first one is the compact-open topology τ_u, generated by all seminorms of the form

$$p \in \mathcal{P}_f(E;F) \rightarrow \sup\{\|p(x)\| : x \in K\}$$

where K runs over the compact subsets of E. The Stone-Weierstrass theorem for polynomial algebras, see [17], says that the space $C(E;F)$, of all continuous functions from E into F, is the τ_u-completion of $\mathcal{P}_f(E;F)$. The second natural topology on $\mathcal{P}_f(E;F)$ is the topology τ_b, of the uniform convergence on bounded subsets of E, which is generated by all seminorms:

$$p \in \mathcal{P}_f(E;F) \rightarrow \sup\{\|p(x)\| : x \in B\}$$

where B is allowed to range over the bounded subsets of E.

2.4. Theorem [5]. *$C_{wbu}(E;F)$ is the τ_b*-completion of $\mathcal{P}_f(E;F)$.

(b) Differentiable case.

Let E, F be real Banach spaces; we endow $C^m(E;F)$, the space of m-times continuously Fréchet differentiable F-valued functions on E, (for $m = 0, 1, \ldots, \infty$) with the compact-open topology of order m. This topology, which we denote by τ_u^m, is generated by all seminorms:

$$f \in C^m(E;F) \rightarrow \sup\{\|f^j(x)\| : x \in K\}$$

where $j \in \mathbf{N}, j \leq m \in \mathbf{N}$, and K is a compact subset of E. The topology τ_u^∞ on $C^\infty(E;F)$ is defined in the obvious way. The Stone-Weierstrass theorem fails for $(C^m(E), \tau_u^m)$ with $m \geq 2$.

2.5. Theorem [15], [16]. *Let H be a real infinite dimensional Hilbert space. Then $\mathcal{P}_f(H)$ is not τ_u^2-dense in $C^2(H)$.*

The main reason of the above result is that generally if $f \in C^m(E)$, then $f^j(x) \in \mathcal{P}({}^jE)\backslash\mathcal{P}_c({}^jE)$, $j \geq 2$. If E' has the approximation property, then $\mathcal{P}_c({}^jE;F) = \mathcal{P}_{wbu}({}^jE;F)$, $(j \leq m)$. Therefore, in order to extend the Stone-Weierstrass theorem for the τ_u^m-topology, it seems logical to establish the following definition.

2.6. Definition. *$C_c^m(E;F)$ is the space of all mappings $f \in C^m(E;F)$ such that for each $x \in E$ and each $n \in \mathbf{N}$, $(1 \leq j \leq m)$, $f^j(x) \in \mathcal{P}_{wbu}({}^jE;F)$.*

The space $C_c^m(E;F)$ is τ_u^m-complete. Note that $C_c^0(E;F) = C(E;F)$ and $C_c^1(E) = C^1(E)$. If $E = C(X)$, with X a dispersed compact Hausdorff space, then $\mathcal{P}({}^jE) = \mathcal{P}_c({}^jE)$; in particular $C_c^m(c_0) = C^m(c_0)$ for all $m = 0, 1, \ldots, \infty$.

2.7. Theorem [5]. *Let E and F be real Banach spaces with E' having the bounded approximation property. Then for all $m = 0, 1, \ldots, \infty$, $C_c^m(E;F)$ is the τ_u^m-completion of $\mathcal{P}_f(E;F)$.*

A classical Bernstein theorem says that for every $f \in C^1(\mathbf{R}^n)$ there is a sequence (p_n) of polynomials such that $p_n \to f$ uniformly on bounded sets and $p_n' \to f'$ uniformly on bounded sets. Restrepo [19] determines the τ_b^1 closure of $\mathcal{P}_f(E)$ for a restricted class of reflexive Banach spaces. We present now an extension of Restrepo's work.

We endow $\mathcal{P}_f(E;F)$ with the locally convex topology of uniform convergence, of order m, on bounded sets of E. This topology, which we denote by τ_b^m, is generated by all seminorms:

$$p \in \mathcal{P}_f(E;F) \to \sup\{\|p^j(x)\| : x \in B, j \leq m\}$$

where $m = 1, \ldots, \infty$, and B runs over the bounded subsets of E.

2.8. Theorem [5]. *Let E and F be two real Banach spaces, with E' having the bounded approximation property, and let $m > 0$. Then, the τ_b^m-completion of $\mathcal{P}_f(E;F)$ is the space $C_{wbu}^m(E;F)$, where:*

$$C_{wbu}^m(E;F) = \{f \in C^m(E;F) : \begin{array}{l} (i)\ f^j(x) \in \mathcal{P}_{wbu}({}^jE;F)\ (x \in E;\ j \leq m), \\ (ii)\ f^j \in C_{wbu}(E;\mathcal{P}_{wbu}({}^jE;F))\ (j \leq m)\}. \end{array}$$

2.9. Remark.

1. *$\mathcal{P}_{wbu}({}^mE;F) \subset C_{wbu}^\infty(E;F)$ for all $m \in \mathbf{N}$, and $C_{wbu}^m(E;F)$ contains all functions $g \circ T$, where $T \in E' \otimes E$ and $g \in C^m(T(E);F)$.*
2. *$C_{wbu}^m(E;F)$ contains no non-zero functions with bounded support, except in the trivial case when $F = \{0\}$ or $dim(E) < \infty$.*

3. *Every $f \in C^m_{wbu}(E;F)$ is uniformly differentiable of order j ($j \le m$), in the following sense: For each $B \subset E$ bounded and each $\varepsilon > 0$, there exists $\delta > 0$ such that if $x \in B, y \in E$ with $\|y\| \le \delta$, then*

$$\|f(x+y) - f(x) - f'(x)(y) - \ldots - \frac{f^j(x)}{j!}(y)\| \le \varepsilon \|y\|^j.$$

3. Interpolation of Bounded Sequences by Weakly Continuous and Weakly Differentiable Functions.

For any Hausdorff real locally convex spaces X and Y, and for $m = 0, 1, \ldots, \infty$, $C^m(X;Y)$ is the space of all m-times continuously Fréchet differentiable mappings from X to Y (see [22], §1.8).

We show now that functions in $C^m_{wbu}(E;F)$ have extensions to functions defined on E'' having the same degree of differentiability. The importance of this result comes from the fact that we can thus obtain a topological and algebraic isomorphism between $C^m_{wbu}(E)$ and $C^m(E''_{bw^*})$.

3.1. Theorem [3]. *Let E be a real Banach space. Every function $f \in C^m_{wbu}(E)$ can be extended in a unique way to a function $\tilde{f} : E'' \to \mathbf{R}$, so that $\tilde{f} \in C^m(E_{bw^*})$. Moreover, the mapping $f \to \tilde{f}$ is a homomorphism and for all $j, n \in \mathbf{N}$ with $j \le m$*

$$\sup\{\|f^j(x)\| : x \in B_n\} = \sup\{\|\tilde{f}^j(x)\| : x \in B''_n\}.$$

3.2. Corollary. *The mapping $f \to \tilde{f}$ is a topological isomorphism, in the sense of Fréchet algebras, between $(C^m_{wbu}(E), \tau^m_b)$ and $(C^m(E''_{bw^*}), \tau^m_b)$ for all $m = 0, 1, \ldots, \infty$.*

Now, we will state a "global" formulation of interpolation problems, in order to apply the results obtained before.

Let E be a real Banach space and (a_n) be a bounded sequence in E, with pairwise different terms. We consider the "restriction homomorphism"

$$\begin{array}{rccl} R: & C^m_{wbu}(E) & \longrightarrow & \ell_\infty \\ & f & \longmapsto & (f(a_n))_{n \in \mathbf{N}} \end{array}$$

Our interest is to determine in which case R is onto ℓ_∞, and in which case it is onto the subspace c of all convergent sequences.

(a) Interpolation of bounded sequences by $C_{wbu}(E)$-functions.

In this case R is induced by a continuous function $\varphi : \beta\mathbf{N} \to E''_{bw^*}$, where $\beta\mathbf{N}$ is the Stone-Čech compactification of $\mathbf{N}$, in the sense that:

$$R(f) = \tilde{f} \circ \varphi, \ \forall f \in C_{wbu}(E).$$

Therefore, R will be onto ℓ_∞ if and only if the composition subalgebra $\{\tilde{f} \circ \varphi : f \in C_{wbu}(E)\} = C(\beta\mathbf{N}) \cong \ell_\infty$.

3.3. Proposition [18]. *The following conditions are equivalent:*

1. *R is onto ℓ_∞.*
2. *$M = \{a_n : n \in \mathbf{N}\}$ is weak-discrete and its weak*-closure $\overline{M}^{w^*}$ is weak*-homeomorphic to $\beta\mathbf{N}$.*

3.4. Proposition [18]. *The range of R is contained in c if, and only if, the sequence (a_n) is weak-Cauchy in E.*

When the range of R is contained in c, R is induced by a continuous function $\varphi : \alpha\mathbf{N} \to E''_{bw^*}$, where $\alpha\mathbf{N}$ represents the Alexandroff compactification of $\mathbf{N}$. R will be onto c if and only if the composition subalgebra $\{\tilde{f} \circ \varphi : f \in C_{wbu}(E)\} = C(\alpha\mathbf{N}) \cong c$.

3.5. Proposition [18]. *The following conditions are equivalent:*

1. *The range of R is c .*
2. *$(a_n)_{n\in\mathbf{N}}$ is weak-Cauchy and, for each $m \in \mathbf{N}, (a_n)_{n\in\mathbf{N}}$ is not weak convergent to a_m.*
3. *$M = \{a_n : n \in \mathbf{N}\}$ is weak-discrete and its weak*-closure $\overline{M}^{w^*}$ is weak* homeomorphic to $\alpha\mathbf{N}$.*
4. *The range of R is contained in c, and there exists a continuous linear section $S : c \to C_{wbu}(E)$ of R.*

Now we present some examples of interpolating and non-interpolating sequences for the case $E = \ell_1$, showing a variety of situations:

3.6. Examples.

1. *For the standard basis $(e_n)_{n\in\mathbf{N}}$ of ℓ_1, the associated homomorphism R is onto ℓ_∞, and there exists a continuous linear section $S : \ell_\infty \to C_{wbu}(\ell_1)$ of R.*
2. *For every bounded unconditional basis $(a_n)_{n\in\mathbf{N}}$ of ℓ_1, the above result (3.6.1) holds.*
3. *The sequence $(a_n)_{n\in\mathbf{N}} \subset \ell_1$ defined by $a_1 = 2e_1, a_n = e_{n-1} - e_n$ (for $n \geq 2$) is a conditional basis of ℓ_1; in this case the associated homomorphism R is onto ℓ_∞ and there exists also a continuous linear section $S : \ell_\infty \to C_{wbu}(\ell_1)$ of R.*

4. *Let $(a_n)_{n\in\mathbf{N}}$ be the sequence in ℓ_1 defined by $a_n = e_n + e_{n+1}$ for each $n \in \mathbf{N}$. In this case the homomorphism R associated to (a_n) is not onto ℓ_∞.*

(b) Interpolation of Schauder bases by $C^m_{wbu}(E)$-functions.

Let E be a real Banach space, (e_n) a normalized Schauder basis in E, with $(e_n) \to 0$ weakly. For each $f \in C^m_{wbu}(E)$, $(f(e_n)) \subset \mathbf{R}$ is a convergent sequence. Our interest now is to study when the associated homomorphism

$$R : C^m_{wbu}(E) \to c, \; R(f) = (f(e_n))_{n\in\mathbf{N}}$$

is onto c and when R admits a continuous linear section.

3.7. Theorem [14].

1. *If E is superreflexive, there exists $m \in \mathbf{N}$ such that $R : \mathcal{P}_{wbu}(^mE) \to c_0$ admits a continuous linear section.*
2. *If E is reflexive and (e_n) is symmetric, then $R : C^\infty_{wbu}(E) \to c$ is onto and admits a continuous linear section.*
3. *If $E = c_0$ then $R : C^1_{wbu}(c_0) \to c$ is onto.*

3.8. Remark. Problem: Is there a Banach space E such that $C^\infty_{wbu}(E)$ is reflexive? Since for each $m \in \mathbf{N}, C^\infty_{wbu}(E)$ induces on $\mathcal{P}_{wbu}(^mE)$ the norm topology, if $C^\infty_{wbu}(E)$ is reflexive then $\mathcal{P}_{wbu}(^mE)$ will be reflexive for all $m \in \mathbf{N}$. In particular, for $m = 1$, $\mathcal{P}_{wbu}(^1E) = E'$ would be reflexive. Therefore we can restrict our attention to reflexive Banach spaces. From 3.7.1 and 3.7.2 it can easily be verified that if E is superreflexive with basis or E is reflexive with symmetric basis, then $C^\infty_{wbu}(E)$ is not reflexive. If T' is the Tsirelson space, T' is reflexive and has unconditional basis. We don't know if $C^\infty_{wbu}(T')$ is reflexive.

REFERENCES

[1] Aron, R.M. *Compact polynomials and compact differentiable mappings*, in: Sém. Pierre Lelong, Lecture Notes in Math. 524, Springer-Verlag (1976) 213-222.

[2] Aron, R.M., Diestel, J. and Rajappa, A.K. *A characterization of Banach spaces containing ℓ_1*, in: Banach spaces, Proc. of the Missouri Conference, Lecture Notes in Math. 1166, Springer-Verlag (1985) 1-3.

[3] Aron, R.M., Gómez, J. and Llavona, J. *Homomorphisms between algebras of differentiable functions in infinite dimensions*, Michigan Math. J. 35 (1988) 163-178.

[4] Aron, R.M., Hervés, C. and Valdivia, M. *Weakly continuous mappings on Banach spaces*, J. Funct. Analysis, vol. 52, No. 2 (1983) 189-203.

[5] Aron, R.M. and Prolla, J.B. *Polynomial approximation of differentiable functions on Banach spaces*, J. Reine Angew. Math. 313 (1980), 195-216.

[6] Collins, H.S. *Completeness and compactness in linear topological spaces*, Trans. Amer. Math. Soc. 79 (1955) 256-280.

[7] Corson, H.H. *The weak topology of a Banach space*, Trans. Amer. Math. Soc. 101 (1961) 1-15.

[8] Day, M.M. Normed linear spaces (Third Edition.) Springer-Verlag, 1973.

[9] Ferrera, J. *Spaces of weakly continuous functions*, Pacific J. Math. 102 (1982) 285-291.

[10] Ferrera, J. Gómez, J. and Llavona, J. *On completion of spaces of weakly continuous functions*, Bull. London Math. Soc. 15 (1983) 260-264.

[11] Gómez, J. *On local convexity of bounded weak topologies on Banach spaces*, Pacific J. Math. 110, No. 1 (1984) 71-75.

[12] Gutiérrez, J. Espacios de funciones continuas y diferenciables en dimensión infinita, Ph.D.Thesis. Universidad Complutense, Madrid (1990).

[13] James, R.C. *Weak compactness and reflexivity*, Israel J. Math. 2 (1964), 101-119

[14] Jaramillo, J.A. and Gómez, J. *Interpolation by weakly differentiable functions on Banach spaces* (Preprint.)

[15] Lesmes, J. *On the approximation of continuously differentiable functions in Hilbert spaces*, Rev. Colombiana de Matemáticas 8 (1974) 217-223.

[16] Llavona, J. *Approximation of differentiable functions*, Advances in Math. Supplementary Studies 4 (1979) 197- 221.

[17] Llavona, J. Approximation of continuously differentiable functions, Math. Studies 130. North-Holland, Amsterdam 1986.

[18] Llavona, J. and Jaramillo, J.A. *Homomorphisms between algebras of continuous functions*, Canadian J. Math. XLI (1) (1989), 132-162.

[19] Restrepo, G. *An infinite dimensional version of a theorem of Bernstein*, Proc. A.M.S. 23 (1969) 193 -198.

[20] Valdivia, M. *Some new results on weak compactness*, J. Funct. Anal. 24 (1977) 1-10.

[21] Wheeler, R.F. *The equicontinuous weak* topology and semi-reflexivity*, Studia Math. XLI (1972) 243-256.

[22] Yamamuro, S. Differential Calculus in Topological Linear Spaces, Lecture Notes in Math. 374, Springer-Verlag, 1974.

Progress in Functional Analysis
K.D. Bierstedt, J. Bonet, J. Horváth & M. Maestre (Eds.)

Factorization of multilinear operators

Jari Taskinen

University of Helsinki, Department of Mathematics, Hallituskatu 15, SF-00100 Helsinki, Finland

Dedicated to Professor M. Valdivia on the occasion of his 60th birthday

Abstract
We show that every complex bilinear partial differential operator with polynomial coefficients acting in some Banach function space can be represented as a product of two abstract linear operators and the multiplication operator.

1. INTRODUCTION.

In this article we consider an approach to multilinear problems which has not yet been studied systematically. We begin with a concrete example. Let $P : C^2(0,1) \to C(0,1)$ be the following bilinear second order differential operator with constant coefficients:

$$P(f) := ff' + (f')^2 + ff'' + f'f'', \tag{1}$$

where $f \in C^2(0,1)$. We consider the equation

$$P(f) = g, \tag{2}$$

where $g \in C(0,1)$ is a given function. Due to the algebraic properties of P it is possible to solve this equation: we can write

$$P(f) = \frac{1}{2}\frac{d((f+f')^2)}{dx} \tag{3}$$

for all f. The equation $f + f' = h$, where $h \in C^1(0,1)$, is solvable so that (2) is reduced to the equation

$$f(x) + f'(x) = 2\Big(\int_0^x g(t)dt + C\Big)^{1/2}, \tag{4}$$

where $C \in \mathbb{R}$ is chosen suitably.

The above method depended on the fact that it was possible to find simple *linear* differential operators B and A such that

$$P(f) = B((Af)^2). \tag{5}$$

If (5) holds for some P, one is naturally led to search for a solution of $P(f) = g$ in the form

$$f = A^{-1}(\sqrt{B^{-1}g}). \tag{6}$$

2. CANONICAL FACTORIZATIONS.

Let us now turn to a more abstract setting. In the following we denote by X a (Lebesgue) measurable subset of $\mathbb{R}^n$, $n \in \mathbb{N}$, such that the vector space $\mathcal{M}_X$ consisting of measurable functions $X \to \mathbb{K}$, $\mathbb{K} = \mathbb{R}$ or $\mathbb{C}$, is infinite dimensional; we identify functions which are equal a.e. in X. If $M \in \mathbb{N}$, $M \geq 2$, and E, F are Banach spaces, we denote by $L(^M E, F)$ the space of continuous M-linear mappings $E^M = E \times \ldots \times E \to F$. We also denote by $\Delta : E \to E^M$ the diagonal mapping $\Delta_M : x \mapsto (x, \ldots, x)$ and by $\psi_M : \mathcal{M}_X^M \to \mathcal{M}_X$ the pointwise multiplication operator $\psi_M : (f_1, \ldots, f_M) \mapsto \prod_{i=1}^{M} f_i$; we also write $\varphi_M := \psi_M \circ \Delta_M$. We define $\bar{e} = (1, \ldots, 1) \in \mathbb{K}^M$. The space of *symmetric* M-linear continuous mappings $G^M \to H$ is denoted by $L^s(^M G, H)$ and the space of M-homogenous continuous polynomials $G \to H$ by $\mathcal{P}(^M G, H)$; recall that $f \in \mathcal{P}(^M G, H)$ if $f = g \circ \Delta_M$ for some $g \in L^s(^M G, H)$.

We state the following

Problem 1. *Let G and H be Banach spaces consisting of elements of $\mathcal{M}_X$ and let $P \in \mathcal{P}(^M G, H)$. Do there exist Banach spaces E and F, which are vector subspaces of $\mathcal{M}_X$, such that φ_M is a well defined and continuous mapping $E \to F$, and continuous linear operators $A : G \to E$ and $B : F \to H$, for which the equality*

$$P(f) = B((Af)^M) \tag{7}$$

holds for all $f \in G$?

Definition 2. If the answer to the preceding problem is positive for some P, we say that P has a *canonical factorization with E, F, A, B*. The operators A and B are called the *inner* and *outer* operators, respectively. Let P, G and H be as above and assume that there exists a Banach space H_1, which is a vector subspace of $\mathcal{M}_X$ containing $H \subset \mathcal{M}_X$, such that the embedding $H \hookrightarrow H_1$ is continuous and such that P has a canonical factorization as an element of $\mathcal{P}(^M G, H_1)$. Then we say that $P \in \mathcal{P}(^M G, H)$ has a *canonical factorization with a weaker norm*.

The motivation of the latter definition are the results of [T] and Theorem 7 below.

Recall that by the polarization formula, [D], Theorem 1.5, we have

$$P(f_1, \ldots, f_M) = \frac{1}{2^M M!} \sum_{\substack{\varepsilon_i = \pm 1 \\ 1 \leq i \leq M}} \varepsilon_1, \ldots, \varepsilon_M (P \circ \Delta_M)(\sum_{j=1}^{M} \varepsilon_j f_j) \tag{8}$$

for all symmetric M–linear mappings P between two Banach spaces and for all f_i. So, assume that $P \in L(^M G, H)$ is symmetric and that (7) holds for $P \circ \Delta_M$. Then we have for all $f_1, \ldots, f_M \in G$

$$\begin{aligned} P(f_1, \ldots, f_M) &= \frac{1}{2^M M!} \sum_{\substack{\varepsilon_i = \pm 1 \\ 1 \le i \le M}} \varepsilon_1 \ldots \varepsilon_M (P \circ \Delta_M)(\sum_{j=1}^{M} \varepsilon_j f_j) \\ &= \frac{1}{2^M M!} \sum \varepsilon_1 \ldots \varepsilon_M B((A(\sum_j \varepsilon_j f_j))^M) \\ &= B(\frac{1}{2^M M!} \sum \varepsilon_1 \ldots \varepsilon_M (\sum_j \varepsilon_j A f_j)^M) \\ &= B(\prod_{i=1}^{M} (A f_i)), \end{aligned} \tag{9}$$

since ψ_M is symmetric. Hence, given a symmetric $P \in L(^M G, H)$, the existence of the factorization (7) for $P \circ \Delta_M$ is equivalent to the existence of the spaces E, F and linear mappings A, B as above such that the equality

$$P(f_1, \ldots, f_M) = B(\prod_{i=1}^{M} A f_i) \tag{10}$$

holds for all $f_1, \ldots, f_M \in G$.

Remark 3. *A canonical factorization need not be unique.* For example, the operator $P(f) = (f')^2$ has a natural factorization with $Af = f'$ and $B = \mathrm{id}$, which can be defined between many different spaces G, H, E and F. However, by the general Theorem 3.1. of [T], P has also a completely different canonical factorization with abstract, non–trivial A and B between some complex Banach function spaces; the natural factorization works also between these spaces.

3. SPECIAL CANONICAL FACTORIZATIONS.

It is elementary to see that the existence of a "good" canonical factorization, e.g., one with surjective multiplication and inner operators, leads to strong conclusions concerning the solutions of equations connected with the given polynomial operator. For the sake of completeness we mention some of these rather trivial facts here.

Remark 4. *Let $P \in \mathcal{P}(^M G, H)$ have such a canonical factorization with E, F, A, B that A and φ_M are surjections.*

a) *If f_i, $i \in I$, with I finite, are solutions of $P(f) = 0$, then all the elements of*

$$A^{-1} \varphi_M^{-1} \{ \sum_{i \in I} c_i (A f_i)^M \mid c_i \in \mathbb{K} \} \tag{10'}$$

are also solutions.

b) *If* $(f_i)_{i\in I}$ *is as in* a) *and* f_0 *is a solution of* $P(f) = g$, $g \in H$, *then also all the elements of*

$$A^{-1}\varphi_M^{-1}\{\sum_{i\in I} c_i(Af_i)^M + (Af_0^M)|c_i \in \mathbb{K}\} \tag{11}$$

are solutions.

Indeed, in the case a) we can calculate

$$B\varphi_M(AA^{-1}\varphi_M^{-1}(\sum_i c_i(Af_i)^M))$$

$$= \sum_i c_i B((Af_i)^M) = 0.$$

The other case is proved similarly.

If the outer operator is an injection, it is easy to see that ker (P) is a linear space: indeed, it is equal to ker (A). Hence, it is very easy to find examples of multilinear operators for which we cannot get a canonical factorization, for example, with a homeomorphism B . We also have

Counterexample 5. *Let* $\mathbb{K} = \mathbb{R}$, $G = C^1(0,1)$, $H = C(0,1)$ *and* $P(f) = (f - f')^2(f + f')$. *Then there does not exist a canonical factorization with* E, F, A, B *where* A *and* φ_3 *are homeomorphisms.*

Proof. It is an elementary fact that

$$G_0 = \{c_1 e^x, c_2 e^{-x}, c_i \in \mathbb{R}\} \subset C^1(0,1) \tag{12}$$

is the solution space of

$$P(f) = 0. \tag{13}$$

Suppose that P has a canonical factorization as above with A and φ_3 homeomorphisms. Then

$$A^{-1}\varphi_3^{-1}(c_1(Ae^x)^3 + c_2(Ae^{-x})^3)$$

is a solution of (13) for all $c_i \in \mathbb{R}$. Let us denote by G_1 the space of all functions of the above form. Since ce^x and ce^{-x} belong to G_1 for all $c \in \mathbb{R}$, we have $G_0 = G_1$.

If $(Ae^x)^3$ and $(Ae^{-x})^3$ linearly dependent, the space G_1 is homeomorphic to $\mathbb{R}$, which leads to a contradiction since G_0 is homeomorphic to the topological space $G = \{(x,y) \in \mathbb{R}^2 \mid x = 0 \quad \text{or} \quad y = 0\}$. ($G$ is not homeomorphic to $\mathbb{R}$ since one can divide it in four disconnected parts by removing a single point, the origin.)

If $(Ae^x)^3$ and $(Ae^{-x})^3$ are linearly independent, G_1 is homeomorphic to $\mathbb{R}^2$, and this is clearly a contradiction since the topological dimension of G_0 is 1. □

4. PRELIMINARIES OF THE MAIN RESULT.

In [T] we proved the existence of a canonical factorization (with a weaker norm, if we consider certain natural norms in image and domain) for all multilinear complex differential operators with constant coefficients. In this paper we generalize this result to all bilinear complex partial differential operators with polynomial coefficients. The idea of the proof is similar to that of [T], but because of the many technical details it is necessary to outline the whole proof. Moreover, it is not a priori clear what kind of norm estimates one can get using the method of [T].

We begin with some notations. By $M \in \mathbb{N}$ we denote the number of variables and by D^M the open unit polydisc of $\mathbb{C}^M$. We use standard multi-index notation: if $z = (z_1, \ldots, z_M) \in \mathbb{C}^M$ and $n = (n_1, \ldots, n_M) \in \mathbb{N}^M$, we denote $z^n = z_1^{n_1} \ldots z_M^{n_M}$ and

$$\frac{\partial^n}{\partial z^n} = \left(\frac{\partial^{n_1}}{\partial z^{n_1}}\right) \cdots \left(\frac{\partial^{n_M}}{\partial z^{n_M}}\right).$$

If $x, y \in \mathbb{R}^M$, then $x \leq y$ means $x_i \leq y_i$ for all i, $1 \leq i \leq M$. We also denote $|x|_\infty = \max_{1 \leq i \leq M} \{|x_i|\}$ for $x = (x_1, \ldots, x_M) \in \mathbb{R}^M$. We write $H^p(D^M)$ for the Hardy space of analytic functions $D^M \to \mathbb{C}$; the norm of $H^p(D^M)$ is defined by

$$\|f\|_p := \sup_{0<r<1} ((2\pi)^{-M} \int_Q |f(re^{i\theta_1}, \ldots, re^{i\theta_M})|^p d\theta)^{1/p}, \tag{14}$$

where $Q = [-\pi, \pi]^M$. We remark that, as in the case of one variable, $\lim_{r \to 1-} f(re^{i\theta_1}, \ldots, re^{i\theta_M}) =: f^*$ exists a.e., if $f \in H^1(D^M)$ and that

$$\|f\|_p^p = (2\pi)^{-M} \int_Q |f^*(e^{i\theta_1}, \ldots, e^{i\theta_M})|^p d\theta_1 \ldots d\theta_M. \tag{15}$$

See [R], Section 3.4, for this and more details on Hardy spaces of M variables. We also define

$$\mathcal{A}_N^p(D^M) := \{f : D^M \to \mathbb{C} \mid f(z_1, \ldots, z_M) = \sum_{m \in \mathbb{N}^M} f_m z^m,$$

$$\|f\|_p^{(N)} := (\sum_{m \in \mathbb{N}^M} (\prod_{i=1}^M (m_i + 1)^{pN_i})|f_m|^p)^{1/p} < \infty\}, \tag{16}$$

where $N = (N_1, \ldots, N_M) \in \mathbb{N}^M$, $m = (m_1, \ldots, m_M)$ and $f_m \in \mathbb{C}$ for all m. The space $\mathcal{A}_N^\infty(D^M)$ is defined in the natural way. We have the norm isomorphism $\mathcal{A}_0^2(D^M) = H^2(D^M)$. We denote by e_m, $m \in \mathbb{N}^M$, the function $z_1^{m_1} \ldots z_M^{m_M} \in \cap_{p \geq 1} H^p(D^M)$, and by e_m^* the functional $\langle f, e_m^* \rangle = e_m$ where $f = \sum_{m \in \mathbb{N}^M} f_m z_1^{m_1} \ldots z_M^{m_M}$ is analytic in D^M. Clearly, $(e_m)_{m \in \mathbb{N}^M}$ (resp. $(e_m^*)_{m \in \mathbb{N}^M}$) is a basis in every $\mathcal{A}_N^p(D^M)$ (resp. $(\mathcal{A}_N^p(D^M))'$).

As a technical tool we need some bijection $\eta : \mathbb{N}^M \to \mathbb{N}$. We choose η such that for all $k \in \mathbb{N}$

$$|\eta^{-1}(k)|_\infty^M \leq k, \; (|\eta^{-1}(k)|_\infty + 1)^M \geq k, \tag{17}$$

where $|m|_\infty = \max_{1\le i\le M} |m_i|$ for $m \in \mathbb{N}^M$. This can be satisfied, since for all $k \in \mathbb{N}$ the set $\mathbb{N}^{M,k} = \{m \in \mathbb{N}^M \mid |m|_\infty \le k\}$ contains $(k+1)^M$ elements: we define η such that for all k we have $\eta^{-1}(t) \in \mathbb{N}^{M,k}$ for all $t, k^M < t \le (k+1)^M$.

We remark that

$$\mathcal{A}_0^\infty(D^M) \subset \mathcal{A}^2_{(-1-\varepsilon)M\bar{e}}(D^M), \tag{18}$$

where $\bar{e}$ is as in the beginning of Section 2, and that the corresponding embedding is continuous. To see this, let $f = \sum_m f_m z^m \in \mathcal{A}_0^\infty(D^M)$. Then we get, using (17),

$$\|f\|_2^{((-1-\varepsilon)M\bar{e})} \le \sum_{m\in\mathbb{N}^M} \Big(\prod_{i=1}^{M}(m_i+1)^{M(-1-\varepsilon)}\Big)|f_m| \le \sum_{m\in\mathbb{N}^M} (|m|_\infty+1)^{M(-1-\varepsilon)}|f_m|$$

$$= \sum_{k\in\mathbb{N}} (|\eta^{-1}(k)|_\infty + 1)^{M(-1-\varepsilon)}|f_{\eta^{-1}(k)}| \le \Big(\sum_{k\in\mathbb{N}}^{\infty} (|\eta^{-1}(k)|_\infty + 1)^{M(-1-\varepsilon)}\Big) \sup_{k\in\mathbb{N}} |f_{\eta^{-1}(k)}|$$

$$\le C\Big(\sum_{k\in\mathbb{N}} (k+1)^{-1-\varepsilon}\Big) \sup_{k\in\mathbb{N}} |f_{\eta^{-1}(k)}| = C' \sup_{m\in\mathbb{N}^M} |f_m| = C'\|f\|_\infty^{(0)} \tag{19}$$

for some positive constants C, C'.

If E and F are Banach spaces, we denote by

$$(E' \otimes E') \otimes_\beta F \quad \text{and} \quad (E' \otimes E')\hat{\otimes}_\beta F$$

the tensor product space and its closure as a subspace of $L(^2E, F)$, the space of continuous bilinear mappings $E \times E \to F$ endowed with the strong operator topology (for definition and more details, see [T]). Let ψ_2 be as in Section 2. We show that

$$\psi_2 = \lim_{m\to\infty} \psi_2^{(m)} \tag{20}$$

in $((H^2(D^M))' \otimes (H^2(D^M))')\hat{\otimes}_\beta H^1(D^M)$, where "$m \to \infty$" means that each $m_i \to \infty$ and where

$$\psi_2^{(m)} = \frac{1}{(m_1+1)\dots(m_M+1)} \sum_{0\le m'\le m} \sum_{0\le k\le m'} \Big(\sum_{0\le t\le k} e^*_{k-t} \otimes e^*_t\Big) \otimes e_k \tag{21}$$

with $m = (m_1, \dots, m_M), m', k, t \in \mathbb{N}^M$. Namely, if $f, g \in H^2(D^M)$ with $f = \sum_m f_m e_m$, $g = \sum_m g_m e_m$, then (21) implies immediately that $\psi_2^{(m)}(f, g)$ is equal to the first arithmetic mean of $fg \in H^1(D^M)$. By [Z], Theorem XVII.1.23 and (15) above we get (20).

We consider now bilinear complex partial differential operators with polynomial coefficients,

$$P(f,g) = \sum_{n,\hat{n}} a_{n,\hat{n}}(z)\Big(\frac{\partial^n f}{\partial z^n}\Big)\Big(\frac{\partial^{\hat{n}} g}{\partial z^{\hat{n}}}\Big) \tag{22}$$

where $n, \hat{n} \in \mathbb{N}^M$, $z = (z_1, \ldots, z_M) \in \mathbb{C}^M$ and $a_{n,\hat{n}}$ is a polynomial of M complex variables. We write

$$a_{n,\hat{n}} = \sum_p a_{n,\hat{n},p} z^p$$

where $a_{n,\hat{n},p} \in \mathbb{C}$ is for all $n, \hat{n}$ different from 0 only for a finite number of multi-indices $p \in \mathbb{N}^M$.

It is elementary to see that $P \in L({}^2\mathcal{A}^2_N(D^M), H^1(D^M))$ if $N = (N_1, \ldots, N_M) \in \mathbb{N}^M$ where N_i is the highest order of the i:th partial derivative occurring in (22); note that multiplication by z^m is for all $m \in \mathbb{N}^M$ a bounded operator in $H^1(D^M)$. The Hausdorff–Young inequality works also in the case of M variables so that we have a continuous canonical inclusion $H^1(D^M) \to \mathcal{A}^\infty_0(D^M)$. Since the identity mapping $\mathcal{A}^\infty_0(D^M) \to \mathcal{A}^2_{(-1-\varepsilon)M\bar{e}}(D^M)$ is continuous for all $\varepsilon > 0$ by (18), we see that $P \in L({}^2\mathcal{A}^2_N(D^M), \mathcal{A}^2_{(-1-\varepsilon)M\bar{e}}(D^M))$. We show that P has even a tensor product representation. So let $f = \sum_m f_m e_m$ and $g = \sum_m g_m e_m$ belong to $\mathcal{A}^2_N$. Then we have

$$\begin{aligned}
P(f,g) &= \sum_{n,\hat{n}} a_{n,\hat{n}}(z) \left(\frac{\partial^n f}{\partial z^n}\right)\left(\frac{\partial^{\hat{n}} g}{\partial z^{\hat{n}}}\right) \\
&= \sum_{n,\hat{n}} (\sum_p a_{n,\hat{n},p} e_p)(\sum_{m \in \mathbb{N}^M} \alpha_{m,n} f_{m+n} e_m)(\sum_{m \in \mathbb{N}^M} \alpha_{m,\hat{n}} g_{m+\hat{n}} e_m) \\
&= \sum_{m \in \mathbb{N}^M} (\sum_{n,\hat{n}} \sum_{0 \le k \le m} \sum_{0 \le p \le k} a_{n,\hat{n},p} \alpha_{m-k,n} \alpha_{k-p,\hat{n}} f_{m-k+n} g_{k-p+\hat{n}}) e_m, \qquad (23)
\end{aligned}$$

where $\alpha_{m,n} \in \mathbb{R}$ is such that $\dfrac{\partial^n e_{m+n}}{\partial z^n} = \alpha_{m,n} e_m$. Note that

$$\begin{aligned}
0 < \alpha_{m,n} &= \prod_{i=1}^M (m_i + n_i)(m_i + n_i - 1) \ldots (m_i + 1) \\
&\le \prod_{i=1}^M (m_i + n_i)^{n_i}. \qquad (24)
\end{aligned}$$

The expression (23) is the Taylor series of the function $P(f,g) \in \mathcal{A}^2_{(-1-\varepsilon)M\bar{e}}(D^M)$; hence, (23) converges for all $f, g \in \mathcal{A}^2_N(D^M)$. This implies that the tensor product representation

$$P = \sum_{m \in \mathbb{N}^M} (\sum_{n,\hat{n}} \sum_{0 \le k \le m} \sum_{0 \le p \le k} a_{n,\hat{n},p} \alpha_{m-k,n} \alpha_{k-p,\hat{n}} e^*_{m-k+n} \otimes e^*_{k-p+\hat{n}}) \otimes e_m \qquad (25)$$

converges in $((\mathcal{A}^2_N(D^M))' \otimes (\mathcal{A}^2_N(D^M))')\hat{\otimes}_\beta \mathcal{A}^2_{(-1-\varepsilon)M\bar{e}}(D^M)$.

5. MAIN RESULT.

Using the notation and preliminary facts presented in the previous section we prove

Theorem 7. *Let P be the symmetric bilinear complex partial differential operator with polynomial coefficients*

$$P(f,g)=\sum_{n,\hat{n}} a_{n,\hat{n}}(z)\Big(\frac{\partial^n f}{\partial z^n}\Big)\Big(\frac{\partial^{\hat{n}} g}{\partial z^{\hat{n}}}\Big), \tag{26}$$

where the notation is as in (22), $f,g\in\mathcal{A}^2_N(D^M)$ and $N=(N_1,\dots,N_M)\in\mathbb{N}^M$ is such that each N_i, $1\le i\le M$, is the highest order of i:th partial derivative occurring in (26). For all $\varepsilon>0$ there exist bounded linear operators $A\in L(\mathcal{A}^2_N(D^M),H^2(D^M))$ and $B\in L(H^1(D^M),\mathcal{A}^2_{(-1-\varepsilon)M\bar{e}}(D^M))$ such that for all $f,g\in\mathcal{A}^2_N(D^M)$

$$P(f,g)=B((Af)(Ag)). \tag{27}$$

Since P is symmetric, we can assume $a_{n,\hat{n},p}=a_{\hat{n},n,p}$ for all $n,\hat{n}$ and p in (26).

Proof. In view of Lemma 2.2 of [T] it is enough to construct bounded linear operators $A'\in L((H^2(D^M))',(\mathcal{A}^2_N(D^M))')$ and B as above such that

$$(A'\otimes A'\otimes B)(\psi_2)=P \tag{28}$$

holds in $((\mathcal{A}^2_N(D^M))'\otimes(\mathcal{A}^2_N(D^M))')\hat{\otimes}_\beta\mathcal{A}^2_{(-1-\varepsilon)M\bar{e}}(D^M)$; see (20), (25) and [T], Section 2, for the notation and definitions connected with (28).

Let us denote by K the number of terms in (26). We define B by

$$Be_m=\begin{cases}0, & \text{if } m\neq(K'3^{\eta(k)+2},0,\dots,0) \text{ for all } k\in\mathbb{N}^M\\ (\eta(k)+1)^{1+\varepsilon}e_k, & \text{if } m=(K'3^{\eta(k)+2},0,\dots,0),\end{cases} \tag{29}$$

where η is as in (17) and $K'>0$ is the number

$$K'=K(2M+1)^M(\rho+1)^M,$$
$$\rho=\max_{n,\hat{n}}\{|p|_\infty \mid a_{n,\hat{n},p}\neq 0\} \tag{30}$$

and extend B linearly to the linear hull of $(e_m)_{m\in\mathbb{N}^M}\subset H^1(D^M)$. Note that B can be factorized as $B=B_2B_1B_0$, where

$$B_0(e_m)=\begin{cases}0, & \text{if } m_i\neq 0 \text{ for some } 2\le i\le m\\ e_m, & \text{if } m=(m_1,0,\dots,0),\end{cases}$$
$$B_1(e_{(m_1,0,\dots,0)})=\begin{cases}0, & \text{if } m_1\neq K'3^{k+2} \text{ for all } k\in\mathbb{N}\\ e_{(k,0,\dots,0)}, & \text{if } m_1=K'3^{k+2},\end{cases} \tag{31}$$

and

$$B_2 e_{(k,0,\ldots,0)} = (k+1)^{1+\varepsilon} e_{\eta^{-1}(k)}$$

for all $k \in \mathbb{N}$; the mappings B_1 and B_2 are defined only on the span of $(e_{(k,0,\ldots,0)})_{k\in\mathbb{N}}$.

We first show that $B_0 : H^1(D^M) \to H^1(D^M)$ is continuous. If $f = \sum_{m\in\mathbb{N}^M} f_m e_m$ with $f_m \neq 0$ only for a finite number of m, we have

$$\int_{-\pi}^{\pi}\cdots\int_{-\pi}^{\pi} f(z_1, e^{i\theta_2}, \ldots, e^{i\theta_M})d\theta_2 \ldots d\theta_M$$

$$= \sum_{m\in\mathbb{N}^M} f_m z_1^{m_1}\Big(\int_{-\pi}^{\pi} e^{im_2\theta_2}d\theta_2\Big)\ldots\Big(\int_{-\pi}^{\pi} e^{im_M\theta_M}d\theta_M\Big)$$

$$= (2\pi)^{M-1}\sum_{m\in\mathbb{N}} f_{(m,0,\ldots,0)} z_1^m.$$

Hence, B_0 has the integral representation

$$(B_0 f)(z) = (2\pi)^{1-M}\int_{-\pi}^{\pi}\cdots\int_{-\pi}^{\pi} f(z_1, e^{i\theta_2}\ldots, e^{i\theta_M})d\theta_2\ldots d\theta_M.$$

Since $B_0 f$ is constant with respect to $z_2, \ldots, z_M$,

$$\|B_0 f\|_1 = (2\pi)^{-M}\int_Q |(B_0 f)(e^{i\theta_1}, \ldots, e^{i\theta_M})|d\theta_1\ldots d\theta_M$$

$$= \frac{1}{2\pi}\int_{-\pi}^{\pi} |B_0 f(e^{i\theta_1}, 0, \ldots, 0)|d\theta_1$$

$$= (2\pi)^{-M}\int_{-\pi}^{\pi}\Big|\int_{-\pi}^{\pi}\cdots\int_{-\pi}^{\pi} f(e^{i\theta_1}, e^{i\theta_2}, \ldots, e^{i\theta_M})d\theta_2\ldots d\theta_M\Big|d\theta_1$$

$$\leq \|f\|_1. \tag{32}$$

We see in fact that B_0 is a continuous projection onto the closed linear subspace of $H^1(D^M)$ spanned by $(e_{(k,0,0)})_{k=0}^{\infty}$. On the other hand this subspace is canonically isometric to $H^1(D)$ and hence we can extend B_1 to $H^1(D^M)$ and prove the continuity of $B_1 : H^1(D^M) \to H^2(D^M)$ using the method of the proof of Theorem 3.1 in [T]. (See (3.5), (3.6) and (3.7)). The mapping B_2 can be continuously extended to $H^2(D^M)$ since $(e_m)_{m\in\mathbb{N}^M}$ is an orthonormal basis and since

$$\|B_2(\sum_{k\in\mathbb{N}} f_k e_{(k,0,\ldots,0)})\|_2^{(-1-\varepsilon)M\bar{e})^2} = \sum_{k\in\mathbb{N}}\Big(\prod_{i=1}^{M}(\eta^{-1}(k)+1)_i^{2M(-1-\varepsilon)}\Big)(k+1)^{2(1+\varepsilon)}|f_k|^2$$

$$\leq \sum_{k\in\mathbb{N}} (|\eta^{-1}(k)|_\infty + 1)^{2M(-1-\varepsilon)}(k+1)^{2(1+\varepsilon)}|f_k|^2$$

$$\leq C\sum_{k\in\mathbb{N}} (k+1)^{2(-1-\varepsilon)}(k+1)^{2(1+\varepsilon)}|f_k|^2 \leq C\|f\|_2^2$$

for some positive constant C, see (17).

The above reasoning shows that we can extend B to a continuous linear mapping $H^1(D^M) \to \mathcal{A}^2_{(-1-\varepsilon)M\bar{e}}(D^M)$.

We define the mapping A' inductively. First of all we set $A'e^*_m = 0$ for all such $m \in \mathbb{N}^M$ that $m_i \neq 0$ for some $2 \leq i \leq M$. We also define $A'e^*_{(m,0,\dots,0)} = 0$ if $0 \leq m \leq 3K'$. Next we assume that $k \in \mathbb{N}^M$ and that $A'e^*_{(m,0,\dots,0)}$ is defined for all m, $0 \leq m \leq K'3^{\eta(k)+1}$; we proceed to define $A'e^*_{(m,0,\dots,0)}$ for $K'3^{\eta(k)+1} < m \leq K'3^{\eta(k)+2}$.

We set

$$A'(e^*_{(m,0,\dots,0)}) = 0 \tag{33}$$

if $K'3^{\eta(k)+2} - K'3^{\eta(k)+1} \leq m \leq K'3^{\eta(k)+2}$. We denote

$$P_k := \sum_{n,\hat{n}} \sum_{0\leq t\leq k} \sum_{0\leq p\leq t} a_{n,\hat{n},p}\alpha_{k-t,n}\alpha_{t-p,\hat{n}} e^*_{k-t+n} \otimes e^*_{t-p+\hat{n}}. \tag{34}$$

Then P_k has at most $K(|k|_\infty + 1)^M(\rho+1)^M$ terms. In view of the definition of B and (25) and (28) we need to define A' such that

$$(A' \otimes A')\Big(\sum_{0\leq m\leq K'3^{\eta(k)+2}} e^*_{(K'3^{\eta(k)+2}-m,0,\dots,0)} \otimes e^*_{(m,0,\dots,0)}\Big) = (\eta(k)+1)^{-1-\varepsilon}P_k. \tag{36}$$

To this end, note that the interval

$$]K'3^{\eta(k)+2}/2, K'3^{\eta(k)+2} - K'3^{\eta(k)+1}[\, \cap \mathbb{N}$$

contains more than $K'3^{\eta(k)} \geq K(|k|_\infty + 1)^M(\rho+1)^M$ numbers (see (17) and (30) for this inequality). Hence, for each term

$$a_{n,\hat{n},p}\alpha_{k-t,n}\alpha_{t-p,\hat{n}} e^*_{k-t+n} \otimes e^*_{t-p+\hat{n}}$$

of (34) we can choose a multi-index

$$\mu(k,t,p,n,\hat{n}) = (\mu(k,t,p,n,\hat{n})_1, 0, \dots, 0)$$

such that $\mu(k,t,p,n,\hat{n})_1 \in\,]K'3^{\eta(k)+2}/2,\ K'3^{\eta(k)+2} - K'3^{\eta(k)+1}[$ and such that $\mu(k,t,p,n,\hat{n}) \neq \mu(k,t_1,p_1,n_1,\hat{n}_1)$ if $t \neq t_1$ or $p \neq p_1$ or $n \neq n_1$ or $\hat{n} \neq \hat{n}_1$. We define

$$A'(e^*_{\mu(k,t,p,n,\hat{n})}) = \frac{1}{2}(\eta(k)+1)^{-\frac{1}{2}-\frac{\varepsilon}{2}} a_{n,\hat{n},p}\alpha_{k-t,n} e^*_{k-t+n},$$

$$A'e^*_{(K'3^{\eta(k)+2}-\mu(k,t,p,n,\hat{n})_1,0,\ldots,0)} = (\eta(k)+1)^{-\frac{1}{2}-\frac{\varepsilon}{2}}\alpha_{t-p,\hat{n}}e^*_{t-p+\hat{n}}. \tag{37}$$

Finally, we set $A'e^*_{(m,0,\ldots,0)} = 0$ in the cases that have not been considered above, if $K'3^{\eta(k)+1} < m \leq K'3^{\eta(k)+2}$. As k runs through $\mathbb{N}^M$, each $A'e^*_m$ becomes defined. We extend A' linearly to sp $(e^*_m \mid m \in \mathbb{N}^M)$.

The fact that (36) holds follows from (37) (as in [T], (3.11)).

We show that A' is bounded on sp $(e^*_m) \subset (H^2(D^M))'$. We denote, for all $t \in \mathbb{N}^M$

$$\mathbb{N}^M_t := \{m \in \mathbb{N}^M \mid A'e^*_m = \beta_m e^*_t \quad \text{for some} \quad \beta_m \neq 0\}.$$

For all t and for all $m \in \mathbb{N}^M_t$ we denote by β_m the number satisfying $A'e^*_m = \beta_m e^*_t$. By (24) and (37) we have for all $t, k \in \mathbb{N}^M$ the estimate

$$\beta_m \leq C(\eta(k)+1)^{-\frac{1}{2}-\frac{\varepsilon}{2}}\alpha_{t,N} \leq C(\eta(k)+1)^{-\frac{1}{2}-\frac{\varepsilon}{2}}\prod_{i=1}^{M}(t_i+N_i)^{N_i} \tag{39}$$

for all $m \in \mathbb{N}^M_{t,k} := \mathbb{N}^M_t \cap \{(p,0,\ldots,0) \mid K'3^{\eta(k)+1} < p \leq K'3^{\eta(k)+2}\}$; here $C = \max\{1, |a_{n,\hat{n},p}|\}$ and N as in (26). We identify $(\mathcal{A}^2_N(D^M))'$ with $\mathcal{A}_{-N}(D^M)$ and $(H^2(D^M))'$ with $H^2(D^M)$ in the canonical way ($e^*_m \sim e_m$) and calculate for $\sum_m x_m e^*_m \in (H^2(D^M))'$ (a finite number of terms):

$$\|A'\sum_m x_m e^*_m\|_2^{(-N)} = \|\sum_{t\in\mathbb{N}^M}\sum_{m\in\mathbb{N}^M_t} x_m A'e^*_m\|_2^{(-N)}$$

$$= (\sum_{t\in\mathbb{N}^M}(\|\sum_{m\in\mathbb{N}^M_t} x_m\beta_m e^*_t\|_2^{(-N)})^2)^{1/2}$$

$$= (\sum_{t\in\mathbb{N}^M}\prod_{i=1}^{M}(t_i+1)^{-2N_i}|\sum_{m\in\mathbb{N}^M_t} x_m\beta_m|^2)^{1/2}$$

$$\leq (\sum_{t\in\mathbb{N}^M}\prod_{i=1}^{M}(t_i+1)^{-2N_i}(\sum_{m\in\mathbb{N}^M_t}|x_m|^2)(\sum_{k\in\mathbb{N}^M}\sum_{m\in\mathbb{N}^M_{t,k}}|\beta_m|^2))^{1/2}. \tag{40}$$

Now we need the fact that card $(\mathbb{N}^M_{t,k}) \leq 2K\rho^{M+1}$ where ρ is as in (30). (See the definition of A'; for each k and $t \in \mathbb{N}^M$ the element e^*_t occurs in P_k, (34), at most $2K\rho^{M+1}$ times, which implies the statement by the definition of $\mathbb{N}^M_{t,k}$.) This and (39) imply, for a constant C',

$$\sum_{k\in\mathbb{N}^M}\sum_{m\in\mathbb{N}^M_{t,k}}|\beta_m|^2 = \sum_{k\in\mathbb{N}^M}2K\rho^{M+1}C^2(\eta(k)+1)^{-1-\varepsilon}\prod_{i=1}^{M}(t_i+N_i)^{2N_i} \leq C'\prod_{i=1}^{M}(t_i+N_i)^{2N_i}. \tag{41}$$

Combining this with (40) yields

$$\|A'\sum_m x_m e_m^*\|_2^{(-N)} \le C_0 \sum_{t\in\mathbb{N}^M} \sum_{m\in\mathbb{N}_t^M} |x_m|^2$$

$$\le C_0\|\sum_m x_m e_m^*\|_2 \tag{42}$$

for a constant C_0. We see that A' can be extended to a bounded linear operator $(H^2(D^M))' \to (\mathcal{A}_N^2(D^M))'$.

Since $A'\otimes A'\otimes B$ is a continuous operator $((H^2(D^M))'\otimes(H^2(D^M))')\hat{\otimes}_\beta H^1(D^M) \to ((\mathcal{A}_N^2(D^M))'\otimes(\mathcal{A}_N^2(D^M))')\hat{\otimes}_\beta \mathcal{A}^2_{(-1-\varepsilon)M\bar{e}}(D^M)$ (see Lemma 2.1 of [T]), (28) is proved as soon as we show

$$\lim_{m\to\infty}(A'\otimes A'\otimes B)\psi_2^{(m)} = P; \tag{43}$$

see (20). First note that for all $m\in\mathbb{N}^M$

$$(A'\otimes A'\otimes B)((\sum_{0\le t\le m} e_{m-t}^*\otimes e_t^*)\otimes e_m)$$

$$= \begin{cases} 0, & \text{if } m\neq (K'3^{\eta(k)+2},0,\ldots,0) \text{ for all } k\in\mathbb{N}^M \\ P_k\otimes e_k, & \text{if } m = (K'3^{\eta(k)+2},0,\ldots,0). \end{cases} \tag{44}$$

see (29) and (36). Using (18) we get

$$(A'\otimes A'\otimes B)\psi_2^{(m)}$$

$$= \frac{1}{(m_1+1)\ldots(m_M+1)} \sum_{0\le m'\le m}\sum_{0\le m''\le m'} (A'\otimes A'\otimes B)$$

$$((\sum_{0\le t\le m''} e_{m''-t}^*\otimes e_t^*)\otimes e_{m''}). \tag{45}$$

In view of (44) this is of the form

$$\sum_{k\in\mathbb{N}^M}(m_1+1)^{-1}\ldots(m_M+1)^{-1}r(k,m)P_k\otimes e_k, \tag{46}$$

where $r(k,m)\in\mathbb{N}$ and for all $m\in\mathbb{N}^M$ we have $r(k,m)\neq 0$ only for a finite number of k, say $|k|_\infty \le k(m)$, where $k(m)\in\mathbb{N}$, $\lim_{m\to\infty}k(m)=\infty$. To estimate $r(k,m)$ note that $P_k\otimes e_k$ occurs in

$$\sum_{0\le m''\le m'}(A'\otimes A'\otimes B)((\sum_{0\le t\le m''} e_{m''-t}^*\otimes e_t^*)\otimes e_{m''}) \tag{47}$$

once, if $m_1' \geq K'3^{\eta(k)+2}$; otherwise it does not occur in (47). Hence,$P_k \otimes e_k$ occurs in (45) $(m_1 - K'3^{\eta(k)+2} + 1)\prod_{i=2}^{M}(m_i + 1)$ times and we deduce for each fixed k

$$\lim_{m \to \infty} (m_1 + 1)^{-1} \dots (m_M + 1)^{-1} r(k, m)$$

$$= \lim_{m \to \infty} (m_1 - K'3^{\eta(k)+2} + 1)(m_1 + 1)^{-1} = 1. \tag{48}$$

This implies for all $f, g \in \mathcal{A}_N^2(D^M)$

$$\|P(f,g) - ((A' \otimes A' \otimes B)\psi_2^{(m)})(f,g)\|_2^{((-1-\varepsilon)M\bar{e})}$$

$$= \|P(f,g) - \sum_{|k|_\infty \leq k(m)} (\prod_{i=1}^{M}(m_i + 1)^{-1}) r(k,m) P_k(f,g) e_k\|_2^{((-1-\varepsilon)M\bar{e})}$$

$$= (\sum_{|k|_\infty \leq k(m)} |\prod_{i=1}^{M}(k_i + 1)^{M(-1-\varepsilon)}(1 - r(k,m)\prod_{i=1}^{M}(m_i + 1)^{-1}) P_k(f,g)|^2)^{1/2} \tag{49}$$

$$+ (\sum_{|k|_\infty > k(m)} |\prod_{i=1}^{M}(k_i + 1)^{M(-1-\varepsilon)} / P_k(f,g)|^2)^{1/2}$$

It is easy to see that (49) tends to 0 as $m \to \infty$: if $\varepsilon > 0$, we can choose a finite set $\mathbb{N}_0$ of indices $k \in \mathbb{N}^M$ such that

$$(\sum_{k \in \mathbb{N}^M - \mathbb{N}_0} |\prod_{i=1}^{M}(k_i + 1)^{M(-1-\varepsilon)} P_k(f,g)|^2)^{1/2} < \varepsilon$$

and $m_0 \in \mathbb{N}^M$ such that $k(m) > \max\{|k|_\infty \mid k \in \mathbb{N}_0\}$ for $m > m_0$ and then use (48). □

The remarks of Section 4 of [T] hold also in the present case.

REFERENCES

[D] Dineen, S: Complex analysis in locally convex spaces. North–Holland Mathematics Studies 57, Amsterdam–New York–Oxford, 1981.
[R] Rudin, W: Function theory in polydiscs. Benjamin, New York–Amsterdam, 1969.
[T] Taskinen, J: Factorization of multilinear differential operators with constant coefficients. To appear in J. Math. Anal. Appl.
[Z] Zygmund, A: Trigonometric series, Vol I–II. Second Edition. Cambridge University Press, 1959.

Progress in Functional Analysis
K.D. Bierstedt, J. Bonet, J. Horváth & M. Maestre (Eds.)

Complex Geodesics on Convex Domains

Seán Dineen[a] and Richard M. Timoney[b]

Dedicated to M. Valdivia on the occasion of his 60th birthday

[a]Department of Mathematics, University College Dublin, Belfield, Dublin 4, Ireland

[b]School of Mathematics, 39 Trinity College, Dublin 2, Ireland

Abstract

Existence and uniqueness of complex geodesics joining two points of a convex bounded domain in a Banach space X are considered. Existence is proved for the unit ball of X under the assumption that X is 1-complemented in its double dual. Another existence result for taut domains is also proved. Uniqueness is proved for strictly convex bounded domains in spaces with the analytic Radon-Nikodym property. If the unit ball of X has a modulus of complex uniform convexity with power type decay at 0, then all complex geodesics in the unit ball satisfy a Lipschitz condition. The results are applied to classical Banach spaces and to give a formula describing all complex geodesics in the unit ball of the sequence spaces ℓ^p $(1 \leq p < \infty)$.

In this article, we discuss the existence, uniqueness and continuity of complex geodesics on a convex domain $\mathcal{D}$ in a complex Banach space X. The term 'complex geodesic' is due to Vesentini [33], although the concept was discussed by Carathéodory [5] and Reiffen [27] under the name 'metric plane'. Recent results on this topic are to be found in [11, 14, 15, 16, 34, 35, 36, 37]. Applications of complex geodesics to the study of biholomorphic automorphisms and to fixed point sets are to be found in [5, 33, 34, 35, 36, 37].

Our results on the existence problem depend on topological properties of the Banach space X, the results on uniqueness depend on the geometry of the boundary $\partial\mathcal{D}$ and on an analytic-geometric property of X (the analytic Radon-Nikodym property), while the continuity (*i.e.* continuous extensions to the boundary) is obtained using complex uniform convexity.

In section 1, we introduce complex geodesics and related concepts and prove some basic results. In section 2, we show that every pair of points in the unit ball B_X of X can be joined by a complex geodesic provided X is 1-complemented in its double dual X^{**}. In proving this we show that the canonical embedding of B_X in $B_{X^{**}}$ is an isometry for the Kobayashi metrics. We also obtain a new simple proof of a result of Davie and Gamelin [7] that bounded analytic functions on B_X extend to $B_{X^{**}}$. In section 3, we use the analytic Radon-Nikodym property and extreme points to obtain uniqueness results and in Section 4 we prove continuity properties of complex geodesics. In Section 5 we apply the results of the preceding three sections to ℓ^p and related spaces.

An examination of the methods of section 2 shows that topological considerations are only used to obtain 'Montel type' theorems for mappings defined on the unit disc in $\mathbb{C}$ with values in the domain $\mathcal{D} \subset X$ under consideration. This led us to consider (in an infinite-dimensional setting) the concepts of taut and complete hyperbolic domains (section 6). In section 6, we also apply our results to show that certain domains have constant negative curvature in the Kobayashi metric.

A number of the results in this paper were announced in [10]. We refer to [10] for background results and further details on complex geodesics.

1 COMPLEX GEODESICS

$\mathbb{D}$ is the open unit disc in $\mathbb{C}$ and ρ will denote the Poincaré distance on $\mathbb{D}$, i.e.

$$\rho(z, w) = \tanh^{-1}\left(\left|\frac{z - w}{1 - wz}\right|\right) \qquad (z, w \in \mathbb{D}).$$

The infinitesimal Poincaré distance α is defined by $\alpha(z, v) = |v|/(1 - |z|^2)$ (for $v \in \mathbb{C}$ and $z \in \mathbb{D}$).

For $\mathcal{D}_1$ and $\mathcal{D}_2$ domains in complex Banach spaces, $H(\mathcal{D}_1, \mathcal{D}_2)$ denotes the set of all $\mathcal{D}_2$-valued holomorphic functions on $\mathcal{D}_1$. For a domain $\mathcal{D}$ in a complex Banach space X, $p, q \in \mathcal{D}$ and $v \in V$,

$$\begin{aligned}
C_{\mathcal{D}}(p, q) &= \sup\{\rho(f(p), f(q)) : f \in H(\mathcal{D}, \mathbb{D})\} \\
c_{\mathcal{D}}(p, v) &= \sup\{|f'(p)(v)|) : f \in H(\mathcal{D}, \mathbb{D})\} \\
\delta_{\mathcal{D}}(p, q) &= \inf\{\rho(u, v) : \exists f \in H(\mathbb{D}, \mathcal{D}), f(u) = p, f(v) = q\} \\
K_{\mathcal{D}}(p, q) &= \inf\left\{\sum_{i=1}^{n} \delta_{\mathcal{D}}(w_{i-1}, w_i) : n \geq 1, p = w_0, w_1, \ldots, w_n = q \in \mathcal{D}\right\} \\
k_{\mathcal{D}}(p, v) &= \inf\{\eta > 0 : \exists f \in H(\mathbb{D}, \mathcal{D}), f(0) = p, f'(0)\eta = v\}.
\end{aligned}$$

$C_{\mathcal{D}}$ is called the Carathéodory distance on $\mathcal{D}$, $K_{\mathcal{D}}$ the Kobayashi distance and $c_{\mathcal{D}}$ and $k_{\mathcal{D}}$ are the corresponding infinitesimal metrics.

For the unit disc, $C_{\mathbb{D}} = K_{\mathbb{D}} = \rho$ and $c_{\mathbb{D}} = k_{\mathbb{D}} = \alpha$. In general $C_{\mathcal{D}} \leq K_{\mathcal{D}} \leq \delta_{\mathcal{D}}$. Unlike $C_{\mathcal{D}}$ and $K_{\mathcal{D}}$, $\delta_{\mathcal{D}}$ does not obey the triangle inequality in general. $K_{\mathcal{D}}$ is the largest distance function on $\mathcal{D}$ smaller than $\delta_{\mathcal{D}}$. Holomorphic mappings are contractions relative to any one of the above distances or infinitesimal metrics.

Definition 1.1 *Let $\mathcal{D}$ be a domain in a complex Banach space X and let d be a distance on $\mathcal{D}$. A mapping $\phi \in H(\mathbb{D}, \mathcal{D})$ is called a complex d-geodesic if*

$$\rho(u, v) = d(\phi(u), \phi(v)) \text{ for } u, v \in \mathbb{D}.$$

If $z, w \in \phi(\mathbb{D})$ are distinct points, then we refer to ϕ as a complex d-geodesic joining z and w.

We will use the term complex geodesic for 'complex $C_{\mathcal{D}}$-geodesic'.

Proposition 1.2 *For a domain $\mathcal{D}$ in a Banach space and $\phi \in H(\mathbb{D}, \mathcal{D})$, the following are equivalent*

(a) ϕ is a complex geodesic.

(b) there exist distinct points $u, v \in \mathbb{D}$ such that $\rho(u, v) = C_{\mathcal{D}}(\phi(u), \phi(v))$.

(c) there exists a point $z \in \mathbb{D}$ such that $\alpha(z, 1) = c_{\mathcal{D}}(\phi(z), \phi'(z))$.

(d) ϕ is biholomorphic from $\mathbb{D}$ to an analytic set $\phi(\mathbb{D})$ and $\phi(\mathbb{D})$ is a holomorphic retract of $\mathcal{D}$ (i.e. there exists $f \in H(\mathcal{D}, \mathcal{D})$ such that $f \circ f = f$ and $f(\mathcal{D}) = \phi(\mathbb{D})$).

(e) ϕ is a complex $K_{\mathcal{D}}$-geodesic and

$$\delta_{\mathcal{D}}|\phi(\mathbb{D}) = K_{\mathcal{D}}|\phi(\mathbb{D}) = C_{\mathcal{D}}|\phi(\mathbb{D}).$$

PROOF: The equivalence of (a), (b) and (c) is due to Vesentini [34] and the fact that (d) is equivalent to (a) is due to Reiffen [27, p. 19] (see also Lempert [24]). (e) $\Rightarrow$ (a) and it remains to show that (a) $\Rightarrow$ (e).

Fix $u \neq v \in \mathbb{D}$ and write $p = \phi(u)$, $q = \phi(v)$. Since $\rho(u, v) = C_{\mathcal{D}}(p, q)$, Montel's theorem implies the existence of $f \in H(\mathcal{D}, \mathbb{D})$ such that $f(p) = u$ and $f(q) = v$. Now $f \circ \phi \in H(\mathbb{D}, \mathbb{D})$ and

$$\rho(f \circ \phi(u), f \circ \phi(v)) = C_{\mathcal{D}}(p, q) = \rho(u, v).$$

By the Schwarz-Pick lemma (see [10, p.5]), $f \circ \phi$ is a biholomorphic automorphism of $\mathbb{D}$ and $\rho(f \circ \phi(z), f \circ \phi(w)) = \rho(z, w)$ for all $z, w \in \mathbb{D}$. Since holomorphic mappings are contractions, we have

$$\begin{aligned} \rho(z, w) &= \rho(f \circ \phi(z), f \circ \phi(w)) \leq C_{\mathcal{D}}(\phi(z), \phi(w)) \\ &\leq K_{\mathcal{D}}(\phi(z), \phi(w)) \leq \delta_{\mathcal{D}}(\phi(z), \phi(w)) \\ &\leq \rho(z, w) \end{aligned} \tag{1.1}$$

and (1.1) consists entirely of equalities. Thus (a) implies (e).

Definition 1.3 *We call a domain $\mathcal{D}$ in a complex Banach space C-connected if every pair of points of $\mathcal{D}$ can be joined by a complex geodesic.*

We remark that it follows from results of Vigué [37] (see also [10, proposition 11.15 and corollary 11.17]) that a subset of a bounded convex finite dimensional domain $\mathcal{D}$ is the range of a complex geodesic if and only if it is a connected one-dimensional analytic subset and a holomorphic retract of $\mathcal{D}$.

An immediate consequence of the equivalence of (a) and (e) in Proposition 1.2 is the following result.

Proposition 1.4 *If $\mathcal{D}$ is a C-connected domain in a Banach space, then*

$$C_{\mathcal{D}} = K_{\mathcal{D}} = \delta_{\mathcal{D}}.$$

This leads to many examples of domains which are *not* C-connected (see for instance [10, p.103]). In particular a proper domain in $\mathbb{C}$ is C-connected if and only if it is simply connected.

On the other hand Lempert [23, 24] (see also Royden-Wong [29]) has shown that $C_{\mathcal{D}} = K_{\mathcal{D}} = \delta_{\mathcal{D}}$ and that $c_{\mathcal{D}} = k_{\mathcal{D}}$ if $\mathcal{D}$ is a convex bounded domain in $\mathbb{C}^n$. This result was extended to convex domains in arbitrary Banach spaces in [11], where it was used to prove that the following are C-connected (see also [10, pp.90–91])

(a) bounded convex domains in reflexive Banach spaces

(b) the open unit ball B_X of a dual Banach space X.

These facts lead us to propose the following conjecture.

Conjecture 1.5 *If $\mathcal{D}$ is a domain in a complex Banach space X which is biholomorphically equivalent to a bounded domain, then the following are equivalent:*

(a) $C_{\mathcal{D}} = K_{\mathcal{D}} = \delta_{\mathcal{D}}$

(b) $\mathcal{D}$ is biholomorphically equivalent to a convex domain

(c) $\mathcal{D}$ is C-connected.

(Note that (c) implies (a) and (b) implies (a).)

2 EXTENSION AND EXISTENCE THEOREMS

In this section, we consider a complex Banach space X as a subspace of its double dual space X^{**} via the natural embedding. B_X denotes the open unit ball of X and $H^\infty(B_X)$ denotes the space of scalar-valued bounded holomorphic functions on B_X, with the supremum norm. $H^\infty(B_X)$ is a Banach algebra. We abbreviate K_{B_X} as K_X and k_{B_X} as k_X from now on.

If $\mathcal{U}$ is an ultrafilter on a set I, then $(X)_\mathcal{U}$ will denote the ultrapower of a Banach space X. More specifically, if $\ell^\infty(I, X)$ denotes the space of bounded X-valued functions on I (with the supremum norm) and

$$N_\mathcal{U} = \{(x_i)_{i\in I} \in \ell^\infty(I, X) : \lim_\mathcal{U} \|x_i\| = 0\},$$

then $(X)_\mathcal{U}$ is the quotient space $\ell^\infty(I, X)/N_\mathcal{U}$.

There is a canonical isometric embedding $j: X \to (X)_\mathcal{U}$ given by

$$j(x) = (x)_{i\in I} + N_\mathcal{U}$$

(where $(x)_{i\in I}$ denotes the constant function x).

The principle of local reflexivity in its ultrapower formulation (see [20]) asserts that given a Banach space X, there exists an ultrapower $(X)_\mathcal{U}$ such that

(i) there is an isometric embedding $J: X^{**} \to (X)_\mathcal{U}$ which extends the canonical embedding $j: X \to (X)_\mathcal{U}$;

(ii) the map

$$\begin{aligned} Q: (X)_\mathcal{U} &\to X^{**} \\ Q((x_i)_{i\in I} + N_\mathcal{U}) &= w^*\text{-}\lim_\mathcal{U} x_i \end{aligned}$$

satisfies $QJ = \mathrm{id}_{X^{**}}$, $\|Q\| = 1$.

(Thus JQ is a contractive projection of $(X)_\mathcal{U}$ onto the isometric copy $J(X^{**})$ of X^{**}.)

The next lemma provides a crude version of the Schwarz inequality, which we prove for completeness.

Lemma 2.1 *Suppose X is a complex Banach space, $f \in H(B_X, \mathbb{D})$, $0 < \varepsilon < 1$ and $a, b \in (1-\varepsilon)B_X$. Then there is a constant A_ε depending only on ε such that*

$$\rho(f(a), f(b)) \le A_\varepsilon \|a - b\|.$$

PROOF: We can first dispose of the case where a and b are far apart, specifically the case $\|a - b\| > \varepsilon$. Define $\phi: \mathbb{D} \to B_X$ by

$$\phi(z) = \frac{z}{1-\varepsilon} a.$$

Then $f \circ \phi \in H(\mathbb{D}, \mathbb{D})$ and it follows from the classical Schwarz-Pick lemma that

$$\rho(f(a), f(0)) = \rho(f \circ \phi(1-\varepsilon), f \circ \phi(0)) \leq \rho(0, 1-\varepsilon).$$

Since the same estimate also applies to b

$$\rho(f(a), f(b)) \leq 2\rho(0, 1-\varepsilon) \leq A_\varepsilon \varepsilon \leq A_\varepsilon \|a - b\|$$

for A_ε chosen suitably large (depending only on ε).

For $\|a - b\| \leq \varepsilon$, we can define $\phi \in H(\mathbb{D}, B_X)$ by

$$\phi(z) = \frac{a+b}{2} + \frac{\varepsilon(a-b)}{\|a-b\|} z$$

and we will then have

$$\phi\left(\frac{\|a-b\|}{2\varepsilon}\right) = a, \quad \phi\left(-\frac{\|a-b\|}{2\varepsilon}\right) = b.$$

Applying the Schwarz lemma to $f \circ \phi$, we find that

$$\rho(f(a), f(b)) \leq \rho\left(\|a-b\|/2\varepsilon, -\|a-b\|/2\varepsilon\right) = 2\rho(\|a-b\|/2\varepsilon, 0) \leq A_\varepsilon \|a-b\|$$

since $\|a - b\|/2\varepsilon \leq 1/2$.

Theorem 2.2 *Suppose $Y = (X)_{\mathcal{U}}$ is an ultrapower of a complex Banach space X, where $\mathcal{U}$ is an ultrafilter on the set I.*

(i) Suppose, for each $i \in I$, $f_i: B_X \to D$ is a holomorphic function from B_X to the unit disc D in the complex plane, and

$$\lim_{\mathcal{U}} |f_i(a_i)| < 1 \tag{2.1}$$

for one point $(a_i)_{i \in I} + N_{\mathcal{U}}$ of B_Y. Then the function

$$\begin{aligned} F: B_Y &\to D \\ F((x_i)_{i \in I} + N_{\mathcal{U}}) &= \lim_{\mathcal{U}} f_i(x_i) \end{aligned}$$

is well-defined and holomorphic.

(ii) The Kobayashi distance K_Y on B_Y is given by

$$K_Y((x_i)_{i \in I} + N_{\mathcal{U}}, (y_i)_{i \in I} + N_{\mathcal{U}}) = \lim_{\mathcal{U}} K_X(x_i, y_i).$$

PROOF: (i) Note first of all that if $(x_i)_{i\in I} + N_{\mathcal{U}} \in B_Y$, then $\lim_{\mathcal{U}} \|x_i\| < 1$ and so there exists $U \in \mathcal{U}$ so that $\|x_i\| < 1$ for all $i \in U$. Thus $f_i(x_i)$ is defined for $i \in U$ and $\lim_{\mathcal{U}} f_i(x_i)$ makes sense. In fact, it is possible to change x_i for $i \notin U$ so as to ensure $\sup_{i\in I} \|x_i\| < 1$ without changing the coset $(x_i)_{i\in I} + N_{\mathcal{U}}$ or the value of the limit. We will make this change without comment from now on for all points in B_Y and, in particular we will assume that $S = \sup_i \|a_i\| < 1$.

By compactness, $\lim_{\mathcal{U}} f_i(x_i)$ certainly exists in the closed unit disc $\mathbb{D}$. To show it is in the open disc fix $(x_i)_i$ and choose $0 < \varepsilon < 1 - S$ so that $\sup_i \|x_i\| < 1 - \varepsilon$. By Lemma 2.1,

$$\rho(f_i(x_i), f_i(a_i)) \leq 2A_\varepsilon \quad (i \in I).$$

Choose now $U \in \mathcal{U}$ so that $T = \sup_{i\in U} |f_i(a_i)| < 1$. It follows by the triangle inequality that

$$\rho(f_i(x_i), 0) \leq 2A_\varepsilon + \rho(0, f_i(a_i)) \leq 2A_\varepsilon + \rho(0, T).$$

Hence $\sup_{i\in U} |f_i(x_i)| < 1$ and F does indeed map B_Y into $\mathbb{D}$.

Next we check that F is continuous. For this, fix $(x_i)_i + N_{\mathcal{U}} \in B_Y$ and choose $\varepsilon > 0$ with $x = \sup_i \|x_i\| < 1 - 2\varepsilon$. For $y = (y_i)_i + N_{\mathcal{U}} \in B_Y$ and $\|x - y\|_Y < \varepsilon$, we have $\|y_i\|_X < 1 - \varepsilon$ and so we can apply Lemma 2.1 to see that

$$\rho(f_i(x_i), f(y_i)) \leq A_\varepsilon \|x_i - y_i\|$$

Taking limits along $\mathcal{U}$, we deduce that

$$\rho(F(x), F(y)) \leq A_\varepsilon \|x - y\|_Y,$$

which is enough to show continuity of F at x.

Finally analyticity of F follows from continuity together with analyticity of the restricton of F on complex lines in Y. For $x = (x_i)_i + N_{\mathcal{U}} \in B_Y$ and $y = (y_i)_i + N_{\mathcal{U}} \in Y$, Montel's theorem shows that

$$F(x + zy) = \lim_{\mathcal{U}} f_i(x_i + zy_i)$$

is an analytic function of z on $\{z \in \mathbb{C} : \|x + zy\|_Y < 1\}$.

(ii) Fix $x = (x_i)_i + N_{\mathcal{U}}, y = (y_i)_i + N_{\mathcal{U}} \in B_Y$ and let $K_{\text{lim}} = \lim_{\mathcal{U}} K_X(x_i, y_i)$. Choose $\varepsilon > 0$ and put $r = \tanh(K_{\text{lim}} + \varepsilon)$. Then $\rho(0, r) = K_{\text{lim}} + \varepsilon$. There exists $U \in \mathcal{U}$ so that $K_X(x_i, y_i) < K_{\text{lim}} + \varepsilon$ for $i \in U$. Thus, since $\delta_X(x_i, y_i) = K_X(x_i, y_i)$, there exists $g_i \in H(\mathbb{D}, B_X)$ satisfying $g_i(0) = x_i$, $g_i(r) = y_i$ (for $i \in U$). For $i \in I \setminus U$, set $g_i = 0$.

Now define $g\colon \mathbb{D} \to B_Y$, by $g(z) = (g_i(z))_i + N_{\mathcal{U}}$. Of course we must check first that $\sup_i \|g_i(z)\|_X < 1$ for all $z \in \mathbb{D}$. To this end, observe that

$$K_X(g_i(z), 0) \leq K_X(g_i(z), g_i(0)) + K_X(g_i(0), 0)$$

$$\begin{aligned} &\leq \rho(z,0) + K_X(x_i,0) \\ &= \rho(z,0) + \rho(\|x_i\|,0) \\ &\leq \rho(z,0) + \sup_{j\in I} \rho(\|x_j\|,0) \\ &< \infty \end{aligned}$$

Here we have used the fact that holomorphic mappings are contractions with respect to the Kobayashi distance and our standing assumption that the x_j are chosen so that $\sup_j \|x_j\| < 1$. The equality $K_X(x,0) = \rho(\|x\|,0)$ (for $x \in B_X$) is elementary. Hence we have

$$\sup_i \rho(\|g_i(z)\|,0) < \infty$$

which implies $\sup_i \|g_i(z)\| < 1$.

Cauchy's formula shows that the functions $\|g_i''(z)\|$ are uniformly bounded on compact subsets of $\mathbb{D}$. This will enable us to show by a direct argument that g is analytic. Fix $z \in \mathbb{D}$ and suppose $|w-z| < (1-|z|)/2$. Then

$$\begin{aligned} \left\| \frac{g_i(w) - g_i(z)}{w-z} - g_i'(z) \right\| &= \left\| \frac{1}{w-z} \int_z^w g_i'(\zeta) - g_i'(z)\, d\zeta \right\| \\ &= \left\| \frac{1}{w-z} \int_z^w \int_z^\zeta g_i''(\eta)\, d\eta\, d\zeta \right\| \\ &\leq |w-z| \sup\{|g_i''(\zeta)| : |\zeta| \leq (1+|z|)/2, i \in I\} \end{aligned}$$

Hence, if $\ell = (g_i'(z))_i + N_{\mathcal{U}}$, we have

$$\left\| \frac{g(w)-g(z)}{w-z} - \ell \right\|_Y \leq M|w-z|$$

for $|w-z| < (1-|z|)/2$ and M a constant depending on z. Taking the limit as $w \to z$, we see that $g'(z)$ exists and is ℓ.

Therefore $g \in H(\mathbb{D}, B_Y)$, $g(0) = x$ and $g(r) = y$. It follows that

$$K_Y(x,y) = K_Y(g(0), g(r)) \leq \rho(0,r) = K_{\lim} + \varepsilon.$$

Since this is true for all $\varepsilon > 0$, we have

$$K_Y(x,y) \leq K_{\lim} = \lim_{\mathcal{U}} K_X(x_i, y_i). \tag{2.2}$$

To establish the reverse inequality, we use the fact that $K_X = C_X$ to select functions $f_i \in H(B_X, \mathbb{D})$ satisfying $f_i(x_i) = 0$, $f_i(y_i) > 0$ and $K_X(x_i,y_i) = \rho(0, f_i(y_i))$. Applying part (i), we get a function $F \in H(B_Y, \mathbb{D})$ which satisfies $F(x) = 0$ and

$$\rho(0, F(y)) = \lim_{\mathcal{U}} \rho(0, f_i(y_i)) = \lim_{\mathcal{U}} K_X(x_i, y_i) = K_{\lim}.$$

But now the distance decreasing property of Kobayashi distances under F allows us to conclude that

$$K_{\lim} = \rho(0, F(y)) = \rho(F(x), F(y)) \leq K_Y(x, y).$$

Combining this with (2.2) completes the proof.

Our next result is a new proof of a result of Davie and Gamelin [7]. It has come to our attention that M. Lindström and R. Ryan have independently obtained a proof of this result using ultrapower techniques.

Theorem 2.3 *For X a complex Banach space, there exists an algebrahomomorphism of norm one,*

$$E : H^\infty(B_X) \to H^\infty(B_{X^{**}})$$

which satisfies

$$Ef|B_X = f$$

for all $f \in H^\infty(B_X)$.

PROOF: Given X and $f \in H^\infty(B_X)$, we choose an ultrafilter $\mathcal{U}$ according to the principle of local reflexivity. To apply Theorem 2.2(i), we take $f_i = f/\|f\|_\infty$ for all i. Unless f is constant, the hypothesis (2.1) is satisfied for $a_i = 0$. In any case, it follows that the function

$$\begin{aligned} F\colon B_Y &\to \mathbb{C} \\ F((x_i)_{i\in I} + N_{\mathcal{U}}) &= \lim_{\mathcal{U}} f(x_i) \end{aligned}$$

is holomorphic and $\|F\|_\infty \leq \|f\|_\infty$. Put $Ef = F \circ J$, where J is as in the principle of local reflexivity. It is straightforward to check that E has the required linearity and multiplicative properties and that $\|E\| \leq 1$. Since J coincides with the canonical embedding $j\colon X \to (X)_{\mathcal{U}}$ on B_X, it is also easy to see that Ef coincides with f on B_X.

Theorem 2.4 *For X a complex Banach space,*

$$K_{X^{**}}(x, y) = K_X(x, y)$$

for all $x, y \in B_X$.

PROOF: Since the canonical inclusion from B_X to $B_{X^{**}}$ is continuous and linear, it is holomorphic and the distance decreasing property of the Kobayashi metric implies

$$K_{X^{**}}(x, y) \leq K_X(x, y)$$

for $x, y \in B_X$.

Fix $x, y \in B_X$. By Montel's theorem, we can find $f \in H(B_X, \mathbb{D})$ so that $f(0) = 0$ and

$$K_X(x, y) = C_X(x, y) = \rho(f(x), f(y)).$$

By Propostion 2.3, we can find an extension $\tilde{f} \in H^\infty(B_{X^{**}})$ of f with $\|\tilde{f}\|_\infty \leq 1$. Since $\tilde{f}(x) = f(x) \in \mathbb{D}$, $\tilde{f}$ has all its values in $\mathbb{D}$. We conclude

$$K_X(x, y) = \rho(f(x), f(y)) = \rho(\tilde{f}(x), \tilde{f}(y)) \leq K_{X^{**}}(x, y).$$

We are now in a position to extend Théorème 4.3 of [11].

Theorem 2.5 *If a complex Banach space X is 1-complemented in its second dual, then B_X is C-connected.*

PROOF: Let P denote a norm 1 projection from X^{**} onto X and let $p \neq q \in B_X$. Since $B_{X^{**}}$ is the unit ball of a dual Banach space, Théorème 4.3 of [11] implies that there exists $\phi \in H(\mathbb{D}, B_{X^{**}})$ and $u, v \in \mathbb{D}$ satisfying $\phi(u) = p$, $\phi(v) = q$ and $\rho(u, v) = K_{X^{**}}(p, q)$.

Now $P \circ \phi \in H(\mathbb{D}, B_X)$ and $P \circ \phi(u) = p$, $P \circ \phi(v) = q$. By Theorem 2.4 and [11, Théorème 2.5],

$$\rho(u, v) = K_{X^{**}}(p, q) = K_X(p, q) = C_X(p, q).$$

Hence $P \circ \phi$ is a complex geodesic in B_X. Since p and q are arbitrary, B_X is C-connected.

Remark 2.6 Preduals of C^*-algebras satisfy the hypotheses of Theorem 2.5 (see [32]) and these include examples which are not covered by the results in [11].

On the other hand, c_0 is well known not to satisfy the hypotheses of Theorem 2.5, although B_{c_0} is C-connected (which follows from homogeneity — see Remarks 5.10).

3 UNIQUENESS RESULTS

If ϕ is a complex geodesic joining the points p and q of a domain $\mathcal{D}$ and f is a biholomorphic automorphism of $\mathbb{D}$, then $\phi \circ f$ is also a complex geodesic joining p and q (because f is a ρ-isometry). Thus there is never a unique complex geodesic joining p and q, because of this possibility of reparametrizing complex geodesics. However, Vesentini [33] has shown that if ϕ and ψ are complex geodesics then they have the same range $\phi(\mathbb{D}) = \psi(\mathbb{D})$ if and only if $\psi = \phi \circ f$ for some biholomorphic automorphism f of $\mathbb{D}$ (this can also be deduced from the global vector-valued subordination theorem of Finkelstein and Whitley [13]). We thus discuss uniqueness of complex geodesics up to reparametrization, by means of the following normalization.

We call a complex geodesic ϕ a *normalized geodesic joining p and q* if $\phi(0) = p$ and $\phi(s) = q$ for some positive real number s. The number s is uniquely determined by p and q — in fact $s = \tanh C_{\mathcal{D}}(p, q)$. By the homogeneity of the unit disc and the result of Vesentini cited above, it follows that there is a unique normalized complex geodesic joining two points $p, q \in \mathcal{D}$ if and only if all complex geodesics joining p and q have the same range.

The following are known results concerning uniqueness.

(a) If B_X is the unit ball of a Banach space X and $x \in B_X$, $x \neq 0$, then there is a unique normalized complex geodesic joining 0 and x if and only if $x/\|x\|$ is a complex extreme point of B_X. (Vesentini [33]).

(b) If $\mathcal{D}$ is a strictly convex domain (i.e. each point of the boundary $\partial\mathcal{D}$ is a (real) extreme point of $\mathcal{D}$) in a finite dimensional space, then there exist unique normalized complex geodesics joining all pairs of points in $\mathcal{D}$. (Lempert [23]).

In this section, we extend (b) to a class of Banach spaces which includes all reflexive Banach spaces and give a general criterion for uniqueness of complex geodesics which highlights the problem of interpolating between the results (a) and (b) above. A more detailed study of non-uniqueness of complex geodesics has been undertaken by Gentili [14, 15, 16] (see also Section 6).

Definition 3.1 *A complex Banach space X has the analytic Radon-Nikodym property (aRNP) if each $f \in H^\infty(\mathbb{D}, X)$ has radial limits almost everywhere on the unit circle.*

$H^\infty(\mathbb{D}, X)$ means those functions in $H(\mathbb{D}, X)$ which have bounded range. If X has aRNP and $f \in H^\infty(\mathbb{D}, X)$, we can extend f to almost all points $e^{i\theta} \in \partial\mathbb{D}$ (almost all with respect to Lebesgue measure on $\partial\mathbb{D}$) by

$$f(e^{i\theta}) = \lim_{r \to 1^-} f(re^{i\theta}).$$

Moreover f is uniquely determined by the boundary values $f(e^{i\theta})$. Reflexive Banach spaces and Banach spaces with the Radon-Nikodym property (e.g separable dual spaces) have aRNP. The Banach space c_0 does not have aRNP. For further details we refer to [10, Chapter 12] and [18].

Now suppose we have a convex bounded domain $\mathcal{D}$ in a Banach space X with aRNP. Let $p \neq q \in \mathcal{D}$ and let $G(p, q)$ denote the set of all normalized complex geodesics joining p and q. If $\phi, \psi \in G(p, q)$ and $0 < \lambda < 1$, and $s = \tanh C_{\mathcal{D}}(p, q)$, then $\lambda\phi + (1 - \lambda)\psi \in G(p, q)$. This follows from convexity of $\mathcal{D}$ (which ensures that $\lambda\phi + (1 - \lambda)\psi(\mathbb{D}) \subset \mathcal{D}$) together with the facts that

$$(\lambda\phi + (1 - \lambda)\psi)(0) = \phi(0) = \psi(0) = p \text{ and } (\lambda\phi + (1 - \lambda)\psi)(s) = \phi(s) = \psi(s) = q.$$

In other words, $G(p, q)$ is a convex subset of $H(\mathbb{D}, \mathcal{D})$

Next, notice that if $\phi \in G(p, q)$, then

$$\lim_{r \to 1-} C_{\mathcal{D}}(\phi(re^{i\theta}), \phi(0)) = \lim_{r \to 1-} \rho(re^{i\theta}, 0) = \infty.$$

It follows that at points $e^{i\theta}$ where $\phi(e^{i\theta})$ is defined (almost all points on $\partial\mathbb{D}$ by aRNP), $\phi(e^{i\theta}) \in \partial\mathcal{D}$. Now if ψ is another element of $G(p, q)$, then for almost all θ,

$$\phi(e^{i\theta}), \psi(e^{i\theta}) \text{ and } \left(\frac{\phi + \psi}{2}\right)(e^{i\theta}) = (1/2)(\phi(e^{i\theta}) + \psi(e^{i\theta}))$$

are all in $\partial\mathcal{D}$. If we now assume that $\mathcal{D}$ is strictly convex, then we must have $\phi(e^{i\theta}) = \psi(e^{i\theta})$ for almost all $e^{i\theta}$. This implies $\phi = \psi$. We have thus proved the following result.

Theorem 3.2 *If X is a complex Banach space with the analytic Radon-Nikodym property and $\mathcal{D} \subset X$ is a strictly convex bounded domain, then there exists at most one normalized complex geodesic joining p and q.*

We now restrict our attention to the case where $\mathcal{D}$ is the open unit ball B_X of X. Let $\phi, \psi \in G(p, q)$ for two points $p, q \in B_X$ and suppose again that X has aRNP. Let $g = \psi - \phi$ and $s = \tanh C_X(p, q)$. Since $\phi(0) = \psi(0) = p$ and $\phi(s) = \psi(s) = q$, $g(0) = g(s) = 0$ and we can therefore write $g(z) = z(z - s)h(z)$ for some $h \in H^\infty(\mathbb{D}, X)$. Using aRNP and the convexity of $G(p, q)$, we can see that the following result holds.

Proposition 3.3 *If X has aRNP and ϕ is a normalized complex geodesic joining two points $p, q \in B_X$, then ϕ is the unique such geodesic if and only if the zero function is the only element $h \in H^\infty(\mathbb{D}, X)$ satisfying*

$$\|\phi(e^{i\theta}) + \lambda e^{i\theta}(e^{i\theta} - s)h(e^{i\theta})\| = 1 \tag{3.1}$$

for almost all θ, all $\lambda \in [0, 1]$ (where $s = \tanh C_X(p, q)$).

Examples 3.4 (a) If X has aRNP and is 1-complemented in X^{**}, and if B_X is strictly convex, then Theorems 2.5 and 3.2 show that there exists a unique normalized complex geodesic joining each pair of points $p, q \in B_X$.

(b) This applies in particular when X is the space ℓ^p of p-summable sequences ($1 < p < \infty$), because ℓ^p is reflexive and has a strictly convex unit ball.

4 CONTINUOUS COMPLEX GEODESICS

In this section we show that complex geodesics can be extended continuously to the boundary under a complex uniform convexity hypothesis.

Definition 4.1 *If $\mathcal{D} \subset X$ is a domain in a complex Banach space X, then we define*

$$\delta_{\mathcal{D}}(z, v) = \sup\{r > 0 : z + rv\mathbb{D} \subset \mathcal{D}\}$$

for $z \in \mathcal{D}$, $v \in X$, $\|v\| = 1$.

We define the modulus of complex convexity of $\mathcal{D}$ to be

$$\delta_{\mathcal{D}}(\varepsilon) = \sup\{\delta_{\mathcal{D}}(z, v) : z \in \mathcal{D}, d(z, \partial\mathcal{D}) \leq \varepsilon, \|v\| = 1\},$$

(where $d(z, \partial\mathcal{D})$ denotes the distance from a point z in $\mathcal{D}$ to the boundary).

The domain $\mathcal{D}$ is called complex uniformly convex if $\delta_{\mathcal{D}}(\varepsilon) \to 0$ as $\varepsilon \to 0$.

Remarks 4.2 For the case where $\mathcal{D} = B_X$ is the unit ball of X, Globevnik [17] introduced

$$\omega_c(\varepsilon) = \sup\{\|y\| : \|x + \zeta y\| < 1 + \varepsilon \text{ for all } \|x\| = 1, \zeta \in \mathbb{D}\}$$

and this function is closely related to $\delta_{\mathcal{D}}(\varepsilon)$. In fact, it is easy to check that

$$\frac{\omega_c(\varepsilon)}{1 + \varepsilon} = \sup\{\delta_{\mathcal{D}}(z, v) : \|v\| = 1, \|z\| = 1/(1 + \varepsilon)\} = \delta_{\mathcal{D}}\left(\frac{\varepsilon}{1 + \varepsilon}\right)$$

and thus that

$$\delta_{\mathcal{D}}(\varepsilon/2) \leq \omega_c(\varepsilon) \leq 2\delta_{\mathcal{D}}(\varepsilon) \quad (0 < \varepsilon \leq 1).$$

Functions which are inverse to $\delta_{\mathcal{D}}(\varepsilon)$ and $\omega_c(\varepsilon)$ were considered by Davis, Garling and Tomczak-Jaegermann [8] and called h_∞^X and H_∞^X (respectively). Dilworth [9, theorem 2.1] has shown that complex uniform convexity of B_X (or uniform H_∞-convexity of X in the notation of [8]) is equivalent to the notion of uniform PL-convexity which was studied intensively in [8].

A result similar to the following one can be obtained for the case where the domain is the unit ball using theorem 2 of [17].

Proposition 4.3 *If $\mathcal{D} \subset X$ is a convex domain, then*

$$c_{\mathcal{D}}(z, v) \geq \frac{\|v\|}{2\delta_{\mathcal{D}}(\varepsilon)}$$

holds for $v \in X$, $z \in \mathcal{D}$, $\varepsilon = d(z, \partial\mathcal{D})$.

PROOF: We will make use of the fact that $c_{\mathcal{D}} = k_{\mathcal{D}}$ for $\mathcal{D}$ convex (see [11]). Fix $z \in \mathcal{D}$, $v \in X$ and consider holomorphic mappings $f: \mathbb{D} \to \mathcal{D}$ with $f(0) = z$, $f'(0) = v/r$, $r > 0$.

Consider the function

$$g(\zeta) = \frac{1}{2\pi} \int_0^{2\pi} f(e^{i\theta}\zeta)(1 + \cos\theta)\, d\theta.$$

Since $(1 + \cos\theta)\, d\theta/(2\pi)$ is a probability measure on $[0, 2\pi]$, we may regard $g(\zeta)$ as a limit of convex combinations of points in $\mathcal{D}$. So g takes values in the closure $\mathcal{D}$. Since $g(0) = f(0) \in \mathcal{D}$ and $\mathcal{D}$ is convex, it follows that g maps $\mathbb{D}$ into $\mathcal{D}$ [33, p. 376].

Working with the power series representation of f, we find that

$$g(\zeta) = f(0) + \frac{1}{2} f'(0)\zeta.$$

(In fact, $g(\zeta)$ is a Cesaro mean of the power series of f.). Thus

$$g(\zeta) = f(0) + \frac{v}{2r}\zeta = z + \frac{\|v\|}{2r}\frac{v}{\|v\|}\zeta$$

maps $\mathbb{D}$ into $\mathcal{D}$, which shows that

$$\frac{\|v\|}{2r} \leq \delta_{\mathcal{D}}(z, v/\|v\|) \leq \delta_{\mathcal{D}}(\varepsilon).$$

Rearranging this, we find $r \geq \|v\|/(2\delta_{\mathcal{D}}(\varepsilon))$. Since $k_{\mathcal{D}}(z, v) = c_{\mathcal{D}}(z, v)$ is the infimum of all possible values of r, the result follows.

Theorem 4.4 *Let $\mathcal{D}$ be the unit ball B_X of a complex Banach space X. If $\mathcal{D}$ is complex uniformly convex and $\delta_{\mathcal{D}}(\varepsilon) \leq A\varepsilon^s$ for some constants $A > 0$, $s > 0$, then all complex geodesics $\phi: \mathbb{D} \to \mathcal{D}$ extend to continuous functions $\phi: \mathbb{D} \to \mathcal{D}$.*

PROOF: By Proposition 4.3, we have (for $\zeta \in \mathbb{D}$)

$$c_{\mathcal{D}}(\phi(\zeta), \phi'(\zeta)) \geq \frac{\|\phi'(\zeta)\|}{2\delta_{\mathcal{D}}(1 - \|\phi(\zeta)\|)}.$$

Using the hypothesis and the fact that ϕ is a complex geodesic, we deduce

$$c_{\mathcal{D}}(\phi(\zeta), \phi'(\zeta)) = c_{\mathbb{D}}(\zeta, 1) = \frac{1}{1 - |\zeta|^2} \geq \frac{\|\phi'(\zeta)\|}{2A(1 - \|\phi(\zeta)\|)^s}$$

or

$$\|\phi'(\zeta)\| \leq \frac{2A(1 - \|\phi(\zeta)\|)^s}{1 - |\zeta|^2}. \tag{4.1}$$

Next observe that

$$C_{\mathcal{D}}(\phi(0), \phi(\zeta)) = \rho(0, \zeta) = \tanh^{-1}(|\zeta|) \leq C_{\mathcal{D}}(0, \phi(\zeta)) + C_{\mathcal{D}}(0, \phi(0)).$$

Recall that $C_{\mathcal{D}}(0, z) = \tanh^{-1} \|z\|$ for $z \in \mathcal{D} = B_X$. Using elementary estimates, we conclude that

$$1 - \|\phi(\zeta)\| \leq A_\phi(1 - |\zeta|)$$

where A_ϕ is a constant depending on $\|\phi(0)\|$.

Combining this observation with (4.1), we see that

$$\|\phi'(\zeta)\| \leq A_1 \frac{(1 - |\zeta|)^s}{1 - |\zeta|^2} \leq A_1 \frac{1}{(1 - |\zeta|)^{1-s}}$$

which implies (see for instance [12, theorem 5.5]) that ϕ satisfies a Lipschitz condition

$$\|\phi(\zeta) - \phi(\eta)\| \leq C|\zeta - \eta|^s.$$

Hence $\phi: \mathbb{D} \to X$ is uniformly continuous and extends continuously to a function $f: \mathbb{D} \to X$.

If a complex geodesic $\phi: \mathbb{D} \to \mathcal{D}$ extends to a continuous function $\phi: \mathbb{D} \to \mathcal{D}$, we call ϕ a *continuous complex geodesic.*

Examples 4.5 (i) For $X = L^1$, $\mathcal{D} = B_X$, Globevnik [17] proved that $\delta_{\mathcal{D}} \leq A\sqrt{\varepsilon}$ (or rather, he proves the equivalent fact that $\omega_c(\varepsilon) \leq A\sqrt{\varepsilon}$). Thus all complex geodesics in B_X are continuous, by Theorem 4.4.

(ii) More generally, if X is the predual of a C^*-algebra and $\mathcal{D} = B_X$, we deduce from a result due to Haagerup (see [8, theorem 4.3]) that $\delta_{\mathcal{D}}(\varepsilon) \leq A\sqrt{\varepsilon}$.

Thus all complex geodesics in B_X are continuous by Theorem 4.4. Existence of complex geodesics in B_X is guaranteed by Theorem 2.5.

Remark 4.6 From corollary 2.5 of [15], it follows that if all complex geodesics in the unit ball B_X are continuous, then all points of ∂B_X are complex extreme points. Theorem 4.4 falls short of being a converse to this.

5 EXAMPLES IN CLASSICAL BANACH SPACES

We apply the results of the preceding sections to give a complete description of the complex geodesics in the unit ball of ℓ^p, $1 \leq p < \infty$. To obtain these examples, we require a Banach space version of a result of Lempert [23, proposition 1] and [29]. Various extensions are possible and we have chosen one which is suitable for the applications we have in mind.

Definition 5.1 *For X a complex Banach space with dual space X^*, we let $H^\infty_*(\mathbb{D}, X^*)$ denote the space of X^*-valued bounded analytic functions on $\mathbb{D}$ which have weak*-radial limits at almost all boundary points.*

In other words $f \in H^\infty_*(\mathbb{D}, X^*)$ means that $f \in H^\infty(\mathbb{D}, X^*)$ and there exists a function $\tilde{f}: \partial\mathbb{D} \to X^*$ so that

$$\lim_{r\to 1^-} \langle x, f(re^{i\theta})\rangle = \langle x, \tilde{f}(e^{i\theta})\rangle \qquad (\text{all } x \in X)$$

holds for almost all $\theta \in \mathbb{R}$.

If X^* has aRNP, then all functions in $H^\infty(\mathbb{D}, X^*)$ have norm radial limits at almost all points of $\partial\mathbb{D}$, and therefore $H^\infty_*(\mathbb{D}, X^*) = H^\infty(\mathbb{D}, X^*)$. By a result of Danilevich [6, theorem 1.4], this equality also holds if X is separable.

In general, the limit function $\tilde{f}(e^{i\theta})$ may only be weak*-measurable, a rather intractable condition. Moreover it is possible that the space $H^\infty_*(\mathbb{D}, X^*)$ depends on the choice of a predual X for X^*. However, the function $\tilde{f}(e^{i\theta})$ determines the holomorphic function $f(\zeta)$ uniquely as can be seen by applying standard results (see [12]) to the scalar-valued bounded analytic functions $\langle x, f(\zeta)\rangle$. We will therefore not cause confusion by using the notation $f(e^{i\theta})$ instead of $\tilde{f}(e^{i\theta})$ for the boundary function.

Lemma 5.2 *Let X be a complex Banach space, $f \in H^\infty_*(\mathbb{D}, X^*)$ and $h: \bar{\mathbb{D}} \to X$ a continuous function which is holomorphic on $\mathbb{D}$. Then*

$$\lim_{r\to r^-} \langle h(re^{i\theta}), f(re^{i\theta})\rangle = \langle h(e^{i\theta}), f(e^{i\theta})\rangle$$

for almost all θ and

$$\langle h(\zeta), f(\zeta)\rangle = \frac{1}{2\pi}\int_0^{2\pi} \langle h(e^{i\theta}), f(e^{i\theta})\rangle \frac{1-|\zeta|^2}{1+|\zeta|^2 - 2\mathrm{Re}\,(e^{-i\theta}\zeta)}\, d\theta$$

for all $\zeta \in \mathbb{D}$.

PROOF: The first assertion follows from the inequality

$$\begin{aligned}
&|\langle h(re^{i\theta}), f(re^{i\theta})\rangle - \langle h(e^{i\theta}), f(e^{i\theta})\rangle| \\
\le\ & |\langle h(re^{i\theta}) - h(e^{i\theta}), f(re^{i\theta})\rangle| + |\langle h(e^{i\theta}), f(re^{i\theta}) - f(e^{i\theta})\rangle| \\
\le\ & \|h(re^{i\theta}) - h(e^{i\theta})\|\, \|f\|_\infty + |\langle h(e^{i\theta}), f(re^{i\theta}) - f(e^{i\theta})\rangle| \\
\to\ & 0
\end{aligned}$$

for almost all θ by continuity of h and the definition of $H^\infty_*(\mathbb{D}, X^*)$. The second assertion follows from the standard fact (see [12]) that scalar-valued bounded analytic functions like $\langle f(\zeta), h(\zeta)\rangle$ are the Poisson integrals of their (almost everywhere) boundary values.

If X is a Banach space and $x \in \partial B_X$, then the Hahn-Banach theorem assures us of the existence of at least one supporting hyperplane for B_X at x. That is, there exists $N_x \in X^*$ such that $\langle x, N_x\rangle = 1$ and $\mathrm{Re}\,\langle p, N_x\rangle < 1$ for $p \in B_X$. We will use the notation N_x for a choice of one such functional, bearing in mind the possibility that it may not be unique.

Proposition 5.3 *Let X be a complex Banach space and $\phi : \mathbb{D} \to B_X$ a continuous map satisfying*

(i) $\phi \mid \mathbb{D}$ is holomorphic and $\phi(\mathbb{D}) \subset B_X$;

(ii) $\phi(\partial\mathbb{D}) \subset \partial B_X$; and

(iii) there exists a choice of $N_{\phi(\zeta)}$ for almost all $\zeta \in \partial\mathbb{D}$ and a measurable function $p : \partial\mathbb{D} \to \mathbb{R}^+$ such that the mapping

$$h(e^{i\theta}) = e^{i\theta} p(e^{i\theta}) N_{\phi(e^{i\theta})}$$

is almost everywhere the weak-radial limit of a function $h \in H^\infty_*(\mathbb{D}, X^*)$.*

Then ϕ is a complex geodesic.

PROOF: Let $g : \mathbb{D} \to B_X$ be a holomorphic mapping with $g(0) = \phi(0)$ and $g'(0) = \lambda\phi'(0)$, $\lambda \geq 0$. Let $g_r(\zeta) = g(r\zeta)$ for $0 < r < 1$ and $\zeta \in \mathbb{D}$. Then g_r is continuous on $\mathbb{D}$ and holomorphic on $\mathbb{D}$, $g_r(0) = \phi(0)$ and $g_r'(0) = \lambda r \phi'(0)$. Moreover $g_r(\zeta) \in B_X$ for $\zeta \in \mathbb{D}$. From the hypotheses, we see that

$$1 = \langle \phi(e^{i\theta}), N_{\phi(e^{i\theta})} \rangle > \mathrm{Re}\,\langle g_r(e^{i\theta}), N_{\phi(e^{i\theta})} \rangle$$

for almost all θ. Hence

$$\mathrm{Re}\left\langle \frac{\phi(e^{i\theta}) - g_r(e^{i\theta})}{e^{i\theta}}, e^{i\theta} p(e^{i\theta}) N_{\phi(e^{i\theta})} \right\rangle = p(e^{i\theta}) \mathrm{Re}\,\langle \phi(e^{i\theta}) - g_r(e^{i\theta}), N_{\phi(e^{i\theta})} \rangle > 0$$

for almost all θ.

Since $\dfrac{\phi(\zeta) - g_r(\zeta)}{\zeta}$ is holomorphic on $\mathbb{D}$ and continuous on $\mathbb{D}$, the function

$$H(\zeta) = \left\langle \frac{\phi(\zeta) - g_r(\zeta)}{\zeta}, h(\zeta) \right\rangle$$

is the Poisson integral of its boundary values $H(e^{i\theta})$ by Lemma 5.2. By the above remark and Lemma 5.2, $\mathrm{Re}\, H(e^{i\theta}) > 0$ for almost all θ and it follows from the Poisson formula that $\mathrm{Re}\, H(0) > 0$, *i.e.*

$$\mathrm{Re}\,\langle \phi'(0) - g_r'(0), h(0) \rangle = (1 - \lambda r) \mathrm{Re}\,\langle \phi'(0), h(0) \rangle > 0.$$

Applying this to the special case where $g(\zeta) = \phi(0)$ is constant (and $\lambda = 0$) we see that $\mathrm{Re}\,\langle \phi'(0), h(0) \rangle > 0$. Thus, returning to the general case, we have $1 - r\lambda > 0$. Since this is true for all $0 < r < 1$, we deduce that $\lambda \leq 1$.

Since this is true for all g, we have established

$$k_X(\phi(0), \phi'(0)) = c_X(\phi(0), \phi'(0)) = 1$$

which shows, by Proposition 1.2, that ϕ is a complex geodesic.

For μ a σ-finite measure on a σ-algebra Σ of subsets of a set Ω and $1 \leq p < \infty$, we use the standard notation $L^p(\mu)$ for the Banach space of (equivalence classes of) p-summable Σ-measurable ($\mathbb{C}$-valued) functions on Ω normed by $\|f\|_p = (\int_\Omega |f|^p \, d\mu)^{1/p}$. $L^\infty(\mu)$ denotes the essentially bounded Σ-measurable functions with the essential sup norm $\|f\|_\infty$. These include as special cases the sequence spaces ℓ^p (where μ is counting measure on the natural numbers) and the finite-dimensional spaces ℓ^p_n (which are $\mathbb{C}^n$ with the norm $\|(z_1, z_2, \ldots, z_n)\|_p = \left(\sum_{j=1}^n |z_j|^p\right)^{1/p}$ for $1 \leq p < \infty$).

In order to discuss complex geodesics in the unit ball B_p of $L^p(\mu)$ for $1 \leq p < \infty$ we will consider nonconstant mappings $\phi\colon \mathbb{D} \to B_p$ of the form

$$\phi(\zeta)(\omega) = c(\omega)\left(\frac{\zeta - \alpha(\omega)}{1 - \overline{\alpha(\omega)}\zeta}\right)^{\beta(\omega)}\left(\frac{1 - \overline{\alpha(\omega)}\zeta}{1 - \gamma\zeta}\right)^{2/p} \tag{5.1}$$

($\zeta \in \mathbb{D}$, $\omega \in \Omega$) where the parameter γ and the measurable functions $\alpha(\omega)$, $\beta(\omega)$ and $c(\omega)$ satisfy

$$\left.\begin{array}{ll} (a) & \gamma \in \mathbb{D}, \quad \alpha \in L^\infty(\mu), \quad \|\alpha\|_\infty \leq 1, \\ & \beta \text{ takes only the values 0 and 1.} \\ (b) & \int_\Omega |c(\omega)|^p (1 + |\alpha(\zeta)|^2)\, d\mu(\omega) = 1 + |\gamma|^2 \\ (c) & \int_\Omega |c(\omega)|^p \alpha(\omega)\, d\mu = \gamma \end{array}\right\} \tag{5.2}$$

Later, we will specialise to the case where $L^p(\mu) = \ell^p$ and then we will start to use subscript notation — $\phi_j(\zeta)$ rather than $\phi(\zeta)(j)$, α_j instead of $\alpha(j)$, *etc.* — and of course summation over j in place of integrals.

Proposition 5.4 *Let B_p denote the open unit ball of $L^p(\mu)$, $1 \leq p < \infty$. Then every nonconstant mapping ϕ of the form (5.1) where γ, $\alpha(\omega)$, $\beta(\omega)$ and $c(\omega)$ satisfy the conditions (5.2) is a complex geodesic in B_p.*

Note that at points where $c(\omega) = 0$, the values of $\alpha(\omega)$ and $\beta(\omega)$ are immaterial. Thus we can suppose if we wish that

$$\text{support}(\beta) = \{\omega : \beta(\omega) = 1\} \subset \{\omega : c(\omega) \neq 0\} = \text{support}(c)$$

and that support$(\alpha) \subset$ support(c). Since $(\zeta - \alpha(\omega))/(1 - \overline{\alpha(\omega)}\zeta)$ is a constant function of ζ when $|\alpha(\omega)| = 1$ (the constant is of modulus 1), we can also assume support$(\beta) \subset \{\omega : |\alpha(\omega)| < 1\}$. (Then there is no problem defining $\phi(\zeta)$ for $|\zeta| = 1$.) With these assumptions, the condition for ϕ to be non-constant is

$$\mu(\{\omega : c(\omega) \neq 0\} \cap \{\omega : \beta(\omega) = 1 \text{ or } \alpha(\omega) \neq \gamma\}) > 0.$$

Note also that condition (5.2)(b) implies that $c \in L^p(\mu)$.

PROOF: Observe that for $\delta = \left(\dfrac{2}{1-|\gamma|}\right)^{2/p}$,

$$|\phi(\zeta)(\omega)| \leq \delta |c(\omega)| \qquad (\zeta \in \mathbb{D}, \omega \in \Omega) \tag{5.3}$$

so that $\phi(\mathbb{D}) \subset L^p(\mu)$. Next if (ζ_n) is a sequence in $\mathbb{D}$ which converges to a point $\zeta \in \mathbb{D}$, then $\phi(\zeta_n)(\omega) \to \phi(\zeta)(\omega)$ for all ω. Hence by the Lebesgue dominated convergence theorem and (5.3), $\|\phi(\zeta_n) - \phi(\zeta)\|_p \to 0$ as $n \to \infty$. This shows that $\phi: \mathbb{D} \to L^p(\mu)$ is a continuous mapping.

For each fixed $\omega \in \Omega$ the map $\zeta \mapsto \phi(\zeta)(\omega)$ is analytic on $\mathbb{D}$ and from (5.3) we see that its Taylor series expansion

$$\phi(\zeta)(\omega) = \sum_n a_n(\omega)\zeta^n \tag{5.4}$$

with coefficients satisfying $|a_n(\omega)| \leq \delta|c(\omega)|$ (all n and all ω). Calculating the $a_n(\omega)$ using the binomial theorem and multiplication of power series, we can check that the functions $a_n(\omega)$ are measurable and then the estimate on the coefficients implies that $a_n \in L^p(\mu)$.

Now, for a fixed $\zeta \in \mathbb{D}$ the sequence $\sum_{j=0}^n a_j(\cdot)\zeta^j$ of measurable functions converges pointwise to $\phi(\zeta)(\cdot)$ as $n \to \infty$ by (5.4). Since

$$\left|\sum_{j=0}^n a_j(\omega)\zeta^j\right| \leq \delta|c(\omega)| \sum_{j=0}^{\infty} |\zeta|^j = \frac{\delta|c(\omega)|}{1-|\zeta|}$$

and $\delta c(\cdot)/(1-|\zeta|) \in L^p(\mu)$, the Lebesgue dominated convergence theorem shows that

$$\phi(\zeta)(\cdot) = \sum_{n=0}^{\infty} a_n(\cdot)\zeta^n$$

in $L^p(\mu)$ for each $\zeta \in \mathbb{D}$. Hence ϕ is holomorphic on $\mathbb{D}$.

For $\theta \in \mathbb{R}$, we have

$$\begin{aligned}
\|\phi(e^{i\theta})\|_p^p &= \int_\Omega |\phi(e^{i\theta})(\omega)|^p \, d\mu(\omega) \\
&= \int_\Omega |c(\omega)|^p \left|\frac{1-\overline{\alpha(\omega)}e^{i\theta}}{1-\gamma e^{i\theta}}\right|^2 d\mu(\omega) \\
&= \frac{1}{1+|\gamma|^2 - 2\mathrm{Re}\,(\gamma e^{i\theta})} \int_\Omega |c(\omega)|^p (1+|\alpha(\omega)|^2 - 2\mathrm{Re}\,(\overline{\alpha(\omega)}e^{i\theta}))\, d\mu(\omega) \\
&= 1
\end{aligned}$$

by conditions (5.2) (b) and (c). Hence $\phi(\partial\mathbb{D}) \subset \partial B_p$ and $\phi(\mathbb{D}) \subset \overline{B}_p$. Since ϕ is non-constant and all unit vectors in $L^p(\mu)$ are complex extreme points of the unit ball the strong maximum modulus theorem of Thorp and Whitley (see for instance [10, proposition 6.19]) implies that $\phi(\mathbb{D}) \subset B_p$.

We consider the dual space $(L^p(\mu))^*$ of $L^p(\mu)$ to be identified with $L^q(\mu)$ ($1/p + 1/q = 1$, $1 < q \leq \infty$) in a complex linear fashion (rather than the conjugate linear identification frequently used) so that $g \in L^q(\mu)$ acts on $f \in L^p(\mu)$ via $\langle f, g\rangle = \int_\Omega f(\omega)g(\omega)\,d\mu(\omega)$. If $f \in \partial B_p$, then one choice of a supporting hyperplane $N_f \in L^q(\mu)$ is given by

$$N_f(\omega) = |f(\omega)|^{p-2}\overline{f(\omega)}$$

(where $0^{p-2} = 0$ for all p).

To complete the proof, we apply Proposition 5.3 with

$$\begin{aligned} p(e^{i\theta}) &= |1 - \gamma e^{i\theta}|^2 \\ h(\zeta)(\omega) &= e(\omega)(1 - \overline{\alpha(\omega)}\zeta)^{2-2/p}(1 - \gamma\zeta)^{2/p}\left(\frac{\zeta - \alpha(\omega)}{1 - \overline{\alpha(\omega)}\zeta}\right)^{1-\beta(\omega)} \end{aligned}$$

with $e(\omega) = |c(\omega)|^{p-2}\overline{c(\omega)}$. Note that $e \in L^q(\mu)$. If $1 < p < \infty$ an argument similar to the one given above to show that ϕ is holomorphic on $\mathbb{D}$ and continuous on $\mathbb{D}$ shows that the same is true of h. This is more than enough to show that $h \in H^\infty_*(\mathbb{D}, (L^p(\mu))^*)$. Since $h(e^{i\theta}) = e^{i\theta}p(e^{i\theta})N_{\phi(e^{i\theta})}$, this shows that ϕ is a complex geodesic when $p > 1$.

When $p = 1$,

$$h(\zeta)(\omega) = (1 - \gamma\zeta)^2 e(\omega)\left(\frac{\zeta - \alpha(\omega)}{1 - \overline{\alpha(\omega)}\zeta}\right)^{1-\beta(\omega)}$$

and $\|e\|_\infty \leq 1$. It is quite easy to see that h is holomorphic on $\mathbb{D}$ and that $\|h(\zeta)\|_\infty \leq (1 + |\gamma|)^2$ for all $\zeta \in \mathbb{D}$. For each $\omega \in \Omega$ and $\theta \in \mathbb{R}$, $\lim_{r\to 1^-} h(re^{i\theta})(\omega) = h(e^{i\theta})(\omega)$. Using the boundedness of h and the Lebesgue dominated convergence theorem it is then easy to see that h has weak* radial limits at all points $e^{i\theta} \in \partial\mathbb{D}$, *i.e.* that

$$\lim_{r\to 1^-}\langle f, h(re^{i\theta})\rangle = \langle f, h(e^{i\theta})\rangle$$

for all $f \in L^1(\mu)$. Thus $h \in H^\infty_*(\mathbb{D}, (L^1(\mu))^*)$ and ϕ is also a complex geodesic in the $p = 1$ case.

Proposition 5.5 *Let B_p denote the open unit ball of ℓ^p, $1 \leq p < \infty$. Then any two distinct points in B_p can be joined by a normalized complex geodesic ϕ of the form given in (5.1).*

That is, there exists a complex geodesic $\phi(\zeta) = (\phi_j(\zeta))_{j=1}^\infty$ joining the points where

$$\phi_j(\zeta) = c_j\left(\frac{\zeta - \alpha_j}{1 - \overline{\alpha_j}\zeta}\right)^{\beta_j}\left(\frac{1 - \overline{\alpha_j}\zeta}{1 - \gamma\zeta}\right)^{2/p} \tag{5.5}$$

$\gamma \in \mathbb{D}$, $\alpha_j \in \mathbb{D}$, $c_j \in \mathbb{C}$, $\beta_j = 0$ or 1, $\sum_j |c_j|^p(1 + |\alpha_j|^2) = 1 + |\gamma|^2$ and $\sum_j |c_j|^p\alpha_j = \gamma$.

PROOF: The existence of some complex geodesic joining the two points follows from Theorem 2.5 (or from [11]). We also know that all complex geodesics in B_p are continuous from Theorem 4.4. For the $p = 1$ case the uniform complex convexity hypothesis was established by Globevnik [17] (see Examples 4.5) and the $p > 1$ case is more straightforward because ℓ^p is uniformly convex in the real sense (see [8]).

Our proof that there exists geodesics of the required form relies heavily on finite-dimensional results (*i.e.* results for the case of ℓ^p_n) of Poletskiĭ [26] and Gentili [16]. For $1 < p < \infty$, Poletskiĭ [26] proved that all geodesics in the unit ball $B_{p,n}$ of ℓ^p_n are of the above form (except that he omitted the possibility that β_j could be 0). Gentili [16, Theorem 6] proved that all continuous complex geodesics in $B_{1,n}$ are of the above form. Now that we know all complex geodesics in $B_{1,n}$ are continuous, it follows that all complex geodesics in $B_{1,n}$ are of the above form. (We could actually circumvent Gentili's result. By taking a limiting argument based on Poletskiĭ's result and the fact that $\bigcap_{p>1} B_{p,n} = B_{1,n}$, we could establish the existence of complex geodesics of the required form joining any pair of points in $B_{1,n}$.)

Now let $x, y \in B_p$ be two distinct points.

We consider ℓ^p_n as being identical with the subspace $\{(z_1, z_2, \ldots, z_n, 0, 0, \ldots)\}$ of ℓ_p and we will use the notation $x^{(n)}$ for the natural projection $(x_1, x_2, \ldots, x_n, 0, \ldots)$ of $x = (x_1, x_2, \ldots)$ into ℓ^p_n. Similarly for $y^{(n)}$. Consideration of the inclusion map : $B_{\ell^p_n} \to B_p$ and the projection $z \mapsto z^{(n)}: B_p \to B_{\ell^p_n}$, which are both holomorphic and therefore contractions with respect to the Carathéodory distance, shows that

$$C_{\ell^p_n}(x^{(n)}, y^{(n)}) = C_{\ell^p}(x^{(n)}, y^{(n)}) \leq C_{\ell^p}(x, y) \quad (x, y \in B_p).$$

Let $\phi^{(n)}$ denote a normalized complex geodesic in the unit ball of ℓ^p_n with $\phi^{(n)}(0) = x^{(n)}$ and $\phi^{(n)}(s_n) = y^{(n)}$, where $s_n = \tanh^{-1} C_{\ell^p}(x^{(n)}, y^{(n)}) > 0$. (Strictly speaking this may not make sense for small n when it may happen that $x^{(n)} = y^{(n)}$.) Let the parameters associated with $\phi^{(n)}$ be denoted $\alpha_j^{(n)}$, $\beta_j^{(n)}$, $\gamma^{(n)}$ and $c_j^{(n)}$ $(1 \leq j \leq n)$.

There is a subsequence of n's along which we have

$$\alpha_j^{(n)} \to \alpha, \quad \beta_j^{(n)} \to \beta_j, \quad \gamma^{(n)} \to \gamma, \quad c_j^{(n)} \to c_j$$

for each j. Since $\sum_j |c_j^{(n)}|^p < 2$ for all n, we have $\sum_j |c_j|^p \leq 2$. We claim that $|\gamma| < 1$. Observe that

$$\phi_j^{(n)}(\zeta) \to \phi_j(\zeta) = c_j \left(\frac{\zeta - \alpha_j}{1 - \alpha_j \zeta}\right)^{\beta_j} \left(\frac{1 - \alpha_j \zeta}{1 - \gamma \zeta}\right)^{2/p}$$

uniformly on compact subsets of $\mathbb{D}$ as $n \to \infty$ along the subsequence. Hence $\phi_j(0) = x_j$ and $\phi_j(s) = y_j$ where

$$s = \lim_n s_n = \lim_n \tanh^{-1} C_{\ell^p}(x^{(n)}, y^{(n)}) = \tanh^{-1} C_{\ell^p}(x, y).$$

Thus we can pick a j with ϕ_j non-constant. If $|\gamma| = 1$, then ϕ_j would be unbounded on $\mathbb{D}$ unless $\alpha_j = \gamma$. ϕ_j unbounded leads to a contradiction since each $\phi_j^{(n)}$ is bounded by 1 and in the case $\alpha_j = \gamma$, ϕ_j would be constant. Hence $|\gamma| < 1$. Using $\sum_j |c_j|^p < \infty$ we can argue as in the proof of Proposition 5.4 to show that

$$\phi = (\phi_1, \phi_2, \ldots) \colon \mathbb{D} \to \ell^p$$

is holomorphic on $\mathbb{D}$ and continuous on $\bar{\mathbb{D}}$.

Since $\sum_j |\phi_j^{(n)}(\zeta)|^p \leq 1$ for all n, we have $\sum_j |\phi_j(\zeta)|^p \leq 1$ for $|\zeta| < 1$. Since $\phi(0) = x$ we have $\phi(\mathbb{D}) \subset B_p$ (see [33, p. 376]) and since $\phi(s) = y$, ϕ must be a complex geodesic. It follows that $\|\phi(\zeta)\|_p^p = 1$ for $|\zeta| = 1$, and expanding this out as in the proof of Proposition 5.4 we find

$$1 + |\gamma|^2 - 2\mathrm{Re}(\gamma\zeta) = \sum_{j=1}^{\infty} |c_j|^p (1 + |\alpha_j|^2 - 2\mathrm{Re}(\alpha\zeta)) \qquad (|\zeta| = 1).$$

As both sides are harmonic for $\zeta \in \mathbb{D}$ and continuous on $\bar{\mathbb{D}}$, the equality remains valid for $\zeta \in \mathbb{D}$. Putting $\zeta = 0$ gives $\sum_j |c_j|^p (1 + |\alpha_j|^2) = 1 + |\gamma|^2$ and then it follows that $\sum_j |c_j|^p \alpha_j = \gamma$.

Lemma 5.6 *Let $f \in H^\infty(\mathbb{D})$, $\gamma \in \mathbb{D}$. Then*

$$\frac{f(e^{i\theta})}{\left(\dfrac{e^{i\theta} - \gamma}{1 - \gamma e^{i\theta}}\right)}$$

is a non-negative real number for almost all $\theta \in \mathbb{R}$ if and only if

$$f(\zeta) = t \left(\frac{\zeta - \alpha}{1 - \alpha\zeta}\right) \left(\frac{1 - \alpha\zeta}{1 - \gamma\zeta}\right)^2$$

for some $t \geq 0$, $|\alpha| \leq 1$.

PROOF: The case $\gamma = 0$ is due to Gentili [16] (see lemma 2 and the proof of theorem 6). The general case follows from Gentili's result by the change of variables $\eta = (\zeta - \gamma)/(1 - \gamma\zeta)$.

Lemma 5.7 *If $X = \ell_n^1$ or $X = \ell^1$ and $x \neq y \in B_X$, then there is a unique normalized complex geodesic in B_X joining x and y.*

PROOF: As already noted, we know that there exists a normalized complex geodesic ϕ in B_X joining x and y. In fact we have an explicit form (5.5) of one such $\phi = (\phi_j)_j$ by Proposition 5.5.

Now suppose ψ is a second normalized complex geodesic joining x and y. Suppose $\phi(0) = \psi(0) = x$ and $\phi(s) = \psi(s) = y$ where $s > 0$. The argument given earlier in the proof of

Theorem 3.2 shows that $f = \lambda\phi + (1-\lambda)\psi$ is also a complex geodesic joining x and y. Thus $\|f(e^{i\theta})\| = 1$ for all $\theta \in \mathbb{R}$. It follows that we must have equality in the triangle inequality

$$1 = \|f(e^{i\theta})\| = \sum_j |\lambda\phi_j(e^{i\theta}) + (1-\lambda)\psi_j(e^{i\theta})| \leq \sum_j \lambda|\phi_j(e^{i\theta})| + (1-\lambda)|\psi_j(e^{i\theta})| = 1.$$

This forces

$$\frac{\psi_j(e^{i\theta})}{\phi_j(e^{i\theta})}$$

to be a non-negative real number for all θ except those for which the denominator is zero.

To prove that $\psi_j = \phi_j$, we consider the case $c_j = 0$ and $c_j \neq 0$ separately. In the first case $\phi_j \equiv 0$ and hence $x_j = y_j = 0$. Since linear isometries of X map complex geodesics in B_X to complex geodesics, ψ and

$$\tilde{\psi} = (\psi_1, \dots, \psi_{j-1}, -\psi_j, \psi_{j+1}, \dots)$$

are both normalized complex geodesics joining x and y. Therefore, so is $g = (\psi + \tilde{\psi})/2$. Since $\|g(e^{i\theta})\| = \|\psi(e^{i\theta})\| = 1$ for all θ, it easily follows that $\psi_j \equiv 0 \equiv \phi_j$.

If $c_j \neq 0$, Lemma 5.6 applied to the function

$$\frac{\phi_j(\zeta)}{c_j}\left(\frac{\zeta - \alpha_j}{1 - \alpha_j\zeta}\right)^{1-\beta_j}$$

shows that

$$\frac{\phi_j(e^{i\theta})}{c_j}\left(\frac{e^{i\theta} - \alpha_j}{1 - \alpha_j e^{i\theta}}\right)^{1-\beta_j} \Big/ \left(\frac{e^{i\theta} - \gamma}{1 - \gamma e^{i\theta}}\right)$$

is non-negative for almost all θ. Therefore the same holds for $\psi_j(e^{i\theta})$ in place of $\phi_j(e^{i\theta})$ and Lemma 5.6 then shows that there exist $t > 0$ and $|\beta| \leq 1$ so that

$$\psi_j(\zeta) = tc_j\left(\frac{\zeta - \alpha_j}{1 - \alpha_j\zeta}\right)^{\beta_j - 1}\left(\frac{\zeta - \beta}{1 - \beta\zeta}\right)\left(\frac{1 - \beta\zeta}{1 - \gamma\zeta}\right)^2.$$

If $\beta_j = 0$ and $|\alpha_j| < 1$, then analyticity of $\psi_j(\zeta)$ forces $\alpha_j = \beta_j$ and two terms cancel in the expression for $\psi_j(\zeta)$. Then from $\phi_j(0) = \psi_j(0) = x_j$, we conclude that $c_j = tc_j$ and $t = 1$. So $\phi_j = \psi_j$ in this situation.

In the remaining cases, $\phi_j(0) = \psi_j(0) = x_j$ yields

$$tc_j(-\alpha_j)^{\beta_j - 1}(-\beta) + c_j(-\alpha_j)^{\beta_j}$$

and hence $\alpha_j = t\beta$. Then $\psi_j(s) = \phi_j(s) = y_j$ and some cancellation of common terms shows that

$$(s - \alpha_j)(1 - \alpha_j s) = t(s - \beta)(1 - \beta s).$$

Combining this with $\alpha_j = t\beta$ yields

$$(1 - t)s(1 - t|\beta|^2) = 0.$$

Hence $t = 1$ or $t = |\beta|^{-2}$. In the second case,

$$\alpha_j = t\beta = 1/\beta.$$

Since $|\alpha_j| \leq 1$ and $|\beta| \leq 1$, we must have $|\alpha_j| = |\beta| = 1$ so that $t = 1$. Hence $\alpha_j = \beta$ and $\psi_j(\zeta) = \phi_j(\zeta)$.

We can summarise our results for ℓ^p as follows.

Corollary 5.8 *Let B_p denote the unit ball of ℓ^p, $1 \leq p < \infty$. Then*

(i) Any two distinct points in B_p can be joined by a unique normalized complex geodesic.

(ii) All complex geodesics in B_p are continuous.

(iii) A map $\phi: \mathbb{D} \to B_p$ is a complex geodesic if and only if it is a non-constant map of the form given in Proposition 5.5

Proof: For all $1 \leq p < \infty$, existence follows from Theorem 2.5 (or from [11]). Uniqueness for $p > 1$ follows from Theorem 3.2 and uniqueness for $p = 1$ has just been established in Lemma 5.7.

(ii) follows from Theorem 4.4. For $p = 1$, this has already been noted in Examples 4.5. It is straightforward that Theorem 4.4 applies to ℓ^p for $1 < p < \infty$ because ℓ^p is uniformly convex in the real sense (see [8]).

(iii) follows from Proposition 5.5 and (i).

We suspect that a more general version of this result holds for $L^p(\mu)$ in place of ℓ^p, but we have not managed to prove L^p-versions of Propositions 5.5 or 5.7.

Example 5.9 Let $X = \ell^{p_1} \oplus_r \ell^{p_2} = \{x = (y, z) : y \in \ell^{p_1}, z \in \ell^{p_2}\}$ normed by $\|x\| = \left(\|y\|_{p_1}^r + \|z\|_{p_2}^r\right)^{1/r}$.

One can check using Proposition 5.3 that for $1 \leq p_i < \infty$, $1 \leq r < \infty$ all nonconstant maps $\phi: \mathbb{D} \to B_X$ of the following form are complex geodesics.

$$\begin{aligned}
\phi(\zeta) &= ((\phi_{1j})_{j=1}^{\infty}, (\phi_{2j})_{j=1}^{\infty}) \\
\phi_{ij}(\zeta) &= c_{ij} \left(\frac{\zeta - \alpha_{ij}}{1 - \bar{\alpha}_{ij}\zeta}\right)^{\beta_{ij}} \left(\frac{1 - \bar{\alpha}_{ij}\zeta}{1 - \bar{\gamma}_i\zeta}\right)^{2/p_i} \left(\frac{1 - \bar{\gamma}_i\zeta}{1 - \bar{\gamma}\zeta}\right)^{2/r}
\end{aligned}$$

where $|\alpha_{ij}| \leq 1$, $|\gamma_i| < 1$, $|\gamma| < 1$, β_{ij} is 0 or 1, and the following relations hold

$$\begin{aligned}
\sum_j |c_{ij}|^{p_i} \alpha_{ij} &= \gamma c_i^{p_i} \qquad (i = 1, 2) \\
c_1^r \gamma_1 + c_2^r \gamma_2 &= \gamma \\
c_1^r(1 + |\gamma_1|^2) + c_2^r(1 + |\gamma_2|^2) &= 1 + |\gamma|^2
\end{aligned}$$

where

$$c_i = \left(\frac{1}{1+|\gamma_i|^2}\sum_j |c_{ij}|^{p_i}(1+|\alpha_{ij}|^2)\right)^{1/p_i} \qquad (i=1,2).$$

The proof of this involves observing first that for $x=(y,z)\in X$ with $\|x\|=1$,

$$N_x = (\|y\|^{r-1}N_{y/\|y\|}, \|z\|^{r-1}N_{z/\|z\|})$$

with $N_{y/\|y\|}$ and $N_{z/\|z\|}$ given as in the proof of Proposition 5.4 for ℓ^p. To apply Proposition 5.3, take $p(\zeta)=|1-\gamma\zeta|^2$ and

$$h(\zeta) = \left(e_{ij}(1-|\alpha_{ij}|^2)^{2-2/p_i}(1-\gamma_i\zeta)^{2/p_i-2/r}(1-\gamma\zeta)^{2/r}\left(\frac{\zeta-\alpha_{ij}}{1-\alpha_{ij}\zeta}\right)^{1-\beta_{ij}}\right)_{ij}$$

where

$$e_{ij} = c_i^{r-p_i}|c_{ij}|^{p_i-2}c_{ij} \quad (i=1,2; 1\le j<\infty).$$

We suspect that all complex geodesics in B_X are of this form. Other examples of complex geodesics in spaces which are direct sums of more than two summands of ℓ^p-type can also be exhibited.

Remark 5.10 The case $p=\infty$ is excluded in all of the previous calculations because it is well known that almost everything is different for ℓ^∞ (and L^∞). Even for the unit ball $B_{\infty,2}=\{(z_1,z_2)\in\mathbb{C}^2 : \max_i|z_i|<1\}$ (polydisc) of ℓ_2^∞, many of the differences are apparent. The only points of $\partial B_{\infty,2}$ that are complex extreme points are those where $|z_1|=|z_2|=1$ and therefore the result of Vesentini [33] cited at the beginning of Section 3 shows that there are many complex geodesics joining 0 to $z=(z_1,z_2)$ if $|z_1|\ne|z_2|$. In fact, if $|z_2|<|z_1|$, the normalized complex geodesics joining 0 to (z_1,z_2) are

$$\phi(\zeta) = \left(\zeta\frac{z_1}{|z_1|}, g(\zeta)\right)$$

where g is any analytic function on $\mathbb{D}$ with $g(0)=0$, $g(|z_1|)=z_2$ and $\sup_{\zeta\in\mathbb{D}}|g(\zeta)|<1$. (This can easily be verified using the fact that the Kobayashi distance between $x,y\in B_{\infty,2}$ is given by $\max_{i=1,2}\rho(x_i,y_i)$.) Thus we see non-uniqueness and discontinuity of complex geodesics. Since $B_{\infty,2}$ is a homogeneous domain, for any pair of points $x,y\in B_{\infty,2}$, we can find a biholomorphic automorphism F with $F(0)=x$ (in fact $F(z_1,z_2)=((x_1-z_1)/(1-x_1z_1),(x_2-z_2)/(1-x_2z_2))$ will do). Then for any normalized complex geodesic ϕ joining 0 and $F^{-1}(y)$, $F\circ\phi$ is a normalized complex geodesic joining x and y.

The infinite dimensional case ℓ^∞ is somewhat similar to the finite dimensional one. We state without proof the following result (B_∞ is the unit ball of ℓ^∞).

A map $\phi = (\phi_1, \phi_2, \ldots): \mathbb{D} \to B_\infty$ *is holomorphic if and only if each coordinate function* $\phi_j(\zeta)$ *is analytic.*

One can check that ϕ is a complex geodesic if and only if either there exists j for which ϕ_j is an automorphism of $\mathbb{D}$ or else there is a subsequence of $(\phi_j)_j$ which converges to an automorphism. This can be done directly or via a result of Gentili [15] who gives a necessary and sufficient conditions for a holomorphic map $\phi: X \to B_X$ to be a complex geodesic in the case when $X = C(K)$ is the space of continuous functions on a compact Hausdorff space K. The condition is that there exists $k \in K$ so that $\zeta \mapsto \phi(\zeta)(k)$ is an automorphism of the unit disc. Since ℓ^∞ is the same as the space of continuous functions on the Stone-Čech compactification of the integers, we can apply his result. There will be more than one complex geodesic joining $x, y \in B_\infty$ unless $|(x_j - y_j)/(1 - x_j y_j)|$ is constant.

Similar remarks apply to c_0. In this case there are no complex extreme points on the unit sphere and there is more than one normalized complex geodesic joining every pair of points in the unit ball.

6 TAUTNESS AND CURVATURE

In previous sections, we have used (for convex domains) convergence principles to establish the existence of holomorphic mappings with certain extremal properties. These convergence properties have been formalised for finite-dimensional domains and manifolds and certain relationships established. In this section, we show that these results do not extend to arbitrary Banach spaces, even for convex bounded domains and give a result on curvature of the Kobayashi infinitesimal metric.

We define hyperbolic, complete hyperbolic and taut domains in a Banach space. The first two definitions are standard, while various versions of the third are possible.

Definition 6.1 *A domain* $\mathcal{D}$ *in a Banach space* X *is called hyperbolic if* $K_\mathcal{D}$ *induces the original topology on* $\mathcal{D}$*. If, moreover,* $(\mathcal{D}, K_\mathcal{D})$ *is a complete metric space, we call* $\mathcal{D}$ *a complete hyperbolic domain.*

If $\mathcal{D}$ is hyperbolic, then $K_\mathcal{D}$ is a distance on $\mathcal{D}$ (*i.e.* $K_\mathcal{D}$ separates the points of $\mathcal{D}$) and the converse is true for finite dimensional domains (see [28, 19]). We do not know if the converse is true for infinite dimensional Banach spaces. Harris [19, theorem 24] has proved that a convex bounded domain in a Banach space is complete hyperbolic and Barth [3] has proved that a convex domain in $\mathbb{C}^n$ which contains no complex lines is hyperbolic (and indeed biholomorphically equivalent to a bounded domain). Again, we do not know if this result extends to infinite dimensional Banach spaces.

Definition 6.2 *A domain $\mathcal{D}$ in a Banach space X is taut if there exists a Hausdorff locally convex topology τ on X such that every net $(f_\alpha)_\alpha$ in $H(\mathbb{D}, \mathcal{D})$ contains either a compactly divergent subnet or a subnet which converges with respect to τ, uniformly on compact subsets of $\mathbb{D}$, to some $f \in H(\mathbb{D}, \mathcal{D})$.*

A net $(f_\alpha)_{\alpha \in \Gamma}$ is compactly divergent if, given K compact in $\mathbb{D}$ and L a norm-compact subset of $\mathcal{D}$, there exists $\alpha_0 \in \Gamma$ such that $f_\alpha(K) \cap L = \emptyset$ for all $\alpha \geq \alpha_0$.

For $\mathcal{D}$ finite dimensional, τ has to be the same as the norm topology and we have

$$\text{complete hyperbolic} \Rightarrow \text{taut} \Rightarrow \text{hyperbolic}$$

(and the converses are both false [22, 2]). Also, for finite dimensional domains the unit disc $\mathbb{D}$ can be replaced by any finite dimensional domain in the definition of tautness [1, 39].

Proposition 6.3 *The unit ball B_X of a Banach space X is taut if and only if X is isometrically isomorphic to a dual space.*

Moreover, if X contains any bounded convex domain $\mathcal{D}$ which is taut (for a locally convex topology τ) then X is isomorphic to a dual Banach space and the topology τ is weaker than the norm topology.

PROOF: We first suppose that B_X is taut and that τ is a locally convex Hausdorff topology associated with tautness.

Let $(x_\alpha)_\alpha$ be a net in B_X and for each α let $f_\alpha(\zeta) = \zeta x_\alpha$ for $\zeta \in \mathbb{D}$. Then f_α is a net in $H(\mathbb{D}, B_X)$ and, since $f_\alpha(0) = 0$ for all α, it contains no compactly divergent subnet. Therefore it must have a subnet $(f_\beta)_\beta$ which converges relative to τ, uniformly on compact subsets of $\mathbb{D}$, to some $f \in H(\mathbb{D}, B_X)$. Since τ is a locally convex topology, it is easily seen that $f(\zeta) = \zeta x$ for some $x \in X$. Because $\|f(\zeta)\| < 1$ for all $\zeta \in \mathbb{D}$, we must have $\|x\| \leq 1$ and moreover $x_\beta \to x$ (with respect to τ). This shows that (B_X, τ) is compact. A result of Ng [25] now implies that X is isometrically isomorphic to a dual space.

Conversely, suppose now that X is isometrically isomorphic to the dual of a Banach space Y. Then B_X is compact in the weak*- or $\sigma(X,Y)$-topology. Let $(f_\alpha)_\alpha$ denote a net in $H(\mathbb{D}, B_X)$. Consider the Taylor series expansions

$$f_\alpha(\zeta) = \sum_n a_{\alpha,n}\zeta^n.$$

From the Cauchy formula, we have $\|a_{\alpha,n}\| \leq 1$ for all α and n. By compactness of the product of infinitely many copies of B_X (for the product $\sigma(X,Y)$-topology) we can find a subnet $(f_\beta)_\beta$ so that $a_{\beta,n} \to a_n \in B_X$ for all n.

If $\|a_0\| = 1$, then for any $0 < r < 1$, $f_\beta(0) = a_{\beta,n} \notin rB_X$ if β is sufficiently large. Thus $\|f_\beta(0)\| \to 1$. Now

$$\begin{aligned} \tanh^{-1}\|f_\beta(\zeta)\| &= K_X(f_\beta(\zeta), 0) \\ &\geq K_X(f_\beta(0), 0) - K_X(f_\beta(\zeta), f_\beta(0)) \\ &\geq \tanh^{-1}\|f_\beta(0)\| - \rho(\zeta, 0) \end{aligned}$$

(here we have used the distance decreasing property for the Kobayashi metric) shows that $\|f_\beta(\zeta)\| \to 1$ uniformly on compact subsets of $\mathbb{D}$. Thus f_β is compactly divergent.

For the case $\|a_0\| < 1$, let $f(\zeta) = \sum_n a_n\zeta^n$. Since $\|a_n\| \leq 1$ for all n, clearly $f \in H(\mathbb{D}, X)$. Fix $y \in Y$, $0 < r < 1$ and $\epsilon > 0$. Choose N so that $\sum_{n=N+1}^{\infty} r^n < \epsilon$. Then, by considering power series, we obtain

$$\begin{aligned} \sup_{|\zeta|\leq r} |\langle y, f_\beta(\zeta) - f(\zeta)\rangle| &\leq \sum_{n=0}^{N} |\langle y, a_{\beta,n} - a_n\rangle| + 2\sum_{n=N+1}^{\infty} r^n\|y\| \\ &\leq \sum_{n=0}^{N} |\langle y, a_{\beta,n} - a_n\rangle| + 2\epsilon\|y\| \end{aligned}$$

for all β. Hence $f_\beta \to f$ with respect to $\sigma(Y, X)$, uniformly on compact subsets of $\mathbb{D}$.

Since $f_\beta(\zeta) \in B_X$ for all ζ and β, it follows that f has values in B_X. Since $f(0) = a_0 \in B_X$, $f \in H(\mathbb{D}, B_X)$. This completes the proof that B_X is taut when X is a dual space.

It remains to show the last part of the proposition. Suppose $\mathcal{D}$ is a taut bounded convex domain in X. We can assume that $0 \in \mathcal{D}$ by a translation. A simple argument based on compactness of the unit circle shows that $\mathcal{D}_0 = \bigcap_{\theta\in\mathbb{R}} e^{i\theta}\mathcal{D}$ is a bounded balanced convex domain. By replacing the norm on X with an equivalent one, we can assume that $\mathcal{D}_0 = B_X$. The same argument as used at the beginning of the proof, together with the observation that a map $f: \mathbb{D} \to X$ of the form $f(\zeta) = \zeta x$ has its values in $\mathcal{D}$ if and only if it has values in $\mathcal{D}_0 = B_X$, shows that X is a dual space. Moreover the proof of Ng [25] implies that τ is weaker than the norm topology. He shows that X is the dual of a certain subspace $Y \subset X^*$ and that the identity map $id: (B_X, \tau) \to (B_X, \sigma(X, Y))$ is a homeomorphism. It follows that $id: (B_X, \|\cdot\|) \to (B_X, \tau)$ is continuous, and hence that $id: (X, \|\cdot\|) \to (X, \tau)$ is continuous.

We remark that we could have appealed to this proposition for the proof of Proposition 5.5 in place of the more direct argument we used.

Example 6.4 For $X = c_0$, B_X is complete hyperbolic (by [19]) but not taut.

The second part of the proof of Proposition 6.3 can easily be modified to show that the following are taut

(a) convex bounded domains in reflexive Banach spaces (for τ the weak topology);

(b) convex bounded domains in dual Banach spaces with the property that their norm closures are weak*-compact (τ the weak*-topology).

Proposition 6.5 *If $\mathcal{D}$ is a convex bounded taut domain in a Banach space X, then $\mathcal{D}$ is C-connected.*

PROOF: Given two points $x, y \in \mathcal{D}$, choose $(f_n)_n$ in $H(\mathbb{D}, \mathcal{D})$ so that $f_n(0) = x$, $f_n(s_n) = y$, $s_n \in (0, 1)$ for all n and $s_n \to s = \tanh K_{\mathcal{D}}(x, y)$ as $n \to \infty$.

Since $\mathcal{D}$ is taut and $f_n(0) = x$ for all n, it follows that $(f_n)_n$ has a τ-convergent subnet (τ being the locally convex topology related to tautness of $\mathcal{D}$). Let $f \in H(\mathbb{D}, \mathcal{D})$ denote the limit of one such subnet. Clearly $f(0) = x$ and we claim $f(s) = y$ (which will show that f is a complex geodesic by convexity and Proposition 1.2).

For $r = (1 + s)/2$, we have

$$f_n(s_n) - f_n(s) = \frac{s_n - s}{2\pi i} \int_{|z|=r} \frac{f_n(z)}{(z - s_n)(z - s)}\, dz.$$

Since $\mathcal{D}$ is bounded and $s_n \to s$, it follows that $f_n(s_n) - f_n(s) \to 0$ in norm (hence also in the topology τ by Proposition 6.3) as $n \to \infty$. Now if $(f_{n_\alpha})_{\alpha \in \Gamma}$ is a subnet of $(f_n)_n$ which converges to f uniformly with respect to τ on compact subsets of $\mathbb{D}$, then

$$f_{n_\alpha}(s_{n_\alpha}) - f(s) = f_{n_\alpha}(s_{n_\alpha}) - f_{n_\alpha}(s) + f_{n_\alpha}(s) - f(s) \to 0$$

in the topology τ. Since $f_{n_\alpha}(s_{n_\alpha}) = y$, it follows that $f(s) = y$.

We observe that one can similarly prove an infinitesimal version of Proposition 6.5: *if $\mathcal{D}$ is a convex bounded taut domain in a Banach space X, $x \in \mathcal{D}$ and $0 \neq v \in X$, then there exists a complex geodesic $f \in H(\mathbb{D}, \mathcal{D})$ such that $f(0) = x$ and $f'(0) = v/k_{\mathcal{D}}(x, v)$.*

Definition 6.6 *If $\mathcal{D}$ is a bounded domain in a Banach space X, then the holomorphic sectional curvature of the Kobayashi infinitesimal metric $k_{\mathcal{D}}$ at a point $x \in \mathcal{D}$ in the (nonzero) direction $v \in X$ is*

$$\kappa_k(x, v) = \sup\left\{ \frac{\triangle \log k_{\mathcal{D}}^2(f(z), f'(z))\,|_{z=0}}{-2k_{\mathcal{D}}^2(f(0), f'(0))} : f \in H(r\mathbb{D}, \mathcal{D}), r > 0, f(0) = x, f'(0) = v \right\}.$$

Since $k_{\mathcal{D}}$ may not even be continuous in general, $\triangle$ above denotes the generalised Laplacian. This is defined (motivated by [21], see [4]) for upper semicontinuous functions u with values in $[-\infty, \infty)$ (but not identically $-\infty$) at points z with $u(z) \neq -\infty$ by

$$\triangle u(z) = 4 \liminf_{r \to 0^+} \frac{1}{r^2} \left\{ \frac{1}{2\pi} \int_0^{2\pi} u(z + re^{i\theta})\, d\theta - u(z) \right\}.$$

At a local maximum point z, $\triangle u(z) \leq 0$ and at a finite local minimum $\triangle u(z) \geq 0$.

The following result for finite-dimensional domains is due to Wong [38], Burbea [4] and Suzuki [30, 31].

Proposition 6.7 *If $\mathcal{D}$ is a taut bounded convex domain in a Banach space X, then*

$$\kappa_k(x, v) = -4 \quad (x \in \mathcal{D}, 0 \neq v \in X).$$

PROOF: We will use the fact that this result is true for the unit disc $\mathbb{D}$ (see for instance [10, §3.4]) — in fact every suitable function f attains the supremum in Definition 6.6 when $\mathcal{D} = \mathbb{D}$ and the supremum is -4.

For general $\mathcal{D}$ and f as in the definition of $\kappa_k(x, v)$, let $\lambda_f(z) = c_{\mathcal{D}}^2(f(z), f'(z))$ (which coincides with $k_{\mathcal{D}}^2(f(z), f'(z))$ by convexity of $\mathcal{D}$). By Montel's theorem, we can find $g \in H(\mathcal{D}, \mathbb{D})$ such that $g(x) = 0$ and

$$c_{\mathcal{D}}(x, v) = |g'(x)(v)| = c_{\mathbb{D}}(g(x), g'(x)(v)) = \alpha(g(x), g'(x)(v)).$$

Let $\beta(z) = \alpha^2(g(f(z)), g'(f(z))(f'(z)))$ for $z \in r\mathbb{D}$. Observe that

$$\beta(z) \leq c_{\mathcal{D}}^2(f(z), f'(z)) = \lambda_f(z)$$

and $\beta(0) = \alpha^2(g(x), g'(x)(v)) = c_{\mathcal{D}}^2(x, v) = \lambda_f(0)$. Since $v \neq 0$ and $\mathcal{D}$ is bounded, there is a neighbourhood of 0 where $\beta(z)$ does not vanish. Hence $\log(\lambda_f/\beta)$ has a local minimum at the origin and therefore

$$\triangle \log \frac{\lambda_f}{\beta}\bigg|_{z=0} = \triangle \log \lambda_f(0) - \triangle \log \beta(0) \geq 0,$$

where we have used the fact that $\log \beta$ is twice continuously differentiable (and hence the lim inf in the definition of $\triangle \log \beta$ is a limit). It follows that

$$\frac{\triangle \log \lambda_f(0)}{-2\lambda_f(0)} \leq \frac{\triangle \log \beta(0)}{-2\beta(0)} = -4$$

and hence that $\kappa_k(x, v) \leq -4$.

Since $\mathcal{D}$ is taut and bounded, there exists a complex geodesic $f \in H(\mathbb{D}, \mathcal{D})$ such that $f(0) = x$ and $f'(0) = v/k_{\mathcal{D}}(x, v)$ (by the infinitesimal version of Proposition 6.5). For this f we have

$$k_{\mathcal{D}}^2(f(z), f'(z)) = \alpha^2(z, 1) = \frac{1}{1 - |z|^2}$$

(see Proposition 1.2) and the one variable result shows that f attains the upper bound of -4.

Corollary 6.8 *If X is a dual Banach space, then the Kobayashi infinitesimal metric k_X on B_X has constant holomorphic sectional curvature -4.*

References

[1] T.J. Barth, *Taut and tight complex manifolds*, Proc. Amer. Math. Soc. **24** (1970) 429–431.

[2] T.J. Barth, *Some counterexamples concerning intrinsic distances*, Proc. Amer. Math. Soc. **66** (1977) 49–53.

[3] T.J. Barth, *Convex domains and Kobayashi hyperbolicity*, Proc. Amer. Math. Soc. **79** (1980) 556–558.

[4] J. Burbea, *On the Hessian of the Carathéodory metric*, Rocky Mountain J. Math. **8** (1978) 555–559.

[5] C. Carathéodory, *Ueber eine spezielle Metrik, die in der Theorie der analytischen Functionen auftritt*, Atti Pontifice Acad. Sc., Nuovi Lincei **80** (1927) 135–141.

[6] A.A. Danilevich, *Some boundary properties of abstract analytic functions and their applications*, Mat. Sbornik **100** (1976) 507–533 [Russian], translated in Math. USSR-Sbornik **29** (1976) 453–474.

[7] A.M. Davie, T.W. Gamelin, *A theorem on polynomial star approximation*, (preprint, 1989).

[8] W.J. Davis, D.J.H. Garling, N. Tomczak-Jaegermann, *The complex convexity of quasi-normed linear spaces*, J. Func. Anal. **55**(1984) 110-150.

[9] S. Dilworth, *Complex convexity and the geometry of Banach spaces*, Math. Proc. Camb. Phil. Soc. **99** (1986) 495–506.

[10] S. Dineen, The Schwarz Lemma, Oxford mathematical monographs (1989).

[11] S. Dineen, R.M. Timoney, J.-P. Vigué, *Pseudodistances invariantes sur les domaines d'un espace localement convexe*, Ann. Sc. Normale Pisa **12** (1985) 515–529.

[12] P.L. Duren, Theory of H^p Spaces, Academic Press, New York and London (1970).

[13] M. Finkelstein, R. Whitley, *Local and global subordination theorems for vector valued analytic functions*, Trans. Amer. Maths. Soc. **144** (1969) 73–76.

[14] G. Gentili, *On non-uniqueness of complex geodesics in convex bounded domains*, Rend. Accad. Naz. Lincei, Serie 8, **79** (1985) 90–97.

[15] G. Gentili, *On complex geodesics of balanced convex domains*, Ann. Mat. Pura Applicata, **144** (1986) 113–130.

[16] G. Gentili, *Regular complex geodesics in the domain* $D_n = \{(z_1, \ldots, z_n) \in \mathbb{C}^n : |z_1| + \cdots + |z_n| < 1\}$, Lecture Notes in Mathematics **1275**, Springer-Verlag (1987) 235–252.

[17] J. Globevnik, *On complex strict and uniform convexity*, Proc. Amer. Math. Soc. **47** (1975) 175–178.

[18] U. Haagerup, G. Pisier, *Factorization of analytic functions with values in non-commutative* L_1*-spaces and applications*, Can. J. Math. **XLI** (1989) 882–906.

[19] L.A. Harris, *Schwarz-Pick systems of pseudometrics for domains in normed linear spaces*, in Advances in Holomorphy (ed. J.A. Barroso), Math. Studies 34, North-Holland, Amsterdam (1979), 345–406.

[20] S. Heinrich, *Ultraproducts in Banach space theory*, J. Reine Angewandte Math. **313** (1980) 72–104.

[21] M. H. Heins, *On a class of conformal metrics*, Nagoya Math. J. **21** (1962) 1–60.

[22] P. Kiernan, *On the relations between taut, tight and hyperbolic manifolds*, Bull. Amer. Math. Soc. **76** (1970) 49–51.

[23] L. Lempert, *La métrique de Kobayashi et la représentation des domaines sur la boule*, Bull. Soc. Math. France **109** (1981) 427–474.

[24] L. Lempert, *Holomorphic retracts and intrinsic metrics in convex domains*, Analysis Mathematica **8** (1982) 257–261.

[25] K.-F. Ng, *On a theorem of Dixmier*, Math. Scand. **29** (1971) 279–280.

[26] E.A. Poletskiĭ, *The Euler-Lagrange equations for extremal holomorphic mappings of the unit disk*, Michigan Math. J. **30** (1983) 317–333.

[27] H.J. Reiffen, *Die differentialgeometrischen Eigenschaften der invarianten Distanzfunktion von Carathéodory*, Schriftenreihe Math. Inst., Univ. Münster **26** (1963).

[28] H.L. Royden, *Remarks on the Kobayashi metric*, Lecture Notes in Mathematics **185** (1971) 125–137.

[29] H.L. Royden, P.M. Wong, *Carathéodory and Kobayashi metric on convex domains*, (preprint).

[30] M. Suzuki, *The holomorphic curvature of intrinsic metrics*, Math. Rep. Toyama Univ. **4** (1981) 107–114.

[31] M. Suzuki, *The intrinsic metrics on the domains in* $\mathbb{C}^n$, Math. Rep. Toyama Univ. **6** (1983) 143-177.

[32] M. Takesaki, *On the conjugate space of operator algebras*, Tôhoku Math. J. **10** (1958) 194–203.

[33] E. Vesentini, *Complex geodesics*, Compositio Math. **44** (1981) 375–394.

[34] E. Vesentini, *Complex geodesics and holomorphic mappings*, Sympos. Math. **26** (1982) 211-230.

[35] E. Vesentini, *Invariant distances and invariant differential metrics in locally convex spaces*, Spectral Theory, Banach Center Publications 8, Warsaw (1982), 211-230.

[36] J.-P. Vigué, *Géodésiques complexes et points fixes d'applications holomorphes*, Advances in Math. **52** (1984) 214–247.

[37] J.-P. Vigué, *Quelques applications des géodésiques complexes aux points fixes d'applications holomorphes*, Sém. P. Lelong, P. Dolbeault, H. Skoda 1983/84, Lecture Notes in Mathematics **1198**, Springer-Verlag (1986) 244–260.

[38] B. Wong, *On the holomorphic curvature of some intrinsic metrics*, Proc. Amer. Math. Soc. **65** (1977) 57-61.

[39] H. Wu, *Normal families of holomorphic mappings*, Acta Math. **119** (1967) 193–233.

Progress in Functional Analysis
K.D. Bierstedt, J. Bonet, J. Horváth & M. Maestre (Eds.)
1992 Elsevier Science Publishers B.V.

Continuity of Tensor Product Operators

Between Spaces of Bochner Integrable Functions

Andreas Defant and Klaus Floret

Dedicated to Manuel Valdivia on the occasion of his sixtieth birthday

For many questions in Analysis it is interesting to know whether or not for given operators $S \in \mathcal{L}(L_q(\mu), L_p(\nu))$ and $T \in \mathcal{L}(E, F)$ (between Banach spaces) there is an operator $U \in \mathcal{L}(L_q(\mu; E), L_p(\nu; F))$ (Bochner integrable functions) such that

$$U(f \cdot x) = Sf \cdot Tx \in L_p(\nu; F)$$

for all $f \in L_q(\mu)$ and $x \in E$. In this survey-like paper some answers to this problem are presented.

1. Introduction

1.1. Throughout this paper all measures μ are supposed to be localizable, i.e. $L_1(\mu)' = L_\infty(\mu)$ holds with the natural duality bracket. For a (real or complex) Banach space E the space $L_p(\mu; E)$ is the Banach space of (classes $\tilde{f}$ of) strongly μ-measurable functions $f : \Omega \to E$ such that $||f(\cdot)||_E^p$ is μ-integrable (for $1 \leq p < \infty$) or μ-essentially bounded $(p = \infty)$;

$$||\tilde{f}||_p := \left(\int_\Omega ||f(t)||_E^p \mu(dt)\right)^{1/p} \quad \text{for } 1 \leq p < \infty$$
$$||\tilde{f}||_\infty := \text{ess} - \sup \ ||f(\cdot)||_E = || \ ||f(\cdot)||_E \ ||_{L_\infty(\mu)}.$$

It is obvious that the tensor product $L_p(\mu) \otimes E$ can be considered as a subspace of $L_p(\mu; E)$ via the canonical bilinear mapping

$$\begin{array}{ccc} L_p(\mu) \times E & \to & L_p(\mu; E) \\ (\tilde{f}, x) & \mapsto & \widetilde{f(\cdot)x} \end{array} .$$

The norm which $L_p(\mu; E)$ induces on $L_p(\mu) \otimes E$ is denoted by Δ_p, the one induced on $E \otimes L_p(\mu)$ (natural transposition) by Δ_p^t:

$$L_p(\mu) \otimes_{\Delta_p} E \overset{1}{\hookrightarrow} L_p(\mu; E) \ ; \ \ E \otimes_{\Delta_p^t} L_p(\mu) \overset{1}{\hookrightarrow} L_p(\mu; E).$$

It is well-known that these isometric embeddings are dense for $1 \leq p < \infty$ but – in general – not for $p = \infty$. So (if $q < \infty$) the problem presented in the abstract reduces to the question under which circumstances the operator

$$S \otimes T : L_q(\mu) \otimes_{\Delta_q} E \longrightarrow L_p(\nu) \otimes_{\Delta_p} F$$

is continuous if S and T are. It is exactly this question (for $p,\ q \in [1,\ \infty]$) which will be investigated in this paper.

Denote by ε the injective and by π the projective tensornorm.

Proposition: *Let $\mu \neq 0$ be a measure and E and F Banach spaces.*

(1) *$\varepsilon \leq \Delta_p \leq \pi$ on $L_p(\mu) \otimes E$; moreover, $\Delta_1 = \pi$ on $L_1(\mu) \otimes E$ and $\Delta_\infty = \varepsilon$ on $L_\infty(\mu) \otimes E$.*

(2) *For each $T \in \mathcal{L}(E, F)$*

$$\| id_{L_p} \otimes T : L_p(\mu) \otimes_{\Delta_p} E \longrightarrow L_p(\mu) \otimes_{\Delta_p} F \| = \|T\|.$$

(3) *If $I : E \overset{1}{\hookrightarrow} F$ is a metric injection (i.e. $\|Ix\| = \|x\|$ for all $x \in E$), then $id_{L_p} \otimes I$ is a metric injection as well.*

(4) *If $Q : E \overset{1}{\twoheadrightarrow} F$ is a metric surjection (i.e. $\| \cdot \|_F$ is the quotient norm of $\| \cdot \|_E$ along Q), then $id_{L_p} \otimes Q$ is also a metric surjection.*

See [10], section 7, for proofs. (1) and the metric mapping property of the tensornorms ε and π show that $S \otimes T$ is always continuous if $q = 1$ (and p arbitrary) or $p = \infty$ (and q arbitrary).

1.2. But (unfortunately or fortunately?) the operator $S \otimes T$ may happen to be uncontinuous. Examples:

(1) For the Fourier transform $\mathcal{F} : L_2(\mathbb{R}) \to L_2(\mathbb{R})$ and a Banach space E the operator

$$\mathcal{F} \otimes id_E : L_2(\mathbb{R}) \otimes_{\Delta_2} E \longrightarrow L_2(\mathbb{R}) \otimes_{\Delta_2} E$$

is continuous if and only if E is isomorphic to a Hilbert space; this is a result due to Kwapien [17] (see also [10], section 30).

(2) For the discrete Hilbert transform $\mathcal{H} : \ell_p(\mathbb{Z}) \to \ell_p(\mathbb{Z})$ (where $1 < p < \infty$) the spaces E for which $\mathcal{H} \otimes id_E$ is continuous are exactly the UMD-spaces (Bourgain [4], Burkholder [5], see also M. Defant [11]).

(3) If $R : L_2[0,1] \to Rad \subset L_2[0,1]$ is the orthogonal projection onto the closed linear span of the Rademacher functions, then the operators T such that $R \otimes T$ is continuous are Pisier's K-convex operators. Denote by $I : \ell_2 \to Rad \subset L_2[0,1]$ the isometric

identification $e_n \mapsto r_n$ (n-th Rademacher function) and by

$$I_p : \ell_p \hookrightarrow \ell_2 \xrightarrow{I} Rad \qquad \text{for} \quad 1 \le p \le 2$$
$$J_q : Rad \xrightarrow{I^{-1}} \ell_2 \hookrightarrow \ell_q \qquad \text{for} \quad 2 \le q < \infty$$

(Rad always considered as an isometric subspace of $L_2[0,1]$). Then $T \in \mathcal{L}(E,F)$ is a type p-operator if

$$I_p \otimes T : \ell_p \otimes_{\Delta_p} E \longrightarrow L_2[0,1] \otimes_{\Delta_2} F$$

is continuous, and a cotype q-operator if

$$J_q \otimes T : Rad \otimes_{\Delta_2} E \longrightarrow \ell_q \otimes_{\Delta_q} F$$

is continuous; here we used for a subspace $G \subset L_p(\mu)$ the notation

$$G \otimes_{\Delta_p} E \overset{1}{\hookrightarrow} L_p(\mu) \otimes_{\Delta_p} E \overset{1}{\hookrightarrow} L_p(\mu;\ E)$$

(see also below 3.2. and 3.3.). Note that the type p-norm of T is just the norm of $I_p \otimes T$ – and analogously in the other cases (for details see [10], sections 7 and 31).

1.3. Three types of problems arise in a natural way. Take $L_q(\mu)$ and $L_p(\nu)$.

(1) Fix $S \in \mathcal{L}(L_q(\mu), L_p(\nu))$. For which operators $T \in \mathcal{L}(E,F)$ is

$$S \otimes T : L_q(\mu) \otimes_{\Delta_q} E \longrightarrow L_p(\nu) \otimes_{\Delta_p} F$$

continuous? The example of cotype q-operators shows that this question is also interesting if S is only defined on a subspace of $L_q(\mu)$.

(2) Which operators $T \in \mathcal{L}(E,F)$ have the property that $S \otimes T$ is continuous for all $S \in \mathcal{L}(L_q(\mu), L_p(\nu))$? Does this depend on the measures μ and ν?

(3) In the case of continuity of $S \otimes T$, is there a good norm-estimate? In terms of the operator norms of S and T or of larger ideal norms?

The last problem, for example, is relevant for the complexification of operators $S : L_q^{\mathbb{R}}(\mu) \to L_p^{\mathbb{R}}(\nu)$ (real functions): since

$$L_q^{\mathbb{R}}(\mu) \otimes_{\Delta_q} \mathbb{R}_2^2 = L_q^{\mathbb{C}}(\mu)$$

holds isometrically (where $\mathbb{R}_2^2$ is the real Euclidean plane) it follows that $S \otimes id_{\mathbb{R}_2^2}$ is just the complexification $S^{\mathbb{C}}$ of S defined by $S^{\mathbb{C}}(f + ig) := Sf + i\,Sg$.

1.4. In the sequel some answers to these questions will be presented – mostly without proofs. The reader will find them in [10] if not otherwise stated.

2. Tensornorms Close to Δ_p and Absolutely p-Summing Operators

2.1. If α is a tensornorm in the sense of Grothendieck ([14], [9], [10]), then, by definition,

$$||S \otimes T : E_1 \otimes_\alpha E_2 \longrightarrow F_1 \otimes_\alpha F_2|| = ||S|| \, ||T||.$$

From the examples in 1.2. it follows that for $1 < p < \infty$ there is no tensornorm α which induces Δ_p (or an equivalent norm) on $L_p(\mu) \otimes E$ for all measures μ and Banach spaces E. But there are tensornorms close to Δ_p – a fact which is very useful for the study of the operators $S \otimes T$:

Take Banach spaces E and F and $z \in E \otimes F$. For $1 \leq p \leq \infty$ define

$$g_p(z;\, E,\, F) := \inf\Big\{ ||(||x_n||)||_{\ell_p} \cdot \sup_{||y'|| \leq 1} ||(\langle y',\, y_n \rangle)||_{\ell_{p'}} \;\Big|\; z = \sum_{n=1}^{N} x_n \otimes y_n \Big\}$$

and d_p its transpose (i.e. $E \otimes_{d_p} F$ and $F \otimes_{g_p} E$ are isometric). Moreover, a tensornorm g_p^* can be defined to be the norm on $E \otimes F$ induced from the embedding

$$E \otimes F \hookrightarrow (E' \otimes_{d_p} F')'$$

(this definition is dangerous: it works since – defined in the correct way – the tensornorm g_p^* is totally accessible; see [10], sections 15 and 21, for details). These tensornorms, due to Saphar, are important for the study of absolutely p-summing operators, since

$$\mathcal{P}_p(E,\, F') = (E \otimes_{d_{p'}} F)'$$

holds isometrically ([10], section 17). The Chevet-Persson-Saphar inequalities say that on $L_p(\mu) \otimes E$

$$d_p \leq \Delta_p \leq g_{p'}^* \quad (\leq g_p)$$

– and there are no (larger or smaller) tensornorms closer to Δ_p; for a more precise statement of this see [10], 15.11..

2.2. The following result, due to Kwapien [16], is striking – later on (4.3) the reader will find a more general version of it.

Theorem: *For each Banach space E and $1 \leq p \leq \infty$ the following two statements are equivalent:*

(a) *For some infinite dimensional $L_p(\mu)$ there is a tensornorm α such that*

$$L_p(\mu) \otimes_{\Delta_p} E = L_p(\mu) \otimes_\alpha E$$

holds isomorphically (resp. isometrically).

(b) *E is isomorphic (resp. isometric) to a quotient of a subspace of some $L_p(\nu)$.*

In this case, the norms α, Δ_p, d_p, and $g^*_{p'}$ are equivalent (resp. equal) on $L_p(\mu) \otimes E$ for all measures μ.

Being a quotient of a subspace of some $L_p(\mu)$ is the same as being a subspace of a quotient of some $L_p(\nu)$; for more details about Kwapien's theorem see [10], 25.10..

2.3. The last results explain why absolutely p-summing operators appear in a natural way in the investigation of the continuity of tensor product operators. To illustrate this, consider the following two results which to a large part are due to Maurey [19]:

Proposition 1: *Take $1 \leq p < q \leq 2$ and E a Banach space. Then the following statements are equivalent:*

(a) *For each Banach space F*

$$\mathcal{P}_q(E', F) = \mathcal{P}_p(E', F).$$

(b) *For all operators $S \in \mathcal{L}(L_q(\mu), L_p(\nu))$ and $T \in \mathcal{L}(E, L_p(\eta))$ the operator*

$$S \otimes T : L_q(\mu) \otimes_{\Delta_q} E \longrightarrow L_p(\nu) \otimes_{\Delta_p} L_p(\eta)$$

is continuous.

(c) *$\mathcal{P}_q(E', F) = \mathcal{P}_1(E', F)$ for each Banach space F.*

(d) *For all $(S,\ T) \in \mathcal{L}(L_q(\mu), L_1(\nu)) \times \mathcal{L}(E, L_1(\eta))$ the operator*

$$S \otimes T : L_q(\mu) \otimes_{\Delta_q} E \longrightarrow L_1(\nu) \otimes_{\Delta_1} L_1(\eta) = L_1(\nu) \otimes_\pi L_1(\eta)$$

is continuous.

The equivalence of (a) and (c) is just the Maurey-Pisier extrapolation theorem (for proofs see [10], 32.8. corollary 2). It is easy to see that one can replace F by ℓ_∞ in (a) and (c).

Proposition 2. *For each Banach space E are equivalent:*

(a) $\mathcal{P}_2(E', \ell_\infty) = \mathcal{P}_1(E', \ell_\infty)$

(b) *For some (and then for all) pairs (p, q) with $1 \leq p < 2 \leq q \leq \infty$ the operator*

$$S \otimes T : L_q(\mu) \otimes_{\Delta_q} E \longrightarrow L_p(\nu) \otimes_{\Delta_p} L_p(\eta)$$

is continuous, whenever S and T are.

(See [10], 32.9. corollary 4.) Note that E' satisfies (a) if it has cotype 2 – and that E'

satisfying (a) implies E' having the Orlicz property (see, e.g., [10], section 32).

These results already show that the continuity behaviour of $S \otimes T$ depends heavily on the indices p and q – and this will be made much more precise.

3. Some General Positive Results

3.1. Now we want to give some more systematic information about the problem. If $S \in \mathcal{L}(L_q(\mu), L_p(\nu))$ is a positive operator, then $S(\chi_A) \geq 0$ for all μ-integrable sets A – and this can be used ([10], 7.3.) to show the

Proposition 1: *If $S \in \mathcal{L}(L_q(\mu), L_p(\nu))$ is positive, then for all $T \in \mathcal{L}(E, F)$*

$$||S \otimes T : L_q(\mu) \otimes_{\Delta_q} E \longrightarrow L_p(\nu) \otimes_{\Delta_p} F|| = ||S||\ ||T||.$$

It follows that for all *regular* S, the operator $S \otimes T$ is continuous. Conversely,

Proposition 2: *Take $S \in \mathcal{L}(L_q(\mu), L_p(\nu))$ and $p < \infty$. If $S \otimes id_G$ is continuous for $G = \ell_1$ or $G = \ell_\infty$, then S is regular and hence $S \otimes T$ is continuous for all $T \in \mathcal{L}(E, F)$.*

PROOF: Assume $S \otimes id_{\ell_1}$ to be continuous, E separable and $T \in \mathcal{L}(E, F)$. Since Δ_p respects the quotient $\ell_1 \overset{1}{\twoheadrightarrow} E$ (see 1.1.(4)) the diagram

$$\begin{array}{ccccc}
L_q \otimes_{\Delta_q} \ell_1 & \overset{S\otimes id}{\longrightarrow} & L_p \otimes_{\Delta_p} \ell_1 & & \\
\Big\downarrow{\scriptstyle 1} & & \Big\downarrow{\scriptstyle 1} & & \\
L_q \otimes_{\Delta_q} E & \overset{S\otimes id}{\longrightarrow} & L_p \otimes_{\Delta_p} E & \overset{id\otimes T}{\longrightarrow} & L_p \otimes_{\Delta_p} F
\end{array}$$

shows that $S \otimes T$ is continuous. The non-separable case follows easily from this. In particular, $S \otimes id_{\ell_\infty}$ is continuous. The continuity of this latter operator applied to $\sum_{k=1}^n f_k \otimes e_k$ implies

$$||\max_k |Sf_k|\ ||_{L_p} \leq c||\max_k |f_k|\ ||_{L_q}$$

which gives that S is order-bounded and hence regular. □

The result on $S \otimes id_{\ell_\infty}$ is due to Virot [23].

3.2. The *continuous Minkowski inequality*

$$\left(\int \left(\int |f(x,y)| \mu(dx)\right)^r \nu(dy)\right)^{1/r} \leq \int \left(\int |f(x,y)|^r \nu(dy)\right)^{1/r} \mu(dx)$$

has an interesting consequence for the norm Δ_p – just take $r := p/q$:

Proposition: *If $1 \le q \le p \le \infty$, then*

$$||id : L_q(\mu) \otimes_{\Delta_q} L_p(\nu) \longrightarrow L_q(\mu) \otimes_{\Delta_p^t} L_p(\nu)|| \le 1 \quad .$$

Since $\mathcal{P}_{p'} \subset \mathcal{P}_{q'}$ implies $d_p \le d_q$, this is also a consequence of the Chevet-Persson-Saphar inequalities:

$$\Delta_p^t \le (g_{p'}^*)^t \le g_p^t = d_p \le d_q \le \Delta_q \quad \text{on} \quad L_q \otimes L_p \quad .$$

Recall from 1.2.(3) the notation $G \otimes_{\Delta_q} E$ for subspaces $G \subset L_q(\mu)$.

Theorem: *Let $1 \le q \le p \le \infty$. If G_1 is a subspace of some $L_q(\mu)$ and G_2 and E Banach spaces, $S \in \mathcal{L}(G_1, G_2)$ and $T \in \mathcal{L}(E, L_p(\eta))$, then*

$$||S \otimes T : G_1 \otimes_{\Delta_q} E \longrightarrow G_2 \otimes_{\Delta_p^t} L_p(\eta)|| = ||S||\,||T||.$$

This result is due to Figiel-Iwaniecz-Pełczyński [12] which generalizes earlier results of Beckner [1] and Bennett [2]. We shall refer to this result as a Beckner-like result. The proof is obvious from the following diagram:

$$\begin{array}{ccc}
G_1 \otimes_{\Delta_q} E & \overset{1}{\hookrightarrow} & L_q \otimes_{\Delta_q} E \\
 & & \downarrow \mathrm{id}_{L_q} \otimes \mathrm{T} \\
\mathrm{id}_{G_1} \otimes \mathrm{T} \,\downarrow & & L_q \otimes_{\Delta_q} L_p \\
 & & \downarrow \text{proposition} \\
G_1 \otimes_{\Delta_p^t} L_p & \overset{1}{\hookrightarrow} & L_q \otimes_{\Delta_p^t} L_p \\
\mathrm{S} \otimes \mathrm{id}_{L_p} \downarrow & & \\
G_2 \otimes_{\Delta_p^t} L_p \ . & &
\end{array}$$

This result *does not hold* for $q > p$: *Maurey's factorization theorem* ([19], see also [10], 18.9.) states, in particular, that for $T \in \mathcal{L}(E,\ L_p(\eta))$ the operator

$$id_{\ell_q} \otimes T : \ell_q \otimes_{\Delta_q} E \longrightarrow \ell_q \otimes_{\Delta_p^t} L_p(\eta)$$

is continuous if and only if T factors through a multiplication operator $L_q(\eta) \to L_p(\eta)$; see [10], 18.9. for more information about this result.

3.3 Some remarks about Δ_p on $G \otimes E \subset L_p \otimes E$:

(1) The theorem implies (take $p = q$) that the definition of Δ_p on $G \otimes E$ does not depend on the measures μ with the property that G is isometrically embedded into $L_p(\mu)$; in other words: $G \otimes_{\Delta_p} E$ is well-defined.

(2) $\pi = \Delta_1$ on $L_1 \otimes E$ – but this does no longer hold on $G \otimes E$: Just take a non-complemented reflexive subspace $G \subset L_1[0,1]$, define $E := G'$ and look at the restriction map

$$(L_1 \otimes_{\Delta_1} G')' = \mathcal{L}(L_1, G) \longrightarrow \mathcal{L}(G, G) = (G \otimes_\pi G')'$$

which is not surjective; the Hahn-Banach theorem gives $\pi \neq \Delta_1$ on $G \otimes G'$.

(3) $\ell_2 \tilde{\otimes}_{\Delta_p} \ell_2 = \ell_2$ holds isomorphically for all $1 \leq p < \infty$.

This result is due to Rosenthal and Szarek [22]. We find it informative to give a proof of this using a tensornorm close to Δ_p: First note that $d^*_{p'} = \Delta_p$ on $L_p \otimes E$ if (and only if) E is isometric to a subspace of some $L_p(\nu)$ (this is also a result of Kwapien's, see [10], 25.10.). Since $d^*_{p'}$ is left-injective ([10], 20.4.) and ℓ_2 is isometric to a subspace of some $L_p(\mu)$ (via Gauss functions; [10], 8.7.) it follows that

$$\ell_2 \otimes_{d^*_{p'}} \ell_2 \overset{1}{\hookrightarrow} L_p \otimes_{d^*_{p'}} \ell_2 = L_p \otimes_{\Delta_p} \ell_2 \quad .$$

Moreover, $d^*_{p'}$ is equivalent to Δ_2 on $\ell_2 \otimes \ell_2$ for $1 \leq p < \infty$ (see [10], 26.6.) and, clearly, $\ell_2 \tilde{\otimes}_{\Delta_2} \ell_2 = \ell_2$ holds isometrically. □

3.4. Take $1 \leq p < q < 2$ and

$$I : \ell_q \overset{1}{\hookrightarrow} L_p(\mu_q)$$

the (isometric) Lévy embedding (where μ_q is the q-stable Lévy measure; see [10], section 24). It was observed by Bennett [2] (see also Rosenthal-Szarek [22] and [10], 26.3.) that

$$I \otimes I : \ell_q \otimes_{\Delta_q} \ell_q \longrightarrow L_p(\mu_q) \otimes_{\Delta_p} L_p(\mu_q)$$

is *not* continuous. This and $I' \otimes I'$ give the

Proposition: *Let $1 \leq p < q \leq \infty$. If for all measures and all $S \in \mathcal{L}(L_q(\mu_1)), L_p(\nu_1))$ and $T \in \mathcal{L}(L_q(\mu_2), L_p(\nu_2))$ the operator*

$$S \otimes T : L_q(\mu_1) \otimes_{\Delta_q} L_q(\mu_2) \longrightarrow L_p(\nu_1) \otimes_{\Delta_p} L_p(\nu_2)$$

is continuous, then $p \leq 2 \leq q$.

It will be seen in the next section (4.2.) that the converse of this result holds as well. Thus it becomes strikingly clear that the continuity behaviour for $q \leq p$ (somehow the "easier" case) and $q > p$ is quite different.

3.5. A careful analysis of Bennett's counterexample (the Lévy embedding) shows the lower estimate of the following finite dimensional result, the upper bound is a consequence of a "mixing" estimate which will be presented in 4.4. (see [22] and [7] for the details):

Proposition: *For $1 \leq p < q < 2$ there is a constant $c_{p,q} > 0$ such that*

$$c_{p,q}(1 + \ln n)^{1/q} \leq \sup_{\substack{\|S\| \leq 1 \\ \|T\| \leq 1}} \|S \otimes T : \ell_q^n \otimes_{\Delta_q} \ell_q^n \longrightarrow \ell_p^n \otimes_{\Delta_p} \ell_p^n\| \leq 5(1 + \ln n)^{1/q}.$$

3.6. The Lévy embedding $\ell_r \hookrightarrow L_p(\mu_r)$ (for $1 \leq p < r < 2$) or Gauss embedding ($r = 2$ and $p \in [1, \infty[$) can be used to show that for every measure η there is a (Borel-Radon) measure $\tilde{\eta}$ and an isometric embedding

$$L_r(\eta) \overset{1}{\hookrightarrow} L_p(\tilde{\eta})$$

if $1 \leq p < r < 2$ or: $r = 2$ and $p \in [1, \infty[$ or (obviously) if $p = r$ (see, e.g., [10], 24.5. for a proof). If follows from the Beckner-like result 3.2. (and 1.1.(3)) that

$$||S \otimes id_{L_r} : L_q \otimes_{\Delta_q} L_r \longrightarrow L_p \otimes_{\Delta_p} L_r|| = ||S||$$

if $q \leq p$ and (p, r) satisfies one of the conditions just stated. For $q = p$ this is an old result of Marcinkiewicz-Zygmund [18]. If $\mathbf{L}_r(T)$ denotes the r-factorable norm of an operator $T \in \mathcal{L}(E, F)$,

$$\mathbf{L}_r(T) := \inf\{||U|| \, ||V|| \;\Big|\; \kappa_F \circ T : E \overset{U}{\longrightarrow} L_r \overset{V}{\longrightarrow} F''\},$$

the following result is easily deduced ([10], 26.3.).

Proposition: *For $p, q, r \in [1, \infty]$*

$$||S \otimes T : L_q(\mu) \otimes_{\Delta_q} E \longrightarrow L_p(\nu) \otimes_{\Delta_p} F|| \leq ||S|| \, \mathbf{L}_r(T)$$

in each of the following four cases:

(1) $q \leq p = r$
(2) $r = 2$ *and* $q \leq p$
(3) $1 \leq q \leq p < r < 2$
(4) $2 < r < q \leq p$.

4. The Operator Ideal Argument

4.1. Fix an $S \in \mathcal{L}(L_q(\mu), L_p(\nu))$ with $||S|| = 1$ and consider the operator ideal $\mathcal{L}_S$ of all operators $T \in \mathcal{L}(E, F)$ such that

$$\mathbf{L}_S(T) := ||S \otimes T : L_q(\mu) \otimes_{\Delta_q} E \longrightarrow L_p(\nu) \otimes_{\Delta_p} F|| < \infty.$$

It can be easily seen ([10], 17.3.) that $(\mathcal{L}_S, \mathbf{L}_S)$ is a Banach operator ideal, injective and surjective (by 1.1.(3) and (4)) – and it is even maximal. Since the ideal $(\mathcal{I}, \mathbf{I})$ of integral operators is contained in every maximal Banach operator ideal, it follows that the injective and surjective hull of $\mathcal{I}$ is contained as well in $\mathcal{L}_S = \mathcal{L}_S^{inj\ sur}$:

$$\mathcal{I}^{inj\ sur} \subset \mathcal{L}_S \quad \text{and} \quad \mathbf{L}_S \leq \mathbf{I}^{inj\ sur}.$$

Clearly, $\mathcal{I}^{inj} = \mathcal{P}_1$, the absolutely summing operators. The following result is proved:

Theorem: *Let $p, q \in [1, \infty]$ and $S \in \mathcal{L}(L_q(\mu), L_p(\nu))$. Then*

$$||S \otimes T : L_q(\mu) \otimes_{\Delta_q} E \longrightarrow L_p(\nu) \otimes_{\Delta_p} F|| \leq ||S|| \, \mathbf{P}_1^{sur}(T : E \to F).$$

4.2. Grothendieck's fundamental theorem of the metric theory of tensor products states that

$$K_G := \mathbf{P}_1^{sur}\,(id_{\ell_2}) < \infty$$

([14], see also [10], 20.17.). It is easy to see that $\mathbf{P}_1^{sur}(id_H) = K_G$ for each infinite dimensional Hilbert space H. The (maximal) Banach operator ideal $(\mathcal{L}_2, \mathbf{L}_2)$ of 2-factorable operators (see 3.6.) is the ideal of those operators factoring through a Hilbert space.

$$\mathbf{L}_2(T : E \longrightarrow F) = \inf\{||U||\,||V|| \;\Big|\; T : E \xrightarrow{U} H \xrightarrow{V} F\};$$

it follows that $\mathbf{P}_1^{sur} \leq K_G\,\mathbf{L}_2$ and hence

Corollary 1: *For $p, q \in [1, \infty]$*

$$||S \otimes T : L_q(\mu) \otimes_{\Delta_q} E \longrightarrow L_p(\nu) \otimes_{\Delta_p} F|| \leq K_G ||S||\,\mathbf{L}_2(T).$$

In particular: $||S \otimes id_H|| \leq K_G||S||$ for every Hilbert space H; for $p, q \in [1, \infty[$ this result is also contained in Marcinkiewicz-Zygmund [18] (with a constant coming from Gauss measures, see below 4.5.). Recall from 3.6. that for $q \leq p$ the Grothendieck constant can be replaced by 1. Kwapien [16] showed that

$$\mathbf{L}_2(T : L_r \longrightarrow L_s) \leq K_G\,||T||$$

if $1 \leq s \leq 2 \leq r \leq \infty$, hence

Corollary 2: For $p, q, r, s \in [1, \infty]$ with $1 \leq s \leq 2 \leq r \leq \infty$

$$||S \otimes T : L_q(\mu_1) \otimes_{\Delta_q} L_r(\mu_2) \longrightarrow L_p(\nu_1) \otimes_{\Delta_p} L_s(\nu_2)|| \leq K_G^2\,||S||\,||T||.$$

More generally, Kwapien's type-cotype theorem ([17], see also [10], section 30) states $\mathcal{C}_2 \circ \mathcal{T}_2 \subset \mathcal{L}_2$ (where $\mathcal{C}_2$ points at the cotype 2-operators and $\mathcal{T}_2$ at the type 2-operators) and $\mathbf{L}_2 \leq \mathbf{C}_2 \circ\, \mathbf{T}_2$. This implies, for example,

$$||S \otimes T : L_q(\mu) \otimes_{\Delta_q} E \longrightarrow L_p(\nu) \otimes_{\Delta_p} F|| \leq K_G\,\mathbf{T}_2(E)\,\mathbf{C}_2(F)\,||S||\,||T||.$$

4.3. So far $p, q \in [1, \infty]$ were arbitrary. It is already known from the Beckner-like result 3.2. (see also 3.6.) that at least in the special case $F = L_p(\eta)$ there is always continuity with $||S \otimes T|| = ||S||\,||T||$ whenever $q \leq p$. Consider now the case $q \geq p$, take $T \in \mathcal{L}(E, F)$ and suppose that T factors as follows

$$\begin{array}{ccccc} E & \xrightarrow{T} & F & \overset{\kappa_F}{\hookrightarrow} & F'' \\ U \big\downarrow & & & \nearrow V & \\ L_q(\eta) & \xrightarrow{I} & L_p(\eta) & & \end{array}$$

where η is a finite measure if $q > p$ and arbitrary if $q = p$. These operators are called (p, q')-factorable, the ideal is denoted by $\mathcal{L}_{p,\,q'}$ and its norm is

$$\mathbf{L}_{p,\,q'}(T) := \inf\{\|U\|\,\|I\|\,\|V\| \,\Big|\, \kappa_F \circ T = V \circ I \circ U \text{ as above}\}.$$

Note $\mathcal{L}_{p,\,p'} = \mathcal{L}_p$. For every $S \in \mathcal{L}(L_q(\mu), L_p(\nu))$ and $T \in \mathcal{L}_{p,\,q'}(E, F)$ one has the following diagram:

$$\begin{array}{ccc} L_q \otimes_{\Delta_q} E & \xrightarrow{\qquad\qquad S\otimes T \qquad\qquad} & L_p \otimes_{\Delta_p} F \overset{1}{\hookrightarrow} L_p \otimes_{\Delta_p} F'' \\ \text{id}\otimes \text{U} \downarrow & & \uparrow \text{id}\otimes \text{V} \\ L_q \otimes_{\Delta_q} L_q = L_q \otimes_{\Delta_q^t} L_q & \overset{S\otimes id}{\longrightarrow} L_p \otimes_{\Delta_q^t} L_q \overset{id}{\longrightarrow} & L_p \otimes_{\Delta_p^t} L_p = L_p \otimes_{\Delta_p} L_p \end{array} .$$

It follows that $\mathbf{L}_S(T) \le \mathbf{L}_{p,\,q'}(T)$ and even

$$\mathbf{L}_S \le \mathbf{L}_{p,\,q'}^{inj\ sur},$$

since the ideal $\mathcal{L}_S$ is injective and surjective. This cannot be improved as the following result from [8] (see also [10], 28.4.) shows

Theorem: *Let* $1 \le p \le q \le \infty$ *and* $L_q(\mu)$ *and* $L_p(\nu)$ *be infinite dimensional. Then*

$$\begin{aligned}\mathbf{L}_{p,\,q'}^{inj\ sur}(T) &= \sup\{\|S \otimes T : L_q(\mu) \otimes_{\Delta_q} E \longrightarrow L_p(\nu) \otimes_{\Delta_p} F\| \,\Big|\, \|S\| \le 1\} = \\ &= \sup\{\|S \otimes T : \ell_q^n \otimes_{\Delta_q} E \longrightarrow \ell_p^n \otimes_{\Delta_p} F\| \,\Big|\, n \in I\!N,\ \|S : \ell_q^n \to \ell_p^n\| \le 1\}.\end{aligned}$$

This implies, that for fixed T the continuity of $S \otimes T$ does not depend on the measures – it is even a "local" problem. Consider the special case $q = \infty$ and $p = 1$: Since $\mathcal{L}_{1,\,1}^{inj} = \mathcal{P}_1$ it follows that

$$K_G = \mathbf{P}_1^{sur}(id_{\ell_2}) = \sup\{\|S \otimes id_{\ell_2} : L_\infty \otimes_\varepsilon \ell_2 \longrightarrow L_1 \otimes_\pi \ell_2\| \,\Big|\, \|S\| \le 1\}.$$

This shows that the appearence of the Grothendieck constant is natural. Note also the relation for the finite dimensional Grothendieck constants $K_G(n)$:

$$K_G(n) := \mathbf{P}_1^{sur}(id_{\ell_2^n}) = \sup\{\|S \otimes id_{\ell_2^n} : L_\infty \otimes_\varepsilon \ell_2^n \to L_1 \otimes_\pi \ell_2^n\| \,\Big|\, \|S\| \le 1\}$$

whenever L_∞ and L_1 are infinite dimensional.

4.4. Recall the maximal Banach operator ideal

$$\mathcal{M}_{q,\,p} = \mathcal{P}_q^{-1} \circ \mathcal{P}_p$$

of (p,q)-mixing operators – its natural quotient ideal norm will be denoted by $\mathbf{M}_{q,p}$. Using traced tensornorms it is possible to prove ([6], see also [10], 32.3.) the

Theorem: *For $S \in \mathcal{L}(L_q(\mu), G)$ and $T \in \mathcal{L}(E, L_p(\eta))$*

$$||S \otimes T : L_q(\mu) \otimes_{\Delta_q} E \longrightarrow G \otimes_{\Delta_p^t} L_p(\eta)|| \leq \mathbf{M}_{s',q'}(S)\, \mathbf{M}_{s,p}(T')$$

whenever $s, p, q \in [1, \infty]$.

This result comprises various of the results which were already mentioned: The implication (a) → (b) of proposition 1 (take $s = q$), one implication of proposition 2 in 2.3., the Beckner-like result from 3.2. (take $q \leq p = s$) and special cases of the results in 4.2. when taking $s = 2$ and the fact that cotype 2-spaces are in space($\mathcal{M}_{2,1}$) $\subset$ space($\mathcal{M}_{2,r}$) for all $r \in [1, \infty]$. This also implies that the operator $S \otimes T$ in the theorem is continuous in each of the following four cases:

(1) $1 \leq q \leq p \leq \infty$
(2) $1 \leq q \leq 2$ and E' has cotype 2
(3) $2 \leq p \leq \infty$ and G has cotype 2
(4) E' and G have cotype 2.

Note that, by Bennett's counterexample in 3.4., the operator $S \otimes T$ is, in general, *not* continuous if, e.g., E and G have cotype 2.

4.5. The Hilbert spaces played a dominant rôle so far – this fact was already known to Marcinkiewicz-Zygmund (see 4.2. above). Using Lévy embeddings (as in 3.6.) they succeeded to replace ℓ_2 by ℓ_r in certain cases and proved, also in [18], the central point of the

Theorem: *If $p, q, r \in [1, \infty]$ satisfy $\max\{p,q\} < r < 2$ or $\min\{p,q\} > r > 2$, then there is a constant $c_{p,q,r}$ such that*

$$||S \otimes T : L_q(\mu) \otimes_{\Delta_q} E \longrightarrow L_p(\nu) \otimes_{\Delta_p} F|| \leq c_{p,q,r} ||S||\, \mathbf{L}_r(T).$$

PROOF: Factorization and $\mathcal{L}_r$-local technique yield that it is enough to take $T = id_{\ell_r}$. Using for $s < r < 2$ the isomorphic Lévy embedding

$$I : \ell_r \hookrightarrow L_s(\mu_r);\ I(\alpha_k) := \sum_{k=1}^{\infty} \alpha_k h_k$$

(with certain functions h_k independent of s) one obtains

$$L_s \otimes_{\Delta_s} \ell_r \overset{\approx}{\hookrightarrow} L_s \otimes_{\Delta_s} L_s(\mu_r) = L_s \otimes_{\Delta_s^t} L_s(\mu_r) \overset{J}{\longrightarrow} L_s \otimes_{\Delta_1^t} L_1(\mu_r)$$

– the latter mapping J having norm ≤ 1 since μ_r is a probability measure. On the other hand,

$$||L_s \otimes_{\Delta_1^t} L_1(\mu_r) \xrightarrow{J^{-1}} L_s \otimes_{\Delta_s} L_1(\mu_r)|| \leq 1$$

(continuous Minkowski inequality 3.2.) and therefore

$$id_{L_s} \otimes I : L_s(\eta) \otimes_{\Delta_s} \ell_r \hookrightarrow L_s(\eta) \otimes_{\Delta_1^t} L_1(\mu_r)$$

is an isomorphic embedding. Now

$$||S \otimes id_{L_1} : L_q(\mu) \otimes_{\Delta_1^t} L_1(\mu_r) \longrightarrow L_p(\nu) \otimes_{\Delta_1^t} L_1(\mu_r)|| = ||S||$$

implies the result for $r < 2$. The case $r > 2$ follows by dualization. □

Note that using the Gauss measure and the embedding

$$\ell_2 \hookrightarrow L_s(\mu_2)$$

for arbitrary $s < \infty$ the arguments in the last proof give corollary 1 in 4.2. for $p,\ q < \infty$. But for $q = \infty$ (and $p = 1$) the result is much stronger: by theorem 4.3. it implies Grothendieck's inequality.

5. Complexification of Operators

5.1. As an application of the results presented, the complexification of operators will be treated. Recall from 1.3. that

$$L_q^{\mathbb{R}} \otimes_{\Delta_q} \mathbb{R}_2^2 = L_q^{\mathbb{C}} \quad \text{and} \quad S^{\mathbb{C}} = S \otimes id_{\mathbb{R}_2^2}$$

whenever $S : L_q^{\mathbb{R}} \to L_p^{\mathbb{R}}$ is a real-linear operator. Since $\mathbb{R}_2^2$ is isometric to a subspace of some L_q (via Gauss functions), theorem 3.2. yields that for every $S \in \mathcal{L}(L_q^{\mathbb{R}}, L_p^{\mathbb{R}})$ and $1 \leq q \leq p \leq \infty$

$$||id \otimes S : \mathbb{R}_2^2 \otimes_{\Delta_q} L_q^{\mathbb{R}} \longrightarrow \mathbb{R}_2^2 \otimes_{\Delta_p^t} L_p^{\mathbb{R}}|| = ||S||.$$

Since $\mathbb{R}_2^2 \otimes_{\Delta_q} L_q^{\mathbb{R}} = \mathbb{R}_2^2 \otimes_{\Delta_q^t} L_q^{\mathbb{R}}$ (isometrically) this implies $||S^{\mathbb{C}}|| = ||S||$ – an observation due to Figiel-Iwaniecz-Pełczyński [12]. In the "bad" case $1 \leq p \leq q \leq \infty$ one obtains directly from theorem 4.3.

$$||S^{\mathbb{C}}|| \leq \mathbf{L}_{p,\,q'}^{inj\ sur}(id_{\mathbb{R}_2^2})||S||.$$

– and the constant is best possible. For the worst case $q = \infty$ and $p = 1$ the considerations in 4.3. on the finite dimensional Grothendieck constant yield

$$||S^{\mathbb{C}}|| \leq K_G^{\mathbb{R}}(2)\ ||S|| \ .$$

Krivine [15] calculated (actually for this purpose) that $K_G^{\mathbb{R}}(2) = \sqrt{2}$. It might be difficult to find the exact value of $\mathbf{L}_{p,\,q'}^{inj\ sur}(id_{\mathbb{R}_2^2})$ in general. For given (p, q) with $1 \leq p \leq q \leq \infty$

it is obvious that $k_{q,\,p} := L^{inj\; sur}_{p,\,q'}(id_{\mathbb{R}^2_2}) \leq \mathbf{L}^{inj}_{p,\,q'}(id_{\mathbb{R}^2_2})$ holds for the best "complexification constant" $k_{q,\,p}$, hence [20], 22.1.5, implies the estimate

$$k_{q,\,p} \leq (\sqrt{\pi})^{1/p-1/q} \left(\frac{\Gamma((1+q)/2)}{\Gamma((2+q)/2)}\right)^{1/q} \left(\frac{\Gamma((2+p)/2)}{\Gamma((1+p)/2)}\right)^{1/p}.$$

Since the operator ideal $\mathcal{L}^{inj}_{p,\,2}$ is surjective and $k_{q,\,p} = k_{p',\,q'}$ by duality, one obtains the

Theorem: *For* $1 \leq p \leq 2$

$$k_{2,\,p} = k_{p',\,2} = \frac{\sqrt{2}}{2}\left(\sqrt{\pi}\frac{\Gamma((2+p)/2)}{\Gamma((1+p)/2)}\right)^{1/p} \quad \textit{and} \quad k_{2,\,1} = k_{\infty,\,2} = \frac{\sqrt{2}}{4}\pi \approx 1,11072.$$

References

[1] Beckner, W.: *Inequalities in Fourier Analysis*; Annals of Math. 102(1975)159-182.

[2] Bennett, G.: *Schur multipliers*; Duke J.Math. 44(1977)603-639.

[3] Bergh, J.-J.Löfström: *Interpolation spaces*; Grundl.math.Wiss. 223, Springer, 1976.

[4] Bourgain, J.: *Some remarks on Banach spaces in which martingale difference sequences are unconditional*; Ark.Mat. 22(1983)163-168.

[5] Burkholder, D.L.: *A geometrical condition that implies the existence of certain singular integrals of Banach space valued functions*; Proc. Conf. Harmonic Analysis in Honour of Antony Zygmund, Chicago (1981)270-286.

[6] Carl, B.-A.Defant: *Tensor products and Grothendieck type inequalities of operators in L_p-spaces*; to appear in Trans.Amer.Math.Soc.

[7] Carl, B.-A.Defant: *An inequality between the p-and $(p,1)$-summing norm of finite rank operators from $C(K)$-spaces*; to appear in Israel J.Math.

[8] Defant, A.: *Produkte von Tensornormen*; Habilitationsschrift, Oldenburg, 1986.

[9] Defant, A.-K.Floret: *Aspects of the metric theory of tensor products and operator ideals*; Note di Matematica 8(1988)181-281.

[10] Defant, A.-K.Floret: *Tensornorms and operator ideals*; to appear in North-Holland Math. Studies/Notas de Matemática.

[11] Defant, M.: *On the vector valued Hilbert transform*; Math.Nachr. 141(1989)251-265.

[12] Figiel, T.-T.Iwaniec - A. Pelczyński: *Computing norms of some operators in L^p-spaces*; Studia Math. 79(1984)227-274.

[13] Garcia-Cuerva, J.-J.L.Rubio de Francia: *Weighted norm inequalities and related topics*; Math. Studies/Notas de Matemática 104, North-Holland, 1985.

[14] Grothendieck, A.: *Résumé de la théorie métrique des produits tensoriels topologiques*; Bol.Soc.Mat.São Paulo 8(1956)1-79.

[15] Krivine, J.L.: *Sur la complexification des opérateurs de L^∞ dans L^1*; C.R.A.S.Paris 284(1977)377-379.

[16] Kwapien, S.: *On operators factoring through L_p-space*; Bull.Soc.Math.France Mémoire 31-32(1972)215-225.

[17] Kwapien, S.: *Isomorphic characterization of inner product spaces by orthogonal series with vector valued coefficients*; Studia Math. 44(1972)583-595.

[18] Marcinkiewicz, J.-A.Zygmund: *Quelques inégalités pour les opérations linéaires*; Fund.Math. 32(1939)115-121.

[19] Maurey, B.: *Théorèmes de factorisation pour les opérateurs linéaires à valeurs dans les espaces L^p*; Astérisque 11, 1974.

[20] Pietsch, A.: *Operator ideals*; Deutscher Verlag der Wiss., 1978 and North-Holland, 1980.

[21] Pisier, G.: *Factorization of linear operators and geometry of Banach spaces*; CBMS Regional Conf. Series 60, Amer.Math.Soc., 1986.

[22] Rosenthal, H.P.-S.J. Szarek: *On tensor products of operators from L^p to L^q*; in: Longhorn Notes (Austin), Funct.Anal.Sem., 1987-89.

[23] Virot, B.: *Extensions vectorielles d'opérateurs linéaires bornés sur L^p*; C.R.A.S.Paris 293(1981)413-415.

A. Defant, K. Floret
Fachbereich Mathematik
der Universität
D-2900 Oldenburg
Germany

K. Floret
IMECC - Unicamp
C.P. 6065
13.081 Campinas, S.P.
Brazil

Progress in Functional Analysis
K.D. Bierstedt, J. Bonet, J. Horváth & M. Maestre (Eds.)

Compact convex sets in the two–dimensional complex linear space with the Yost property

Ehrhard Behrends

Freie Universität Berlin, Institut für Mathematik I, Arnimallee 2 - 6, D - 1000 Berlin 33

Abstract

Let $K \subset \mathbb{C}^2$ be a compact convex set such that $f(K)$ is a disk for every complex linear functional f. Such sets occurred in D. Yost's investigation of certain approximation problems in Banach spaces.

Here we present some general results concerning sets K with the above property. Also we sketch an example showing that the K under consideration do not necessarily have a center of symmetry.

1 History

Let J be a closed subspace of a Banach space X and $n \in \mathbb{N}$. J is said to satisfy the *n–ball property* if, for all open balls $B_1, \ldots, B_n$ with $\cap B_i \neq \emptyset$ and $B_i \cap J \neq \emptyset$ for all i, one has $\cap B_i \cap J \neq \emptyset$. This property has been introduced in [1] for the case of real scalars. There it was also shown that J satisfies the n–ball property for every n provided it satisfies the 3–ball property (in which case it is called an *M–ideal)*, but that the 2–ball property does not imply the 3–ball property. Later, Lima [3] investigated the case of complex scalars. Whereas he was able to prove counterparts of most of the results in [1] one problem remained open: does the 2–ball property imply the 3–ball property for complex spaces? (We note in passing that when dealing with intersection properties the behaviour for the real case can be strikingly different from the complex situation; see e. g. [4].)

It was D. Yost in [5] who was able to show that this question of maybe very restricted interest is equivalent with a surprisingly elementary problem: The 2–ball property implies the 3–ball property for arbitrary complex spaces iff every compact convex set $K \subset \mathbb{C}^2$ for which all $f(K)$ are disks for $f \in (\mathbb{C}^2)^*$ necessarily has a center of symmetry.

Compact convex sets K for which all $f(K)$ are disks will be said to satisfy the *Yost property* here. Recently it was shown that they need not have a center of symmetry ([2], see also section 4) so that by the above results there is in fact also in complex spaces a difference between the 2–ball property and the 3–ball property. However, the

Yost property implies some other interesting consequences for K which we are going to describe in this paper. They concern strong regularity properties of the topological and convex structure; details are provided in section 3 after some preparations in section 2.

In section 4 we sketch an example of a set with the Yost property but without points of symmetry, and finally in section 5 we indicate some open problems.

2 Preliminaries

Let $K \subset \mathbb{C}^2$ be a compact, convex and nonvoid set. What are the interior properties of K which are special for the K with the Yost property? In order to treat this problem it is desirable to have the following description of K by means of lower dimensional objects:

2.1 Lemma: *Let $\pi : \mathbb{C}^2 \longrightarrow \mathbb{C}$ be the projection $(z, w) \longmapsto z$.*

(i) *Let the family $(K_z)_{z\in\pi(K)}$ be defined by $K_z := \{w | (z, w) \in K\}$. Then*
   (1) *$K_{tz+(1-t)w} \supset tK_z + (1-t)K_w$ for $z, w \in \pi(K)$ and $0 \leqslant t \leqslant 1$.*
   (2) *$z \longmapsto K_z$ is upper semicontinuous, i. e. if a K_{z_0} is contained in an open set O then so is K_z for z close to z_0.*

(ii) *Conversely, let π_K and $(\widetilde{K}_z)_{z\in\pi_K}$ be nonvoid compact convex subsets of $\mathbb{C}$ with (i), (1), and (2). Then $K := \{(z, w) | z \in \pi_K, w \in \widetilde{K}_z\}$ defines a compact convex subset of $\mathbb{C}^2$ with $\pi(K) = \pi_K$ and $K_z = \widetilde{K}_z$ for all $z \in \pi_K$.*

The elementary proof is omitted.

Now let $\mathcal{K}$ be the collection of nonvoid compact convex subsets of $\mathbb{C}$, provided with the Hausdorff metric:

$$d(L, M) = \inf\{\varepsilon | L \subset M + D(0, \varepsilon),\ M \subset L + D(0, \varepsilon)\}$$

(where $D(0, \varepsilon)$ denotes the closed disk with center 0 and radius ε).

2.2 Lemma: *Let K be as above. Then $z \longmapsto K_z$ is continuous on the interior of $\pi(K)$.*

Proof: Let z_0 be in the interior of $\pi(K)$ and $\varepsilon > 0$. We have to show that there is a neighbourhood U of z_0 such that $K_z \subset K_{z_0} + D(0, \varepsilon)$, $K_{z_0} \subset K_z + D(0, \varepsilon)$ for $z \in U$. By 2.1(i)(2) we have $K_z \subset K_{z_0} + D(0, \varepsilon)$ for z close to z_0. For the proof of the other inclusion we first show that $z \longmapsto K_z$ is lower semicontinuous at z_0, i.e. $K_z \cap O \neq \emptyset$ for z in a neighbourhood of z_0 if $K_{z_0} \cap O \neq \emptyset$ and O is open.

Let such an O be given and choose $w_0 \in K_{z_0} \cap O$, $\eta > 0$ with $D(w_0, \eta) \subset O$. Since $z_0 \in \pi(K)^0$ there is a $\delta > 0$ such that the following holds: every z with $|z - z_0| \leqslant \delta$ has a representation $z = tz_0 + (1-t)\hat{z}$ with $\hat{z} \in \pi(K)$ and $t \in [1 - \frac{\eta}{2R}, 1]$ (here R denotes any number such that $D(0, R)$ bounds all K_z). We claim that K_z meets O for these z.

In fact, if $\hat{w} \in K_{\hat{z}}$ is arbitrary we know from 2.1(i)(1) that $w := tw_0 + (1-t)\hat{w} \in K_z$, so that, since $|w - w_0| \leqslant |1 - t|(|w_0| + |\hat{w}|) \leqslant \eta$, $w \in O \cap K_z$.

To conclude the proof let $w_1, \ldots, w_r$ be an $\varepsilon/2$–net in K_{z_0}. By the preceding step there is a neighbourhood $U$of z_0 such that $K_z \cap D(w_\rho, \varepsilon/2) \neq \emptyset$ for $\rho = 1, \ldots, r$ and $z \in U$. But this yields immediately $K_{z_0} \subset K_z + D(0, \varepsilon)$ for these z.

Note: The proof shows even more: one has continuity at every z_0 such that

$$\underset{\varepsilon>0}{\forall} \ \underset{\delta>0}{\exists} \ \underset{\substack{z\in K \\ |z-z_0|\leqslant\delta}}{\forall} \ \underset{\substack{\widehat{z}\in K \\ t\in[1-\varepsilon,1]}}{\exists} \quad z = tz_0 + (1-t)\widehat{z}\,.$$

This is satisfied, e.g., for all z_0 if $\pi(K)$ is a square but only for the z_0 in the interior of $\pi(K)$ in the case of a disk.

It should be stressed that continuity can not be guaranteed at all z_0 in general: Consider, for example any $\tau : \{z \mid |z| = 1\} \longrightarrow \mathcal{K}$ which is upper semicontinuous but not continuous and put $K_\tau := co\,\{(z, w) \mid |z| = 1,\ w \in \tau(z)\}$. Then K_τ is a compact convex set for which $\pi(K_\tau) = D(0, 1)$, and $\pi^{-1}(z) = \tau(z)$ for $|z| = 1$. Thus $z \longmapsto K_z$ is discontinuous at every point where τ is.

3 Properties of compact convex sets with the Yost property

By the preceding results compact convex subsets of $\mathbb{C}^2$ are in one-to-one correspondence with families $(K_z)_{z\in\pi_K}$, where π_K and the K_z are compact and convex in $\mathbb{C}$ and for which the convexity and continuity properties 2.1(i), (1) and (2), hold.

Now suppose that K has the Yost property. To avoid trivial cases we will assume that the disks $f(K)$ are all non-degenerate for $f \neq 0$: if $f(K)$ is a singleton for some $f \neq 0$ then K is a disk in a suitable hyperplane, a situation which obviously does not give rise to interesting questions. Thus in particular $\pi(K)$ and $\bigcup_{z\in\pi(K)} K_z$ are nontrivial disks, and we may assume in the sequel that both are just $D(0, 1)$. This follows from the fact that like K also TK has the Yost property for every $T : \mathbb{C}^2 \longrightarrow \mathbb{C}^2$ of the form linear plus constant.

Let, for $\lambda \in \mathbb{C}$, $f_{(1,\lambda)}$ be the functional $(z, w) \longmapsto z + \lambda w$. The fact that all $\Delta_\lambda := f_{(1,\lambda)}(K) = \bigcup_{|z|\leqslant 1} (z + \lambda K_z)$ are disks will play an important role in the following considerations.

At first we investigate *continuity properties* of the map $z \longmapsto K_z$. It will turn out that in contrast to the general situation described in 2.2 here $z \longmapsto K_z$ is *everywhere* continuous which means that the Yost property implies a certain regularity at the boundary.

3.1 Lemma: *Suppose that K satisfies the Yost property with $\pi(K) = \cup K_z = D(0, 1)$. Let $\beta \in [-1, 1[$ be such that $\sup\{Re\, w \mid w \in K_1\} > \beta$. Then there is an $\eta > 0$ such that $\sup\{Re\, w \mid w \in K_z\} > \beta$ if $|z - 1| \leqslant \eta$.*

Proof: At first we will concentrate on the z close to 1 with $|z| = 1$ which we will write as $e^{i\theta}$ with small real θ. The following elementary fact will be of importance: if θ and θ_0 are given such that $0 < |\theta_0| \leqslant 0.01$ and $|\theta| \leqslant 0.01$, $|\theta - \theta_0| \leqslant |\theta_0|$, then $|\cos\theta - \cos\theta_0|$ $(= |\sin\widehat{\theta}|\,|\theta - \theta_0| \leqslant |\widehat{\theta}|\,|\theta - \theta_0|) \leqslant 2|\theta_0|\,|\theta - \theta_0|$.

Now suppose that $Re\, K_{e^{i\theta_0}}$ is strictly bounded by some $\alpha_0 \in [-1, +1]$, i.e. $Re\, w < \alpha_0$ for $w \in K_{e^{i\theta_0}}$. By 2.1(i)(2) the same is then true for the $Re\, K_z$ with $|z - e^{i\theta_0}| \leqslant 2\delta$, say, where $\delta > 0$ and w. l. o. g. $2\delta \leqslant |\theta_0|$. If then θ satisfies $|e^{i\theta} - e^{i\theta_0}| \leqslant 3\varepsilon$ for an $\varepsilon \leqslant \delta/3$ then $|\theta - \theta_0| \leqslant 6\varepsilon \leqslant |\theta_0|$ so that $\cos\theta \leqslant \cos\theta_0 + 12|\theta_0|\varepsilon$; more generally $Re\, z \leqslant \cos\theta_0 + 12|\theta_0|\varepsilon$ whenever $|z| \leqslant 1$ and $|z - e^{i\theta_0}| \leqslant 3\varepsilon$ (note that the maximum of the function $z \longmapsto Re\, z$ on $\{z \mid |z \leqslant 1,\ |z - e^{i\theta_0}| \leqslant 3\varepsilon\}$ is achieved at a point of the form $e^{i\theta}$).

Let any z with $|z| \leqslant 1$ be given and $\varepsilon \leqslant \delta/3$.

Case 1: $|z - e^{i\theta_0}| \leqslant 3\varepsilon$.
Then, for $w \in K_z$, we know that $Re\, w < \alpha_0$ so that $Re\,(z + \varepsilon w) < Re\, e^{i\theta_0} + 12\varepsilon|\theta_0| + \varepsilon\alpha_0$; in particular, $e^{i\theta_0} + \varepsilon(\alpha_0 + 12|\theta_0|) \notin z + \varepsilon K_z$.

Case 2: $|z - e^{i\theta_0}| > 3\varepsilon$.
We have $|\alpha_0| + 12|\theta_0| \leqslant 2$. Hence $|z - e^{i\theta_0} + \varepsilon(\alpha_0 + 12|\theta_0|)| > \varepsilon$, and thus also in this case $z + \varepsilon K_z$ does not contain $e^{i\theta_0} + \varepsilon(\alpha_0 + 12|\theta_0|)$.

Summing up, we have shown so far that

$$(1) \qquad \underset{\substack{\theta_0 \\ 0<|\theta_0|\leqslant 0.01}}{\forall} \quad \underset{\substack{\alpha_0\in[-1,+1] \\ Re\, K_{e^{i\theta_0}}<\alpha_0}}{\forall} \quad \underset{\varepsilon_0>0}{\exists} \quad \underset{\varepsilon\leqslant\varepsilon_0}{\forall} \; e^{i\theta_0} + \varepsilon(\alpha_0 + 12|\theta_0|) \notin \Delta_\varepsilon\,.$$

We now turn to the proof of 3.1. Fix any $\beta_0 > \beta$ such that there exists a $w \in K_1$ with $Re\, w = \beta_0$ and choose α_0 with $\beta_0 > \alpha_0 > \beta$. We claim that $\sup\{Re\, w \mid w \in K_z\} \geq \alpha_0$ for all z with $|z| = 1$ and $|z - 1|$ small. To prove this we have to make use of

(2) There is a constant $M > 0$ with the following property: Whenever θ_0 and ε are given such that $0 < \varepsilon \leqslant |\theta_0| \leqslant 0.01$ one has $m_1 + \sqrt{r^2 - (\sin\theta_0 - m_2)^2} \geq (\cos\theta_0)(1 + \varepsilon\beta_0) - M\varepsilon|\sin\theta_0|$; here m_1, m_2, r denote the real numbers such that $\Delta_\varepsilon = D(m_1 + im_2, r)$.

(This will be proved below.)

Put $\tilde{\eta} := \frac{1}{2}\left(\frac{\beta_0 - \alpha_0}{12 + 2M}\right)$, we will show that $\sup Re\, K_{e^{i\theta}} \geq \alpha_0$ whenever $|e^{i\theta} - 1| \leqslant \tilde{\eta}$. Suppose that this were not true. We then could, by (1), find a $\theta_0 \neq 0$ with $|e^{i\theta_0} - 1| \leqslant \tilde{\eta}$ such that $e^{i\theta_0} + \varepsilon(\alpha_0 + 12|\theta_0|) \notin \Delta_\varepsilon$ for sufficiently small ε. But on the other hand, by (2),

$$([(1 + \varepsilon\beta_0)\cos\theta_0 - \varepsilon M|\sin\theta_0|] - m_1)^2 + (\sin\theta_0 - m_2)^2 \leqslant r^2\,,$$

i.e. $e^{i\theta_0} + (\beta_0\cos\theta_0 - M\sin\theta_0)\varepsilon \in \Delta_\varepsilon$.

Since $\Delta_\varepsilon \cap (e^{i\theta_0} + \mathbb{R})$ is convex this implies

$$(\beta_0\cos\theta_0 - M\sin\theta_0)\varepsilon < (\alpha_0 + 12|\theta_0|)\varepsilon\,,$$

and this leads after some elementary calculation to a contradiction to our choice of θ_0.

It remains to consider *all* z in $D(0,1)$ with $|z-1|$ small (and not only those with $|z|=1$) and to prove (2). To fill the first gap let $z \in D(0,1)$ with $|z-1| \leqslant \eta := \tilde{\eta}/2$. We may write $z = |z|z_0 + (1-|z|)0$, where $z_0 := \frac{z}{|z|}$ is $\tilde{\eta}$–close to 1. By the preceding investigations we have $Re\, K_{z_0} \geq \alpha_0$. Thus, with an arbitrary $\tilde{w} \in K_0$ and any $w_0 \in K_{z_0}$ with $Re\, w \geq \alpha_0$, $w := |z|w_0 + (1-|z|)\hat{w} \in K_z$ (2.1(i)(1)) and $Re\, w \geq (1-\eta)\alpha_0 - \eta$ or $Re\, w \geq \alpha_0 - \eta$ if $\alpha_0 \geq 0$ or $\leqslant 0$ respectively. Since both $(1-\eta)\alpha_0 - \eta$ and $\alpha_0 - \eta$ are larger than β_0 by the choice of η this leads to $\sup Re\, K_z \geq \beta_0$.

Finally we prove (2). Let ε and θ_0 be given as in this assertion and write Δ_ε as $D(m_1 + im_2, r)$. Then, since $\sup\{Re\, w \,|\, w \in K_1\} \geq \beta_0$ and $D(0, 1-\varepsilon) \subset \Delta_\varepsilon \subset D(0, 1+\varepsilon)$ it follows that

(3) $|m_1|, |m_2| \leqslant \varepsilon,\ r \in [1-\varepsilon, 1+\varepsilon],\ m_1 + r \geq 1 + \varepsilon\beta_0.$

Choose an $M_1 > 0$ with

$$\sqrt{1 - \left(\frac{|t|+\varepsilon}{1-\varepsilon}\right)^2} \Big/ \sqrt{1-t^2} \geq 1 - \varepsilon |t| M_1$$

for $0 \leqslant \varepsilon \leqslant |t| \leqslant 0.01$. The following steps are easy consequences of (3):

$$r\sqrt{1 - \left(\tfrac{|t|+\varepsilon}{1-\varepsilon}\right)^2} \Big/ \sqrt{1-t^2} \geq r - \varepsilon |t| M_2$$

(for a suitable M_2 and ε, t as above);

$$m_1 + r\sqrt{1 - \left(\tfrac{|t|+\varepsilon}{1-\varepsilon}\right)^2} \Big/ \sqrt{1-t^2} \geq m_1 + r - \varepsilon |t| M_2\,;$$

$$\left(m_1 + r\sqrt{1 - \left(\tfrac{|t|+\varepsilon}{1-\varepsilon}\right)^2}\right) \Big/ \sqrt{1-t^2} \geq m_1 + r - \varepsilon |t| M_3$$

(for a suitable M_3 and ε, t as above);

$$m_1 + r\sqrt{1 - \left(\tfrac{|t|+\varepsilon}{1-\varepsilon}\right)^2} \geq (m_1 + r) - \varepsilon |t| M_4$$

(for a suitable M_4 and ε, t as above),

and this yields (2) if we put $t = \sin\theta_0$ (note that, by (3), $\left(\frac{|t|+\varepsilon}{1-\varepsilon}\right)^2 \geq \left(\frac{t-m_2}{r}\right)^2$).

3.2 Theorem: *Let K be a compact convex set with the Yost property. Then, with the notation of section 2, $z \longmapsto K_z$ is continuous on $\pi(K)$.*

Proof: As before we may assume that $\pi(K) = \cup K_z = D(0,1)$. Since with $(K_z)_{|z|\leqslant 1}$ also $(K_{(ze^{i\rho})})_{|z|\leqslant 1}$ gives rise to a set with the Yost property it suffices to prove continuity at $z = 1$.

Lemma 3.1 states that with K_1 also the K_z with small $|z-1|$ meet the halfspace $\{w \,|\, Re\, w > \beta\}$, and by applying 3.1 for the family $(e^{i\rho}K_z)_{|z|\leqslant 1}$ we conclude that the same is true for all halfspaces.

Now let $\varepsilon > 0$ be arbitrary. Since upper semi-continuity always holds (2.1(i)(2)) we only have to find a $\delta > 0$ such that $D(0,\varepsilon) + K_z \supset K_1$ for $|z-1| \leqslant \delta$. To this end choose an open set $O \supset K_1$ and open halfspaces $H_1, \ldots, H_n$ such that the following holds:

(4) $H_i \cap K_1 \neq \emptyset$, and whenever $y_j \in H_j \cap O$ are given then $D(0,\varepsilon) + co\{y_1, \ldots, y_n\} \supset K_1$.

(Choose e.g. strongly exposed points $z_1, \ldots, z_n \in K_1$ with $K_1 = co\{z_1, \ldots, z_n\}$, $\tau > 0$ and exposing normalized functionals $f_1, \ldots, f_n$ such that $s_j := \sup Re\, f_j(K) = Re\, f_j(z_j)$, diam $\{z | z \in K_1,\ s_j - \tau \leqslant Re\, f_j(z) \leqslant s_j\} \leqslant \varepsilon$; then $O := K_1 + D(0, \tau/2)^0$ and $H_j := \{z | Re\, f_j(z) > s_j - \tau/2\}$ have the required properties.)

By the preceding considerations and 2.1(i)(2) we find a $\delta > 0$ such that, for $|z-1| \leqslant \delta$, $K_z \subset O$ and $K_z \cap H_j \neq \emptyset$ for all j. (4) immediately implies that then $K_1 \subset K_z + D(0,\varepsilon)$.

It is clear from the proofs of 3.1 and 3.2 that in order to show 3.2 it is only necessary to use a local variant of the Yost property: Δ_λ has to be a disk for small $|\lambda|$.

The next lemma shows that it is generally sufficient to consider the behaviour of $\lambda \longmapsto \Delta_\lambda$ for λ close to zero when dealing with the Yost property. Since, as it will be made precise in lemma 3.4, this behaviour is essentially determined by the K_z with $|z| = 1$ it follows that in order to get general assertions one needs results providing information on these K_z. This motivates the importance of proposition 3.5. and theorem 3.6 below.

3.3 Lemma: *Let K be as in 3.1 and $\lambda_0 \in \mathbb{C}$. Then there is a K^{λ_0} with the Yost property such that the local behaviour of $\lambda \longmapsto f_{(1,\lambda)}(K) =: \Delta_\lambda$ at λ_0 is precisely the local behaviour of $\lambda \longmapsto f_{(1,\lambda)}(K^{\lambda_0}) =: \Delta_\lambda^{\lambda_0}$ at zero:*

$$\Delta_{\lambda_0+\lambda} = \Delta_\lambda^{\lambda_0} \quad \textit{for all} \quad \lambda \in \mathbb{C}.$$

Proof: Put, for $|w| \leqslant 1$, $\widetilde{K}_w := \{z | (z,w) \in K\}$. From the definition it follows immediately that $\bigcup_{|z|\leqslant 1} z + \lambda K_z = \bigcup_{|w|\leqslant 1} \lambda w + \widetilde{K}_w$, and this gives

$$\begin{aligned} \Delta_{\lambda_0+\lambda} &= \cup z + (\lambda_0+\lambda)K_z \\ &= \cup(\lambda_0+\lambda)w + \widetilde{K}_w \\ &= \cup \lambda w + (\lambda_0 w + \widetilde{K}_w) \\ &= \cup z + \lambda K_z^{\lambda_0}, \end{aligned}$$

where $K_z^{\lambda_0} = \{w | z \in \lambda_0 w + \widetilde{K}_w\}$. It is routine to show that $K^{\lambda_0} := \{(z,w) | w \in K_z^{\lambda_0}\}$ has the claimed properties.

Thus we may restrict ourselves to the study of the behaviour of $\lambda \longmapsto \Delta_\lambda$ for λ close to zero, and this is – up to an error of second order – determined by the K_z with $|z| = 1$:

3.4 Lemma: *Let (K_z) be as before. For $\theta \in [0, 2\pi]$ put $\alpha(\theta) := \sup Re\, e^{-i\theta} K_{e^{i\theta}}$. Then, for every $\delta > 0$ there is an $\varepsilon_0 > 0$ such that, for $0 \leqslant \varepsilon \leqslant \varepsilon_0$,*

$$S_\varepsilon^{\alpha,-} \subset \Delta_\varepsilon \subset S_\varepsilon^{\alpha,+};$$

the $S_\varepsilon^{\alpha,\pm}$ are defined by $\{te^{i\theta} | 0 \leqslant \theta \leqslant 2\pi,\ 0 \leqslant t \leqslant 1 + \varepsilon\alpha(\theta) \pm \varepsilon\delta\}$.

Proof: Let $\delta > 0$ be given. Choose $\varepsilon_0 \in]0, \delta/2]$ so small that $d(K_{z_1}, K_{z_2}) \leqslant \delta/3$ whenever $|z_1 - z_2| \leqslant 3\varepsilon_0$ (this is possible by theorem 3.2).

Now let $0 \leqslant \varepsilon \leqslant \varepsilon_0$; we will show that $S_\varepsilon^{\alpha,-} \subset \Delta_\varepsilon \subset S_\varepsilon^{\alpha,+}$.

1. $\underline{S_\varepsilon^{\alpha,-} \subset \Delta_\varepsilon:}$
By symmetry it suffices to check a particular θ-direction, e.g. $\theta = 0$; is it true that $1 + \varepsilon\alpha(0) - \varepsilon\delta \in \Delta_\varepsilon$?

Consider $z_\pm = e^{\pm 2\varepsilon i}$. We have $|z_\pm - 1| \leqslant 3\varepsilon$ so that there are $w_\pm \in K_{z_\pm}$ with $Re\, w_\pm / z_\pm \leqslant \alpha(0) - \delta/2$. Thus $Re\, w_\pm \geq \alpha(0) - \frac{2}{3}\delta$, and also $z_\pm + \varepsilon w_\pm \in \Delta_\varepsilon$. By convexity therefore Δ_ε must contain a real number $\geq \cos 2\varepsilon + \varepsilon(\alpha(0) - \frac{2}{3}\delta) \geq 1 + \varepsilon\alpha(0) - \delta\varepsilon$ which shows that $1 + \varepsilon\alpha(0) - \delta\varepsilon \in \Delta_\varepsilon$.

2. $\underline{\Delta_\varepsilon \subset S_\varepsilon^{\alpha,+}:}$
We have to show that $z + \varepsilon K_z \subset S_\varepsilon^{\alpha,+}$ for $|z| \leqslant 1$ which is obvious for $|z| \leqslant 1 - \varepsilon$. So suppose that $|z| > 1 - \varepsilon$, and we assume for convenience that $z = |z|$. Thus $|1 - z| \leqslant \varepsilon$, and in this case we have $K_z \subset K_1 + D(0, \delta/3)$.

Let $w \in K_z$ be given. We may write $z + \varepsilon w = e^{i\theta} r$ with $|\theta| \leqslant 2\varepsilon$ and $r \leqslant 1 + \varepsilon\alpha(0) + \varepsilon\frac{\delta}{3}$. But for the θ with $|\theta| \leqslant 2\varepsilon$ we have $|e^{i\theta} - 1| \leqslant 3\varepsilon$ so that again $d(K_{e^{i\theta}}, K_1) \leqslant \delta/3$ and consequently $\alpha(\theta) \geq \alpha(0) - \delta/3$.

$$\text{From}\quad \begin{aligned} 1 + \varepsilon\alpha(\theta) + \varepsilon\delta &\geq 1 + \varepsilon(\alpha(0) - \delta/3) + \varepsilon\delta \\ &= 1 + \varepsilon\alpha(0) + \tfrac{2}{3}\varepsilon\delta \geq r \end{aligned}$$

we now see that $z + \varepsilon w \in S_\varepsilon^{\alpha,+}$.

From 3.4 it follows that α must be a very special function:

3.5 Proposition: *There are $\mu_0 \geq 0$, $\theta_0 \in [0, 2\pi]$, and $r \in \mathbb{R}$ such that $\alpha(\theta) = \mu_0 \cos(\theta - \theta_0) + r =: \beta_{\mu_0,\theta_0,r}(\theta)$ for $\theta \in [0, 2\pi]$.*

Proof: Let $\mathcal{F} \subset C[0, 2\pi]$ be the collection of all $\beta_{\mu_0,\theta_0,r}$. It is easy to see that $\mathcal{F}$ is closed, and hence it suffices to show that α can be arbitrarily well approximated by the elements of $\mathcal{F}$.

To this end, let $\delta > 0$ be given. We choose $\varepsilon_0 > 0$ as in 3.4 and fix any $\varepsilon \in]0, \varepsilon_0]$. By 3.4 the disk $\Delta_\varepsilon = D(m(\varepsilon), r(\varepsilon))$ lies between $S_\varepsilon^{\alpha,-}$ and $S_\varepsilon^{\alpha,+}$, and it follows from elementary considerations that then, with $m(\varepsilon) = \widehat{\mu_0} e^{i\theta_0}$, $1 + \varepsilon\alpha(\theta) - 2\varepsilon\delta \leqslant r(\varepsilon) + \widehat{\mu_0} \cos(\theta - \theta_0) \leqslant 1 + \varepsilon\alpha(\theta) + 2\varepsilon\delta$. Hence $\|\alpha - \beta_{\mu_0,\theta_0,r}\| \leqslant 2\delta$, where $\mu_0 := \frac{\widehat{\mu_0}}{\varepsilon}$, $r := (r(\varepsilon) - 1)/\varepsilon$.

The consequences of proposition 3.5 are collected in

3.6 Theorem: Let K be a compact convex subset of $\mathbb{C}^2$ with the Yost property such that $\pi(K) = \cup K_z = D(0,1)$.

(i) *The midpoint– and the radiusfunction m and r, which are defined by $f_{(1,z)}(K) = D(m(z), r(z))$ have one–sided derivatives at every z_0 into all directions.*

(ii) *Suppose that K_z is known for three different z_1, z_2, z_3 with $|z_j| = 1$. Then all K_z with $|z| = 1$ can be constructed.*

Proof: (i) Let α be written as $\beta_{\mu_0,\theta_0,r}$ for suitable μ_0, θ_0, r. Then, by 3.5, it follows that $m(\varepsilon)$ and $r(\varepsilon)$ equal $\varepsilon\mu_0 e^{i\theta_0}$ and $1 + \varepsilon r$, respectively, up to an error term of second order in ε. Hence $\lim\limits_{\varepsilon\to 0+} \dfrac{m(\varepsilon) - m(0)}{\varepsilon} = \mu_0 e^{i\theta_0}$ and $\lim\limits_{\varepsilon\to 0+} \dfrac{r(\varepsilon) - r(0)}{\varepsilon} = r$, and this is just the assertion for the case $z_0 = 0$ and the direction $t \geq 0$. The case of arbitrary directions follows from this result by considering the family $(e^{i\rho} K_z)_{|z|\leqslant 1}$, and for an investigation of the z_0 which are not at the origin one only has to apply lemma 3.3.

(ii) This will be derived from the observations that

1. a function $\beta_{\mu_0,\theta_0,r}$ can be reconstructed from its values at three different $\theta_1, \theta_2, \theta_3$ in $[0, 2\pi[$; and

2. a compact convex subset L in $\mathbb{C}$ is uniquely determined by the family of numbers $(\sup\, Re\, e^{i\rho} L)$ (this is a consequence of the separation theorem).

Let $(K_{z_j})_{j=1,2,3}$ be given for different z_j. From 3.5 and "1." we can derive the values of all $\sup\, Re\, e^{-i\theta} K_{e^{i\theta}}$. Applying this observation to the family $(e^{i\rho} K_z)_{|z|\leqslant 1}$ we get the $\sup\, Re\, e^{i(\rho-\theta)} K_{e^{i\theta}}$ for all ρ, and by "2." this determines the $K_{e^{i\theta}}$.

Notes:

1) The function m is necessarily bounded, and if it were *complex* differentiable it would be constant by the Liouville theorem. This, however, need not be the case as the counterexample to be sketched in the next section shows. Therefore only comparitively weak differentiability properties like the one in 3.6 (i) are to be expected.

2) Since all functions $\beta_{\mu_0.\theta_0,r}$ are continuous one could easily derive from 3.5 that $z \longmapsto K_z$ is continuous on $\partial D(0,1)$. Unfortunately this observation cannot be taken as a first step to avoid the tedious proof of theorem 3.2 since this theorem was essential for the proof of 3.5.

4 A counterexample to the symmetry problem

One of the fundamental questions in connection with the Yost property was whether or not all sets with this property have a center of symmetry (cf. section 1). The anwer is "no":

4.1 Theorem: *There is a continuous concave function $\varphi : D(0,1) \longrightarrow \mathbb{R}^+$ such that*

$$K_\varphi := \{(z,w) \mid |z| \leqslant 1,\ |w| \leqslant \varphi(z)\}$$

has the Yost property, but K_φ has no center of symmetry.

Sketch of proof: (Details are given in [2].)
Fix a, b with $1/2 < a < b$. It can be shown that there is a $\delta > 0$ such that the following holds: Whenever $m : \mathbb{R}^+ \longrightarrow \mathbb{R}$ is three times continuously differentiably with support in $[a,b]$ and $\|m\|_3 := \|m\|_\infty + \|m'\|_\infty + \|m''\|_\infty + \|m'''\|_\infty \leqslant \delta$, then for every $z \in \mathbb{C}$, $0 < |z| \leqslant 1$ there are precisely one $t \geq 1/2$ and one $e^{i\theta}$ ($\theta \in [0, 2\pi]$) such that

$$z = z_{t,\theta} := m(t) + e^{i\theta}(r(t) - th'_\theta(t));$$

here r denotes the function $r(t) = t + \frac{1}{4t}$, and $h_\theta(t) := r(t) + m(t)\cos\theta$.

Define

$$\varphi(z) := \begin{cases} 1 & z = 0 \\ h'_\theta(t) & z = z_{t,\theta}\,. \end{cases}$$

Then it can be shown that – at least for even smaller $\|m\|_3$ – the associated set K_φ has the claimed properties.

5 Some open problems

Once having a counterexample it is not hard to prove that in fact "most" sets with the Yost property fail to have a center of symmetry: in $\mathcal{K}_{Y.P.} = \{$ the K with the Yost property$\}$ the subset $\mathcal{K}_{S.P.}$ of the K with a point of symmetry is a closed set with empty interior (this follows immediately from the fact that with $K \in \mathcal{K}_{Y.P.}\backslash\mathcal{K}_{S.P.}$ and $\widehat{K} \in \mathcal{K}_{S.P.}$ one has $\varepsilon K + (1-\varepsilon)\widehat{K} \in \mathcal{K}_{Y.P.}\backslash\mathcal{K}_{S.P.}$ for $\varepsilon \in]0,1]$). However, it would be interesting to know some qualitative estimates, e.g. to provide a "small" universal constant $\eta > 0$ such that the $K \in \mathcal{K}_{Y.P.}$ which are bounded in norm by one, say, can be η–approximated by a set in $\mathcal{K}_{S.P.}$ (or, what amounts to be the same, the diameter of the range of the function $z \longmapsto$ midpoint of $f_{1,z}(K)$ is always small).

A second problem should be mentioned. Since by 4.1 the Yost property does not imply symmetry one would like to have an additional "natural" sufficient condition which could easily be formulated in terms of the convex geometry of K.

References

[1] E. M. ALFSEN AND E. G. EFFROS. *Structure in real Banach spaces. Part I and II.* Ann. of Math. **96** (1972), 98 - 173.

[2] E. BEHRENDS. *Points of symmetry of convex sets in the two–dimensional complex space – a counterexample to D. Yost's problem.* (Preprint).

[3] Å. LIMA. *Intersection properties of balls and subspaces in Banach spaces.* Trans. Amer. Math. Soc. **227** (1977), 1 - 62.

[4] Å. LIMA. *Complex Banach spaces whose duals are L^1–spaces.* Israel Journal of Math. **24**, 1976, 59 – 72.

[5] D. YOST. *Semi-M-ideals in complex Banach spaces.* Rev. Roum. Math. Pures Appl. **29** (1984), 619 - 623.

Progress in Functional Analysis
K.D. Bierstedt, J. Bonet, J. Horváth & M. Maestre (Eds.)

SOME REMARKS ON A LIMIT CLASS OF APPROXIMATION IDEALS

Fernando Cobos* and Thomas Kühn**

*Departamento de Matemáticas, Facultad de Ciencias, Universidad Autónoma de Madrid, 28049 Madrid. Spain.

**Sektion Mathematik, Universität Leipzig, 7010 Leipzig. Germany.

Dedicated to Professor M. Valdivia on his 60th birthday

Abstract

We give some results that show the difference between approximation ideals $\mathcal{L}_{\infty,q}$ and classical Lorentz ideals $\mathcal{L}_{p,q}$.

Let H be a Hilbert space and let $T:H \longrightarrow H$ be a compact linear operator. Denote by $|T| = [T^*T]^{1/2}$ the unique positive operator such that $|T|^2 = T^*T$. Then $|T|$ is compact and positive. Consequently the spectrum of $|T|$ is non-negative, and all its non-zero spectral values are eigenvalues, having no accumulation point except possibly zero. We can order the eigenvalues with respect to decreasing values, repeating each one as many times as its (algebraic) multiplicity indicates,

$$\lambda_1(|T|) \geq \lambda_2(|T|) \geq \dots \geq 0.$$

If $|T|$ has only a finite number N of eigenvalues, then we complete the sequence by setting $\lambda_n(|T|) = 0$ for $n > N$.

The n-th singular number of T is given by

$s_n(T) = \lambda_n(|T|)$.

In the theory of operators on a Hilbert space, a major role is played by operator spaces defined by means of summability conditions on the singular numbers. So Lorentz operator spaces are defined as follows

$$S_{p,q} = \{T \in \mathcal{L}(H) : \sigma_{p,q}(T) = [\sum_{n=1}^{\infty} (n^{1/p} s_n(T))^q n^{-1}]^{1/q} < \infty\}.$$

Here $0 < p < \infty$ and $0 < q \leq \infty$. Note that for $p = q$, $S_{p,p}$ coincides with the Schatten-von Neumann p-class S_p. In particular, S_1 is the collection of all nuclear operators on H, and S_2 the set of Hilbert-Schmidt operators (see,e.g.,[4]).

In order to treat certain problems of perturbation theory and invariant subspaces, Macaev [8] introduced in 1961 the following space

$$S_{\infty,1} = \{T \in \mathcal{L}(H) : \sigma_{\infty,1}(T) = \sum_{n=1}^{\infty} s_n(T)\, n^{-1} < \infty\}$$

(see also [4],[5]). Much more recently, spaces

$$S_{\infty,q} = \{T \in \mathcal{L}(H) : \sigma_{\infty,q}(T) = [\sum_{n=1}^{\infty} s_n(T)^q n^{-1}]^{1/q} < \infty\} \quad (0 < q < \infty)$$

appeared naturally in the work of Peller [10] on the averaging projection onto the set of Hankel matrices.

Observe that formally $S_{\infty,q}$ is defined by taking $p = \infty$ in the expression of Lorentz spaces $S_{p,q}$. A first relationship between these two scales of spaces is that for $0 < p,q < \infty$

and $0 < r \leqq \infty$ it holds

$S_{p,r} \hookrightarrow S_{\infty,q}$ (continuous inclusion).

Take now any bounded linear operator $T \in \mathcal{L}(E,F)$ acting between the Banach spaces E and F. The approximation numbers of T are given by

$$a_n(T) = \inf\{\|T-R\| : R \in \mathcal{L}(E,F) \text{ , rank } R < n\} \quad n=1,2,...$$

It is well-known that approximation numbers coincide with singular numbers for operators acting on a Hilbert space (see, e.g., [4]). So we can extend the operator spaces described above to the class of all Banach spaces by setting for $0 < p < \infty$ and $0 < q \leqq \infty$

$$\mathcal{L}_{p,q}(E,F) = \{T \in \mathcal{L}(E,F) : \sigma_{p,q}(T) = [\sum_{n=1}^{\infty} (n^{1/p} a_n(T))^q n^{-1}]^{1/q} < \infty\}$$

and for $0 < q < \infty$

$$\mathcal{L}_{\infty,q}(E,F) = \{T \in \mathcal{L}(E,F) : \sigma_{\infty,q}(T) = [\sum_{n=1}^{\infty} a_n(T)^q n^{-1}]^{1/q} < \infty\} .$$

Lorentz ideals $\mathcal{L}_{p,q}$ are considered nowadays indispensable tools in the modern theory of eigenvalue distributions (see, e.g., [7], [13]). Some important properties of Lorentz ideals follow from the fact that they can be regarded as approximation spaces. In particular, this is the case of the following representation theorem in terms of finite rank operators:

<u>Theorem 1.</u>- Let $0 < p < \infty$, $0 < q \leqq \infty$ and let $T \in \mathcal{L}(E,F)$. Then the following conditions are equivalent.

(i) $T \in \mathcal{L}_{p,q}(E,F)$.

(ii) There is a sequence $(T_n)_{n=0}^{\infty} \subset \mathcal{L}(E,F)$, such that

rank $T_n \leq 2^n$, $T = \sum_{n=0}^{\infty} T_n$ converges in the operator norm

and $\left[\sum_{n=0}^{\infty} (2^{n/p}\|T_n\|)^q\right]^{1/q} < \infty$.

The general theory of Lorentz ideals does not cover the case of ideals $\mathcal{L}_{\infty,q}$ by taking $p=\infty$. In fact, Theorem 1 does not hold for $p=\infty$. Cobos and Resina [2] gave an example of an operator T from ℓ_∞ into ℓ_1 which does not belong to $\mathcal{L}_{\infty,q}(\ell_\infty,\ell_1)$ but that can be written in the form (ii) with $p=\infty$.

From the point of view of "Geometry of Banach Spaces", spaces ℓ_1 and ℓ_∞ are "bad" ones. So one might expect that Theorem 1 remains true for $p=\infty$ when E and F are Banach spaces having "good" geometrical properties; in particular, when E and F coincide with a Hilbert space.

As a matter of fact, in the literature one frequently finds examples of operator ideals whose behaviour improves notably when the underlying spaces are Hilbert spaces (see, e.g., [6]).

Nevertheless, our next example shows that even in the class of Hilbert spaces, conditions (i) and (ii) in Theorem 1 are no longer equivalent if $p=\infty$.

Let $(\delta_n)_{n=0}^{\infty}$ be the sequence given by $\delta_0 = 1$ and $\delta_n = n^{-1-1/q}$ for $n=1,2,\ldots$ Define operators D_n in ℓ_2 by $D_n(\xi_m) = (\mu_m)$ where

$$\mu_m = \begin{cases} \delta_n \xi_m & \text{if} \quad 1 \leqq m \leqq 2^n \\ 0 & \text{otherwise} \end{cases}$$

Clearly rank $D_n = 2^n$. Since $\|D_n\| = \delta_n$, we have

$$\sum_{n=0}^{\infty} \|D_n\|^q = 1 + \sum_{n=1}^{\infty} n^{-q-1} < \infty .$$

Moreover, it follows from $\sum_{n=0}^{\infty} \delta_n < \infty$ that $D = \sum_{n=0}^{\infty} D_n$ is a well-defined bounded operator in ℓ_2. Therefore, the operator D can be represented in the form (ii) with $p = \infty$ by means of the sequence $(D_n)_{n=0}^{\infty}$.

But D does not belong to $\mathcal{L}_{\infty,q}(\ell_2, \ell_2)$. To check this, first observe that

$$\sigma_{\infty,q}(D)^q \leqq \sum_{n=0}^{\infty} s_{2^n}(D)^q \leqq \frac{2}{\log 2} \sigma_{\infty,q}(D)^q .$$

Furthermore, taking into account that D is a diagonal operator, we get

$$s_{2^n}(D) = \sum_{k=n}^{\infty} \delta_k = \sum_{k=n}^{\infty} k^{-1-1/q} \geqq q\, n^{-1/q} .$$

Hence

$$\sigma_{\infty,q}(D)^q \geqq q^q \frac{\log 2}{2} \sum_{n=1}^{\infty} n^{-1} = \infty .$$

The representation theorem that holds in this limit case is the following (see [2], Thm.1.2):

<u>Theorem 2.-</u> An operator $T \in \mathcal{L}(E,F)$ belongs to $\mathcal{L}_{\infty,q}(E,F)$ if and only if there exists a sequence $(T_n)_{n=0}^{\infty} \subset \mathcal{L}(E,F)$ with

rank $T_n \leqq 2^{(2^n)}$ such that $T = \sum_{n=0}^{\infty} T_n$ converges in the operator norm and $[\sum_{n=0}^{\infty} 2^n \| T_n \|^q]^{1/q} < \infty$.

Another point where the gap between ideals $\mathcal{L}_{p,q}$ and $\mathcal{L}_{\infty,q}$ becomes obvious is the behaviour under tensor products. Pietsch [12](case $p < q$) and König [6](case $p \geqq q, 1/q<1/p+1/2$) have shown that Lorentz ideals $\mathcal{L}_{p,q}$ fail to be tensor-stable. Nevertheless, ideals $\mathcal{L}_{\infty,q}$ are stable under tensor norms. To see it, we first describe the approximation structures that underlie these ideals and then we prove a result on "bilinear" operators. The stability property follows from this last result.

Subsequently $(X, \| \ \|_X)$ denotes a quasi-normed Abelian group and $(G_n)_{n=0}^{\infty}$ is a sequence of subsets of X such that

(1) $G_0 = \{0\}$;

(2) $G_n \subset G_{n+1}$, $n=0,1,...$;

(3) $G_n \pm G_m \subset G_{n+m}$, $n,m=0,1,...$

Given any $f \in X$, put

$$E_n(f) = \inf\{\|f-g\|_X : g \in G_{n-1}\} \quad , \quad n=1,2,...$$

In this way we associate to each $f \in X$ the sequence of non-negative numbers $(E_n(f))$ whose n-th coordinate is the error of approximation of f by elements of G_{n-1}.

Let $0 < \alpha < \infty$ and $0 < q \leqq \infty$. The classical approximation space X_q^{α} consists of all $f \in X$ which have a finite quasi-norm

$$\|f\|_{X_q^{\alpha}} = [\sum_{n=1}^{\infty} (n^{\alpha} E_n(f))^q n^{-1}]^{1/q} \quad .$$

Spaces X_q^α have been investigated by many authors and have important applications in operator theory and function theory. Let us refer, for example, to the papers by Peetre and Sparr [9], Pietsch [11] and DeVore and Popov [3].

The relationship between X_q^α and $\mathcal{L}_{p,q}(E,F)$ becomes clear when taking $X = \mathcal{L}(E,F)$ and

$$G_n = \{R \in \mathcal{L}(E,F) : \text{rank } R \leq n\} \quad , \quad n=0,1,\ldots$$

Then

$$X_q^\alpha = \mathcal{L}_{1/\alpha,q}(E,F) .$$

There is also a scale of approximation spaces associated to ideals $\mathcal{L}_{\infty,q}$. Namely, for $0 < q < \infty$, define X_q as the collection of all f in the group X which have a finite quasi-norm

$$\|f\|_{X_q} = [\sum_{n=1}^{\infty} E_n(f)^q \, n^{-1}]^{1/q} .$$

Spaces X_q were studied in [1]. Observe that once more, when $X = \mathcal{L}(E,F)$ and

$$G_n = \{R \in \mathcal{L}(E,F) : \text{rank } R \leq n\} \quad , \quad n=0,1,\ldots$$

we have

$$X_q = \mathcal{L}_{\infty,q}(E,F) .$$

Next we describe the behaviour of "bilinear" operators under X_q-spaces.

Take three quasi-normed Abelian groups X,Y,Z and let $(G_n)_{n=0}^\infty$, $(F_n)_{n=0}^\infty$, $(H_n)_{n=0}^\infty$ be sequences of subsets in X,Y,Z respectively, satisfying conditions (1) to (3). Let

$$J: X \times Y \longrightarrow Z$$

be an operator. We assume that there is $M < \infty$ such that for all $f, f_0, f_1 \in X$ and all $g, g_0, g_1 \in Y$ the following holds:

(I) $\quad J(f_0 \pm f_1, g) = J(f_0, g) \pm J(f_1, g)$;

(II) $\quad J(f, g_0 \pm g_1) = J(f, g_0) \pm J(f, g_1)$;

(III) $\quad \|J(f,g)\|_Z \leq M \|f\|_X \|g\|_Y$.

<u>Theorem 3.</u>- Let X,Y,Z and J be as above, and suppose also that

(IV) $\quad J(G_m, F_n) \subset H_{mn} \quad$ for $m,n=0,1,\ldots$

Then

$$J: X_q \times Y_q \longrightarrow Z_q$$

is bounded for all $0 < q < \infty$.

<u>Proof</u>. First we show that there is a constant $C < \infty$ independent of f,g and n such that

$$E_{2^{2n}}(J(f,g)) \leq C[E_{2^n}(f)\|g\|_Y + \|f\|_X E_{2^n}(g)] \qquad (*)$$

For this we use a similar argument to one in [14]. Let c_1, c_3 be the constants of the quasi-triangle inequality in X,Z respectively. Given any $\varepsilon > 0$, find

$$f_0 \in G_{2^n-1} \quad \text{with} \quad \|f-f_0\|_X < E_{2^n}(f) + \varepsilon$$

and

$$g_0 \in F_{2^n-1} \quad \text{with} \quad \|g-g_0\|_Y < E_{2^n}(g) + \varepsilon \ .$$

By (IV),

$J(f_0,g_0) \in H_{2^{2n}-1}$.

Hence, using (I),(II) and (III), we get

$$E_{2^{2n}}(J(f,g)) \leq \|J(f,g) - J(f_0,g_0)\|_Z$$

$$= \|J(f-f_0,g) + J(f_0,g-g_0)\|_Z$$

$$\leq c_3M[\|f-f_0\|_X\|g\|_Y + \|f_0-f+f\|_X\|g-g_0\|_Y]$$

$$\leq c_3M[(E_{2^n}(f) + \varepsilon)\|g\|_Y + c_1(E_{2^n}(f) + \varepsilon + \|f\|_Y)(E_{2^n}(g) + \varepsilon)]$$

$$\leq c_3M[(E_{2^n}(f) + \varepsilon)\|g\|_Y + c_1(2\|f\|_Y + \varepsilon)(E_{2^n}(g) + \varepsilon)] \quad .$$

Passing to the limit when $\varepsilon \to 0$ we obtain ($\ast$) with $c=2Mc_1c_3$.

Now we have

$$\|J(f,g)\|_{Z_q} = [\sum_{n=1}^{\infty} E_n(J(f,g))^q\, n^{-1}]^{1/q}$$

$$\leq [\sum_{n=0}^{\infty} E_{2^n}(J(f,g))^q]^{1/q}$$

$$\leq 2^{1/q}[\sum_{n=0}^{\infty} E_{2^{2n}}(J(f,g))^q]^{1/q}$$

$$\leq 2^{1/q}C[\sum_{n=0}^{\infty}(E_{2^n}(f)\|g\|_Y + \|f\|_X E_{2^n}(g))^q]^{1/q}$$

$$\leq 2^{1/q}C\max\{1,2^{(1/q)-1}\}[(\sum_{n=0}^{\infty} E_{2^n}(f)^q)^{1/q}\|g\|_Y + \|f\|_X(\sum_{n=0}^{\infty} E_{2^n}(g)^q)^{1/q}]$$

$$\leq M_1\|f\|_{X_q}\|g\|_{Y_q}$$

where M_1 is a constant independent of f and g. This completes the proof. //

Theorem 3 improves [1], Thm.3.2, where the same conclusion was obtained assuming that the group Z is normed and that $0 < q \leqq 1$.

As a direct consequence of Theorem 3 we have that ideals $\mathcal{L}_{\infty,q}$ are tensor stable for any $0 < q < \infty$. In other words, if τ is any tensor norm, then $S \in \mathcal{L}_{\infty,q}(E,F)$ and $T \in \mathcal{L}_{\infty,q}(E_0,F_0)$ imply that

$$S \hat{\otimes}_\tau T \in \mathcal{L}_{\infty,q}(E \hat{\otimes}_\tau E_0 \, , \, F \hat{\otimes}_\tau F_0) \, .$$

References

1 F.Cobos and M.Milman, On a limit class of approximation spaces, Numer.Funct. Anal. and Optimiz.11 (1990) 11-31.

2 F.Cobos and I.Resina, Representation theorems for some operator ideals, J. London Math. Soc. 39 (1989) 324-334.

3 R.A.DeVore and V.A.Popov, Interpolation and non-linear approximation, in "Function Spaces and Applications",p. 191 to 205, Springer L.N.M. 1302, Berlin 1988.

4 I.C.Gohberg and M.G.Krein,"Introduction to the Theory of Linear Nonselfadjoint Operators", Amer. Math. Soc., Providence 1969.

5 I.C.Gohberg and M.G.Krein, "Theory and Applications of Volterra Operators on Hilbert Spaces", Amer. Math. Soc., Providence 1970.

6 H.König, On the tensor stability of s-numbers ideals, Math. Ann. 269 (1984) 77-93.

7 H.König, "Eigenvalue Distribution of Compact Operators", Birkhäuser, Basel 1986.

8 V.I.Macaev, A class of completely continuous operators, Soviet Math. Dokl. 2 (1961) 972-975.

9 J.Peetre and G.Sparr, Interpolation of normed abelian groups, Ann. Mat. Pura et Appl. 92 (1972) 217-262.

10 V.V.Peller, Metric properties of an averaging projection onto the set of Hankel matrices, Soviet Math. Dokl. 30 (1984) 362-368.

11 A.Pietsch, Approximation spaces, J. Approx. Th. 32 (1981) 115-134.

12 A.Pietsch, Tensor products of sequences, functions, and operators, Arch. Mat. (Basel) 38 (1982) 335-344.

13 A.Pietsch, "Eigenvalues and s-numbers", Cambridge Univ. Press, Cambridge 1987.

14 N.Tita, On a class of $\ell_{\Phi\phi}$ operators, Collect. Math. 32 (1981) 275-279.

The second author was supported in part by DGICYT (SAB-90-0033).

Progress in Functional Analysis
K.D. Bierstedt, J. Bonet, J. Horváth & M. Maestre (Eds.)

Some Factorization Properties of Composition Operators

Hans Jarchow*
Mathematisches Institut
Universität Zürich
Rämistrasse 74
CH - 8001 Zürich / Switzerland

Dedicated to Professor Manuel Valdivia
on his sixtieth birthday

Introduction

Let D be the open unit disk in the complex plane, and let φ be an analytic function on D such that $|\varphi(z)| < 1$ for each $z \in D$. It is a well-known consequence of Littlewood's subordination principle (see e.g. [Du] p.10) that $f \mapsto f \circ \varphi$ defines a bounded linear operator $C_\varphi : H^p \to H^p$ on each of the classical Hardy spaces $H^p = H^p(D)$ $(0 < p \leq \infty)$, the so-called *composition operator* induced by φ.

A central topic in the study of composition operators on Hardy spaces (as well as on various other function spaces) is to investigate the interplay between properties of the generating function φ and properties of the operator C_φ on the spaces under consideration. See C.C. Cowen's recent survey article [Co] for a discussion of these and related aspects.

For example, it was shown in 1973 by J.H. Shapiro and P.D. Taylor [ST] that $C_\varphi : H^2 \to H^2$ is a Hilbert-Schmidt operator of and only if $\frac{1}{1-|\varphi|^2}$ belongs to $L^1(m)$. Here m denotes normalized Lebesgue measure on ∂D; we also follow the common habit to denote by φ not only the analytic function $D \to D$ but also its (m-*a.e.* defined) extension $\partial D \to \overline{D}$ to the boundary. The question of characterizing φ such

* Supported by the Swiss National Science Foundation

that $C_\varphi : H^2 \to H^2$ belongs to the Schatten-von Neumann class $\mathfrak{S}_p(0 < p < \infty)$ was settled only recently by D.H. Luecking and K. Zhu [LZ], extending a corresponding description of compactness of C_φ obtained earlier by J.H. Shapiro [Sh].

We shall concentrate on composition operators C_φ whose symbol φ satisfies the integrability condition $\frac{1}{1-|\varphi|^2} \in L^\beta(m)$ for some $0 < \beta < \infty$. Various questions in connection with such operators have been investigated in [Hr], [HJ], [HJM]. We shall supplement the corresponding results by some additional characterizations. For example, if $1 \leq \beta < \infty$, then the above integrability condition is equivalent to saying that C_φ admits a factorization $H^{2/\beta} \xrightarrow{v} l^2 \xrightarrow{u} H^2$ where u is a Hilbert-Schmidt operator, whereas in case $0 < \beta \leq 1$ one has to require the existence of a factorization $H^2 \xrightarrow{v} l^2 \xrightarrow{u} H^{2\beta}$, with v being Hilbert-Schmidt.

Order Bounded Operators

Let X be a quasi-Banach space, let (Ω, Σ, μ) be any measure space, and let $0 < p < \infty$. We say that an operator (= bounded linear map) $u : X \to L^p(\mu)$ is *order bounded* if there exists a non-negative member g of $L^p(\mu)$ such that $|ux| \leq g$ μ-*a.e.* whenever x belongs to B_X, the unit ball of X.

It is easily seen that $u : X \to L^p(\mu)$ is order bounded if and only if there is a factorization $u : X \xrightarrow{b} L^\infty(\nu) \xrightarrow{i_p} L^p(\nu) \xrightarrow{a} L^p(\mu)$ where ν is a probability measure on (Ω, Σ), a is a multiplication operator induced by a suitable measurable function, i_p is the formal identity, and b is a bounded operator. In a particular, such an operator is always (absolutely) p-summing. We refer to A. Pietsch's monograph [Pi] for details on summing operators and related operator ideals.

Back to L. Schwartz and S. Kwapień (cf. [Kw]) goes the result that if X is a Banach space and $1 \leq p < \infty$, then $u \in \mathcal{L}(X, L_p(\mu))$ is order bounded whenever its adjoint, $u^* : L_{p^*}(\mu) \to X^*$, is p-summing ([Pi], 24.5.8). Since u is then even p-integral, we obtain an easy consequence:

(1) PROPOSITION: *Let $1 \leq p < \infty$, and let μ and ν be any measures. The following statements about an operator $u : L_{p^*}(\mu) \to L_p(\mu)$ are equivalent*

(i) u is order-bounded.

(ii) *u is p-integral.*

(iii) *u is p-summing.*

(iv) *u^* is order-bounded.*

(v) *u^* is p-integral.*

(vi) *u^* is p-summing.*

If $p > 1$, then p-nuclearity for u and u^* may be added to this list. It is actually a result of A. Grothendieck ([Gr], I, §4, n°3) that $L_1(\mu)$-valued operators are order bounded precisely when they are 1-integral.

Let now $0 < \beta < \infty$ and an analytic function $\varphi : D \to D$ be given. It is known that, regardless of our choice of $0 < p < \infty$, C_φ induces an operator $H^p \to H^{\beta p}$ which is order bounded when considered as a map into $L^{\beta p}(m)$ if and only if $\frac{1}{1-|\varphi|^2} \in L^\beta(m)$; see e.g. [HJ]. Let us agree to simply say that C_φ is *β-order bounded* in this case.

The appearance of $|\varphi|^2$ in these matters is of no significance (it has its roots in working with normalized point evaluations). Any positive exponent can be used instead of 2. The above condition is also equivalent to m-integrability of $(\log \frac{1}{|\varphi|})^{-\beta}$. More important for us, however, will be the fact that C_φ is β-order bounded if and only if $(\|\varphi^n\|_{H^r})_{n=1}^\infty$ belongs to the Lorentz sequence space $l^{r/\beta,r}$, no matter how we choose $0 < r < \infty$. Before we use this, let us state an immediate corollary to the above proposition.

(2) COROLLARY: *Let $\varphi : D \to D$ be an analytic function and let $0 < \beta < \infty$. Then β-order boundedness of the associated composition operator C_φ is equivalent to C_φ inducing a $(1+\beta)$-summing operator $C_\varphi : H^{1+1/\beta} \to H^{1+\beta}$.*

Of course, we could as well have asked $C_\varphi : H^{1+1/\beta} \to H^{1+\beta}$ (or its adjoint) to be $(1+\beta)$-integral, or $(1+\beta)$-nuclear.

A related, but somewhat deeper characterization is as follows:

(3) PROPOSITION: *Suppose that $\beta \geq 1$ and that $\varphi : D \to D$ is analytic. Then C_φ is β-order bounded if and only there are a bounded operator $v : H^1 \to l^{\beta^*}$ and a β-summing operator $u : l^{\beta^*} \to H^\beta$ such that $uv(f) = f \circ \varphi \;\; \forall f \in H^1$.*

So, briefly and diagrammatically, C_φ factorizes as $C_\varphi : H^1 \xrightarrow{v} l^{\beta^*} \xrightarrow{u} H^\beta$, with u and v as announced.

Proof. If C_φ admits such a factorization, then u is order-bounded as an operator into $L^\beta(m)$, by (1), hence C_φ is β-order bounded.

Suppose conversely that C_φ is β-order bounded, so that $\sum_{n=0}^\infty (n+1)^{\beta-1}\|\varphi^n\|_{H^1}^\beta < \infty$. By Hardy's inequality ([Du], p.48), if $f \in H^1$, then

$$\left(\sum_n \frac{|\hat{f}(n)|^{\beta^*}}{n+1}\right)^{1/\beta^*} \leq \sup_n |\hat{f}(n)|^{1/\beta} \cdot \left(\sum_n \frac{|\hat{f}(n)|}{n+1}\right)^{1/\beta^*} \leq \pi^{1/\beta^*} \cdot \|f\|_{H^1},$$

so that $v : H^1 \to l^{\beta^*} : f \mapsto \left(\frac{\hat{f}(n)}{(n+1)^{1/\beta^*}}\right)_n$ is a well-defined bounded linear map [$\hat{f}(n)$ is the n-th Fourier coefficient of $f \in H^1$].

Thanks to our hypothesis, $u((\xi_n)_n) := \sum_{n=1}^\infty (n+1)^{1/\beta^*}\xi_n\varphi^n$ defines a bounded linear map $u : l^{\beta^*} \to H^\beta$ with $\|u\|^\beta \leq \sum_n (n+1)^{\beta-1}\|\varphi^n\|_{H^\beta}^\beta$. u is not only bounded : u^* is even β-nuclear, and so u is β-summing by (1).

If $f = \sum_n \hat{f}(n)z^n$ is in H^1, then $uv(f) = \sum_n \hat{f}(n)\varphi^n = C_\varphi(f)$, and so the proof is complete.

Hilbert-Schmidt Factors of Composition Operators

As we shall see, β-order bounded composition operators can be characterized by means of a factorization scheme in which one of the factors is a Hilbert-Schmidt operator. We shall have to distinguish between the cases $1 \leq \beta < \infty$ and $0 < \beta < 1$. Throughout, $\varphi : D \to D$ will be an analytic function.

(4) Proposition: *If $1 \leq \beta < \infty$, then C_φ is β-order bounded if and only if there is a factorization $C_\varphi : H^{2/\beta} \xrightarrow{v} l^2 \xrightarrow{u} H^2$ where v is a bounded operator and u is a Hilbert-Schmidt operator.*

Proof. Since Hilbert-Schmidt operators have 2-summing adjoints, the existence of such a factorization implies that C_φ is β-order bounded.

Suppose that conversely C_φ is β-order bounded. We know that this is equivalent to $\sum_n (n+1)^{\beta-1}\|\varphi^n\|_{H^2}^2 < \infty$. Accordingly, $u : l^2 \to H^2$ is well-defined by $u((\xi_n)) := \sum_n (n+1)^{\frac{1}{2}(\beta-1)}\xi_n\varphi^n$; it is even Hilbert-Schmidt since $\sum \|ue_n\|_{H_2}^2 < \infty$ holds for the standard unit vector basis (e_n) of l^2.

If we can show that $f \mapsto \left((n+1)^{-\frac{1}{2}(\beta-1)}\hat{f}(n)\right)_n$ defines a bounded operator $v : H^{2/\beta} \to l^2$, then we will again be done.

Since $0 < \frac{2}{\beta} \leq 2$, we may apply a classical result of Hardy and Littlewood ([Du], p. 95) to find a constant C such that

$$\left(\sum_{n=0}^{\infty} \frac{|\hat{f}(n)|^{2/\beta}}{(n+1)^{2/\beta^*}}\right)^{\beta/2} \leq C \cdot \|f\|_{H^{2/\beta}} \qquad \forall f \in H^{2/\beta}.$$

If moreover $2/\beta \geq 1$, then the Hausdorff-Young theorem ([Du], p. 94) guarantees that $(\hat{f}(n))_n$ belongs to $l^{2/(2-\beta)}$, with $\|(\hat{f}(n))_n\|_{l^{2/(2-\beta)}} \leq \|f\|_{H^{2/\beta}}$ for $f \in H^{2/\beta}$. Consequently, if $f \in B_{H^{2/\beta}}$,

$$\begin{aligned}\sum_n \frac{|\hat{f}(n)|^2}{(n+1)^{\beta-1}} &\leq \left(\sum_n \Big(\frac{|\hat{f}(n)|}{(n+1)^{\beta-1}}\Big)^{2/\beta}\right)^{\beta/2} \cdot \Big(\sum_n |\hat{f}(n)|^{2/(2-\beta)}\Big)^{(2-\beta)/2} \\ &\leq \left(\sum_n \frac{|\hat{f}(n)|^{2/\beta}}{(n+1)^{2/\beta^*}}\right)^{\beta/2} \leq C,\end{aligned}$$

and we are done in this case.

If $0 < \frac{2}{\beta} < 1$, we may apply another result of Hardy and Littlewood ([Du], p. 98): There is a constant c such that, for each $f \in H^{2/\beta}$ and each $n = 0, 1, 2, \ldots$,

$$|\hat{f}(n)| \leq c \cdot (n+1)^{\frac{\beta}{2}-1} \cdot \|f\|_{H^{2/\beta}}$$

For $f \in B_{H^{2/\beta}}$ and with $\gamma = c^{2-2/\beta}$ we thus get

$$\begin{aligned}\sum_n \frac{|\hat{f}(n)|^2}{(n+1)^{\beta-1}} &\leq \gamma \cdot \sum_n \frac{|\hat{f}(n)|^{\frac{2}{\beta}} \cdot (n+1)^{(\frac{\beta}{2}-1)(2-\frac{2}{\beta})}}{(n+1)^{\beta-1}} \\ &\leq \gamma \cdot \sum_n \frac{|\hat{f}(n)|^{2/\beta}}{(n+1)^{2/\beta^*}} \leq \gamma \cdot C^{2/\beta},\end{aligned}$$

which was what we wanted.

Let us denote by

$$\mathfrak{S}_{p,q}(H,K)$$

the class of all compact operators u from a Hilbert space H to a Hilbert space K such that the sequence of singular numbers

$$a_n(u) \qquad (n = 1,2,3,\cdots)$$

(= eigenvalues of $(u^*u)^{1/2}$) belongs to the Lorentz space $l^{p,q}$ $(0 < p < \infty,$ $0 < q \leq \infty)$. Of course, $\mathfrak{S}_{p,p}(H,K)$ is the classical Schatten-von Neumann class $\mathfrak{S}_p(H,K)$.

(5) COROLLARY: *If C_φ is β-order bounded, $\beta \geq 1$, then $C_\varphi \in \mathfrak{S}_{\frac{2}{\beta},2}(H^2,H^2)$.*

Proof. According to (4), C_φ has a decomposition $C_\varphi : H^2 \xrightarrow{v} l^2 \xrightarrow{u} H^2$, where the operator v is given by $vf = \sum_n (n+1)^{(1-\beta)/2}\hat{f}(n)$, $\forall f \in H^2$ and u is a Hilbert-Schmidt operator. Thus, in other notation, $v = \sum_{n=0}^{\infty}(n+1)^{(1-\beta)/2} z^n \otimes e_n$, where (z^n) and (e_n) are the canonical orthonormal bases in H^2 and l^2, respectively. We conclude that v belongs to $\mathfrak{S}_{\frac{2}{\beta-1},\infty}(H^2,l^2)$ and $a_n(v) = n^{\frac{1}{2}(1-\beta)}$ for $n = 1,2,3,\cdots$.

Using the so-called multiplicativity of $a_n(\cdot)$ ([Pi], 11.9.1 [and 11.3.4]) we obtain

$$\begin{aligned}\sum_{n=1}^{\infty} n^{\beta-1} a_n(C_\varphi)^2 &\leq 2 \cdot \sum_{n=1}^{\infty}(2n)^{\beta-1} a_{2n-1}(C_\varphi)^2 \\ &\leq 2^\beta \cdot \sum_{n=1}^{\infty} n^{\beta-1} \cdot a_n(u)^2 \cdot a_n(v)^2 = 2^\beta \cdot \sum_{n=1}^{\infty} a_n(u)^2 < \infty.\end{aligned}$$

Hence $\left(a_n(C_\varphi)\right)_n \in l^{\frac{2}{\beta},2}$, and so $C_\varphi \in \mathfrak{S}_{\frac{2}{\beta},2}(H^2,H^2)$.

We doubt that the converse of (5) is true. Nevertheless, if $p \leq 2$ and $\beta > 2/p$, then β-order boundedness of C_φ implies $C_\varphi \in \mathfrak{S}_p(H^2,H^2)$. In particular, when $\beta > 2$, then β-order bounded composition operators are in the trace class $\mathfrak{S}_1(H^2,H^2)$.

The situation changes drastically if $\beta < 1$. The corresponding result appears now as a corollary of a general statement about order bounded operators. In the sequel, μ will again be any positive measure.

(6) PROPOSITION: *Let H be a Hilbert space, $0 < p \leq 2$, and $u : H \to L^p(\mu)$ an operator. Order boundedness of u is equivalent to the existence of a factorization $u : H \xrightarrow{w} H_0 \xrightarrow{v} L_p(\mu)$, where H_0 is a Hilbert space, v is a bounded operator, and w is a Hilbert-Schmidt operator.*

Proof. Suppose first that u is order bounded. As was indicated before, u then admits a factorization $u : H \xrightarrow{b} L^\infty(\nu) \xrightarrow{i_p} L^p(\nu) \xrightarrow{a} L^p(\mu)$, where a and b are suitable operators, ν is a probability measure, and i_p is the formal identity. On its way to $L^p(\nu)$, this map inevitably encounters $L^2(\nu)$ (since $p \leq 2$), and so the announced decomposition occurs as a consequence of the fact that a Hilbert space operator is Hilbert-Schmidt if (and only if) it factors through an L_∞-space; see e.g. [Pi], 22.1.8 and 22.4.2.

We settle the converse first for $1 \leq p \leq 2$. In that case, w^*v^* is p-summing since one of its factors is Hilbert-Schmidt, and this implies as before that $u = vw$ is order bounded ([Pi], 24.5.8).

Next we turn to the case $0 < p < 1$. Since H_0 is a Hilbert space, 2-summing operators with domain H_0 are automatically p-summing ([Pi], 22.4.1). According to a factorization theorem of B. Maurey ([Ma], Théorème 23), there exists thus a factorization $v : H_0 \xrightarrow{v_2} L^1(\mu) \xrightarrow{v_1} L_p(\mu)$ where v_1 is a multiplication operator induced by a suitable function in $L^r(\mu)$ ($\frac{1}{r} = \frac{1}{p} - 1$). By what we have seen before, $v_2 w$ is order bounded, and $u = vw$ inherits this property, thanks to the nature of v_1.

In particular:

(7) COROLLARY: *If $0 < \beta \leq 1$, then C_φ is β-order bounded if and only if it factors $C_\varphi : H^2 \xrightarrow{v} l^2 \xrightarrow{u} H^{2\beta}$, where u is a bounded operator and v is a Hilbert-Schmidt operator.*

Of course, $\beta = 1$ is just the case of Hilbert-Schmidt composition operators. But in contrast to (5), and possibly surprising at first sight, there is, for each $0 < \beta < 1$, an analytic function $\varphi : D \to D$ such that C_φ is β-order bounded but fails to be compact as an operator $H^2 \to H^2$ (equivalently, as an operator $H^p \to H^p$, regardless of $0 < p < \infty$). In fact, it was shown by H. Hunziker [Hu] that, for $\beta < \alpha < 1$, a non-compact composition operator C_φ is generated by

$$\varphi(z) = \exp\left(\frac{1}{2\pi} \int_{-\pi}^{\pi} \frac{e^{it} + z}{e^{it} - z} \cdot \log\left[1 - \left(\frac{|t|}{2\pi}\right)^{1/\alpha}\right] dt \right).$$

But this function still satisfies $\lim_{n\to\infty} n^\alpha \cdot \|\varphi^n\|_{H^1} = 0$, so that $(\|\varphi^n\|_{H^1})_n$ belongs to $l^{1/\alpha,\infty}$, a subset of $l^{1/\beta,1}$.

References

[Co] C.C. Cowen: Composition operators on Hilbert spaces of analytic functions: a status report. Proc. AMS 1988 Summer Res.Inst. on Operator Theory and Operator Algebras.

[Du] P.L. Duren: Theory of H^p spaces. Academic Press 1970.

[Gr] A. Grothendieck: Produits tensoriels topologiques et espaces nucléaires. Mem. AMS *16*, 1955.

[Hu] H. Hunziker: Kompositionsoperatoren auf klassischen Hardyräumen. Dissertation Universität Zürich 1989.

[Hu] H. Hunziker, H. Jarchow: Composition operators which improve integrability. Math. Nachr., to appear.

[HJM] H. Hunziker, H. Jarchow, V. Mascioni: Some topologies on the space of analytic self-maps on the unit disk. Proc. Conf. Geometry of Banach Spaces, Strobl 1989, Camb. Univ. Press, to appear.

[Kw] S. Kwapień: On a theorem of L. Schwartz and its applications to absolutely summing operators. Studia Math. *38* (1970) 193-201.

[LZ] D.H. Luecking, K. Zhu: Composition operators belonging to the Schatten ideals. Preprint.

[Ma] B. Maurey: Théorèmes de factorization pour les opérateurs linéaires à valeurs dans les espaces L^p. Soc. Math. France, Astérisque *11*, 1974.

[Pi] A. Pietsch: Operator ideals. VEB Deutscher Verlag der Wissenschaften 1978; North-Holland 1980.

[Sh] J. H. Shapiro: The essential norm of a composition operator. Ann.Math. *12* (1987) 375-404.

[ST] J. H. Shapiro, P.D. Taylor: Compact, nuclear, and Hilbert-Schmidt composition operators on H^2. Indiana Univ.Math.J. *23* (1973) 471-496.

Progress in Functional Analysis
K.D. Bierstedt, J. Bonet, J. Horváth & M. Maestre (Eds.)

Eigenvalues of nuclear operators on Tsirelson spaces

Albrecht Pietsch

Department of Mathematics, Friedrich Schiller University,
0-6900 Jena, Germany

Abstract

For $1 \le r < 2$, let $\mathbb{E}_r$ and $\mathbb{E}_{r,\infty}$ denote the classes of all complex Banach spaces E such that the eigenvalue sequences of all nuclear operators on E belong to l_r and $l_{r,\infty}$, respectively. Then $\mathbb{E}_r \subseteq \mathbb{E}_{r,\infty}$. Our main result states that this inclusion is strict. The case r = 1 was already treated by G. Pisier.

1. INTRODUCTION

For every complex Banach space E the n-th Grothendieck number is defined by

$$\Gamma_n(E) := \sup\left\{ |\det(\langle x_i, a_j\rangle)|^{1/n} : \begin{matrix} x_1,\dots,x_n \in U_E \\ a_1,\dots,a_n \in U_E^o \end{matrix} \right\},$$

where U_E and U_E^o denote the closed unit balls of E and the dual E', respectively.

Using these quantities, for $0 \le \rho \le 1/2$, we let

$$\Gamma_\rho := \{E : (n^{-\rho}\Gamma_n(E)) \in l_\infty\} .$$

Note that $\Gamma_{1/2}$ consists of *all* Banach spaces, while Γ_o is the class of *weak Hilbert spaces* in the sense of G. Pisier ([11] and [12], p. 189). Various characterizations and properties of the class Γ_ρ are to be found in [9] and [10]. It follows from

$$\Gamma_n(l_q) = n^{1/2-1/q} \quad \text{for} \quad 2 \le q \le \infty$$

that

$$l_q \in \Gamma_\rho \setminus \Gamma_{\rho-\varepsilon}, \text{ where } \rho = 1/2 - 1/q > \varepsilon > 0 .$$

Let $\mathfrak{N}(E)$ denote the set of all nuclear operators acting on a Banach space E. We agree that the eigenvalues of any operator $T \in \mathfrak{N}(E)$ are arranged as a sequence $(\lambda_n(T))$ such that

$$|\lambda_1(T)| \geq |\lambda_2(T)| \geq \cdots \geq 0,$$

counting multiplicities; see ([7], 3.2.20). The formula

$$(\lambda_n(T)) \in l_{r,\infty} \quad \text{means that} \quad (n^{1/r}\lambda_n(T)) \in l_\infty .$$

The following criterion is due to G. Pisier ([11], Theorem 2.9) for $\rho = 0$ and to S. Geiss ([2], Prop. 3.5) for $0 < \rho < 1/2$; see also ([9], Theorem 5.5).

Theorem. *Let* $0 \leq \rho < 1/2$ *and* $1 \leq r < 2$ *such that* $1/r = 1 - \rho$. *Then* $E \in \Gamma_\rho$ *if and only if*

$$T \in \mathfrak{N}(E) \quad \textit{implies} \quad (\lambda_n(T)) \in l_{r,\infty} .$$

Applying this result to l_q with $2 \leq q < \infty$ and letting $1/r = 1/2 + 1/q$, we see that

$$T \in \mathfrak{N}(l_q) \quad \text{implies} \quad (\lambda_n(T)) \in l_{r,\infty} .$$

However, W. B. Johnson, H. König, B. Maurey and J. R. Retherford ([3], Theorem 2.6) proved a stronger conclusion, namely,

$$T \in \mathfrak{N}(l_q) \quad \text{implies} \quad (\lambda_n(T)) \in l_r .$$

Thus one may ask whether, even for all Banach spaces $E \in \Gamma_\rho$,

$$T \in \mathfrak{N}(E) \quad \text{implies} \quad (\lambda_n(E)) \in l_r \ ?$$

Using the 2-convexified Tsirelson space, G. Pisier ([12], Chap. 13) showed that this conjecture fails for $\rho = 0$. It is the purpose of the present paper to extend his observation to the case $0 < \rho < 1/2$. To this end, we give an *explicit* example of a nuclear operator T on some Banach space $E \in \Gamma_\rho$ for which $(\lambda_n(T)) \in l_{r,\infty} \setminus l_r$.

Concerning the standard definitions, the reader is referred to my monographs [6] and [7]. The notation is adopted from the latter.

Throughout, we let $2 \leq q < \infty$, $1 \leq r < 2$ and $0 \leq \rho < 1/2$ such that $1/r = 1/2 + 1/q$, $1/r = 1 - \rho$ and $\rho = 1/2 - 1/q$.

2. PRELIMINARIES

Let E and F be complex Banach spaces. For every (bounded linear) operator $T \in \mathfrak{L}(E,F)$ and $n = 1,2,\dots$, we define the following quantities:

The *Grothendieck numbers*

$$\Gamma_n(T) := \sup\left\{ |\det(\langle Tx_i, b_j\rangle)|^{1/n} : \begin{array}{l} x_1,\dots,x_n \in U_E \\ b_1,\dots,b_n \in U_F^o \end{array} \right\},$$

the *z-numbers*

$$z_n(T) := \sup\left\{ a_n(BTX) : \begin{array}{l} X \in \mathfrak{L}(l_2^n, E),\ \| X \mid \mathfrak{P}_2 \| \le 1 \\ B \in \mathfrak{L}(F, l_2^n),\ \| B \mid \mathfrak{P}_2' \| \le 1 \end{array} \right\},$$

where $\| X \mid \mathfrak{P}_2 \|$ and $\| B \mid \mathfrak{P}_2' \|$ denote the 2-summing norms of X and B', respectively, and $a_n(BTX)$ is the n-th approximation number of BTX,

the *ideal norms*

$K_n(T) := \inf c$,

where the infinum is taken over all constants $c \ge 0$ such that

$$\left(\sum_{i=1}^{n} \| \sum_{j=1}^{n} \sigma_{ij} T x_j \|^2 \right)^{1/2} \le c \left(\sum_{j=1}^{n} \| x_j \|^2 \right)^{1/2}$$

for $x_1,\dots,x_n \in E$ and all unitary (n,n)-matrices (σ_{ij}).

In the special case when we consider the identity map of a Banach space E, the above quantities are simply denoted by $\Gamma_n(E)$, $z_n(E)$ and $K_n(E)$, respectively.

For proofs of the following inequalities, we refer to ([9], Lemmas 1.11, 1.13, 2.11) and ([10], Prop. 6.2):

$$n z_n(T) \le e^2 \Gamma_n(T), \quad \Gamma_n(T) \le K_n(T) \quad \text{and} \quad \Gamma_n(T) \le n \left(\prod_{k=1}^{n} z_k(T) \right)^{1/n}.$$

All algebraic operations for sequences are defined coordinatewise. For example, given $x = (\xi_k)$ and $y = (\eta_k)$, we let

$xy = (\xi_k \eta_k)$ and $(|x|^2 + |y|^2)^{1/2} = ((|\xi_k|^2 + |\eta_k|^2)^{1/2})$.

3. THE q-CONVEXIFIED TSIRELSON SPACE

3.1. On the linear space f, the set of all finite sequences $x = (\xi_k)$, we inductively define a sequence of norms by letting

$$\|x\|_1 := \sup_k |\xi_k|$$

and

$$\|x\|_{m+1} := \max\left\{ \|x\|_m , \tfrac{1}{2} \sup\left(\sum_{i=1}^{n} \|a_i x\|_m^q \right)^{1/q}\right\}$$

where the inner supremum is taken over all finite families of sequences $a_1 = (\alpha_k^{(1)}), \dots, a_n = (\alpha_k^{(n)}) \in f$ such that

$$\alpha_k^{(i)} = 0 \quad \text{for} \quad k \le n \quad \text{and} \quad \sum_{i=1}^{n} |\alpha_k^{(i)}|^q \le 1 \quad \text{for} \quad k > n .$$

Note that

$$\|x\|_1 \le \|x\|_2 \le \dots \le \|x\|_m \le \dots \le \|x| l_q\| \quad \text{for all} \quad x \in f .$$

The q-*convexified Tsirelson space* t_q is defined to be the completion of f with respect to the norm

$$\|x|t_q\| := \lim_m \|x\|_m .$$

It turns out that t_q can be identified with a subset of c_o, the set of zero sequences. More precisely, we have $l_q \subset t_q \subset l_{q,\infty}$.

For further information about Tsirelson spaces, we refer the reader to [1].

3.2. The basic idea of the following proof is taken from ([12], Chap. 13). However, using the concept of z-numbers, we were able to simplify Pisier's reasoning.

Proposition. $t_q \in \Gamma_\rho$.

P r o o f . Write t_q as a direct sum $t_q = u_q^{(n)} \oplus v_q^{(n)}$, where

$$u_q^{(n)} := \{x = (\xi_k) \in t_q : \xi_k = 0 \quad \text{for} \quad k > n\}$$

and

$$v_q^{(n)} := \{x = (\xi_k) \in t_q : \xi_k = 0 \quad \text{for} \quad k \le n\} .$$

The projections from t_q onto $u_q^{(n)}$ and $v_q^{(n)}$ are denoted by P_n and Q_n, respectively. Note that $\text{rank}(P_n) = n$ and $\|Q_n\| = 1$.

By ([9], Prop. 2.2), we have

$$z_{2n}(t_q) \le z_{n+1}(P_n) + z_n(Q_n) \le z_n(v_q^{(n)}) \, \|Q_n\| .$$

Hence in view of

$n z_n(T) \le e^2 \Gamma_n(T) \le e^2 K_n(T)$ for all $T \in \mathfrak{L}(E,F)$,

it follows that

$z_{2n}(t_q) \le \frac{e^2}{n} K_n(v_q^{(n)})$.

Next, following G. Pisier ([12], Chap. 13), we estimate $K_n(v_q^{(n)})$. To this end, let $x_j = (\xi_k^{(j)}) \in v_q^{(n)} \cap f$ for $j = 1, \dots, n$. Given any unitary (n,n)-matrix (σ_{ij}), we define $y_i = (\eta_k^{(i)}) \in v_q^{(n)}$ for $i = 1, \dots, n$ by

$$y_i = \sum_{j=1}^{n} \sigma_{ij} x_j .$$

Then the sequences

$$x := \Big(\sum_{j=1}^{n} |x_j|^2 \Big)^{1/2} \quad \text{and} \quad y := \Big(\sum_{i=1}^{n} |y_i|^2 \Big)^{1/2}$$

coincide.

It follows by induction that all norms $\|\cdot\|_m$ are q-convex. Thus the Tsirelson norm $\|\cdot | t_q\|$ is q-convex, as well. Hence, by ([5], Prop. 1. d. 5), it is 2-convex. This yields

$$\| x | t_q \| \le \Big(\sum_{j=1}^{n} \| x_j | t_q \|^2 \Big)^{1/2}. \tag{1}$$

Define $b_i = (\beta_k^{(i)})$ for $i = 1, \dots, n$ by

$$\beta_k^{(i)} := \begin{cases} \eta_k^{(i)} \Big(\sum_{h=1}^{n} | \eta_k^{(h)} |^2 \Big)^{-1/2} & \text{if } \eta_k^{(i)} \ne 0 , \\ 0 & \text{if } \eta_k^{(i)} = 0 . \end{cases}$$

Then

$$\beta_k^{(i)} = 0 \quad \text{for} \quad k \le n \quad \text{and} \quad \sum_{i=1}^{n} | \beta_k^{(i)} |^q \le \Big(\sum_{i=1}^{n} | \beta_k^{(i)} |^2 \Big)^{q/2} \le 1 \quad \text{for} \quad k > n .$$

Taking into account that $y_i = b_i y$ and using the definition of $\| y \|_{m+1}$, we obtain

$$\Big(\sum_{i=1}^{n} \| y_i \|_m^q \Big)^{1/q} \le 2 \| y \|_{m+1} .$$

Passing to the limit as $m \to \infty$ yields

$$\Big(\sum_{i=1}^{n} \| y_i | t_q \|^q \Big)^{1/q} \le 2 \| y | t_q \| .$$

Hence, by Hölder's inequality,

$$\Big(\sum_{i=1}^{n} \| y_i | t_q \|^2 \Big)^{1/2} \le 2 n^{1/2 - 1/q} \| y | t_q \| . \tag{2}$$

In view of $x = y$, the inequalities (1) and (2) imply

$$\Big(\sum_{i=1}^{n} \|\sum_{j=1}^{n} \sigma_{ij} x_j | t_q \|^2\Big)^{1/2} \le 2n^{1/2-1/q} \Big(\sum_{j=1}^{n} \| x_j | t_q \|^2\Big)^{1/2}$$

for $x_1, \ldots, x_n \in v_q^{(n)} \cap f$. Since $v_q^{(n)} \cap f$ is dense in $v_q^{(n)}$, we obtain

$$z_{2n}(t_q) \le \frac{e^2}{n} K_n(v_q^{(n)}) \le 2e^2 n^{-1/2-1/q}.$$

Therefore

$$\Gamma_n(t_q) \le cn^{1/2-1/q}.$$

This completes the proof.

4. NUCLEAR OPERATORS ON TSIRELSON SPACES

4.1. The use of twisted operators in the theory of eigenvalue distributions goes back to [4] and [8]; see also ([7], 3.8.2).

Proposition. Given $s = (\sigma_k) \in l_2$ *and* $t = (\tau_k) \in t_q$, *there exists a nuclear operator on* $t_q \oplus t_q'$ *with eigenvalues* $\pm\sigma_1\tau_1, \pm\sigma_2\tau_2, \ldots$

P r o o f . Write s in the form $s = ab$ with $a = (\alpha_k) \in l_2$ and $b = (\beta_k) \in c_o$ such that $\alpha_k \neq 0$ for $k = 1,2,\ldots$. Let $A \in \mathfrak{L}(c_o, l_1)$, $B \in \mathfrak{L}(l_1, c_o)$ and $T \in \mathfrak{L}(c_o, t_q)$ be the diagonal operators associated with the sequences $a^2 \in l_1$, $b^2 \in c_o$ and $t \in t_q$, respectively. Clearly, A is nuclear. Concerning the nuclearity of B, we refer to ([6], 6.3.5) or ([7], 1.7.19). The fact that T acts from c_O (even from l_∞) into t_q is an immediate consequence of the following property of t_q:

Let $x = (\xi_k)$ and $y = (\eta_k)$. If $|\xi_k| \ge |\eta_k|$ for $k = 1,2,\ldots$, then $x \in t_q$ implies $y \in t_q$ and $\|x|t_q\| \ge \|y|t_q\|$.

Finally, we let I denote the embedding map from t_q into c_o.

It follows from the preceding observations that the operators

$$U: \ t_q \xrightarrow{I} c_o \xrightarrow{A} l_1 \xrightarrow{I'} t_q'$$

and

$$V: \ t_q' \xrightarrow{T'} l_1 \xrightarrow{B} c_o \xrightarrow{T} t_q$$

are nuclear. Hence the twisted operator

$$W: (x,y) \longrightarrow (Vy, Ux)$$

is nuclear on $t_q \oplus t_q'$.

Since the finite sequences are dense in t_q, the dual t_q' can be viewed as a sequence space via the duality

$$\langle x,y\rangle = \sum_{k=1}^{\infty} \xi_k \eta_k \quad \text{for} \quad x = (\xi_k) \in t_q \quad \text{and} \quad y = (\eta_k) \in t_q'.$$

Then W acts as follows:

$$W: ((\xi_k),(\eta_k)) \longrightarrow ((\beta_k^2 \tau_k^2 \eta_k),(\alpha_k^2 \xi_k)).$$

Let e_n denote the n-th unit sequence. It follows from

$$W(\beta_n \tau_n e_n, \pm\alpha_n e_n) = (\pm\alpha_n \beta_n^2 \tau_n^2 e_n, \alpha_n^2 \beta_n \tau_n e_n) = \pm\sigma_n \tau_n(\beta_n \tau_n e_n, \pm\alpha_n e_n)$$

that $\pm\sigma_n\tau_n$ is an eigenvalue of W.

4.2. We are now in a position to state the main result of this paper.

Theorem. *There exists a nuclear operator* T *on some Banach space* $E \in \Gamma_\wp$ *for which* $(\lambda_n(T)) \in l_{r,\infty} \setminus l_r$.

P r o o f . The crucial point is the observation that t_q is strictly larger than l_q. The proofs of this fact given in ([1], Cor. IV.c.E) for q = 1 and in ([12], Chap. 13) for q = 2 can be adapted to the case $2 \le q < \infty$ we are interested in.

Choose $t = (\tau_k) \in t_q \setminus l_q$. By the classical Landau theorem, there must exist $s = (\sigma_k) \in l_2$ such that $st = (\sigma_k \tau_k) \in l_{r,\infty} \setminus l_r$. From Proposition 3.2 and ([9], Theorem 5.2) we see that $t_q \oplus t_q' \in \Gamma_\wp$. Thus the nuclear operator constructed in Proposition 4.1 has the desired properties.

Remark. The construction of an explicit sequence $t = (\tau_k) \in t_1 \setminus l_1$ given in ([1], pp. 40-45) is horrible. Who can do better?

5. OPEN PROBLEMS

5.1. As already mentioned, we have $l_q \subset t_q \subset l_{q,\infty}$. There even holds $l_q \subset t_q \subset l_{q,q+\varepsilon}$ for all $\varepsilon > 0$, and this is only an epsilon of the truth. In fact, the deviation of t_q from l_q is unimaginably small. Thus, it can be conjectured that the eigenvalue sequences of operators $T \in \mathfrak{N}(t_q)$ are "almost" r-summable. In particular, we formulate the

PROBLEM. *Is it true that*

$T \in \mathfrak{N}(t_q)$ *implies* $(\lambda_n(T)) \in l_{r,r+\varepsilon}$

for all $\varepsilon > 0$?

5.2. For arbitrary Banach spaces $E \in \Gamma_\rho$ the statement

$T \in \mathfrak{N}(E)$ implies $(n^{1/r} \lambda_n(T)) \in l_\infty$

can easily be improved to

$T \in \mathfrak{N}(E)$ implies $(n^{1/r} \lambda_n(T)) \in c_o$.

However, it is unknown whether the latter conclusion is the best possible.

PROBLEM. *Let* $r < w < \infty$. *Does there exist a nuclear operator* T *on some Banach space* $E \in \Gamma_\rho$ *for which* $(\lambda_n(T)) \in l_{r,\infty} \setminus l_{r,w}$?

6. REFERENCES

[1] P. C. Casazza and T. J. Shura, Tsirelson's space, Lecture Notes in Math. 1363, Springer, Berlin-Heidelberg-New York 1989.

[2] S. Geiss, Grothendieck numbers of linear und continuous operators on Banach spaces, Math. Nachr. 148 (1990) 65-79.

[3] W. B. Johnson, H. König, B. Maurey and J. R. Retherford, Eigenvalues of p-summing and l_p-type operators in Banach spaces, J. Funct. Anal. 32 (1979) 353-380.

[4] R. J. Kaiser and J.R. Retherford, Eigenvalue distribution of nuclear operators, Proc. Conf. Functional Analysis/Banach Space Geometry, 245-287, Univ. Essen 1982.

[5] J. Lindenstrauss and L. Tzafriri, Classical Banach spaces, Vol. II, Springer, Berlin-Heidelberg-New York 1979.

[6] A. Pietsch, Operator ideals, North-Holland, Amsterdam-New York-Oxford 1980.

[7] A. Pietsch, Eigenvalues and s-numbers, Cambridge Univ. Press. 1987.

[8] A. Pietsch, Distribution of eigenvalues and nuclearity, Banach Center Publ. 8, 361-365, Warsaw 1982.

[9] A. Pietsch, Eigenvalue distribution and geometry of Banach spaces, Math. Nachr.

[10] A. Pietsch, Sequences of ideal norms, Note di Mat. (Lecce), Köthe memorial volume.

[11] G. Pisier, Weak Hilbert spaces, Proc. London Math. Soc. (3) 56 (1988), 547-579.

[12] G. Pisier, The volume of convex bodies and Banach space geometry, Cambridge Univ. Press 1989.

Progress in Functional Analysis
K.D. Bierstedt, J. Bonet, J. Horváth & M. Maestre (Eds.)

Absolutely summing surjections from Sobolev spaces in the uniform norm

A. Pełczyński and M. Wojciechowski

Dedicated to Professor Manuel Valdivia on the occasion of his 60th birthday

Introduction

In the present paper we continue the study of translation invariant projections on anisotropic Sobolev spaces on tori. We shall show (Theorem 1.1) that for arbitrary smoothness S (see Section 1 for the definition) every infinite set of characters of the d-dimensional torus group $\mathbb{T}^d$ contains an infinite subset, say W, such that the orthogonal projection on the Sobolev Hilbert space $L^2_S(\mathbb{T}^d)$ regarded as operator from $C_S(\mathbb{T}^d)$ into a subspace of $L^2_S(\mathbb{T}^d)$ generated by W has closed range, equivalently it is a surjection onto this subspace. This is an analogue of the classical result for spaces of continuous functions on arbitrary compact abelian group. Combining Theorem 1.1 with our earlier result [P-W] we give a necessary and sufficient condition for arbitrary smoothness $S \subset \mathbb{R}^d$ in order that there exist an absolutely summing translation invariant surjection from $C_S(\mathbb{T}^d)$ onto an infinite dimensional subspace of $L^2_S(\mathbb{T}^d)$. (The simplest example when such a surjection exists is the 2-dimensional smoothness generated by the partial derivatives D_{xx} and D_y.) Our last result has been motivated by similar property of the Disc Algebra discovered by Fournier [F] and Paley [Pa].

1 Preliminaries and statement of the main result

A finite subset S of $\mathbb{R}^d$ (= d-dimensional real vector space) consisting of vectors with integer-valued non-negative coordinates is called a *smoothness* provided the origin $O \in S$ and if $a = (a(j)) \in S$ and $b = (b(j))$ has non-negative integer-valued coordinates such that $0 \le b(j) \le a(j)$ for $j = 1, 2, \ldots, d$ then $b \in S$. With each $a \in S$ we associate the partial derivative

$$D^a = \frac{\partial^{a(1)+a(2)+\ldots+a(d)}}{\partial_1^{a(1)} \partial_2^{a(2)} \ldots \partial_d^{a(d)}}$$

whose symbol $\sigma(a)\colon \mathbb{R}^d \to \mathbb{C}$ ($\mathbb{C}$ - is the complex plane) is defined by $\sigma(a)(\xi) = \prod_{j=1}^{d} (i\xi(j))^{a(j)}$ for $\xi = (\xi(j)) \in \mathbb{R}^d$. To the origin O we correspond the identity operator with symbol $\sigma(O) \equiv 1$. The function

$$Q_S = \sum_{a \in S} |\sigma(a)|^2$$

is called the *fundamental polynomial* of smoothness S.

$\mathbb{T}$ - stands for the d-dimensional torus identified with the cube $[-\pi,\pi)^d$. The torus $\mathbb{T}^d$ is a compact abelian group whose dual group is identified with the integer-valued lattice $\mathbb{Z}^d \subset \mathbb{R}^d$. To each $n \in \mathbb{Z}^d$ corresponds the character $t \to \exp(i\langle t,n\rangle) = \exp\left(i\sum_{j=1}^{n} t(j)n(j)\right)$ for $t=(t(j))$, $n=(n(j))$. The complex linear combinations of characters are called trigonometric polynomials.

For fixed smoothness $S \subset \mathbb{Z}^d$ we define on trigonometric polynomials the Sobolev norms $\|\cdot\|_{S,p}$ for $1 \le p \le \infty$ by

$$\|f\|_{S,p} = \left(\int_{\mathbb{T}^d}\left(\sum_{a\in S}|D^a f(t)|^2\right)^{p/2} dt\right)^{1/p} \qquad (1 \le p < \infty),$$

$$\|f\|_{S,\infty} = \sup_{t\in\mathbb{T}^d}\left(\sum_{a\in S}|D^a f(t)|^2\right)^{1/2}.$$

The integration is taken against the normalized Haar measure of $\mathbb{T}^d$.

The Sobolev space $L^p_S(\mathbb{T}^d)$ (resp. $C_S(\mathbb{T}^d)$) is the completion of the trigonometric polynomials in the norm $\|\cdot\|_{S,p}$ for $1 \le p < \infty$ (resp. $\|\cdot\|_{S,\infty}$). To define the space $L^\infty_S(\mathbb{T}^d)$ note that $C_S(\mathbb{T}^d)$ can be regarded as a subspace of the space $C(\mathbb{T}^d;\, l^2_s)$ of all continuous functions on $\mathbb{T}^d$ with values in s-dimensional complex Hilbert space l^2_s where s is the number of elements of S. The space $L^\infty_S(\mathbb{T}^d)$ is defined to be the closure of this subspace in the $L^1(\mathbb{T}^d;\, l^2_s)$ topology of $L^\infty(\mathbb{T}^d;\, l^2_s)$.

For $f \in L^1(\mathbb{T}^d)$ and $n \in \mathbb{Z}^d$ we put

$$\hat{f}(n) = \int_{\mathbb{T}^d} f(t)\exp(-i\langle t,n\rangle)\, dt.$$

For $f \in L^2_S(\mathbb{T}^d)$ we have

$$\|f\|_{S,2} = \left(\sum_{n\in\mathbb{Z}^d}|\hat{f}(n)|^2 Q_S(n)\right)^{1/2},$$

equivalently the multiple of characters $\{Q_S^{-1/2}(n)\exp(i\langle\cdot,n\rangle)\}_{n\in\mathbb{Z}^d}$ form an orthonormal basis in $L^2_S(\mathbb{T}^d)$. Note that $\|\exp(i\langle\cdot,n\rangle)\|_p = Q_S^{1/2}(n)$ for $1 \le p \le \infty$. Given a subset $\{n_1,n_2,\dots\} \subset \mathbb{Z}^d$ we consider the orthogonal projection $R\colon L^2_S(\mathbb{T}^d) \to L^2_S(\mathbb{T}^d)$ onto the closed linear subspace generated by the characters $\{\exp(i\langle\cdot,n_k\rangle)\colon k=1,2,\dots\}$. By $\hat{R}$ we denote the characteristic function $1_{\{n_1,n_2,\dots\}}$ of the subset $\{n_1,n_2,\dots\}$ of $\mathbb{Z}^d$.

Recall (cf. [P-W]) that an orthogonal projection $R\colon L^2_S(\mathbb{T}^d) \to L^2_S(\mathbb{T}^d)$ onto a subspace generated by some characters is called a *Paley projection* provided it has infinite dimensional range and there exists $C > 0$ such that $\|Rf\|_{S,2} \le C\|f\|_{S,1}$ for every trigonometric polynomial f. In fact every Paley projection uniquely extends to a projection on $L^1_S(\mathbb{T}^d)$.

Every orthogonal projection $R: L^2_S(\mathbb{T}^d) \to L^2_S(\mathbb{T}^d)$ onto a subspace generated by the characters $\{\exp(i\langle\cdot, n_k\rangle): k = 1, 2, \ldots\}$ uniquely determines translation invariant operators from $C_S(\mathbb{T}^d)$ into $L^2_S(\mathbb{T}^d)$ and from $L^\infty_S(\mathbb{T}^d)$ into $L^2_S(\mathbb{T}^d)$ which are formally defined by

$$f \to \sum_k \hat{f}(n_k) \exp i\langle\cdot, n_k\rangle$$

where the series in the right hand side unconditionally converges in $L^2_S(\mathbb{T}^d)$. We use the same letter R to denote these operators.

Now we are ready to state the main result of this paper.

1.1 Theorem. *For every infinite subset $V \subset \mathbb{Z}^d$ there exists an infinite subset $W \subset V$ such that the operator $R: C_S(\mathbb{T}^d) \to L^2_S(\mathbb{T}^d)$ defined by*

$$R(f) = \sum_{n \in W} \hat{f}(n) \exp(i\langle\cdot, n\rangle)$$

has closed range.

The proof of this result is presented in the next section. It heavily depends on the following technical fact.

1.1 Proposition. *Let $S \subset \mathbb{Z}^d$ be a smoothness. Then given ε with $0 < \varepsilon < 1$ and $D = 1, 2, \ldots$ there exists $\varrho = \varrho(D, \varepsilon) > 1$ such that for every n, m in $\mathbb{Z}^d$ the condition*

$$\min_{1 \le j \le d} |n(j)| > \varrho \quad \text{and} \quad \sum_{j=1}^{d} |n(j) - m(j)| \le D$$

implies

$$\sum_{a \in S} |\sigma(a) Q_S^{-1/2}(n) - \sigma(a) Q_S^{-1/2}(m)|^2 < \varepsilon^2.$$

For the proof cf. [P-W], Proposition 1.1.

2 Proof of the main result

First we recall a well known construction essentially due to Fournier [F].

2.1 Lemma. *Let $(m_k) \subset \mathbb{Z}^d$ satisfy*

(2.1) $$m_1(1) \ge 1; \quad m_{k+1}(1) m_k^{-1}(1) \ge 2 \quad \text{for} \quad k = 2, 3, \ldots$$

Then for every positive integer N and for arbitrary complex numbers $c_1, c_2, \ldots, c_N$ there exists a trigonometric polynomial $F\colon \mathbb{T}^d \to \mathbb{C}$ such that

$$\hat{F}(m_k) = c_k \quad \text{for} \quad k = 1, 2, \ldots, N, \tag{2.2}$$

$$|F(t)|^2 \le e \sum_{k=1}^{N} |c_k|^2 \quad \text{for } t \in \mathbb{T}^d, \tag{2.3}$$

$$\text{if } \hat{F}(m) \neq 0 \text{ then } m = \sum_{k=1}^{N} \eta_k m_k \text{ for some} \tag{2.4}$$

$\eta_1, \eta_2, \ldots, \eta_N$ such that $\eta_k \in \{0, -1, +1\}$ for $k = 1, 2, \ldots, N$ and if $k_0 = \max\{k\colon \eta_k \neq 0\}$ then $\eta_{k_0} = 1$.

Outline of the construction.

If $\sum_{k=1}^N |c_k|^2 = 0$ we put $F = 0$. Otherwise put $\gamma_k = c_k \left(\sum_{j=1}^N |c_j|^2\right)^{-1/2}$ for $k = 1, 2, \ldots, N$ and define functions $F_0, F_1, \ldots, F_n$ and $G_0, G_1, \ldots, G_n$ by the recursive formulae

$$\begin{aligned} F_0 &= 0; \quad G_0 = 1 \\ F_k &= F_{k-1} + \gamma_k \exp(i\langle \cdot, m_k\rangle) G_{k-1} \\ G_k &= G_{k-1} - \overline{\gamma}_k \exp(-i\langle \cdot, m_k\rangle) F_{k-1} \quad (k = 1, 2, \ldots, N). \end{aligned}$$

Then

$$|F_k|^2 + |G_k|^2 = (1 + |\gamma_k|^2)(|F_{k-1}|^2 + |G_{k-1}|^2) \quad (k = 1, 2, \ldots, N).$$

Therefore $|F_N|^2 \le \prod_{k=1}^{N} (1 + |\gamma_k|^2) \le e$. Finally we put

$$F = \left(\sum_{j=1}^{N} |c_j|^2\right)^{1/2} F_N.$$

Then F satisfies (2.2) – (2.4). For details see [F].

In the next proposition we overcome essential technical difficulties

2.1 Proposition. *Let $S \subset \mathbb{Z}^d$ be an arbitrary smoothness and let a sequence $(n_k) \subset \mathbb{Z}^d$ satisfy*

$$\lim_k \inf_{1 \le j \le d} |n_k(j)| = \infty. \tag{2.5}$$

Then there exists an infinite subsequence (m_k) of the sequence (n_k) such that the translation invariant operator $R\colon C_S(\mathbb{T}^d) \to L^2_S(\mathbb{T}^d)$ with $\hat{R} = 1_{\{m_1, m_2, \ldots\}}$ has closed range.

PROOF. Let s be the number of elements of S. In the s-dimensional Hilbert space l^2_s consider the vectors

$$x_m = \left([\sigma(b) Q_S^{-1/2}](m)\right)_{b \in S} \quad \text{for} \quad m \in \mathbb{Z}^d.$$

Clearly $\|x_m\|_2 = 1$ for all m. Using the compactness of the unit sphere of l_s^2 and passing to a subsequence of (n_k), if necessary, one can assume without loss of generality that for some $x = (x(b))_{b \in S} \in l_s^2$

$$\lim_k \|x_{n_k} - x\|_2 = 0.$$

Applying Proposition 1.1 and using (2.5) we define inductively a subsequence (m_k) of (n_k) and an increasing sequence (D_k) of positive integers so that

$$D_1 = 1,$$

m_1 is the smallest element in the set $\{n_1, n_2, \ldots\}$ such that

$$\min_{1 \le j \le d} |m_1(j)| > \varrho(D_1, 4^{-1}) > 1,$$

$$\|x - x_{m_1}\|_2 < 4^{-1};$$

next for $k = 2, 3, \ldots$ we put

$$D_k = \sum_{r=1}^{k-1} \max_{1 \le j \le d} |m_r(j)|,$$

m_k is the smallest element in the set $\{n_1, n_2, \ldots\}$ such that

$$m_k(1) \ge 2m_{k-1}(1),$$

$$\min_{1 \le j \le d} |m_k(j)| > \varrho(D_k, 4^{-k}(2D_k - 1)^{-d}),$$

$$\|x - x_{m_k}\| < 4^{-k}(2D_k - 1)^{-d}.$$

Let us put $B_1 = \{m_1\}$,

$$B_k = \{m \in \mathbb{Z}^d : \sum_{j=1}^{d} |m(j) - m_k(j)| < D_k\} \quad (k = 2, 3 \ldots).$$

Clearly the number of elements of B_k does not exceed $(2D_k - 1)^d$. Moreover it follows from the definitions of B_k and m_k that

$$\|x_m - x_{m_k}\|_2 < 4^{-k}(2D_k - 1)^d \quad \text{whenever } m \in B_k.$$

Therefore

$$\sum_{k=1}^{\infty} \sum_{m \in B_k} \|x - x_m\|_2 < \sum_{k=1}^{\infty} 2 \cdot 4^{-k}. \tag{2.6}$$

Now we are ready to show that the translation invariant operator $R: C_S(\mathbb{T}^d) \to L^2_S(\mathbb{T}^d)$ with $\hat{R} = 1_{\{m_1, m_2, \ldots\}}$ has closed range.

To this end it is enough to show that a dense set of the unit sphere of the closure of the range of R is contained in the image of a fixed ball of the domain of R. Specifically we shall show that if $h = \sum_{k=1}^N d_k \exp(i\langle \cdot, m_k\rangle)$ satisfies $\|h\|_{S,2} = \left(\sum_{k=1}^N |d_k|^2 Q_S(m_k)\right)^{1/2} = 1$ then there exists an f in $C_S(\mathbb{T}^d)$ such that $h = R(f)$ and $\|f\|_{S,\infty} \le 2\sqrt{e}$.

Put $c_k = d_k Q_S(m_k)$ for $k = 1, 2, \ldots, N$. Then $\sum_{k=1}^N |c_k|^2 = 1$. Clearly our sequence (m_k) satisfies (2.1). Thus, by Lemma 2.1, there exists a trigonometric polynomial $F: \mathbb{T}^d \to \mathbb{C}$ satisfying (2.2), (2.3), (2.4). We define $f: \mathbb{T}^d \to \mathbb{C}$ by

$$f = \sum_{m \in \mathbb{Z}^d} \hat{F}(m) Q_S^{-1/2}(m) \exp(i\langle \cdot, m\rangle).$$

In view of (2.2), we have $R(f) = h$. To complete the proof we show that $\|f\|_{S,\infty} \le 2\sqrt{e}$ or equivalently

$$\left(\sum_{b \in S} |D^b f(t)|^2\right)^{1/2} = \|(D^b[f(t)])_{b \in S}\|_2 \le 2\sqrt{e} \quad \text{for } t \in \mathbb{T}^d.$$

We have

$$\|(D^b(f)(t))_{b \in S}\|_2 \le \|(D^b(f)(t))_{b \in S} - xF(t)\|_2 + \|xF(t)\|_2.$$

Since $\|x\|_2 = 1$, it follows from (2.3) that $\|xF(t)\|_2 \le \sqrt{e}$.

On the other hand we have

$$\begin{aligned} y &= (D^b(f)(t))_{b \in S} - xF(t) \\ &= \left(\sum_{m \in \mathbb{Z}^d} \left(\hat{F}(m)[\sigma(b) Q_S^{-1/2}](m) - \hat{F}(m) x(b)\right) \exp(i\langle t, m\rangle)\right)_{b \in S} \\ &= \sum_{m \in \mathbb{Z}^d} \hat{F}(m)(x_m - x) \exp(i\langle t, m\rangle). \end{aligned}$$

Next observe (important!) that (2.4) and the definition of B_k imply

$$\text{if } \hat{F}(m) \ne 0 \text{ then } m \in \bigcup_{k=1}^N B_k.$$

Thus

$$y = \sum_{k=1}^N \sum_{m \in B_k} \hat{F}(m) \exp(i\langle t, m\rangle)(x_m - x).$$

Therefore, by (2.6),

$$\|y\|_2 \leq \sum_{k=1}^{N} \sum_{m \in B_k} \|x_m - x\|_2 \cdot \sup_{m \in \mathbb{Z}^d} |\hat{F}(m)| \leq \sup_{m \in \mathbb{Z}^d} |\hat{F}(m)|.$$

Since $|\hat{F}(m)| \leq \|F\|_\infty$ for $m \in \mathbb{Z}^d$, the condition (2.3) yields $\|y\|_2 \leq \sqrt{e}$. Thus $|f(t)| \leq 2\sqrt{e}$ for $t \in \mathbb{T}^d$.

PROOF OF THEOREM 1.1. First we pick from V a sequence (n_k) so that for some non-empty set $A \subset \{1, 2, \ldots, d\}$ one has

$$\liminf_{k} \inf_{j \in A} |n_k(j)| = \infty,$$

$$n_k(j) = c_j \quad \text{for } j \notin A.$$

Put $c = (c(j)) \in \mathbb{Z}^d$ where $c(j) = 0$ for $j \in A$ and $c(j) = c_j$ for $j \notin A$. Let

$$S_A = \{a \in S : a(j) = 0 \quad \text{for } j \notin A\}.$$

Applying Proposition 2.1 to the δ-dimensional smoothness S_A (where δ denotes the number of elements of A) and to the sequence (n_k^A) (where $\xi^A = (\xi(j))_{j \in A}$ for $\xi = (\xi(j))_{1 \leq j \leq d} \in \mathbb{R}^d$) we extract an infinite subsequence (m_k) of (n_k) so that the translation invariant operator $R_A : C_{S_A}(\mathbb{T}^\delta) \to L^2_{S_A}(\mathbb{T}^\delta)$ with $\hat{R}_A = 1_{\{m_1^A, m_2^A, \ldots\}}$ has closed range.
Finally we define $R : C_S(\mathbb{T}^d) \to L^2_S(\mathbb{T}^d)$ by

$$R = M_c I_A R_A E_A M_{-c}$$

where M_a is the operator of multiplication by $\exp(i\langle\cdot, a\rangle)$ for $a \in \mathbb{Z}^d$; the averaging $E_A : C_S(\mathbb{T}^d) \to C_{S_A}(\mathbb{T}^\delta)$ is defined by $E_A(f)(x^A) = \int_{\mathbb{T}^{d'}} f((x^A, x^{A'}) d(x^{A'})$ for $f \in C_S(\mathbb{T}^d)$ (we represent each $x \in \mathbb{T}^d$ as the pair $x = (x^A, x^{A'})$ where $A' = \{1, 2, \ldots, d\} \setminus A$); the natural isometric embedding $I_A : L^2_{S_A}(\mathbb{T}^\delta) \to L^2_S(\mathbb{T}^d)$ is defined by $I_A(g)(x) = g(x^A)$ for $g \in L^2_{S_A}(\mathbb{T}^d)$ and for $x \in \mathbb{T}^d$ almost everywhere with respect to the Haar measure.

3 Existence of absolutely summing surjections

The reader is referred to [Pi], chap. 17 and [Wt], III.F. for the definition and basic properties of absolutely summing operators.

A direct application of Theorem 1.1 yields

3.1 Corollary. *Let $S \in \mathbb{Z}^d$ be an arbitrary smoothness. Then there exists 1-absolutely summing translation invariant surjection from $C_S(\mathbb{T}^d)$ onto a (translation invariant) infinite dimensional subspace of $L^2_S(\mathbb{T}^d)$ iff there exists a Paley projection on $L^1_S(\mathbb{T}^d)$.*

PROOF. Assume that $P: L^1_S(\mathbb{T}^d) \to L^1_S(\mathbb{T}^d)$ is a Paley projection with $\hat{P} = 1_{\{n_1,n_2,\dots\}}$. Applying Theorem 1.1 we can extract an infinite subsequence (m_k) of (n_k) so that $R: C_S(\mathbb{T}^d) \to L^2_S(\mathbb{T}^d)$ with $\hat{R} = 1_{\{m_1,m_2,\dots\}}$ has closed range. Since P is a Paley projection and (m_k) is a subsequence of (n_k), there is $C > 0$ such that $\|Rf\|_{S,2} \le C\|f\|_{S,1}$ for $f \in C_S(\mathbb{T}^d)$. This shows that R admits the factorization $R = \tilde{R}I_S$ where $I_S: C_S(\mathbb{T}^d) \to L^2_S(\mathbb{T}^d)$ is the natural embedding (= the formal identity map) and $\tilde{R}: L^1_S(\mathbb{T}^d) \to L^2_S(\mathbb{T}^d)$ is the Paley projection regarded as an operator from $L^1_S(\mathbb{T}^d)$ into $L^2_S(\mathbb{T}^d)$ (with $\|\tilde{R}\| \le C$). Note that I_S is 1-absolutely summing because it is the restriction to $C_S(\mathbb{T}^d)$ regarded as a subspace of $C(\mathbb{T}^d;\, l^2_s)$ of the natural embedding $I: C(\mathbb{T}^d;\, l^2_s) \to L^1(\mathbb{T}^d;\, l^2_s)$ and I is 1-absolutely summing because l^2_s is finite dimensional. Hence R is 1-absolutely summing. Obviously R having closed range is a surjection onto the closed linear subspace of $L^2_S(\mathbb{T}^d)$ spanned by the characters $\{\exp(i\langle\cdot, m_k\rangle): k = 1, 2.\dots\}$.

Conversely if $R: C_S(\mathbb{T}^d) \to L^2_S(\mathbb{T}^d)$ is a 1-absolutely summing surjection then using the fact that the space $C(\mathbb{T}^d;\, l^2_s)$ is naturally isomorphic with the space $C(\bigcup_{a\in S} \mathbb{T}^d)$ of scalar valued continuous function on the direct union of s copies of $\mathbb{T}^d$ and the Pietsch Factorization Theorem (cf. e.g. [Wt], III.F.8.) there exists a probability Borel measure $\mu = (\mu_a)_{a\in S}$ on $\bigcup_{a\in S} \mathbb{T}^d$ and a positive K such that

$$\|Rf\|_{S,2} \le K \sum_{a\in S} \int |D^a f(t)| \mu_a(dt). \tag{3.1}$$

For $b \in \mathbb{T}^d$ let τ_b denote the operator of translation by b defined for a (continuous) function f by $(\tau_b f)(t) = f(t+b)$ for $t \in \mathbb{T}^d$ and for a measure μ by $\tau\mu(A) = \mu(A+b)$ for Borel $A \subset \mathbb{T}^d$. Averaging against the normalized Haar measure of $\mathbb{T}^d$ and using the relation $\tau_b R = R\tau_b$ for $b \in \mathbb{T}^d$ we get from (3.1)

$$\|Rf\|_{S,2} \le K \sum_{a\in S} \int_{\mathbb{T}^d} |D^a f(t)| \left(\int_{\mathbb{T}^d} \tau_b \mu_a \tau_{-b}\, db \right)(dt).$$

By the uniqueness of the (translation invariant) Haar measure on $\mathbb{T}^d$ each of the measures $\int_{\mathbb{T}^d} \tau_b \mu_a \tau_{-b}\, db$ is a multiple of the normalized Haar measure of $\mathbb{T}^d$. Thus there are positive K_a for $a \in S$ such that

$$\begin{aligned} \|Rf\|_{S,2} &\le K \sum_{a\in S} K_a \int_{\mathbb{T}^d} |D^a f(t)|\, dt \\ &\le K\Big(\sum_{a\in S} K_a^2\Big)^{1/2} \int \Big(\sum_{a\in S} |D^a f(t)|^2\Big)^{1/2} dt \\ &= C\|f\|_{S,1} \end{aligned}$$

with $C = K\left(\sum_{a\in S} K_a^2\right)^{1/2}$. Hence R is a Paley projection.

Remark. A similar argument works if in Corollary 3.1 $C_S(\mathbb{T}^d)$ is replaced by $L_S^\infty(\mathbb{T}^d)$.

Recall (cf. [P-W], Theorem 2.1 and Corollary 2.1) that given $S \subset \mathbb{Z}^d$ there exists a Paley projection on $L_S^1(\mathbb{T}^d)$ iff either S contains all the unit vectors (corresponding to the partial derivatives of the first order) and contains points a and b such that $\sum_{j=1}^d [a(j) - b(j)] \equiv 1 \bmod 2$ which belong to a $(d-1)$-dimensional hyperplane which supports the convex hull of S and is not parallel to any axis of $\mathbb{R}^d$, or the same property has one of the lower dimensional smoothnesses being the intersection of S with some numbers of coordinate hyperplanes.

Thus combining Corollary 3.1 with the above result of [P-W], in a special case, we get

3.2 Corollary. *The 2-dimensional smoothness generated by the partial derivatives D_{xx} and D_y, i.e. the smoothness $S = \{(0,0),(1,0),(2,0),(0,1)\}$ has the property that there exists a translation invariant 1-absolutely summing surjection from $C_S(\mathbb{T}^2)$ onto an infinite dimensional subspace of $L_S^2(\mathbb{T}^2)$.*

References

[F] J. J. F. Fournier: An interpolation problem for coefficients of H_∞ functions, Proc. Amer. Math. Soc. 42 (1974), 402–407.

[Pa] R. E. A. C. Paley: On the lacunary coefficients of power series, Ann. of Math. 34 (1933), 615–616.

[Pi] A. Pietsch: Operator ideals, VEB Deutscher Verlag der Wissenschaften, Berlin 1978.

[P-W] A. Pełczyński and M. Wojciechowski: Paley projections on anisotropic Sobolev spaces on tori, submitted for publication.

[Wt] P. Wojtaszczyk: Banach spaces for analysts, Cambridge University Press. Cambridge 1991.

Institute of Mathematics
Polish Academy of Sciences
Śniadeckich 8
00-950 Warszawa Poland